EVOLUTION AND DIFFERENTIATION OF THE CONTINENTAL CRUST

The evolution and differentiation of the continental crust pose fundamental questions that are being addressed by new research concerning melting, melt extraction and transport through the crust, and the effect of melt on crustal rheology, in addition to new advances involving geophysics and geochemistry. Many new insights into crustal processes have been triggered by combined field observations and laboratory experiments, supported by developments in numerical modeling.

The first three chapters deal with the structure of the continents, controls on heat production and the composition, differentiation and evolution of continental crust. The role of arc magmatism in the Phanerozoic and crustal generation in the Archean are addressed next. To understand the modification and differentiation of continental crust we first consider two regional examples, one of the lower crust and one of the middle crust. There follows a series of process-oriented chapters involving melting, melt extraction and migration and crustal rheology. The final two chapters review the emplacement and growth of plutons and outline a modeling approach to the physical controls on crustal differentiation.

Written by experts active in the field, this book provides a valuable summary of recent advances for graduate students and research workers.

MICHAEL BROWN has held faculty positions at Oxford Brookes and Kingston universities in the UK, and at the University of Maryland in the USA, where he is currently Professor of Geology and Director of the Laboratory for Crustal Petrology. He is the founding editor of the *Journal of Metamorphic Geology* and is editor of books on *High-temperature Metamorphism and Crustal Anatexis* (Unwin Hyman, 1990) and *The Origin of Granites and Related Rocks* (Geological Society of America Special Paper).

Michael Brown and his colleagues and students investigate the pressure–temperature–time–deformation evolution of metamorphic belts, and the generation, segregation, transfer, and emplacement of granite within the Earth's crust. His work involves integration between field and laboratory studies, and theoretical analysis and modeling. Michael Brown was awarded the Coke Medal for 2005 by the Geological Society.

TRACY RUSHMER is currently a faculty member at the University of Vermont and teaches introductory and advanced courses in petrology and tectonics. Her research has focused on the physical and chemical processes of partial melting and differentiation. She integrates both igneous and metamorphic rock processes and explores melting processes mainly through experimentation. The rock deformation laboratory at the University of Vermont has been built to investigate the chemical signatures of magmas derived in active environments and compares them with partial melt compositions when deformation is not present. Experimentally investigating this interactive process, deformation combined with partial melting in both silicate and metal–silicate systems, has resulted in multiple papers and book chapters over the past decade. She was the Mineralogical Society of America's Distinguished Lecturer in 1999–2000.

EVOLUTION AND DIFFERENTIATION OF THE CONTINENTAL CRUST

Edited by

MICHAEL BROWN
University of Maryland

TRACY RUSHMER
University of Vermont

CAMBRIDGE
UNIVERSITY PRESS

CAMBRIDGE UNIVERSITY PRESS
Cambridge, New York, Melbourne, Madrid, Cape Town, Singapore, São Paulo

Cambridge University Press
The Edinburgh Building, Cambridge CB2 8RU, UK

Published in the United States of America by Cambridge University Press, New York

www.cambridge.org
Information on this title: www.cambridge.org/9780521782371

© Cambridge University Press 2006

First published 2006
This digitally printed version (with corrections) 2008

A catalogue record for this publication is available from the British Library

ISBN 978-0-521-78237-1 hardback
ISBN 978-0-521-06606-8 paperback

At the time of going to press the original colour images for figures 2.1,
2.3, 2.6, 2.7, 2.8, 2.9, 7.6, 7.10, and 8.7 were available for download from
www.cambridge.org/9780521066068

Contents

List of Contributors   *page* vi

1   Introduction   1

2   Structure of the continental lithosphere   21

3   Thermo-mechanical controls on heat production distributions
    and the long-term evolution of the continents   67

4   Composition, differentiation, and evolution of continental crust:
    constraints from sedimentary rocks and heat flow   92

5   The significance of Phanerozoic arc magmatism in generating
    continental crust   135

6   Crustal generation in the Archean   173

7   Structural and metamorphic process in the lower crust: evidence
    from a deep-crustal isobarically cooled terrane, Canada   231

8   Nature and evolution of the middle crust: heterogeneity
    of structure and process due to pluton-enhanced tectonism   268

9   Melting of the continental crust: fluid regimes, melting
    reactions, and source-rock fertility   296

10 Melt extraction from the lower continental crust of orogens:
    the field evidence   331

11 The extraction of melt from crustal protoliths and the flow
    behavior of partially molten crustal rocks: an experimental
    perspective   384

12 Melt migration in the continental crust and generation of lower
    crustal permeability: inferences from modeling and
    experimental studies   430

13 Emplacement and growth of plutons: implications for rates
    of melting and mass transfer in continental crust   455

14 Elements of a modeling approach to the physical controls
    on crustal differentiation   520

Index   550

Contributors

Richard J. Arculus
Department of Geology
The Australian National University
Canberra, ACT 0200, Australia

Scott A. Barboza
ExxonMobil Upstream Research Co.
3319 Mercer Street
Houston, TX 77027–6019

George W. Bergantz
Department of Earth and Space
Sciences
63 Johnson Hall, Box 351310
Seattle, WA 98195

Michael Brown
Department of Geology
University of Maryland
College Park, MD 20742

John D. Clemens
Centre for Earth and Environmental
Sciences Research
Kingston University
Penrhyn Road, Kingston-upon-
Thames
Surrey, KT1 2EE, UK

Alexander R. Cruden
Department of Geology
University of Toronto
22 Russell Street
Toronto, Ontario M5S 3B1, Canada

Jon P. Davidson
Department of Earth Sciences
University of Durham
Durham, DH1 3LE, UK
and formerly
Department of Earth and Space
Sciences
University of California at Los
Angeles
Los Angeles, CA 90095, USA

Simon Hanmer
Natural Resources Canada
615 Booth Street
Ottawa, Ontario K1A 0E9, Canada

Sidney R. Hemming
Lamont–Doherty Earth Observatory
Rt. 9W
Palisades, NY 10964

Karl E. Karlstrom
Department of Earth and Planetary
Sciences
University of New Mexico
Albuquerque, NM 87131

Adrian Lenardic
Department of Earth Science
Rice University
Houston, TX 77005

Alan Levander
Department of Earth Science
Rice University
Houston, TX 77005

Sandra McLaren
Research School of Earth Sciences
The Australian National University
Canberra, ACT 0200, Australia

Scott M. McLennan
Department of Geosciences
State University of New York at
Stony Brook
Stony Brook, NY 11794–2100

Julian Mecklenburgh
Rock Deformation Laboratory
Department of Earth Sciences
University of Manchester
Manchester, M13 9PL, UK

Stephen A. Miller
Department of Geodynamics
and Geophysics
University of Bonn
Nussallee 8
D-53115 Bonn
Germany

Hugh Rollinson
Department of Earth Sciences
Sultan Qaboos University
PO Box 36, Al Khodh
Postal Code 123
Muscat, Oman
and formerly
Geography and Environmental
Management Unit
Cheltenham and Gloucester College
of Higher Education
Francis Close Hall, Swindon Road
Cheltenham, GL50 4AZ, UK

Tracy Rushmer
Department of Geology
University of Vermont
Burlington, VT 05405

Ernest H. Rutter
Rock Deformation Laboratory
Department of Earth Sciences
University of Manchester
Manchester, M13 9PL, UK

Mike Sandiford
School of Earth Sciences
University of Melbourne
Victoria, 3010, Australia

Stuart R. Taylor
Department of Geology
Australian National University
Canberra, ACT 2601, Australia

Michael L. Williams
Department of Geosciences
University of Massachusetts
Amherst, MA 01003, USA

1

Introduction

MICHAEL BROWN AND TRACY RUSHMER

1.1 Rationale

Understanding the evolution of the continental crust poses many fundamental geological questions that are being addressed currently by several different groups within the community of solid earth scientists. The general approach is one in which the aim is to understand the processes of generation, modification and destruction of continental crust in relation to tectonics on modern Earth, and then to apply this understanding to the formation and evolution of the continental crust throughout Earth's history. However, the Earth is a self-organized body with an irreversible evolution dissipating energy into space, so that applying what we learn from modern Earth to the past must be moderated by a consideration of secular evolution. One way to do this is by tracking geological variables as a function of time, which is referred to as "secular analysis." For example, Earth's radiogenic heat production has declined exponentially with time. If this evolution translates into a warmer mantle in the past, which is a reasonable inference, this will have had an effect on mantle dynamics and degree of melting during decompression, which, in turn, has implications for the formation and evolution of continental crust. For this reason, the full characterization of the continental crust as it is preserved in the exposed geology of the continents remains an important area of research, for without these data we cannot undertake a complete secular analysis using, for example, proxies for subduction- or plume-related activities.

New advances in geophysics and geochemistry have led to ever more detailed information concerning the structure and composition of the continental crust, which, in turn, has shed light on additions and losses of mass to the continental crust during subduction, the nature of the Moho, the role of crust–mantle interactions, and the processes involved in crust–mantle evolution. Research concerning the mechanisms of chemical and physical

Evolution and Differentiation of the Continental Crust, ed. Michael Brown and Tracy Rushmer. Published by Cambridge University Press. © Cambridge University Press 2005.

differentiation of the continental crust has been invigorated by integrative studies that combine field observations with laboratory experiments and numerical modeling. In particular, new results from such studies address crustal melting, melt and fluid extraction and transport through the crust, and the effect of melt and fluid generation and migration on crustal rheology. However, even with this renewed attention on the continental crust, basic questions continue to linger in regard to many vital processes, particularly with respect to secular evolution, and it is these topics that are addressed in this book.

There are three fundamental questions that drive research into the origin and evolution of Earth's continental crust. First, by what process or processes has the continental crust been extracted from Earth's mantle, how have these processes changed through time, and has extraction been a continuous or episodic activity? Second, how much of the continental crust has been recycled back into Earth's mantle, by what processes has this recycling taken place, and what has been the net rate of continental growth through time? Third, in stabilized continental crust, upper, middle and lower crust (and lithospheric mantle) are distinguished on the basis of chemical composition and geophysical properties, but by what processes has the continental crust been differentiated, have these processes changed with time, and what are the consequences for evolution of the Moho?

We have made significant advances in characterizing the petrological and geochemical composition of the mantle-derived products that form the primitive continental crust. However, determining when the bulk of the continental crust was generated and what was the balance between extraction from the mantle in subduction settings versus plume activity in the past remain largely unresolved. Arc and plume magmatism contrast not only in the composition of the primary addition to the continental crust (predominantly andesite in continental arcs versus basalt above plumes), but also because subduction is an ongoing process that is argued to produce essentially quasi-continuous additions to the continental crust, whereas superplume events are inferred to be rare global-scale episodes that imply non-continuous additions to the continental crust.

Arcs are traditionally viewed as sites of crustal growth, and in many models of crustal evolution the subduction factory is regarded as the principal site of production of continental crust. Even in modern arcs, though, it is challenging to quantify the mass of material being transferred from mantle to crust, and the mass of material as sediment or crust being transferred back to the mantle by sediment subduction and by subduction erosion and delamination. Furthermore, although we make a conceptual distinction between recycling crust into the mantle and reworking crust leading to differentiation, it remains problematic in ancient crust to determine the relative importance of these processes in the generation of particular rock suites exposed at Earth's surface.

Alternative models for crustal growth place greater importance on basaltic magmatism related to superplume events and the accretion of oceanic plateaus to continents. Oceanic plateaus cover about 3% of modern Earth's sea floor, and if the Ontong Java Plateau collision with the Solomon Islands Arc is typical, about 80% of the thickness of the plateau is subducted and recycled into the mantle (Mann & Taira, 2004). Although crustal growth by this mechanism appears minimal on modern Earth, it is interesting to speculate whether superplume events were more important earlier in Earth history, and this is one question that will only be answered by full characterization of Earth's continental crust and a careful secular analysis of appropriate proxies for mantle plume magmatism. For example, with higher degrees of mantle melting earlier in Earth history, oceanic plateaus may have been thicker and more buoyant, and consequently could have made a greater contribution to the mass of the continental crust, particularly if subduction was absent from the tectonics of early Earth.

With respect to differentiation of primitive crust that has been extracted from the mantle, we have a good understanding of the melting relations of crustal protoliths (Chapter 9) but we need to develop a better understanding of the dynamics of partial melting in the crust, and to characterize the rheological response of the crust to partial melting and melt transfer. Sources of heat to drive crustal melting processes and the specific links between the petrologic and structural, and the kinematic and dynamic expressions of crustal differentiation by intra-crustal magmatism, continue to be highly debated topics in which it is common for case studies to reach as many different interpretations as there are tectonic settings on modern Earth. As a result, numerical modeling has been used increasingly in crustal studies to determine the primary drivers and elemental properties of melt flow through the crust and magma emplacement in the upper crust.

In this first chapter, we outline briefly our understanding of the structure of the continental crust and its chemical composition, and we consider the distribution of heat sources and their role in crustal evolution. We speculate about the extraction and turnover of crust, and the generation of an evolved "continental" crust on early Earth, and we consider the subduction factory as the primary source of continental crust production on modern Earth. Thus, we recognize that secular change has been important, and we touch upon the contrasts between modern arc magmatism and magmatism in the Archean. We introduce convergent margin processes and accretionary versus collisional orogenesis, and this leads into a discussion about the geologic materials that melt; we consider outstanding questions in regard to melt segregation and transport through the crust. We discuss the driving forces for melt migration in relation to different tectonic regimes, because these are the physical controls on crustal differentiation. Finally, we review the important issue of fluids in the crust, both aqueous and silicate-melt, and we introduce recent advances in modeling the modes of transport of fluids and mechanisms of magma ascent and emplacement.

1.2 Structure, composition and heat production
in the continental lithosphere

The continental crust is highly heterogeneous, both in terms of chemical composition and geophysical properties (Chapters 2–8). Indeed, the hetero-geneity of the continental crust–mantle lithosphere system itself is one of its most important properties. Lithospheric heterogeneity is described and docu-mented by geophysicists and geologists alike who have studied the seismic structure and the potential strength during large-scale tectonic deformation of exposed lower crustal sections and crust–mantle sections reconstructed from xenolith suites (e.g., Rudnick & Fountain, 1995; Karlstrom & Williams, 1998). This heterogeneity is probably a product of the buoyancy forces associated with Earth's convection cycle, heat and mass transfer associated with plumes, and the large-scale horizontal translation of crustal plates and their interac-tions during convergence (Davies, 1999; Turcotte & Schubert, 2002). These geodynamic processes have led to redistribution of mass and differentiation of crust from mantle.

Horizontal translation of crustal material has become increasingly identi-fied as of fundamental importance in the development of the overall hetero-geneity of continental crust, mainly through integrated seismic studies such as Lithoprobe, INDEPTH and the US Continental Dynamics Working Groups (e.g., Cook, 1995; Brown *et al.*, 1996; Fliedner *et al.*, 1996; Karlstrom & CDROM Working Group, 2002). One conclusion derived from interpretation of these data sets is that whole crustal detachments are common, even in cratonic portions of the continents, which suggests that metamorphic and igneous rock packages are thrust up from lower crustal depths to become interlayered with more recently deposited sedimentary rocks, thereby increas-ing local crustal heterogeneity (Chapter 2). Indeed, recently tectonic interleav-ing by ductile thrusting of oceanic and continental materials has been identified as an important process in the Isua Greenstone belt of west Greenland (Fedo *et al.*, 2001; Hanmer & Greene, 2002), which suggests that the deformational behavior, rheological constitution and overall strength of Paleoarchean and modern continental crust were similar.

It is also clear that crustal heterogeneity and variation in seismic character are functions of continental antiquity and recent tectonic regime. For example, shields and platforms, such as those present in North and South America, Eurasia and Africa, contrast significantly with Phanerozoic orogenic belts and convergent margins with accretionary orogenic belts. Orogenic belts undergoing net contraction, such as those in Tibet, the Altiplano–Puna and the Canadian Cordillera, show variations in crustal structure, including bright middle and lower crustal reflectivity and deformation fabrics, in comparison with those belts undergoing extension and collapse, such as the Basin and Range and the Colorado Plateau in the southern Rocky Mountains. Variation in seismic character also applies to rifted margins such as North America and Europe, and South America and Africa, and to the significant, but smaller-scale examples of rifting, such as the Rio Grande Rift and the North Sea, where ocean basin formation processes apparently have stalled (Burke, 1977; Klemperer *et al.*, 1986). Considering the heterogeneity observed both compositionally and structurally in the continental crust, it may be surprising that estimates of the average chemical composition of continental crust derived from seismic velocity variation and thicknesses of upper, middle and lower continental crust are in reasonable agreement with estimates derived from other approaches to assessing the average chemical composition of continental crust (Taylor & McLennan, 1981, 1985; Rudnick & Fountain, 1995).

Determining the bulk composition of the crust is a formidable, but important, task because the composition and element distribution in the crust may be used to develop models of crust–mantle evolution, and to test models of crustal production and crustal evolution through time. The crust is about 0.6% by mass of the silicate Earth, but it is the major reservoir for the incompatible elements (e.g., Taylor, 1992; Taylor & McLennan, 1985). There are several methods to estimate upper crustal elemental abundances. One successful approach that has been used by several researchers is the chemical composition of well-mixed clastic sedimentary rocks. The natural sampling of the exposed continental crust of different ages provided by erosion, transport and deposition of sediments allows the composition of upper crust to be estimated over geological time, because there is a sedimentary record dating back to at least 3.8 Ga (Taylor & McLennan, 1981; Chapter 4). With this approach, constraints may be placed on the composition of the bulk continental crust and possible recycling of continental crust. The chemical composition of the sedimentary rock record supports growth of the continental crust in an episodic manner, but with a major increase in the growth rate in the Late Archean (Chapters 4 and 6). Continental crust generated during the Archean has been suggested to be more mafic in overall composition compared to the composition of younger crust, but the difference may not be as profound as once thought (Rudnick & Gao, 2003). On modern Earth, the crust continues to grow mainly by island-arc volcanism and related magmatism (Chapters 4 and 6), but today this appears to be approximately balanced by crustal recycling through sediment subduction and subduction erosion (Scholl & von Huene, 2004).

Crustal differentiation is mainly achieved through partial melting processes and segregation of more felsic melt from the mafic residue, which means that heterogeneity in the crust, both local and regional, is an inevitable result. Thus, magmas that crystallize after migration to shallower levels of the crust contribute to a more evolved upper crust, whereas the lower crust becomes dominated by residue. However, the process by which sufficient heat is introduced into the crust to promote partial melting is a major area of research, because the thermal regimes for the crust calculated using numerical models (e.g., Connolly & Thompson, 1989; Jamieson *et al.*, 1998) at present do not achieve temperatures required for extensive melting of crustal protoliths as inferred from experiments (Chapter 9).

One significant source of heat in addition to underplating of mafic magma to the base of the crust is the presence and distribution of heat-producing elements (U, Th, K). A better understanding of crustal differentiation and the origin of crustal heterogeneity may be obtained by investigating the relationship between observed heat flow and possible heat sources and crustal heat production. Modern heat-flow observations suggest that heat-producing elements in the crust contribute on average about 50–66% of the observed surface heat flow and that the modern continental crust is strongly differentiated with respect to the heat-producing elements (McLennan & Taylor, 1996; Chapters 3 and 4). The observed distribution of crustal heat sources in stable continental crust may be understood in a similar way to explaining crustal structure and compositional heterogeneity, in that deformation, magmatism, and erosion redistribute the heat-producing elements. In general, these tectonic and petrologic processes increase the long-term thermal and mechanical stability of the continental crust by moving the heat-producing elements upwards in the crust through magmatism and reducing the total budget of heat-producing elements in the lithospheric column (Jaupart, 1983; Chapters 3 and 4). This process was particularly important in the evolution of cratons in the Archean, where tectonic processes ultimately achieved a stable ordering of the heat-producing elements.

1.3 Generating continental crust

It is of fundamental importance to identify when crust was first differentiated from the mantle and whether oceans were present at the same time, but this is a challenging task, for the evidence is rare and fragmentary and our knowledge of the earliest differentiation processes on Earth is limited because information retrieved from the earliest crust is punctuated by large gaps. Earth's earliest primitive crust most probably was mafic, and formed on top of a previously convecting magma ocean of uncertain depth (Boyet *et al.*, 2003). However, no outcrop evidence of this earliest primitive crust is known to have survived, probably due to the intense meteor bombardment that affected Earth between its formation and *c.* 4.0 Ga, the age of the oldest known components in outcrops of the Acasta gneisses of northwestern Canada (Bowring & Williams, 1999). Thus, the discovery of detrital zircons with U–Pb ages that are older than 4.0 Ga and that are characterized by high $\delta^{18}O$ isotopic compositions provides evidence that both continents and liquid water were surface features of the earliest Earth (Wilde *et al.*, 2001; Mojzsis *et al.*, 2001), and it is now considered possible that differentiation processes that produce a felsic evolved crust have been active since very close to the birth of the planet (e.g., Amelin *et al.*, 1999). This requires the planet to have formed an embryonic, differentiated, silicic crust during the Hadean (the time period between the formation of the Earth at *c.* 4.56 and *c.* 4.0 Ga, corresponding to the first preserved terrestrial rocks (Gradstein *et al.*, 2004)), and has reinvigorated the discussion on continental growth through time (e.g., Armstrong, 1991; Chapter 6).

Archean Earth was a different place from modern Earth, and the formation and evolution of the lithosphere and the tectonic interactions between segments of lithosphere will have been affected by secular change due to cooling from the Hadean magma ocean to an Earth at *c.* 2.5 Ga with a significant proportion of the volume of continental crust that we have today. Herein is one of the important questions in Earth Science: was crustal growth progressive and stepwise through time, or was it primarily an Archean phenomenon with significant crustal recycling since that time?

One view from isotope geochemistry has been summarized recently by Kramers *et al.* (2004). These authors consider it unlikely that before 4.0 Ga there was much continental crust, and whatever continental crust that may have existed most likely was largely recycled into the mantle so fast that with present techniques we cannot identify any isotopic trace of it. However, we do have those enigmatic detrital zircons! These authors argue that from *c.* 4.0 to *c.* 2.0 Ga, net growth of continental crust was rapid, resulting in about 75% of the present mass at 2.0 Ga, and from *c.* 2.0 Ga to the present, net growth of continental crust has continued at a slower pace. Kramers *et al.* (2004) argue that rates of recycling by continental erosion have remained approximately proportional to continent mass throughout Earth history.

This view is broadly consistent with the data of Scholl and von Huene (2004), who argue that on modern Earth crustal recycling to the mantle at subduction zones and new additions from the mantle at arcs are similar at approximately $70 \, km^3 \, Myr^{-1}/km$. By uniformitarian extrapolation this implies that the mass of recycled continental crust since the beginning of the Proterozoic is approximately equivalent to the mass of the present continental crust, which is less than 1% of the mass of the mantle. However, in an alternative view, Condie (2000) argues for episodic crustal formation with significant peaks in production at 2.7, 1.9 and 1.2 Ga for periods up to 100 Myr long related to probable superplume events. These peaks appear to correspond to periods of crustal amalgamation in the supercontinent cycle. Thus, the increased rate of crustal production and/or growth may reflect the mantle dynamics responsible for supercontinent formation (Condie, 2004). Condie (2002) argues that growth of juvenile crust occurred chiefly in the form of new arc systems accreted to continents during collision, but net growth may have been increased by reduced sediment subduction rates due to higher sea levels during these periods. According to Scholl and von Huene (2004), about $30 \, km^3 \, Myr^{-1}/km$ of crustal recycling is accomplished by sediment subduction on modern Earth, so that any reduction in this rate in the past is likely to have led to a significant increase of continental mass.

Currently, magmatism at arcs is viewed as representative of the main long-term flux of material from the mantle to the crust, but plume activity and construction of oceanic plateaus may have been more significant in the past (Chapter 6). As a result, the continental crust that is being formed today is generated in arc settings. However, assessing how important this process has been throughout geological time also addresses whether or not the plate tectonic paradigm for modern Earth is an adequate tectonic model for the generation and evolution of continental crust throughout Earth history (DePaolo, 1988). A comparison of estimated average compositions of the continental crust with the compositions of material associated with modern mantle–crust arc fluxes will enable an evaluation of arc magmatism in the generation of continental crust earlier in Earth history. Evaluation of data from specific subduction zones shows that primary arc magmas are high MgO basalts, and that the composition of magmas emplaced in the near surface environment (volcanic and plutonic crust) is actually a reflection of protracted intra-crustal assimilation and differentiation (Davidson, 1996).

But, what happened in the Archean? Although it is debated whether or not the Archean crust is more mafic than modern-day crust (Hamilton, 2003), studies of Archean crust have led many workers to the conclusion that the mechanisms by which the Earth's continental crust formed during the Archean were fundamentally different than those processes forming the continental crust today (e.g., Condie, 1981, 1990, 2000). Contrasts exist between the average composition of the continental crust, the estimated contemporary fluxes from the mantle via multistage evolution in arcs on modern Earth and typical Archean felsic crust. Models based on modern Earth and extrapolation of present-day arc volcanic processes to a hotter early Earth have not adequately explained the observations in these terranes, particularly the predominance of the tonalite–trondhjemite–granodiorite (TTG) suite of rocks and high-grade (amphibolite and granulite facies) metamorphism.

Zegers and van Keken (2001) have suggested that the earliest continental crust may have been differentiated and stabilized by delamination of the lower part of an oceanic plateau-like protocrust equilibrated at eclogite facies. Such delamination could have resulted in uplift, extension and the production of TTG suites as recorded in Archean cratons. Indeed, this process may not be that dissimilar to the process of foundering of a continental arc root that has been proposed recently to have occurred beneath the southern part of the Sierra Nevada in California (Ducea, 2002; Saleeby *et al.*, 2003; Zandt *et al.*, 2004). The removal of the root is hypothesized to have occurred by a Rayleigh–Taylor-type instability, with a nearly horizontal shear zone accommodating the detachment of the ultra-mafic root from its overlying granitic crust, and such a process could have been more common in Archean time. Ultimately, discerning the balance of contribution among melts derived from over-ridden "oceanic" lithosphere (so-called "slab" melts), mantle that might separate over-ridden "oceanic" lithosphere (so-called mantle "wedge") from continental lithosphere, mafic underplate and older continental crust to the process of Archean TTG suites and crust-formation, is of prime importance if we are to understand Archean geodynamics (Kramers, 1987; Kramers & Tolstikhin, 1997; Chapter 6).

An evaluation of modern arc magmatism suggests that either arc magmatism is not representative of the process that has generated the bulk of the continental crust in the past, or if it is an important crust-building process, it must be accompanied by an additional differentiation process that returns mafic materials to the mantle (Arculus, 1999). Initial mass additions to the crust through arc magmatism are basaltic. These compositions generate the distinctive trace element signatures that are observed in all continental crust by subsequent modification of this primary arc material through intra-crustal differentiation or contamination by the deep mafic lower crust (Chapter 5).

Differentiation may involve the formation of cumulates derived from fractional crystallization of magma located close to the Moho or residues derived from later crustal melting (Rudnick, 1995). Partial melting residues and cumulates from the original mantle influx may have been recycled into the mantle (Jull & Kelemen, 2001) or reside below the Moho (e.g., Ducea & Saleeby, 1998; Cook & Vasudevan, 2003), and this suggests that the Moho itself changes position (Arndt & Goldstein, 1989; Chapter 5).

1.4 Intracrustal differentiation at convergent margins

Studies of modern convergent plate margins have led to a better understanding of tectonic settings associated with most observed crustal melting (Chapter 10). In general, convergent margins generate two types of orogen. Accretionary orogens are produced along convergent plate margins characterized by long periods of ocean plate subduction (e.g., the Phanerozoic–Mesozoic–Cenozoic evolution of the Pacific Margin, e.g., the Japanese Islands, Tasmanides of eastern Antarctica and Australia). They form by creation and destruction of large arc systems, with growth of accretionary wedges in the fore arc, and extension and sedimentation in the back arc during prolonged slab roll-back. Episodic basin inversion occurs during short-lived events, probably driven by subduction of ocean floor features such as plateaus (Collins, 2002). Shortening and thickening of behind-arc basin sequences formed on thin lithosphere characterized by high heat flow give rise to typical low-pressure–high-temperature metamorphism, with counterclockwise pressure–temperature–time paths that may involve melting at the metamorphic peak and synchronous granite magmatism (e.g., Johnson & Vernon, 1995; Brown & Solar, 1999; Johnson et al., 2003; Brown, 2003). Additionally, subduction of an active ocean ridge may be an important component of the thermal evolution of some accretionary orogens (e.g., Brown, 1998).

In contrast, collisional orogens form by closure of an ocean basin (e.g., the Cenozoic Himalayas of Asia, the Paleozoic Variscides of Europe). Collisional orogenesis involves a period of ocean plate subduction followed by an arc or continent collision, so that in most circumstances the effects of collision will be superimposed on an older accretionary orogen. The geotherm and the rheology of the lower crust, the convergence rate and any metamorphic changes that occur in the crust if it is subducted determine what happens during the collision period (e.g., Toussaint et al., 2004). Arc or continent subduction may occur if the lithosphere is strong and the convergence rate is high, in which case the trench absorbs the convergence until it locks, after which ongoing convergence is accommodated by internal deformation.

Conversely, if the lithosphere is weak and the convergence rate is low, the convergence may be accommodated by pure shear thickening and Rayleigh–Taylor instabilities. Pressure–temperature–time paths in subducted continental material generally are clockwise, and crustal melting may occur (e.g., Brown, 1993, 2002). Extensional collapse of overthickened crust has been inferred for parts of the European Variscides (e.g., Burg *et al.*, 1994), where collapse is associated with widespread crustal melting.

Provided that a free aqueous fluid phase is present, common crustal rock types may be partially melted at temperatures as low as 700 °C. However, geologic evidence suggests that fluid-undersaturated conditions predominate during most high-grade metamorphic events (Chapter 9). Under these conditions, major melt production only occurs at temperatures above 800 °C, and temperatures need to be > 900 °C if hornblende-bearing metabasalts are involved. These thermal requirements generally demand that extra-crustal heat sources be available and, as described above, apart from some migmatite terranes, magmatic underplating and concentration of heat-producing elements are of major importance to the occurrence of crustal melting.

The volume of fluid-absent melt will vary as a function of *P*, *T*, and hydrous mineral content in the protolith, but common rock types (metapelites, metagreywackes, meta-andesites and some amphibolites) may yield 10 to 50% by volume of H_2O-undersaturated melt at attainable crustal temperatures (Clemens & Vielzeuf, 1987; Thompson, 1990; Chapter 9). The compositions of the melts may range from leucogranites to trondhjemites and granodiorites to tonalites, depending on the chemistry and mineralogy of the source rocks and the temperature of melting. The origin of high-K calc-alkaline magmas, which are of significant volumetric importance in orogenic belts, has been debated; in particular, how to form partial melts of high-K content. One possible source for these magmas involves AFC processes and mantle-derived magmas and assimilated crust; or, mixing between crustal and mantle melts. However, these compositions may be produced by crustal melting without the need for hybridization with mantle magmas, as suggested by Roberts and Clemens (1993, 1995). Melting experiments have shown that meta-tonalitic and high-K andesitic protoliths produce melts of high-K, calc-alkaline composition and the results suggest that K-rich meta-andesitic rocks are viable source rocks for producing high-K, calc-alkaline granite magmas in modern arcs (Roberts & Clemens, 1995).

1.5 Fluids in the crust

Fluids in the crust, both aqueous- and silicate-melt, have long been a focus of geologic studies because the porosity induced by mineralogical reaction, coupled with fluid flow and chemical transport and/or deposition processes, give rise to many features observed in the crust. These include metamorphic differentiation, differentiated cleavage and schistosity, lineation defined by mineral differentiation and, in higher-temperature regimes, where melt generation occurs, gneissosity and migmatitic layering. The importance of determining the mechanisms of fluid distribution and extraction at the grain scale is fundamental for understanding crustal differentiation processes in different tectonic environments.

A basis for understanding fluid migration in the crust is to determine the controls on the development of permeable networks, and several basic modes of fluid distribution and transport may be identified (Miller *et al.*, 2003; Chapter 12). These include: initial permeability (primary); deformation-induced permeability (secondary); chemical reaction-induced permeability; and coupling between deformation and chemical reactions (Connolly, 1996, 1997; Miller *et al.*, 2003). Do different patterns of fluid flow result from these different modes? How does feedback between fluid flow and chemical reactions result in different patterns of deformation and mineralogical distribution? What kinds of structures characterize these modes of transport? These questions reflect that the fundamental modes of transport of fluids through the crust involve the generation or removal of porosity.

Fluid flow in rocks with an initial intrinsic permeability is essentially driven by Darcy flow, e.g., by compaction. However, permeability may also be produced through deformation and/or chemical reactions. Connolly (1997) has shown how deformation-induced permeability in the vicinity of a devolatilization front (positive ΔV of metamorphic reactions) may lead to circumstances in which zones of highly localized porosity are generated in the crust. While the ΔV of metamorphic reactions is of importance in determining whether reactions will generate over-pressuring and embrittlement, the more important variables are, in fact, connected porosity, changes in rock and fluid pressure gradients, and compaction. Porosity induced by chemical reaction, coupled with fluid flow and chemical transport/deposition processes may also lead to the propagation of multiple reaction fronts to "fingering" or "channelling" (Connolly, 1996, 1997).

The results from these studies, which are relevant to both metamorphic dehydration reactions and partial melting reactions in the mid to lower crust, suggest that in the lower crust there should exist zones or domains of high permeability. The segregation of partial melt, from a fluid dynamics point of view, requires that viscosity does not play a significant role and that compaction is possible for segregation at geologically relevant time scales. For the most part, this treatment is illustrative and explains permeability features observed in partially molten terrains; however, the rheological changes that occur in rocks undergoing partial melting require some additional consideration.

1.6 Extraction of crustal melts and physical controls on crustal differentiation

Once partial melt is generated, it needs to leave its source if the crust is to be efficiently differentiated. Understanding what drives the extraction of crustal melts requires knowledge of the mechanical properties of the partially molten protoliths. Extraction of crustal melts, when aided by tectonic deformation, may occur by gravity and deformation. Several different deformation mechanisms may dominate: dislocation creep of the grain matrix, diffusion creep enhanced by the presence of the high diffusivity melt, or granular flow of the matrix of grains, with or without fracturing of the grains. Experimental data suggest that any of these mechanisms may be dominant under particular conditions (Rutter & Neumann, 1995; Rushmer, 1995; Holyoke & Rushmer, 2002).

Experiments on partially molten crustal rocks have all shown that as melt fraction increases, the rock undergoes significant weakening. Mecklenburgh and Rutter (2003) have shown that partially molten granitic rocks are extremely weak and able to flow at geologically relevant strain rates at differential stresses on the order of 1 MPa or less. This suggests that as melt fraction increases, strain will be localized and shearing will be focused. As a result, even thin melt-bearing horizons in the crust may be effective tectonic detachment horizons (see Chapter 11). In addition, field and experimental studies of partially molten crustal rocks suggest that they may fracture even under conditions in which ductile behavior should dominate (melt-enhanced embrittlement or ductile fracture; Chapter 10) or they may develop melt-filled cracks as a result of volume change on the onset of melting (Davidson *et al.*, 1994; Klepeis *et al.*, 2003). Also, they may develop ductile, dilatant features such as interboudin partitions and shear bands, forming vein networks, that may feed dikes or shear zones to facilitate melt drainage from the interconnected melt-filled networks (see Chapter 10).

Gravity-driven drainage of a rock-melt system has been shown to separate melt in the mantle in a geologically reasonable time frame (the "compaction" model (McKenzie, 1984)) due to the density difference between melt and its source. However, granite melt that has low melt water content may be too viscous for homogeneous intergranular porous flow to cause efficient segregation of granitic liquids. Therefore, mechanisms such as pressure-gradient driving forces that are effective in moving more viscous liquids are likely to prevail in the crust, where differential stress gradients may develop as melting begins (see Chapter 12). The role of pore fluid pressure gradients in driving melt migration may work locally to interconnect sites of melting and may lead to rapid melt segregation (Miller & Nur, 2000).

One of the key questions in crustal differentiation is the nature of melt transport as it occurs over kilometers in the crust. Melt transport is often considered episodic, especially as we continue to collect better and more accurate geochronology data. However, what controls the episodicity? Is it the rate of heat input or something inherent in the system itself? If melt migration is continuous, how is the thermal balance maintained and what is the ascent mechanism? As pointed out in Chapter 14, the main difficulty in modeling multiphase behavior is the relative motion between adjacent phases at many scales (Bergantz, 1992).

Many of the available analog and numerical models of crustal differentiation are difficult to test with the complex data sets from natural systems, so models are driven by simplicity and appeal to the mixture assumption (Yuen *et al.*, 2000). Although the mixture approach is useful in some geological applications, it is not robust enough for general application to problems in crustal differentiation (Bergantz, 2000). The quantitative description of the transport of mixed-phase materials is especially problematic but of great importance in understanding crustal differentiation processes. Both buoyancy and tectonic processes must be active, but the relative importance in time and space is unclear. Magma mixing is controlled by both the intrinsic density differences as well as the aspect ratio of the magma chamber, which itself is controlled by the cooling history (Chapters 13, 14). Ultimately, both well-chosen field examples and new continuum theories and numerical models play an important role in understanding the best application of current models of multiphase behavior to magma transport and mixing (Chapter 14).

1.7 Emplacement and growth of plutons

Finally, the emplacement of granites in the middle and upper crust and the processes involved in the extraction of melt from partially molten source regions in the lower crust need to be linked. As discussed in Chapter 9, it is generally agreed that granitic magmas are generated mainly by crustal melting at high temperature, and that there exists a genetic connection between granitic magma generation and granulite-facies metamorphism. Insight into granite emplacement will be greatly enhanced with the development of integrated models that use a variety of approaches to explore the relationships between all aspects of crustal melting and granitic magmatism (e.g., Petford *et al.*, 2000; Chapter 13). Most recent studies find the limiting controls on pluton growth, size and spacing to be most closely related to the degree of melting and the mechanisms of melt extraction in the lower crust (Vigneresse, 1995; Brown & Solar, 1999; Cruden & McCaffrey, 2001; Chapter 13). However, tectonic setting also has a major influence on the distribution, shape and form of plutons, with field studies showing the importance of the interaction and feedback between active regional deformation and pluton growth, melt transfer and source extraction processes over local structural controls on pluton emplacement (Paterson & Schmidt, 1999).

1.8 *Quo vadimus*

In reading this book, we ask that you keep in mind the scientific method, particularly the principle of falsification, and two specific issues concerning the formation and evolution of the continental crust.

In her seminal book on *The Rejection of Continental Drift: Theory and Methods in American Earth Science* (1999), Naomi Oreskes writes (p. 3):

History is littered with the discarded beliefs of yesteryear, and the present is populated by epistemic resurrections. This realization leads to the central problem of the history and philosophy of science: How are we to evaluate contemporary science's claims to truth given the perishability of past scientific knowledge? The question is of considerable philosophic interest and of practical import as well. If the truths of today are falsehoods of tomorrow, what does this say about the nature of scientific truth?

In this book, we make no claim about the truth of any beliefs held by the contributing authors; however, we do believe in the longevity of data collected in the field and in the laboratory, and the contribution to understanding that may be gleaned from modeling studies. Thus, the book is a statement about our present understanding and, of course, it will be overtaken by new research – this is the nature of scientific inquiry, and why it is so much fun to be a participant.

In putting together this volume, we have kept in mind two particular issues; although these issues are not discussed in the same level of detail in each chapter, they crop up in most and we suggest them as themes to consider as you read. The first issue concerns the formation and stabilization of continental crust. This may be summarized in two questions. Are oceanic plateaus or arcs the seeds of the continents, and are continental arcs the recycling machine? How has secular change affected the tectonic behavior of the lithosphere? The second issue is the response of the crust to melting during orogenesis, and the process of extraction and transport of melt from lower to upper crust. Here, a principal concern remains the source of the heat, or perhaps more correctly how asthenospheric heat is transferred to the lower crust to facilitate melting and differentiation of the continental crust.

We have come far in our understanding of the origin and evolution of the continental crust, but much remains to be done; this book represents but one stepping stone on the path to a better understanding.

References

Amelin, Y., Lee, D.-C., Halliday, A. and Pidgeon, R. T. (1999). Nature of the Earth's earliest crust from hafnium isotopes in single detrital zircons. *Nature*, **399**, 252–5.

Arculus, R. J. (1999). Origins of the continental crust. *Journal and Proceedings of the Royal Society of New South Wales*, **132**, 83–110.

Armstrong, R. L. (1991). The persistent myth of crustal growth. *Australian Journal of Earth Science*, **38**, 613–30.

Arndt, N. T. and Goldstein, S. L. (1989). An open boundary between lower continental crust and mantle; its role in crust formation and crustal recycling. *Tectonophysics*, **161**, 201–12.

Bergantz, G. W. (1992). Conjugate solidification and melting in multicomponent open and closed systems. *International Journal of Heat Mass Transfer*, **35**, 533–43.

Bergantz, G. W. (2000). On the dynamics of magma mixing by reintrusion: implications for pluton assembly processes. *Journal of Structural Geology*, **22**, 1297–1309.

Bowring, S. A. and Williams, I. S. (1999). Priscoan (4.00 – 4.03) orthogneiss from northwestern Canada. *Contributions to Mineralogy and Petrology*, **134**, 3–16.

Boyet, M., Rosing, M., Blichert-Toft, J., Storey, M. and Albaréde, F. (2003). Nd evidence for early Earth differentiation. *Earth and Planetary Science Letters*, **214**, 427–42.

Brown, L. D., Zhao, W., Nelson, K. D., Hauck, M., Alsdof, D., Ross, A., Cogan, M., Liu X. and Che, J. (1996). Bright spots, structure and magmatism in Southern Tibet from Tibet from INDEPTH Seismic Reflection Profiling. *Science*, **274**, 1688–90.

Brown, M. (1993). *P-T-t* evolution of orogenic belts and the causes of regional metamorphism. *Journal of the Geological Society London*, **150**, 227–41.

Brown, M. (1998). Ridge-trench interactions and high-*T* – low-*P* metamorphism, with particular reference to the Cretaceous evolution of the Japanese Islands. In *What Drives Metamorphism and Metamorphic Reactions*, ed. P. J. Treloar and P. J. O'Brien. Geological Society Special Publication No. 138, pp. 131–63.

Brown, M. (2002). Prograde and retrograde processes in migmatites revisited. *Journal of Metamorphic Geology*, **20**, 25–40.

Brown, M. (2003). Hot orogens, tectonic switching, and creation of continental crust: comment. *Geology: Online forum, June 2003*, DOI 10.1130/0091–7613, **31**, e9.

Brown, M. and Solar, G. S. (1999). The mechanism of ascent and emplacement of granite magma during transpression: a syntectonic granite paradigm. *Tectonophysics*, **312**, 1–33.

Burg, J.-P. van den Driessche, J. and Brun, J. -P. (1994). Syn- to post-thickening extension in the Variscan Belt of Western Europe: modes and structural consequences. *Géologie de la France*, **1994**, 33–51.

Burke, K. (1977). Aulacogens and continental breakup. *Annual Reviews of Earth and Planetary Science Letters*, **5**, 371–96.

Clemens, J. D. and Vielzeuf, D. (1987). Constraints on melting and magma production in the crust. *Earth and Planetary Science Letters*, **86**, 287–306.

Collins, W. J. (2002). Nature of extensional accretionary orogens. *Tectonics*, **21**, doi:10.1029/2000TC001272.

Condie, K. C. (1981). *Archean Greenstone Belts*. Amsterdam: Elsevier.

Condie, K. C. (1990). Growth and accretion of continental crust: inferences based on Laurentia. *Chemical Geology*, **83**, 183–94.

Condie, K. C. (2000). Episodic continental growth models: afterthoughts and extensions. *Tectonophysics*, **322**, 153–62.

Condie, K. C. (2002). Continental growth during a 1.9-Ga superplume event. *Journal of Geodynamics*, **34**, 249–64.

Condie, K. C. (2004). Supercontinents and superplume events: distinguishing signals in the geologic record. *Physics of the Earth and Planetary Interiors*, **146**, 319–32.

Connolly, J. A. D. (1996). Mid-crustal focused fluid movement in the lower crust: thermal consequences and silica transport. In *Fluid Flow and Transport in Rocks: Mechanisms and Effects*, ed. B. Jamtviet and B. W. D. Yardley. London: Chapman and Hall, pp. 235–50.

Connolly, J. A. D. (1997). Devolatilization-generated fluid pressure and deformation-propagated fluid flow during prograde regional metamorphism. *Journal of Geophysical Research*, **102**, 18,149–73.

Connolly, J. A. D. and Thompson, A. B. (1989). Fluid and enthalpy production during regional metamorphism. *Contributions to Mineralogy and Petrology*, **102**, 346–66.

Cook, F. A. (1995). Lithospheric processes and products in the southern Canadian Cordillera: a Lithoprobe perspective. *Canadian Journal of Earth Sciences, Journal Canadien des Sciences de la Terre*, **32**, 1803–24.

Cook, F. A. and Vasudevan, K. (2003). Are there relict crustal fragments beneath the Moho? *Tectonics*, **22**, 1026, doi:10.1029/2001TC001341.

Cruden, A. R. and McCaffrey, K. J. W. (2001). Growth of plutons by floor subsidence: implications for rates of emplacement, intrusion spacing and melt-extraction mechanisms. *Physics and Chemistry of the Earth, Part A, Solid Earth and Geodesy*, **26**, 303–15.

Davidson, J. P. (1996). Deciphering mantle and crustal signatures in subduction zone magmatism. In *Subduction: Top to Bottom*, ed. G. E. Bebout, D. W. Scholl, S. H. Kirby and J. P. Platt. American Geophysical Union Monograph 96, pp. 251–62.

Davidson, C., Schmid, S. M. and Hollister, L. S. (1994). Role of melt during deformation in the deep crust. *TERRA Nova*, **6**, 133–42.

Davies, G. F. (1999). *Dynamic Earth: Plates, Plumes and Mantle Convection*. Cambridge: Cambridge University Press.

DePaolo, D. J. (1988). Age dependence of the composition of the continental crust: evidence from Nd isotopic variations in granitic rocks. *Earth and Planetary Science Letters*, **90**, 263–71.

Ducea, M. N. (2002). Constraints on the bulk composition and root foundering rates of continental arcs: a California arc perspective. *Journal of Geophysical Research*, **107**, 2304, doi:10.1029/2001JB000643.

Ducea, M. N. and Saleeby, J. B. (1998). The age and origin of a thick mafic–ultramafic keel from beneath the Sierra Nevada batholith. *Contributions to Mineralogy and Petrology*, **133**, 169–85.

Fedo, C. M., Myers, J. S. and Appel, P. W. U. (2001). Depositional setting and
    Paleogeographic implications of Earth's oldest supracrustal rocks, the >3.7 Ga
    Isua Greenstone belt, west Greenland. *Sedimentary Geology*, **141**, 61–77.
Fliedner, M. M., Ruppert, S. D., Malin, P. E., Park, S. K., Jiracek, G. R., Phinney,
    R. A., *et al.* (1996). Three-dimensional crustal structure of the southern Sierra
    Nevada from seismic fan profiles and gravity modeling: U.S. Southern Sierra
    Nevada Continental Dynamics Working Group, United States (USA). *Geology*,
    24, 367–70.
Gradstein, F. M., Ogg, J. G., Smith, A. G., Bleeker, W. and Lourens, L. J. (2004).
    A new geologic timescale with special reference to Precambrian and Neogene.
    Episodes, **27**, 83–100.
Hamilton, W., (2003). An alternative Earth. *GSA Today*, **13**(11), 4–12.
Hanmer, S. and Greene, D. C. (2002). A modern structural regime in the Paleoarchean
    (~3.64 Ga); Isua Greenstone belt, southern west Greenland. *Tectonophysics*, **346**,
    201–22.
Holyoke, C. III and Rushmer, T. (2002). An experimental study of grain-scale melt
    segregation mechanisms in crustal rocks. *Journal of Metamorphic Geology*, **20**,
    493–512.
Jamieson, R. A., Beaumont, C., Fullsack, P. and Lee, B. (1998). Barrovian regional
    metamorphism: where's the heat? In *What Drives Metamorphism and
    Metamorphic Reactions?*, ed. P. J. Treloar and P. J. O'Brien. London: Geological
    Society, Special Publication No. 138, pp. 23–45.
Jaupart, C. (1983). Horizontal heat transfer due to radioactivity contrasts: causes and
    consequences of the linear heat flow relation: *Geophysical Journal of the Royal
    Astronomical Society*, **75**, 411–35.
Johnson, S. E. and Vernon, R. H. (1995). Stepping stones and pitfalls in the
    determination of an anticlockwise P–T–t-deformation path in the low-P, high-T
    Cooma Complex, Australia. *Journal of Metamorphic Geology*, **13**, 165–83.
Johnson, T. E., Brown, M. and Solar, G. S. (2003). Low-pressure subsolidus and
    suprasolidus phase equilibria in the MnNCKFMASH system: constraints on
    conditions of regional metamorphism in western Maine, northern Appalachians.
    American Mineralogist, **88**, 624–38.
Jull, M. and Kelemen, P. B. (2001). On the conditions for lower crustal convective
    stability. *Journal of Geophysical Research*, **106**, 6423–46.
Karlstrom, K. E. and CDROM Working Group (2002). Structure and evolution of
    the lithosphere beneath the Rocky Mountains. *GSA Today*, **12**, 4–10.
Karlstrom, K. E. and Williams, M. L. (1998). Heterogeneity of the middle crust;
    implications for strength of continental lithosphere. *Geology*, **26**, 815–18.
Klemperer, S. L., Hauge, T. A., Hauser, E. C., Oliver, J. E. and Potter, C. J. (1986).
    The Moho in the northern Basin and Range Province, Nevada, along the
    COCORP 40 degrees N seismic-reflection transect. *Geological Society of America
    Bulletin*, **97**, 603–18.
Klepeis, K. A., Clarke, G. L. and Rushmer, T. (2003). Magma transport and coupling
    between deformation and magmatism in the continental lithosphere. *GSA Today*,
    13, 4–11.
Kramers, J. D. (1987). Link between Archean continent formation and anomalous
    sub-continental mantle. *Nature*, **325**, 47–50.
Kramers, J. D. and Tolstikhin, I. N. (1997). Two terrestrial lead isotope paradoxes,
    forward transport modelling, core formation and the history of the continental
    crust. *Chemical Geology*, **139**, 75–110.

Kramers, J. D., Kleinhanns, I. C., Nägler, Th. F. and Tolstikhin, I. N. (2004). Continental crust growth through geological time: an isotope geochemical perspective. In *Geoscience Africa 2004 abstracts*, ed. L. D. Ashwal. School of Geosciences, University of the Witwatersrand, pp. 360–1.

Mann, P. and Taira, A. (2004). Global tectonic significance of the Solomon Islands and Ontong Java Plateau convergent zone. *Tectonophysics*, **389**, 137–90.

McKenzie, D. (1984). The generation and compaction of partially molten rock. *Journal of Petrology*, **25**, 713–65.

McLennan, S. M. and Taylor, S. R. (1996). Heat flow and the chemical composition of continental crust. *Journal of Geology*, **104**, 369–77.

Mecklenburgh, J. and Rutter, E. H. (2003). On the rheology of partially molten synthetic granite. *Journal of Structural Geology*, **25**, 1575–85.

Miller, S. A. and Nur, A. (2000). Permeability as a toggle-switch in fluid-controlled crustal processes. *Earth and Planetary Science Letters*, **183**, 133–46.

Miller, S. A., van der Zee, W., Olgaard, D. L. and Connolly, J. A. D. (2003). A fluid-pressure-controlled feedback model of dehydration reactions: experiments, modeling, and application to subduction zones. *Tectonophysics*, **370**, 241–51.

Mojzsis, S. J., Harrison, T. M. and Pidgeon, R. T. (2001). Oxygen-isotope evidence from ancient zircons for liquid water at the Earth's surface 4,300 Myr ago. *Nature*, **409**, 178–81.

Oreskes, N., (1999). *The Rejection of Continental Drift: Theory and Methods in American Earth Science*. New York and Oxford: Oxford University Press.

Paterson, S. R. and Schmidt, K. L. (1999). Is there a close spatial relationship between faults and plutons? *Journal of Structural Geology*, **21**, 1131–42.

Petford, N., Cruden, A. R., McCaffrey, K. J. W. and Vigneresse, J.-L. (2000). Granite magma formation, transport and emplacement in the Earth's crust. *Nature*, **408**, 669–73.

Roberts, M. P. and Clemens, J. D. (1993). Origin of high-potassium, calc-alkaline, I-type granitoids. *Geology*, **21**, 825–8.

Roberts, M. P. and Clemens, J. D. (1995). Feasibility of AFC models for the petrogenesis of calc-alkaline magma series. *Contributions to Mineralogy and Petrology*, **121**, 139–47.

Rudnick, R. L. (1995). Making continental crust. *Nature*, **378**, 571–8.

Rudnick, R. L. and Fountain, D. M. (1995). Nature and composition of the continental crust; a lower crustal perspective. *Reviews of Geophysics*, **33**, 267–309.

Rudnick, R. L. and Gao, S. (2003). Composition of the continental crust. In *The Crust*, ed. R. L. Rudnick. Treatise on Geochemistry, Amsterdam: Elsevier, vol. 3, pp. 1–64.

Rushmer, T. (1995). An experimental deformation study of partially molten amphibolite: applications to low-fraction melt segregation. In *Journal of Geophysical Research* Special Section "Segregation of melts from crustal protoliths: mechanisms and consequences", ed. M. Brown, T. Rushmer and E. W. Sawyer. *Journal of Geophysical Research*, **100**, 15,681–96.

Rutter, E. H. and Neumann, D. H. K. (1995). Experimental deformation of partially molten Westerly granite, with implications for the extraction of granitic magmas. *Journal of Geophysical Research*, **100**, 15,697–715.

Saleeby, J., Ducea, M. and Clemens-Knott, D. (2003). Production and loss of high-density batholithic root, southern Sierra Nevada, California. *Tectonics*, **22**, 1064, doi:10.1029/2002TC001374.

Scholl, D. W. and von Huene, R. (2004). *Continental Crust Recycling at Convergent Ocean Margins Compared to Arc Magmatic Additions – Implications for the*

Growth of Continental Crust and Evolution of Mantle Geochemistry. 32nd IGC, Florence 2004 Scientific Sessions: Abstracts (Part 2), p. 1334.

Taylor, S. R. (1992). *Solar System Evolution: A New Perspective*. New York: Cambridge University Press.

Taylor, S. R. and McLennan, S. M. (1981). The composition and evolution of the continental crust: rare earth element evidence from sedimentary rocks. *Philosophical Transactions of the Royal Society of London*, **A301**, 381–99.

Taylor, S. R. and McLennan, S. M. (1985). *The Continental Crust: Its Composition and Evolution*. Oxford: Blackwell Scientific.

Thompson, A. B. (1990). Heat, fluids and melting in the granulite facies. In *Granulites and Crustal Differentiation*, ed. D. Vielzeuf and P. Vidal. Dordrecht: Kluwer Academic Publishers, pp. 37–58.

Toussaint, G., Burov, E. and Jolivet, L. (2004). Continental plate collision: unstable vs. stable slab dynamics. *Geology*, **32**, 33–6.

Turcotte, D. L. and Schubert, G. (2002). *Geodynamics*. Cambridge: Cambridge University Press.

Vigneresse, J.-L. (1995). Control of granite emplacement by regional deformation. *Tectonophysics*, **249**, 173–86.

Wilde, S. A., Valley, J. W., Peck, W. H. and Graham, C. M. (2001). Evidence from detrital zircons for the existence of continental crust and oceans on the Earth 4.4 Gyr ago. *Nature*, **409**, 175–78.

Yuen, D., Vincent, A. P., Bergeron, S. Y., Dubuffet, F., Ten, A. A., Steinbach, V. C. and Starin, E. (2000). Crossing of scales and non-linearities in geophysical processes. In *Problems in Geophysics in the New Millennium*, ed. E. Boschi, G. Ekstrom and A. Morelli. Bologna: Editrice Compositori. pp. 403–62.

Zandt, G., Gilbert, H., Owens, T. J., Ducea, M. F., Saleeby, J. and Jones, C. H. (2004). Active foundering of a continental arc root beneath the southern Sierra Nevada in California. *Nature*, **432**, 41–6.

Zegers, T. E. and van Keken, P. E. (2001) Middle Archean continent formation by crustal delamination. *Geology*, **29**, 1083–6.

2

Structure of the continental lithosphere

ALAN LEVANDER, ADRIAN LENARDIC AND KARL KARLSTROM

2.1 Introduction

More than forty years have passed since the beginning of the revolution in global tectonics, ushered in by the recognition of seafloor spreading. The success of plate tectonics in describing the evolution of the ocean basins lies in the regular progression of oceanic lithosphere from its generation at spreading ridges to its consumption at subduction zones, and the simple description of the tectonics of oceanic lithosphere in terms of translation and rotation of generally large "rigid" nearly uniform plates obeying fairly simple geometric laws at their boundaries. This process dominates global tectonics at scales of hundreds to thousands of kilometers and durations of $\sim$0.2 Ga and is often taken as the surface expression of mantle convection. That is, the oceanic thermal lithosphere as a whole, both its crustal and mantle components, forms the active upper thermal boundary layer of mantle convection (Turcotte & Oxburgh, 1967; Parsons & Sclater, 1977; Parsons & McKenzie, 1978). Differentiation processes at ocean ridges are also relatively simple, creating plate-wide layered structures that are largely recycled by subduction, and relationships between age of lithosphere, its thickness, thermal state, and density are well known (Turcotte & Schubert, 1982).

Earth, unique among the planets, has a bimodal distribution of topography, with $\sim$40% of its surface made up of continents whose surfaces reside above, to slightly below, sea-level and $\sim$60% of its surface comprised of ocean with average elevation of $\sim$ -4 km. This is the most obvious difference between the continents and oceans, but there are equally profound differences extending to depths of 200–400 km. In contrast to the 60% of Earth forming the oceanic plates, the 40% comprising the continents has grown over the last $\sim$4.5 Ga through a complex sequence of differentiation and deformation processes that operate on scales of hundreds of kilometers and smaller, including highly

Evolution and Differentiation of the Continental Crust, ed. Michael Brown and Tracy Rushmer. Published by Cambridge University Press. © Cambridge University Press 2005.

localized phenomena such as faulting and pluton emplacement. The resulting continental lithosphere consists of a complex collage of fragments of various tectonic affinities. Whereas these fragments may be collected to form super-continents or dispersed by the same convection system that creates and recycles oceanic plates, the development of continental lithosphere itself, unlike the oceanic lithosphere, cannot be viewed in terms of creation of a thermal boundary layer in the convecting mantle system, and continental tectonics cannot be simply described by translation, rotation, and consump-tion of large "rigid," nearly uniform plates obeying fairly simple geometric laws at their boundaries. In general, continental crust does not participate in global mantle overturn and this also appears to be the case for the subconti-nental mantle beneath many cratons. Thus, the continental lithosphere is not a simple active upper thermal boundary layer within the convecting mantle system, and understanding the structure of continents requires understanding of the cumulative products of magmatic and deformational processes that have acted over billions of years of Earth history.

This chapter provides an overview of the continental crust and aspects of the continental upper mantle within a largely geophysical frame of reference. In addition to the vast store of geological and geochemical data, a large part of our knowledge of continents is derived from seismic, magnetotelluric, potential field, heat flow, and topographic data. Among the most important methods for understanding the structure of continents are the various branches of seismology (reflection, refraction, and earthquake seismology), which provide images of the elastic properties or their changes in the Earth. Another important geophysical method is magnetotellurics, which provides images of electrical properties.

It is beyond the scope of this chapter to review the many geophysical methods used to investigate the crust, or all the areas of their applications. Reviews on seismology may be found in Mooney and Brocher (1987), Mooney (1989) and Braile *et al.* (1995). For the magnetotelluric methods, Jiracek *et al.* (1995) provide an instructive overview. We have selected some representative or illustrative examples from the literature in an attempt to summarize globally important attributes of the continental crust, and dwell on resolution issues in seismology, as seismic images are currently the most widely used and reliable indicators of the structure and composition of rocks in the subsurface.

Potential field studies (gravity and magnetics) investigate potential energy differences; the potential field method is a powerful tool for examining lateral density variations and geodynamics of stress distribution in the crust. Magnetics looks at the uppermost part of crystalline crust and is good for examining fine-scale variations in the crystalline basement in platform areas that are concealed by thin sedimentary cover. Magnetotellurics may be used to image conductive layers (usually fluids and melts, but also conductive compositions such as graphite and sulphides) and hence help us to understand temperature, chemical composition, and physical state variations. Heat flow measurements and modeling are important in terms of understanding variations in continental temperature, rheology, and seismic velocity with depth, as well as inferring crustal, as opposed to mantle, contributions to heat flow.

The term "remote sensing" may be used to describe structural seismology, since the development and interpretation of seismic images is very similar to the imaging science used in medicine and elsewhere (e.g., Blackledge, 1989). There is one important difference – Earth's vertical density and rheological stratification lead to a nearly continuous increase in seismic velocity from the surface to the core–mantle boundary. The exceptions to this continuous velocity increase with depth are important mineralogical, chemical, or rheological layers in Earth. For example, the crust–mantle boundary is defined as the place where seismic velocities increase discontinuously from $< 7.6\,\mathrm{km\ s^{-1}}$ to $> 7.6\,\mathrm{km\ s^{-1}}$, a low velocity channel at depths of 100–150 km is synonymous with the asthenosphere beneath the oceans, the 410 km discontinuity is associated in part with the phase change of olivine to β-spinel, and the 670 km discontinuity with the phase change from β-spinel to perovskite (Anderson, 1989). The consequence of the velocity variation with depth for imaging is that seismic energy travels along curved, sometimes quite complicated, paths, and estimation of the path is required to properly focus seismic images. In the crystalline crust, the path estimation may be complicated, due to both the large- and small-scale lateral heterogeneity of the seismic velocity field. This lateral heterogeneity is apparent in most surface exposures of formerly middle and lower crust, e.g., the Ivrea Zone, Italy (Rutter *et al.*, 1999), where lithologies having velocity contrasts exceeding 10% are mixed at the reflection–refraction seismic wavelength scale (100–5000 m). None the less, images from modern deep crustal reflection data sets are now quite good, frequently identifying features far deeper than the crust itself. For example, recent Lithoprobe data show reflections to depths of $\sim$100 km beneath the Wopmay orogen in northwestern Canada (Cook *et al.*, 1999).

Structural seismology describes the use of seismic methods to produce images of the mechanical properties of Earth's interior, and falls into two classes of imaging methods: direct imaging – which uses the recorded wavefield in image formation, and travel time methods – which use the derived quantity travel time to make an image. The former includes reflection seismology, and teleseismic receiver-function seismology. The latter includes much of refraction seismology and tomography. Resolution is an inherent property of the seismic experiment design and imaging system used, and includes the bandwidth of the recorded signals. Generally, one may write the overall resolution in an image as $R_d = \lambda/2$ for direct imaging methods and $R_t = (L\lambda)^{0.5}$ for tomographic methods, where R is the resolution radius, λ is the wavelength of the illuminating field, and L is the number of wavelengths the waves travel in the imaging volume. For a pathlength of 100 wavelengths – a typical value – direct imaging methods have 20 times the resolving power of tomographic ones. However, the spatially averaged velocity model provided by the tomographic methods is necessary to make a highly resolved direct image, so the two methods are complementary.

Active source seismology, which makes use of man-made (anthropogenic) energy sources, produces tomographic velocity models and reflection images. Passive or natural seismology relies on earthquake sources and produces local, regional, or global tomography images of velocity structure, receiver function images of velocity and density perturbations, and shear wave-splitting measurements of anisotropy. Refraction studies produce information on velocity structure of the crust and mantle, giving vertical layering and long wavelength lateral changes, and identifying regionally important interfaces such as the Moho. Reflection seismology, an analog to ultrasound imaging, is sensitive to the impedance contrasts in Earth, but only advanced applications of this technique image vertical boundaries or velocity changes in weakly layered crust and mantle.

Seismic refraction measurements are typically made along profiles 200–500 km long. Modern refraction surveys resolve average velocities over distances of 10–50 km, and depth intervals of 5–10 km. More simply put, generally we may estimate an average velocity in a "pixel" 5 km deep and 10–50 km long with accuracy of a few percent. Reflection images provide resolution of structures with impedance contrasts[1] at the scale of hundreds of meters vertically and laterally, meaning that we may see details in larger structures as small as ~100 m horizontally and vertically. In modern practice then, reflection images are about 2 orders of magnitude higher resolution than refraction images. Whereas reflection seismology gives us spectacular images of geologic structures, it provides poorly constrained information on seismic velocity of the crystalline crust using present imaging technologies; therefore, refraction and reflection are complementary in providing lithological inference from seismic velocity, and structural geology geometries. Modern crustal

---

[1] Impedance is the product of velocity and density: $I = V_p*\rho$. Impedance contrast is the difference in impedance between two units, usually corresponding to a lithologic, chemical, or structural boundary. The reflection coefficient from such a boundary is the impedance contrast divided by twice the average impedance.

surveying sometimes merges the reflection and refraction methods into a single experiment (e.g., Fuis *et al.*, 1995, 1997; Levander *et al.*, 1994c; Wissinger *et al.*, 1997).

Similarly, earthquake tomography provides broad averages of crust and mantle velocities, with averaging along ray directions, near vertical for teleseismic signals, more nearly horizontal for local and regional earthquakes. Seismic tomography arrays vary in size from tens of kilometers to the entire globe. For travel-time tomography, resolution laterally is somewhat less than station spacing; vertical resolution is generally worse due to narrow ranges of incidence angles. Receiver functions provide velocity and density contrast images similar to reflection seismograms, with a factor of 5–20 greater resolution than tomography (e.g., Rondenay *et al.*, 2001; Bostock *et al.*, 2002; Wilson & Aster, 2003; Levander, 2003). Receiver function images are about an order of magnitude lower in resolution than reflection seismology, as seismic reflection methods normally record in the 8–80 Hz band, and teleseismic arrays record signals at less than about 2.5 Hz.

To summarize, we learn different things about the structure of the continents from different imaging methods, and our understanding is enhanced if we merge different geophysical data sets and study these in conjunction with available geologic data. Modern seismic imaging makes use of all of the available seismic probes to produce multiscale images of the crust and upper mantle. For example, Fig. 2.1 shows a combined reflection–refraction–receiver function-tomography image from the western US Rocky Mountains (Karlstrom *et al.*, 2002). The combination of different imaging methods provides identification of features in the crust as fine as 100 m to identification of features in the mantle as large as ~100 km, a three order-of-magnitude scale range. Interpretation of the images relies heavily on geological inference.

Seismic velocity in rocks depends on a number of factors, including bulk chemistry and mineralogy (and, hence, metamorphic grade), temperature, pressure, porosity, fracturing, and the presence or absence of fluids. Seismic velocities increase rapidly with depth through the first 5 km due to the closure of fractures and pore space. At depths greater than 10 km for most crustal lithologies in a normal continental geotherm, the increase in velocity due to increasing pressure is closely matched or exceeded by the decrease in velocity due to increasing temperature.

Rocks with velocities below about 7.6 km s^{-1} measured at crustal depths (<80 km) and temperatures (<800 °C) are, by seismic definition, crustal rocks. Large contiguous bodies of higher-velocity rocks for which an unbroken path may be drawn into the deeper Earth are part of the mantle under these *P–T* conditions. From seismic velocity one may make inferences about

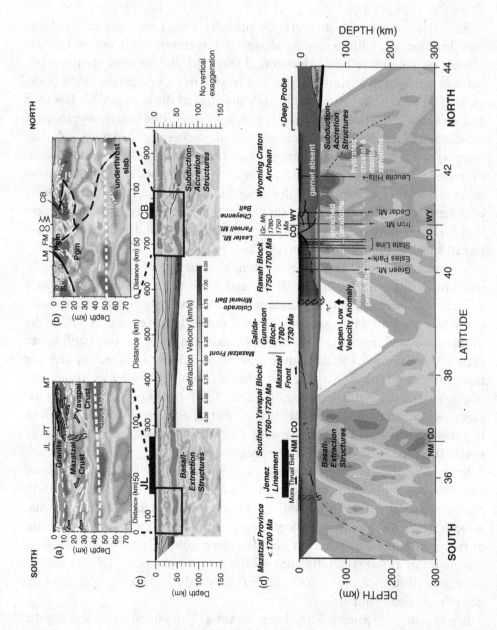

Fig. 2.1. Synthesis of seismic results from the southern Rocky Mountains. (a) Seismic reflection image (black and white lines) superimposed on depth migrated receiver function image of Jemez lineament in northern New Mexico. (b) Seismic reflection image (black and white lines) superimposed on Cheyenne Belt receiver function image in southern Wyoming. (c) Receiver function images superimposed on refraction velocity model extending from central New Mexico to southern Wyoming. (d) Mantle *P*-wave tomography velocity model, interpreted crustal section, and locations of mantle xenoliths. The combined images show upper crustal features as small as folds and thrusts in crystalline nappes ((a) between PT and MT), intermediate-scale crustal features such as the high velocity lower crustal layer, and broad-scale mantle anomalies associated with surface structures. A north-dipping positive *P*-wave velocity anomaly is correlated with the Cheyenne belt crustal structures. A south dipping negative *P*-velocity anomaly lies beneath the Jemez lineament, and is roughly correlative with a south dipping suture imaged by the seismic reflection data. (Modified after Karlstrom *et al.* (2002), Structure and evolution of the lithosphere beneath the Rocky Mountains. *GSA Today*, **12**, 4–10. © 2002 Geological Society of America. Modified by permission of the Geological Society of America.) At the time of going to press the original colour version was available for download from www.cambridge.org/9780521066068

lithology and, hence, chemistry. Crustal lithologies include felsic and mafic rocks; mantle lithologies are ultramafic. An important exception for the upper mantle is mafic rocks in the eclogite stability field that acquire velocities $(8.0–8.4 \, km \, s^{-1})$ and densities $(3300–3400 \, kg \, m^3)$ typical of mantle rocks. A second exception is serpentinite, hydrated peridotite, which may have unusually low seismic velocities $(5.5 \, km \, s^{-1})$ and densities $(2500 \, kg \, m^3)$. That high-velocity ultramafic rocks $(\sim 8 \, km \, s^{-1})$ are found exposed at the surface, and presumably low-velocity sedimentary rocks $(< 6 \, km \, s^{-1})$ are subducted beneath continental margins, points out the problems associated with simple definitions of crust and mantle based on seismic velocity alone. The adherence to a strict definition of the crust–mantle boundary, the Moho, based on seismic velocities is undoubtedly exasperating to petrologists, whose view of the crust and mantle is based on a differentiation model. We discuss this further below, as this leads to conceptual as well as semantic problems.

2.2 Crustal differentiation, melt generation and melt migration in the continents: basic considerations

Continents represent the products of repeated differentiation cycles whereby less dense, intermediate to felsic crust is upwardly stratified by combinations of tectonomagmatic processes. Accretion of uniquely continental materials (anything more felsic than basalt) takes place above subduction zones in magmatic arcs and orogenic plateaus, but formation of true continental lithosphere requires multiple differentiation events to reach the average composition of continents (which is near to andesite; see Table 2.1). These events take place during plate convergence, which may lead to collisions of arcs with arcs, arcs with continents, and continents with continents. Such events may also take place during incipient rifting of continents via interactions between lithosphere and asthenosphere.

In terms of the differentiation processes that make continental crust over time, the important physical quantities are heat sources, both mantle and crustal, magma sources within the mantle (i.e., fertile mantle), other fluid sources in the crust and mantle (e.g., providing H_2O and CO_2), and migration pathways provided by existing fractures and shear zones, or in the case of over-pressured magmas, the regional stress field that controls the orientations and locations of hydrofractures. It is generally believed that the radiogenic heat production in the crust is inadequate to produce crustal magmas, except in the greatly thickened, probably hydrated, crust occurring near or within orogenic plateaus. Most crustal melting results from heat originating in the mantle that is advectively transferred to the crust in the form of mantle-derived mafic

Table 2.1. *Chemical composition of the continental crust in weight percent*

	Continental crust (T & M, 1985)	Continental crust (R & F, 1995)	Continental crust (C & M, 1995)
SiO_2	58.0	59.1	61.7
TiO_2	0.8	0.7	0.9
Al_2O_3	18.0	15.8	14.7
Fe_2O_3			1.9
FeO	7.5	6.6	5.1
MnO	0.1	0.11	0.1
MgO	3.5	4.4	3.1
CaO	7.5	6.4	5.7
Na_2O	3.5	3.2	3.6
K_2O	1.5	1.9	2.1

T & M: Taylor and McLennan (1985).

R & F: Rudnick and Fountain (1995).

C & M: Christensen and Mooney (1995).

melts, with the melting process facilitated by other fluids residing in the crust, or transferred to the crust from the mantle, e.g., hydration melting above subduction zones.

2.3 Continental subdivisions

Traditionally, the continents have been divided into shields, platforms (shield regions covered by sedimentary deposits), Paleozoic and Mesozoic–Cenozoic orogenic belts, rifts, and extensional terranes. Continents are bounded by convergent, transcurrent (transpressional and transtensional) or passive margins. Both convergent and transcurrent margins adjacent to or within continents are generally associated with orogenic belts. Notable features within orogenic belts are Earth's three great orogenic plateaus in Tibet, western North America, and western South America, all regions of high topography but relatively low relief that are the result of plate collisions lasting throughout the Cenozoic.

Globally, of the ~40% area of Earth's surface underlain by the continents, roughly 24% of the area forms shields or platforms, about 8% Paleozoic orogenic belts, about 6% Mesozoic–Cenozoic orogenic belts, and approximately 1% rifts. Systematic variations in crustal thickness and velocity in each continental province have been reviewed by several authors (Christensen & Mooney, 1995; Rudnick & Fountain, 1995). Mean crustal thickness is 41.0 ± 6.2 km (Christensen & Mooney, 1995), with an average velocity structure implying a bulk composition close to andesite (Christensen & Mooney, 1995; Rudnick & Fountain, 1995), in agreement with estimates made from sedimentary rocks (Taylor & McLennan, 1985; Table 2.1; Chapter 4). Average crustal compressional velocity is 6.45 km s^{-1} ± 0.21 km s^{-1}; average upper mantle velocity is 8.09 ± 0.20 km s^{-1}. The crust in cratonic regions is ~41–43 km thick on average, but varies in thickness from 32 to 55 km (Table 2.2), in Mesozoic–Cenozoic orogenic belts the crust is 41–78 km thick, in Cenozoic extensional provinces the crust is ~28–35 km thick, and passive continental margins are simply stretched continents, so the crust varies from oceanic to the thin side of continental crustal thickness, from ~10 to ~35 km thick. Regions of Cenozoic tectonics generally have detectably lower seismic velocity in the crust and mantle than older regions (6.21 and 8.02 km s^{-1} compared to 6.45 and 8.09 km s^{-1}, respectively), due to increased temperatures.

The only pervasive seismic boundaries of the continental crust are its surface and the Moho. In areas with sedimentary cover, the crystalline basement–sediment interface, referred to as the "acoustic basement," is usually identifiable. In some provinces a high-velocity lower crustal layer has been identified, and in Europe this interface has been referred to as the "Conrad discontinuity." In some areas, a refraction arrival is clearly associated with a high velocity (7.00 km s^{-1} +) lower crustal layer indicative of a mafic lower crust (e.g., Henstock *et al.*, 1998; Snelson *et al.*, 1998). In many areas seismic fabric changes corresponding to the onset of bright reflectivity or its disappearance, rather than bulk velocity changes, produce wide-angle reflection arrivals of the Conrad discontinuity (Levander & Holliger, 1992). In this case, the boundary in question identifies a change in lithologic fabric that scatters seismic waves rather than a change in bulk velocity between two layers.

Discussions of the meaning of the Moho are extensive, and frequently semantically complicated. Defined seismically as the boundary separating velocities less than about 7.6 km s^{-1} from layers with greater velocities, most Earth scientists equate Moho with crust–mantle boundary, a meaning that implies different things to different Earth scientists. For instance, Griffin and O'Reilly (1987) have noted that crustal rocks in the granulite and eclogite metamorphic fields may have high seismic velocities that place them in the mantle by the seismic Moho definition, but are still crustal rocks based on petrology. Similarly, where the Moho is a detachment surface, upper mantle and lower crustal rocks may be tectonically interfingered, putting ultramafic rocks in the lower crust, so that average velocities are lower than 7.6 km s^{-1}, but individual slivers have higher velocity. The conceptual dilemma arises because the means of remotely measuring the location of the continental Moho based on its definition is relatively simple, but the crust–mantle boundary itself commonly is not a simple boundary, nor even the result of a single process at one location.

The differences between the continents and oceans extend much deeper than their crusts, and it is difficult to discuss the continental crust in any modern sense without reference to the continental mantle. The concept of the tectosphere, a long-lived high velocity but low-density mantle root beneath the cratons that extends deep into the upper mantle (200–350 km), appears well established (Jordan, 1975, 1988), although its depth extent is still debated (e.g., Polet & Anderson, 1995; James *et al.*, 2001). The tectosphere is thought to be comprised of a highly iron-depleted peridotitic residuum, at least some of which dates to continental formation in the Archean. The fertile mantle system is unusual in that extracting the light component (basalt or komatiite under high temperatures) produces two materials that are both less dense than the starting material. The greater the depletion in iron – brought on by higher temperatures – the greater the corresponding density reduction in the residuum. The tectospheric mantle is from ~1 to 2% faster, and from ~1 to 2% less dense than the corresponding upper mantle of the oceanic lithosphere (Jordan, 1988; Lee, 2003). Distributed uniformly through a 400 km column, one may account for the bulk difference in elevation of a 120 Ma oceanic lithosphere (~−4 km) and >2 Ga continent (~+250 m) entirely with buoyancy forces.

Detailed studies of tectospheric mantle beneath the continents form a large part of modern structural seismology, particularly using dense arrays of earthquake recorders. Recent large-scale seismic studies in Australia and North America have shown a surprising complexity in ancient and modern mantles (e.g., Van der Hilst *et al.*, 1998; Karlstrom *et al.*, 2002). Here we refer to an ancient mantle as depleted tectospheric mantle with limited or no mobility, whereas modern mantle may be fertile or recently depleted and implies mobility. The Canadian Lithoprobe results from the Proterozoic Wopmay Orogen/Slave Province suggest that at least in the Proterozoic, the tectospheric mantle was formed in part by accretion of subducted lithosphere around the peripheries of the cratons, i.e., by largely horizontal advection of mass (Cook *et al.*, 1999), a model proposed by Jordan (1988).

Crust–mantle interactions constitute a large part of modern solid Earth science investigation of the continents. The regularity of the oceanic lithosphere in formation, thickening, subsidence, and subduction is in stark contrast to the complexity of continental crust–mantle interactions. The continental crust is characterized by lateral and vertical heterogeneity in bulk chemistry, mineralogy, lithology, fracture density, and fluid content, and therefore in rheology. The recent investigations of the continental lithospheric mantle suggest that it may be also far more complex than its oceanic counterpart (e.g., Thybo & Perchuc, 1997; Ryberg *et al.*, 2000; Dueker *et al.*, 2001).

2.4 Continental rheology

Important as observational seismology is to understanding the continental crust, equally important are the developments in understanding and inferring the rheology of the continental crust from geologic studies, laboratory experiments, lithospheric flexure studies, gravity and geodetic data, and modeling. Kohlstedt *et al.* (1995) provide a review of continental rheology based on laboratory-derived constitutive laws for typical mineral species and rocks, notably quartz, calcite, feldspar aggregates, pyroxenes, olivine, and peridotite. The now standard continental crustal strength profile shows strength of the upper crust dominated by brittle frictional deformation – a pressure-sensitive phenomenon – and lower crustal deformation governed by temperature-sensitive plastic flow. The region of the crust where plastic failure occurs at approximately the same stresses as frictional failure is termed the "brittle–ductile transition zone" where semi-brittle behavior occurs, and generally coincides closely with the bottom of the seismogenic zone. Below this to the top of the strong mantle is the regime of plastic flow. The resultant crustal rheology profile is euphemistically referred to as the "jelly sandwich-" type continental lithosphere, with the strongest parts in the middle crust (just above the brittle–ductile transition) and in the uppermost mantle (Fig. 2.2; Chapter 8.)

In simple strength–depth profiles, the upper crust and upper mantle deform by brittle failure on pre-existing fractures, and yield strength increases linearly with depth. This relation, known as Byerlee's Law (Kohlstedt *et al.*, 1995), is expressed as the following for upper crust and mantle, respectively:

$$\tau = 0.85\sigma_n \quad 3 < \sigma_n < 200 \text{ MPa}$$

$$\tau = 60 + 0.6\sigma_n \quad 200 \text{ MPa} < \sigma_n$$

where σ_n is effective normal stress, and τ is the strength. The lower crust and lower parts of the mantle deform predominantly via power law creep of quartz and/or feldspar and olivine, respectively, where strain rates increase, i.e., strength decreases, exponentially with increasing temperature

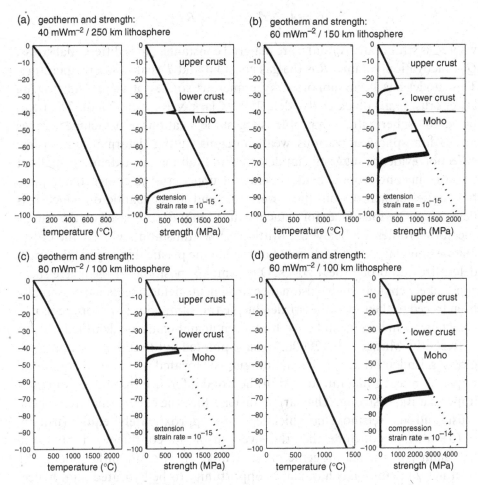

Fig. 2.2. Geotherms and lithospheric strength profiles. In all plots the crust is 40 km thick, has a granite rheology in the upper 20 km, and a diorite rheology in the lower 20 km (Hansen & Carter, 1982). Different curves in the upper mantle (solid and dashed) correspond to different values of the material constants for upper mantle rocks (see Evans & Kohlstedt (1995), and references therein) (a)–(c) are for a lithosphere in extension with a strain rate of 10^{-15} s^{-1}; (d) is for a lithosphere in compression with a strain rate of 10^{-14} s^{-1}. (a) A tectospheric lithosphere 250 km thick with surface heat flow 40 m Wm^{-2}. (b) An intermediate thickness lithosphere, 150 km thick, with a surface heat flow of 60 m Wm^{-2}. (c) A thin, 100 km, lithosphere with a surface heat flow of 80 m Wm^{-2}. (d) A thin, 100 km thick lithosphere with a surface heat-flow of 60 m Wm^{-2} under compression. In moderate to high heat-flow regimes, the lower crust undergoes varying degrees of plastic flow. In compression, the lower crust may act as a broad zone of detachment.

$$\dot{\varepsilon} = A_d \Delta\sigma^n \exp(-Q_d/RT)$$

where $\dot{\varepsilon}$ is strain rate, A_d and n are material constants, $\Delta\sigma$ is differential stress, Q_d is activation volume, R is the gas constant and T is absolute temperature. This model has been supported by numerous studies of weak and flowing lower crust (e.g., Block & Royden, 1990; Clark & Royden, 2000). The combined lithospheric strength profiles suggest the locations of detachment zones in the lithosphere at various weak horizons during deformation. Seismic reflection evidence suggests that detachment faulting may extend throughout the crust in contractional belts, making the upper crust in contractional plate boundaries detached from the lower crust and upper mantle on which it is deforming. Obvious zones for detachments to root or sole based on the crustal rheology profiles are below the brittle–ductile transition and near the crust–mantle boundary. As detachment faults enter the plastic flow regime they may delocalize into broad shear zones. Over time the positions of the active shear zones likely change. In extensional environments detachments have generally been traced to the top of the brittle–ductile transition, but few thoroughgoing detachments in continental crust have been unambiguously identified.

Recently, Maggi *et al.* (2000a, b) have pointed out that there is almost no global seismicity that may be unambiguously located within the continental upper mantle, calling into question the usual inference that the continental upper mantle is strong. They argue instead that the elastic thickness of the crust and its seismogenic thickness are approximately equal (usually 8–15 km), and conclude that the strength of the continental lithosphere lies mainly in the seismogenic crust. Unlike the oceanic lithosphere, the continental mantle has had ample opportunity to be hydrated and, hence, weakened since its formation. In this case the "jelly sandwich" is open-faced, down.

Complications to the simple 1-D rheological model arise from variations in fluid and magma pressures, thermal weakening adjacent to localized heat sources such as intrusions, marked heterogeneity in rock type, and localization of high strain rates. This latter factor may cause large lateral variations to form within a deforming section of the crust even if its pre-deformation structure was laterally uniform. Finite-element models of crustal deformation driven by imposed basal boundary conditions meant to mimic the subducting portion of the mantle lithosphere have shown this quite clearly (e.g., Willett *et al.*, 1993). Figure 2.3b shows the results from such a model. The large lateral variations in deformational response, due to dynamic strain-rate localization, clearly point to the limits of using 1-D strength profiles in tectonically active regions. More recently, models have been

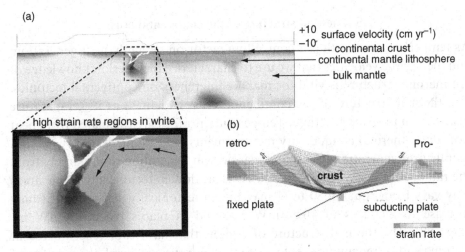

Fig. 2.3. (a) A model of coupled mantle convection and continental tectonics. Continental mantle lithosphere (green) subducts causing the crust (brown) to deform. The non-linear crustal rheology leads to localization in shear zones (from Lenardic *et al.* (2003), Longevity and stability of cratonic lithosphere: insights from numerical simulations of coupled mantle convection and continental tectonics, *Journal of Geophysical Research*, doi: 10.1029/2002JB001859. © 2003 American Geophysical Union. Reproduced by permission of American Geophysical Union). (b) Model of crustal deformation driven by a static velocity field associated with subcontinental subduction. The Pro- and Retro-refer to localized high strain-rate zones that form in the model and that are observed in several orogens. (From Willet *et al.* (1993), Mechanical model for the tectonics of doubly vergent compressional orogens. *Geology*, **21**, 371–4. © 1993 The Geological Society of America. Reproduced by permission of the Geological Society of America.) The 2-D numerical simulations identify regions of strain localization in the crust and mantle. At the time of going to press the original colour version was available for download from www.cambridge.org/9780521066068

developed that do not require mantle flow to be imposed as a boundary condition on the crust, but that rather couple crustal deformation and mantle convection (Fig. 2.3b; see Lenardic *et al.*, 2000, 2003). The model of Fig. 2.3a confirms a key assumption inherent to the model of Fig. 2.3b, i.e., the idea that detachments may form in the lower crust that allow for a decoupling between the subducting mantle component of the lithosphere and the crust above. As computer capabilities increase, and the sophistication of the models themselves increase, these simulations will undoubtedly have greater predictive value at the field geology level. This will require geometric improvements, 3-D modeling and improvements in physical processes that are allowed within the models, e.g., the allowance for lateral variations due to rheologically and/or chemically distinct rock types.

2.5 Thermal structure of the continental crust

As temperature in Earth is a controlling factor on strength, melt generation, and the seismic velocities that we use to characterize the interior, knowledge of the thermal structure is vital for making inferences on continental tectonics. The thermal structure of the continental crust is defined as its temperature distribution in space and time. Temperature profiles are calculated from heat flow data, thermal conductivity measurements, and estimates of crustal and mantle heat generation. Crust and mantle heat generation result largely from the radioactive decay of U, Th, and K. Heat flow in tectonically stable areas may vary from roughly 30 to 60 mW m^2; in tectonically active areas it may increase to averages of 70–90 mW m^2, and in places reach 150 mW m^2 (Fig. 2.2). The thermal structure of regions that have experienced a recent tectonic and/or magmatic event will contain a transient signal, the specifics of which will depend on the nature of the event, making it difficult to generalize about tectonically active regions. For this reason, we focus on stable regions that have not experienced a major tectonothermal event within the last several 100 Myr, after which heat flow decays to local equilibrium values.

The thermal structure of stable crust will depend on its heat generation and thermal conductivity structure and on the mantle heat flow into its base. The concentration of radiogenic elements within a crustal column depends on the rock types that compose it. Thus, the thermal and chemical structures are connected. Another connection that must be considered is the thermal coupling between the crust and the mantle, which must be constrained to characterize the thermal structure of the continental lithosphere as a whole. This may be done if the mantle and crustal contribution to continental heat flow may be separated. Based on analysis of global heat flow data, mean surface heat flow values from Archean, Proterozoic, and late Paleozoic regions have been determined to be 41 mW m^2, 55 mW m^2, and 61 mW m^2, respectively (Pollack, 1982; Morgan & Sass, 1984; Nyblade & Pollack, 1993a, b). Efforts to get at the nature of crust–mantle thermal coupling have focused on extracting the crustal contribution from this integrated signal. The thermal conductivity of crystalline rocks has a relatively small range for temperatures below 1300 K. For thermal modeling purposes, constant values of 2.5–3.0 W m^{-1}K for the crust and 3.0 to 3.5 W m^{-1}K for the mantle have been adopted by many workers and more elaborate conductivity functions tend to produce second order effects (Morgan & Sass, 1984).

The distribution of heat-producing elements within different continental regions is more difficult to constrain. The earliest efforts to do so relied on an empirical relationship between surface heat flow and near-surface heat production (Birch *et al.*, 1968), suggesting that crustal heat production decreased exponentially with depth and that lower crustal heat production was very low. However, the significance of this relationship was questioned early on (Jaupart, 1983), and it has been shown not to hold in many regions (Guillou *et al.*, 1994). Further, a simple exponential dependence does not hold in regions in which full crustal columns are exposed, so that direct heat production measurements may be made (e.g., Ashwal *et al.*, 1987; Ketcham, 1996). These studies confirmed that lithology plays the dominant role in determining heat production, and they showed that heat production within the lower crust may be significant.

The failings noted above have led to the general abandonment of seeking any one generic function to describe the distribution of heat sources within stable continental crust worldwide. A less global, although not truly regional, approach has been to seek a representative chemical structure, and by association, a heat source distribution, for Archean, Proterozoic, and Paleozoic crust (e.g., Rudnick & Fountain, 1995). This approach has been driven by the observed globally averaged variation of surface heat flow already noted for these age provinces. Globally averaged trends may not hold locally. For example, global data suggest that the surface heat flow of Proterozoic terrains is higher than that of Archean (Nyblade & Pollack, 1993a), but heat-flow measurements from the Canadian Shield do not show such a trend (Jaupart *et al.*, 1998). Given that the component of crustal heat depends on local crustal thickness and lithology, it is not surprising that regional trends may differ from globally averaged trends. Thus, care must be taken in applying global estimates to a specific geologic region.

Tectonic activity has exposed crustal cross-sections in several regions worldwide allowing direct crustal composition and heat source distribution models to be made (Nicolaysen *et al.*, 1981; Weaver & Tarney, 1984; Ashwal *et al.*, 1987; Fountain *et al.*, 1987; Pinet & Jaupart, 1987; Ketcham, 1996). No section discovered to date is fully complete. Most commonly, the lowermost crust is missing. This problem is alleviated to a degree as a variety of studies suggest that some granulite facies terrains may be representative of the lower crust, and the measured heat production of such terrains worldwide clusters at a value of approximately 0.4 $\mu W\ m^{-3}$ (Pinet & Jaupart, 1987). There is also agreement that much of the middle crust contains rocks in the amphibolite facies that have an average heat production between 1.0–1.2 $\mu W\ m^{-3}$ (Rudnick & Fountain, 1995; Jaupart *et al.*, 1998). Heat production within the uppermost crust may be measured directly. It depends strongly on rock type with values ranging from 0.2–3 $\mu W\ m^{-3}$ (e.g., Pollack, 1982; Wollenberg & Smith, 1987; Guillou *et al.*, 1994). In practice, crustal models are constructed allowing for a degree of variation in the heat source concentration of lower, middle, and upper crustal layers with values of the type above providing estimates of likely mean values.

Although seismic data may constrain the thickness of crustal layers within given regions, ~10% uncertainty will still exist. Taken with the uncertainty in crustal heat production values, this means that thermal models cannot be fully constrained. None the less, at least one very robust conclusion has been drawn for Archean terrains, i.e., it is now generally agreed that the mantle heat-flow into Archean crust is relatively low and uniform with values that fall in the approximate range 10–$20 \, mW \, m^{-2}$ (Nicolaysen et al., 1981; Weaver & Tarney, 1984; Ashwal et al., 1987; Fountain et al., 1987; Pinet & Jaupart, 1987; Gupta et al., 1991; Pinet et al., 1991; Guillou et al., 1994). This does not mean that a single thermal structure may be applied to all sections of Archean crust, as lateral variations in crustal composition within Archean cratons may lead to variations in thermal structure. The key point is that the variations are dominantly controlled by crustal heterogeneity as opposed to deeper mantle effects. Further conclusions are less well agreed upon.

The smaller number of exposed Proterozoic crustal section and the larger range of surface heat flow values in Proterozoic relative to Archean terrains makes it a more difficult task to extract the crustal and mantle components of heat flow in Proterozoic terrains. The difficulty may be made clear by noting that one of the simplest questions related to the thermal structure of stable continental regions is still not fully answered. That being so, what is principally responsible for surface heat flow variations in stable continental regions: is it variations in crustal heat production (e.g., Morgan, 1985) or variations in mantle heat flow (e.g., Ballard & Pollack, 1987)? The issue has centered largely on heat flow variations from Archean to Proterozoic terrains and, from a global viewpoint, it remains open. However, progress has been made from a more regional viewpoint.

Addressing crustal thermal structure in a regional framework involves comparing surface heat flow and crustal structure in a specific well-mapped geologic province. To date, the most complete regional study is from central and eastern Canada (Mareschal et al., 1989, 1997, 1999; Pinet et al., 1991; Guillou et al., 1994; Guillou-Frottier et al., 1995, 1996; Jaupart et al., 1998). The key result, related to the issue at hand, is that surface heat flow variations, from stable regions spanning Archean to Paleozoic ages, may be accounted for by known variations in crustal composition with the component of mantle heat flow having a low, near uniform, value of 10–$15 \, mW \, m^{-2}$. This conclusion need not hold for stable continental regions worldwide. Indeed, the Canadian data point to the potential problems of global generalizations, as they show that significant heat flow variations may exist within a province of a single geologic age. None the less, the Canadian data do suggest that the thermal structure of stable continental crust, in one large geologic region at least, is principally determined by its thickness and chemical structure, including lateral variations, with mantle effects playing a lesser role. Similar detailed regional studies are needed to confirm whether this conclusion applies to other stable geologic provinces.

As noted at the beginning of this section, our discussion has focused on regions that have not experienced relatively recent (within 100 Myr) thermo-tectonic disruptions. As such, it does not apply to regions of active rifting, volcanism, and/or contractional tectonics. The thermal structure of such regions cannot be separated from their tectonic and/or magmatic history, which requires that coupled thermal–tectonic models be explored. The range of literature for such efforts is as large as the number of regions with distinct tectonic histories. We will focus on the more recent efforts that seek to couple thermal and tectonic modeling of continental lithosphere within a mantle convection framework in Section 2.7.

2.6 Structural seismology of the continental crust and its provinces

Several authors have compiled average seismic velocity structures of the different tectonic provinces of the continental crust based on the regional averaging made by seismic refraction measurements (Smithson *et al.*, 1981; Rudnick & Fountain, 1995; Christensen & Mooney, 1995). Rudnick and Fountain (1995) and Christensen and Mooney (1995) provide large compilations of the seismic refraction measurements of the continental crust, as well as the petrophysical properties of its constituent rock types. With some small differences, the two groups subdivide the continents into shields and platforms, (Phanerozoic) orogenic belts, continental arcs, extended crust, and rifts.

The 1-D continental velocity structure of Christensen and Mooney (1995) is shown in Fig. 2.4, along with their average inferred nonlinear velocity–density relationship. The crustal chemistry estimates given from the seismic velocity analysis, along with estimates from several other studies, are given in Table 2.1. Pressure and temperature derivatives of most crystalline rocks nearly cancel one another for an average geotherm (and, in fact, the velocity decrease associated with temperature is somewhat greater than the pressure-induced increase), so the general increase of velocity with depth in the CM1 model is interpreted as an increase in mafic content and metamorphic grade in the crust.

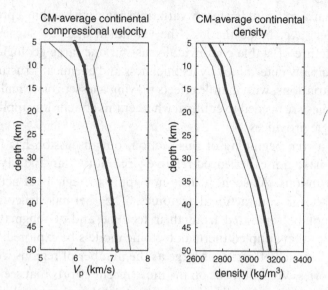

Fig. 2.4. Average crustal velocity–depth and density–depth functions (after Christensen & Mooney, 1995). The average velocity profiles are based on a worldwide compilation of seismic refraction profiles. The density profile is based on a non-linear velocity–density relationship. The thin lines show the first standard error on each measurement.

It is interesting to note that the CM1 average velocity profile probably does not exist anywhere on the continents. We will discuss this more fully below. It is also important to note that the average velocity–depth functions for the different provinces, shields, orogens, continental arcs, extended crust, and rifts, only differ by 2.5–4.5% over the whole depth range – smaller than the formal errors associated with the continental averages. The average density profile derived from the velocity–density relationship is an important reference to consider when making arguments of melt emplacement at neutral density levels in the crust (Fig. 2.4).

In contrast to the fairly smooth average refraction velocity function (Fig. 2.4), field geology and reflection seismology (Figs. 2.1, 2.5–2.8) have shown the continental crust to be highly heterogeneous, with evidence for structural complexity from the largest scales, (e.g., crustal thickness), to the hand sample, a scale range $>10^4$. (At smaller scales than a hand sample, mineral chemistry imposes a different set of scales.) However, the reflection and refraction observations are not in contradiction with one another – refraction measurements are lateral averages. The first standard deviation on the average continental velocity depth function (Fig. 2.4) permits reflection coefficients of ~0.1 throughout the crust. This is a substantial reflection coefficient, and one commonly inferred for the brightest reflections observed in deep crustal data (Warner, 1990). The extensive rock physics tabulations presented in Christensen and Mooney (1995) show broad ranges of velocity for most rock types constituting the middle and lower crust, including bimodal distributions of velocity in rock types expected at a given depth in the crust (Christensen & Mooney, 1995, Fig. 15). This laboratory result underscores the point that the manner in which different lithologies are interlayered at scales discernible by reflection seismics determines crustal reflectivity (e.g., Holliger *et al.*, 1993; Holliger & Levander, 1994a, b; Levander *et al.*, 1994a, b; Hurich, 1996; Hurich & Kocurko, 2000). The field, petrophysical, and seismic reflection evidence points to a highly heterogeneous, as opposed to well-mixed, crust.

Models for complex crustal reflectivity have been based on detailed outcrop studies of middle and lower crustal exposures that are treated deterministically (e.g., Rutter *et al.*, 1999; Khazenehdari *et al.*, 2001) or stochastically (e.g., Holliger & Levander, 1992; Holliger *et al.*, 1993; Levander *et al.*, 1994a; Goff & Levander, 1996). The former attempt to predict exact reflections, the latter to reproduce pervasive reflection patterns. The goal of both methods is to contribute to the structural interpretation of the crust, provide quantitative estimates of the lithologies present, and constrain developmental models of crustal formation (e.g., Smithson & Johnson, 2003). Interestingly, all geologic maps analyzed to date, as well as most well logs (e.g., Holliger & Goff, 2003) in the crystalline crust have self-affine (i.e., fractal) spatial correlation properties (Fig. 2.8).

2.6.1 Cratons and shields

A number of surprises in deep crustal structure come from the detailed work of the Canadian Lithoprobe program in cratonic North America. Recognized as a collage of microcontinental granite–greenstone belts welded together by Proterozoic deformed belts (Hoffman, 1988), the cratons have proven to be structurally as complex and not dissimilar from modern collision zones. Paleosubduction structures have been identified in the crust and upper mantle in the 2.69 Ga Superior Province (Fig 2.5; Calvert *et al.*, 1995) beneath the Abitibi granite–greenstone belt and the plutonic, arc-related Opatica Belt. When examined in aggregate, many of the Precambrian crustal structures of the Canadian Shield are remarkably similar to those of modern orogenic belts (but lacking both surface and Moho topography), with seismic images showing that the whole crust was involved in large-scale horizontal detachment and translation at collision zones between island arcs and/or microcontinents. Crustal reflectivity is often similar in the upper and lower crust.

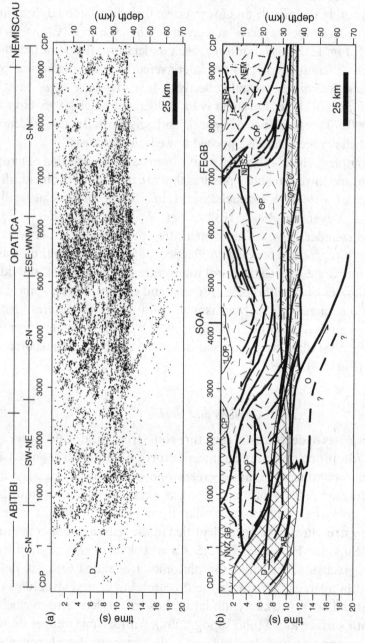

Fig. 2.5. Seismic reflection section and interpretation from the Archean Superior Province. (From Calvert *et al.* (1995), Archaean subduction inferred from seismic images of a mantle suture in the Superior Province. *Nature*, **375**, 670–4. © 1995 Nature Publishing Group. Reproduced by permission of Nature Publishing Group.) The seismic image is shown at 1:1 scale. The Moho is at 37–42 km depth. The bright sub-Moho reflection is interpreted as a relict 2.69 Ga subduction zone resulting from collision of the Abitibi granite–greenstone subprovince with the Opatica belt, implying that tectonic processes were involved in Archean continental formation.

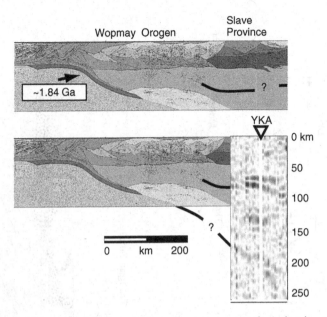

Fig. 2.6. Interpreted deep crustal and upper mantle seismic reflection profile (top and bottom) and receiver functions (bottom) through the Proterozoic Wopmay orogen and Archean Slave Province showing spectacular mantle reflectivity. (From Cook *et al.* (1999), Frozen subduction in Canada's Northwest Territories: lithoprobe deep lithospheric reflection profiling of the western Canadian Shield. *Tectonics*, **18**, 1–24. © 1999 American Geophysical Union. Reproduced by permission of American Geophysical Union.) Receiver function images made at the Yellowknife (YKA) array show PdS converted events at the same depths as the vertical incidence reflections (H, and possibly X and L). The Moho is relatively flat under the Wopmay orogen. The profile shows an east dipping 1.84 Ga subduction zone in which the subducting oceanic lithosphere appears to be tectonically accreted to the underside of the Wopmay orogen and the Slave craton. Crustal shortening is accommodated by a succession of west vergent crustal scale thrusts and duplexes.
At the time of going to press the original colour version was available for download from www.cambridge.org/9780521066068

What appear to be paleo-detachments extending from the upper crust to the lower crust and Moho are common, and are shown spectacularly in the Slave profile crossing the Proterozoic Wopmay Orogen and Slave Province (Fig. 2.6; Cook *et al.*, 1999). The Moho is at a near constant depth of 33–37.5 km. This section shows both impressive crustal deformation involving the entire crust, and a paleo-subduction zone in the upper mantle to depths of 95–100 km that may be traced over distances of 200 km. The latter includes fragments of presumably Proterozoic oceanic lithosphere that have contributed to the growth of cratonic mantle, i.e., the subducting lithosphere accretes to the underside of the continental tectosphere.

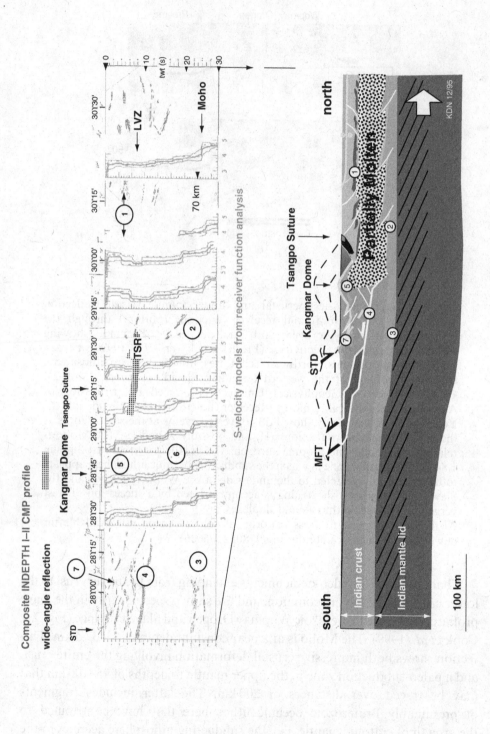

Fig. 2.7. Geophysical observations (top) from Project INDEPTH in the southern Tibetan Plateau with interpretation (bottom). (From Nelson *et al.* (1996), Partially molten middle crust beneath southern Tibet: synthesis of Project INDEPTH results. *Science*, **274**, 1684–7. © 1996 AAAS. Reproduced by permission of AAAS). Top: Blue (see color plate) are 1-D receiver function *S*-velocity-depth profiles. Red (see color plate) are events from a vertical-incidence seismic reflection profile. Blue stipple (see color plate) is wide-angle reflections through a gap in the vertical-incidence profile. The combined measurements suggest that the middle crust north of the Kangmar Dome from 20–40 km depths is partially molten, as evidenced by the low velocity S-wave velocity zone (LVZ) and the bright reflections that delineate its top (1). South of the Kangmar Dome the measurements image the thrust belt at the Indian-Asian collision zone. The Moho is at ~70 km depth. STD is the Southern Tibetan Detachment system; (4) is the Main Frontal Thrust. At the time of going to press the original colour version was available for download from www.cambridge.org/9780521066068

2.6.2 Orogens

Modern orogens and collapsed orogens, i.e., extended or extending terranes, may have the simplest reflection patterns if the most recent tectonics produces a strong overprinting on whatever pre-existing reflection fabric existed. In provinces that have experienced repeated tectonic events, unambiguous interpretation of seismic reflection data from the crust may be difficult or impossible without corroborating evidence provided by wells, geochronology/thermometry/barometry and structural geology. A form of seismic stratigraphy for the crystalline crust based on cross-cutting structural relations has been proposed by Brown (1998) as a means of providing relative ages for reflections.

Whole crustal detachment faulting related to foreland fold and thrust belts has been identified as a dominant deformation mode in modern contractional orogens, as seen in the southern Appalachians (Cook *et al.*, 1981), the Alps (e.g., Valasek *et al.*, 1991; Ansorge *et al.*, 1991), the Brooks Range (Wissinger *et al.*, 1997; Fuis *et al.*, 1997) and the Canadian Cordillera (e.g., Cook, 1995). Crustal thickness of orogenic belts is 30–80 km, with the latter thickness generally being obtained only in the great orogenic plateaus. The basal detachment surfaces root within the lower crust or at the Moho, often forming a province wide zone of orogenic float (Oldow *et al.*, 1990). These structures have been imaged at depths greater than 40 km; they allow subduction of a large volume of oceanic lithosphere while accommodating an order of magnitude less crustal shortening. When viewed in total cross-section, many of the orogens are bivergent (Willet *et al.*, 1993) but asymmetric, forming on thrust faults of opposite vergence on either side of a mantle root zone. The root zone may be overlain by a complicated, melt-depleted crust (e.g., the Alps and the Canadian Cordillera).

The orogenic plateaus are of particular note. The Indian–Eurasian collision is the only large-scale continent–continent collision acting today, although similar collisions occurred in the past. Deep reflection profiling in the Tibetan plateau has identified the Main Himalayan Thrust extending to > 45 km depth beneath southern Tibet and the Moho at 75 km depth (Fig. 2.7; Nelson *et al.*, 1996; Brown *et al.*, 1996). Seismic analyses of many wave types and magnetotelluric observations have led to an interpretation of a series of bright reflections at ~15–20 km depth in south-central Tibet as the tops of partially molten granitic plutons extending through the mid-crust over a distance > 200 km from approximately the Zangbo suture to the north (Nelson *et al.*, 1996; Makovsky *et al.*, 1996; Kind *et al.*, 1996; Chen *et al.*, 1996). Although argument persists as to whether the bright reflections result from steam-water systems (Makovsky & Klemperer, 1999) or from ponding silicate melts, there is compelling evidence and consensus that a zone within the crust extending from 15–30 km depth is in a partially molten state, and is a source of granitic melt.

In the southern plateau melting is thought to arise from heating during crustal thickening due to the concentration of large volumes of radiogenic elements in the abnormally thickened crust, rather than from advection of mantle melts. This zone of partial melt mechanically separates the upper crust from the lower, acting as a crustal asthenosphere, and provides an explanation for the absence of appreciable relief on the Tibetan plateau, i.e., the flowing middle crust is too weak to support much topography. Mechie *et al.*, (2004) identify a velocity contrast at 18–32 km depth in the crust beneath central Tibet that they identify as the phase transition from α to β quartz, one of the few phase transitions observed seismically in Earth's crust resulting from high temperature and a quartz-rich felsic crustal column.

Western North America has been the site of ocean–continent convergence since the Cretaceous, transitioning to transform motion beginning at ~28.5 Ma. The plateau has a variety of different structural styles and crustal thicknesses. The western plateau is experiencing extension that initiated at 55 Ma and now has thin, ~30 km crust, but is still elevated (~2 km) due to the chemically depleted and hot upper mantle (Humphreys & Dueker, 1994a, b; and see following section on extended crust). In contrast, the eastern intact plateau has much thicker crust (40 to > 50 km), and higher elevations. It also has remarkably different crust and mantle structures along the strike, changing fairly abruptly across the Paleoproterozoic Cheyenne Belt in southern Wyoming (Fig. 2.1). To the north, in the deformed Wyoming Archean craton, the crust is ~50 km thick, the bottom half of which is a 25-km thick layer with high compressional velocity. Beneath the southern Rocky Mountains and the Colorado Plateau, crustal structure is more varied, with the crust ranging from 40 to > 50 km thick, and the high-velocity lower crust is less well-developed and in places absent (Henstock *et al.*, 1998; Snelson *et al.*, 1998; Rumpel *et al.*, 2005; Karlstrom *et al.*, 2002; Levander *et al.*, 2005).

Upper mantle velocities undergo a profound change across this boundary as well. Beneath the Wyoming and northern cratonic blocks, compressional velocities are high and increase with depth (from 8.1–8.2 at the Moho to 8.6 km s^{-1} at 125 km depth) signifying a tectosphere-like mantle, whereas to the south they are low (8.0 km s^{-1}) with a low velocity channel, ~7.7 km s^{-1}, at relatively shallow depths (65 km), suggesting a young, or at least rejuvenated, and mobile upper mantle (Henstock *et al.*, 1998; Yuan & Dueker, 2005). Parts of Proterozoic southwestern North America are supported isostatically by the low-density upper mantle, rather than by crustal roots (Sheehan *et al.*, 1995), and parts by crustal roots.

In western South America, a zone of modern ocean–continent convergence, the orogenic plateau, consisting of the continental arc, the Altiplano–Puna plateau, and the eastern cordillera, is relatively narrow, nowhere being greater than ~600 km, in contrast to the much greater widths (~1200 km) of the Tibetan and North American orogenic plateaus. Standing at about 4 km elevation with crustal thicknesses up to ~75 km, the South American plateau has many features in common with the Tibetan Plateau, despite being the result of ocean–continent convergence rather than continent–continent collision, and having largely developed in the past 20 Myr. Most striking are the receiver function and wide-angle reflection observations of a plateau-wide 10–20-km thick mid-crustal low-velocity zone, inferred to be partially molten and a source for granitic magmas (Yuan *et al.*, 2000). This low-velocity layer is interpreted as a ductile zone separating brittle deformation in the upper crust from a stronger residuum in the lower crust. Moho depths vary from 75 km under the Andean volcanoes to as little as 40 km under the plateau, representing a high degree of lithospheric heterogeneity. Another important observation is the image of the subducting oceanic crust, which may be traced to depths of 80 km in reflection data (ANCORP Working Group, 1999), and 120 km on receiver function images (Yuan *et al.*, 2000), at which depths the basalt-to-eclogite phase transition is believed to have reached completion, eliminating the seismic velocity contrasts between oceanic crust and mantle that give rise to the seismic signals.

2.6.3 Rifts

Rift zones represent a small fraction of the continental crust, yet have assumed large importance because of the repeated construction of supercontinents followed by their subsequent dispersal by formation of new ocean basins. Most extended terranes are presumed to have started as rifts. Numerous continental rifts have been studied, including the Mid-Continent Rift and the Rio Grande Rift in North America, the East African and Red Sea Rifts in Africa, the European Cenozoic Rift system, and the Baikal Rift in Asia. Dead rifts, such as the Mid-Continent Rift in the US, have been described as "failed ocean basins," prompting K. Burke (personal communication) to comment that they are quite successful rifts, with the implication that a rift and an ocean basin are not necessarily two consecutive parts of the same process. A continental rift has been defined by Burke (1977) and modified by Olsen and Morgan (1995) as an "elongate tectonic depression associated with which the entire lithosphere has been modified in extension." Consequently, continental rifts are characterized by a distinct, elongated topographic low normally flanked by uplifts, high heat flow ($>100\,mW\,m^{-2}$), thin crust (20–36 km), and unusually low upper mantle seismic velocities. Along the strike of the Kenya Rift, the upper mantle compressional velocities are $\sim$7.6 km s^{-1}, with $\sim$12 km of Moho relief. Surprisingly, across strike the low upper mantle velocities extend little further than the surface expression of the rift zone (Fig. 2.9).

2.6.4 Extended crust

Extended terranes such as the US Basin and Range Province and parts of the Western European Variscides are thought to be an end member in the progression of a continental rift, having experienced more than 50% extension. Like rifts, modern extended terranes also have relatively thin crust (30 km) and high heat flow, low upper mantle seismic velocities, and a characteristic seismic signature associated with crustal extension – an upper crust of moderate to low reflectivity cut by brittle normal faults overlies a brightly reflective lower crust dominated by subhorizontal reflections. The Moho in extended terranes typically has little topography at large wavelengths (e.g., Klemperer *et al.*, 1986; Larkin *et al.*, 1997). Seismic fabric analysis of data from the Basin and Range and from offshore southwest Britain (Fig. 2.8) show that the crust has statistical scale lengths[2] of 250–600 m (Larkin, 1996; LaFlame, 1999; Pullammanappallil *et al.*, 1997). Crustal thicknesses are low; the base of the crust may or may not have a mafic underplate layer.

The bright lower and middle crustal reflectivity has been attributed to basaltic intrusions that result in melting, segregation, and concentration of the felsic components of country rock, as metamorphic fabrics, as ductile flow features, and as distributed lower crustal fluids. Surface exposures of extended deep crust favor the intrusion and melt segregation hypotheses as the likely cause of most of the deeper reflectivity (e.g., Rutter *et al.*, 1999; Khazenehdari *et al.*, 2001), whereas metamorphic fabrics may be the cause of middle crustal reflectivity (Holbrook *et al.*, 1991). In regions undergoing extension, fluid sills, believed to be basaltic magma have been identified seismically at mid-crustal levels (DeVoogd *et al.*, 1988), and at Moho levels (Jarchow *et al.*, 1993). The division between a largely unreflective upper crust and a highly reflective lower

---

[2] The characteristic scale of a fractal medium is roughly 0.4 the visual scale; i.e., a geologist's observation of a unit 1000 m in length corresponds to a statistically measured length of 400 m.

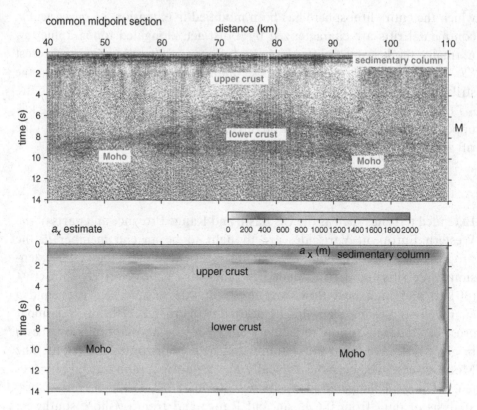

Fig. 2.8. Top: SWAT seismic profile from the extended terrane offshore southwest Great Britain shown in a power display. This section is typical of data from extended terranes, in that the lower crust is highly reflective and the upper crystalline crust is weakly reflective. From Klemperer & Hobbs, 1991, *The BIRPS Atlas: Deep Seismic Reflection Profiles around the British Isles.* Cambridge: Cambridge University Press. © 1991. Reproduced by permission of Cambridge University Press.) Bottom: Statistical analysis of the lower crustal lateral scale parameter based on a self-affine (fractal) reflectivity model (LaFlame, 1999). The scale analysis provides a seismic attribute that maps the large lateral scales in the sedimentary section, short scales in the upper crystalline crust, and intermediate to large scales in the lower crust. In the lower crust, a large antiformal feature of large scales is outlined by the scale analysis. The geologic or visual scale is about 2.2 times greater than the statistical scale. The moderate- to large-scale parameters in the lower crust are the result of intrusives and/or flow patterns in the extended crust. At the time of going to press the original colour version was available for download from www.cambridge.org/9780521066068

one coincides nicely with division of the crust rheologically into a brittle upper and a ductile lower layer with the temperature at the top of the reflective zone, 300–400 °C, roughly corresponding to the onset of ductile behavior in intermediate composition rocks (Klemperer, 1987).

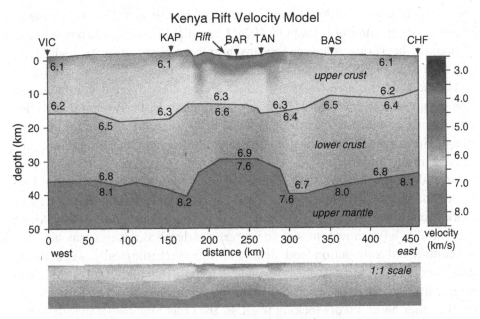

Fig. 2.9. East–west (dip direction) compressional wave seismic velocity profile across the Kenya Rift. This profile is a composite of a crust–upper mantle (to ~50 km depth) refraction profile and teleseismic *P*-velocity tomography. The crust is thinned and the mantle exhibits slow velocities only under the region of active surface rifting. Similarly, lower crustal velocities are somewhat higher directly under the rift zone than to either side, suggesting intrusion by mantle derived basaltic rocks. (From Braile *et al.* (1995) Seismic techniques. In *Continental Rifts: Evolution, Structure and Tectonics*, ed. K. H. Olsen, Amsterdam: Elsevier, pp. 61–87. © 1995 Braile *et al.* Elsevier. Reproduced by permission of Elsevier.) At the time of going to press the original colour version was available for download from www.cambridge.org/9780521066068

2.7 The continental crust and mantle dynamics

The thickness and chemical buoyancy of continental crust makes continents resistant to subduction. That this could affect large-scale mantle dynamics was proposed almost 70 years ago (Pekeris, 1935). Since then, a number of investigations have shown that the local insulating effects associated with non-subducting continents could have low-order (i.e., long wavelength) effects on the large-scale temporal and spatial patterns of mantle convection. The implications of this have been explored for several specific issues, including the mechanisms of continental drift (e.g., Elder, 1967), the dynamics of super-continent assembly and dispersal (e.g., Gurnis, 1988; Lowman & Jarvis, 1993), the cause of long-wavelength mantle structure (e.g., Zhong & Gurnis, 1993), the effects of continents on spatial patterns of mantle heat flow (Nyblade & Pollack, 1993b; Guillou & Jaupart, 1995; Lenardic, 1997), and the partitioning of mantle heat flow between oceans and continents over geologic time (e.g., Lenardic, 1998).

The bulk of the investigations, noted above, have treated continents as homogenous, non-deformable bodies. The interaction of continental tectonics and mantle convection and the effects of both vertical and horizontal continental heterogeneity on issues of large-scale continent mantle interactions are only beginning to be explored.

Several groups have recently examined the interaction of continental tectonics and mantle dynamics within a self-consistent mantle convection framework (Lenardic et al., 2000; Pysklywec et al., 2000; Schott et al., 2000). All of these studies have shown the strong role crustal depth variations may have on subcontinental subduction and delamination. A rheologically weak lower crust favors detachment surfaces forming at its base that allows subcontinental mantle lithosphere to be recycled into the deeper mantle. A strong lower crust, on the other hand, favors locking together the crust and deeper mantle lithosphere, which tends to suppress recycling of subcontinental mantle lithosphere by increasing its integrated buoyancy. This suggests an interesting connection between crustal structure and the structure of deeper continental lithosphere. A strong lower crust is favored in regions of low heat flow and associated low crustal heat generation. That is, it is most favored in Archean regions and it is these regions that are associated with thick, subcrustal lithosphere, i.e., tectosphere. The degree to which the depth-variable properties of the continental crust do truly influence the stability and longevity of deeper lithosphere remains to be determined, but the studies cited suggest that a connection exists and that the specific properties of the lower continental crust may have effects that extend beyond its own depth.

Continental geologists have long known that lateral variations in crustal properties may influence patterns of tectonic deformation (e.g., Smith & Mosley, 1993). This issue has been explored in geodynamic models of the lithosphere and it has been shown that lateral variations within the crust may determine where continental rifting will be most likely to occur (Dunbar & Sawyer, 1989), may have first-order influence on the dynamics of continental collision (Ellis et al., 1998) and may influence first-order strain distribution within continents as a whole (Tommasi et al., 1995). More recently, models of the full mantle convection system have shown how lateral variations within the continental crust may play a strong role in determining the deep structure of continents (Lenardic et al., 2000). The important, overreaching point from all of these studies is a point that has been made throughout this chapter – the heterogeneity of continental crust has first-order effects on the dynamics of the continental lithosphere as a whole. Fully quantifying this statement from a global to a regional scale will probably form an active area of geodynamics research in the near future.

2.8 Discussion and major unanswered questions about continental lithosphere

Study of crust–mantle interactions constitutes a large part of modern solid Earth science investigations of the continents. The regularity of the oceanic lithosphere in formation, thickening, subsidence, and subduction is in counterpoint to the complexity of continental crust–mantle interactions. As we have noted above, the continental crust is characterized by lateral and vertical heterogeneity in bulk chemistry, mineralogy, lithology, fracture density, and fluid content. The recent investigations of the continental lithospheric mantle suggest that it is also far more complex than its oceanic counterpart (e.g., Ryberg *et al.*, 2000, Dueker *et al.*, 2001), and far more complex than was previously appreciated. Moreover, the upper mantle plays a role in crustal deformation and magmatism – mafic magmas extracted from shallow, fertile asthenosphere invade the crust in rifts and extended terranes. Delamination of lower crust and upper mantle appears to either initiate or accompany orogenic collapse and the formation of extensional structures. Stacking of subducting lithosphere around the peripheries of growing continents may be a means of stabilizing cratons.

Understanding the various ways in which the crust–mantle boundary forms and evolves is one of the outstanding unanswered problems in Earth Science. Evidence from reflection seismology suggests that, in contractional belts, the Moho, or perhaps the entire lowermost crust is a detachment surface, is quite strong (Cook, 1995; Wissinger *et al.*, 1997, 1998). In these areas it is likely that ultramafic and crustal rocks become intermixed at a relatively fine (100–1000 m) scale.

As orogens deepen, the more mafic lower crust may be converted to eclogite, resulting in a negatively buoyant layer that may delaminate from the remaining crust. The ecoligitization process itself resets the Moho to a higher level, by the seismic definition. The delamination process also will reset it from the petrological viewpoint. In some areas, reflection and refraction data suggest that the Moho evolves as a downward-moving boundary through the addition to the crust of an underplated mafic layer provided by a fertile mantle (Karlstrom *et al.*, 2002). In other areas, the Moho is just a place in the Earth where the seismic velocity exceeds a certain value, and the crust–mantle

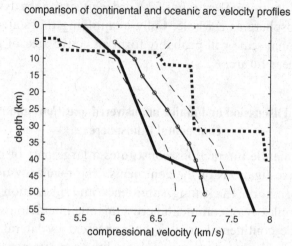

Fig. 2.10. Velocity depth profiles through the Aleutian island arc (short dashes; Holbrook *et al.*, 1999), the Sierra Nevada continental arc (solid line; Fliedner & Klemperer, 1999), and the Christensen–Mooney (1995) average crustal velocity model (Christensen & Mooney 1995) (solid line with symbols; first standard error shown as thin dashed lines). The Aleutian arc is too mafic to be easily transformed to an average continental crust profile, whereas the Sierran profile is too felsic.

boundary petrologically lies far below the seismically determined depth (e.g., Griffin & O'Reilly, 1987). In this case, metamorphic reactions within crustal rocks have proceeded well into the granulite or eclogite facies, producing high-velocity crustal rocks. Thus, Moho cannot be considered a datum; instead, it evolves through time and has different significance in different tectonic settings.

The infusion of heat and fluids, both mafic and aqueous, results in the melting of existing crustal rocks. In some cases, an existing intermediate-composition rock is segregated into silicic and mafic components by complete melting and mobilization of the lighter component. The mafic and silicic segregation structure is common in exposures of the lower crust (e.g., Khazenehdari *et al.*, 2001 and references therein). Under the right physical conditions, the silicic component may accumulate to form large igneous bodies (Chapters 10, 11, 13).

Significant chemical problems occur if the continental crust is simply viewed as being constructed of island arcs. The most recent seismic investigations through a modern island arc, the Aleutians (Holbrook *et al.*, 1999), indicate that it is too mafic to form continental crust, and it is unlikely that a simple chemical or mechanical refining process could push the composition to intermediate, even if large-scale lower crustal delamination were to occur. Similarly, recent studies of the Sierra Nevada continental arc (e.g., Fliedner *et al.*, 1996) indicate that it is too felsic to form continental crust without an influx of mafic material (Fig. 2.10). The formation of new continental crust on the large scale involves either the amalgamation of both mafic and felsic terranes, e.g., continental and oceanic arcs, or oceanic crust or plateaus and the accretionary rocks that overlie them in subduction zones. Continental–crust is certainly recycled in ocean–continent and continent–continent collisions, but evidence of formation of new crust in these tectonic settings is not yet available.

2.9 Conclusions

The worldwide geophysical database on crustal structure allows for generalizing seismic structure of the continental crust. The average continental crustal thickness and velocities permit estimates of the average crustal chemistry (Table 2.1), and the seismically derived estimates are in reasonable agreement with estimates derived from sediment compositions (Chapter 4). The range of crustal thicknesses within a given province type is quite large (Table 2.2). Differences in average velocity between the different province types is less than the formal error on the continental average velocity, suggesting that the velocity structures of the provinces have nearly as much inter-province variation as intra-province variation. Most tectonic provinces, e.g., the Phanerozoic extended terranes, have common seismic signatures that characterize deformational styles. Despite a high degree of success, the detailed analysis of seismic reflections for physical properties is at an intermediate stage of development. Most analysis methods are not automated and therefore may be done only on a small data volume. This analysis is often done only on the most interesting and therefore anomalous reflection events.

As the examples shown in this chapter make clear, discussion of the deformation of the continental crust is virtually impossible without consideration of the mantle underneath it. In both cratonic and modern provinces, surface structures appear to be highly correlated with mantle structures. Heterogeneities observed in the mantle may be used to infer some of the forces driving crustal deformations. The heterogeneity of the continental crust–mantle system is perhaps one of its most important properties. The deviations from the mean, and the spatial and temporal correlation functions of the deviations, are as important as the average values. The heterogeneity that arises in chemistry and lithology, fracture density, fluid content, and the geothermal field produce seismic velocity and impedance anomalies that

Table 2.2. *Crustal thickness of continental provinces in kilometers*

	Extensional terranes	Continental arcs	Active rifs	Paleozoic orogens	Cenozoic-Mesozoic orogens	Shields and platforms	Average continent
C & M(1995)	30.5 ± 5.3	38.7 ± 9.6	36.2 ± 7.9	46.3 ± 9.5		41.5 ± 5.8	41.0 ± 6.2
R & F(1995)	31.13 ± 2.42	40.00 ± 5.87	27.63 ± 5.66	34.84 ± 5.04	52.43 ± 13.03	43.37 ± 5.19	
Max[1]	35.00[2]	46.00[3]	36.00[4]	48.00[5]	78.00[6]	55.00[7]	
Min[1]	28.00[8]	33.00[9]	20.00[10]	27.00[11]	41.00–42.00[12]	32.00–33.00[13]	

C & M: Christensen & Mooney (1995).

R & F: Rudnick & Fountain (1995).

Error estimates on each individual refraction measurement from which the compilations are made are ± 10%.

[1] From Rudnick and Fountain (1995).
[2] Omineca Belt, Canada.
[3] Cascades, Oregon, USA.
[4] Southern Kenya Rift.
[5] Valley and Ridge, USA.
[6] Himalaya, S Lhasa Block, Tibet.
[7] Svecokarelides.
[8] Basin and Range, USA.
[9] Northern Honshu, Japan.
[10] Salton Sea, USA; Kenya Rift North.
[11] Swabian Jura, Germany.
[12] East Himalaya, Southern Alps.
[13] Kaapvaal, Africa; East Superior, USA; SvecoFennides.

may be imaged over a broad range of scales. The material and physical state heterogeneity influences strength, strain rates, and strain localization.

Heterogeneities in the crust influence tectonic style observed in the field. Numerical models of orogen-wide deformation are now detailed enough and utilize realistic enough rheologies to provide broad-scale predictions of field observations. At present, the bulk of such models allows mechanical heterogeneities to form in largely uniform starting models in response to deformation and strain localization (Fig. 2.3), but it is clear that in the future they will need to incorporate longer-lived heterogeneities, e.g., different lithologies, long-lived faults and shear zones, etc. This corresponds to the geologists' often-cited zones of pre-existing weakness or strength, in explaining tectonic models.

Crustal heterogeneity has been pointed out many times by petrologists and structural geologists working on crystalline rocks in reference to extant lithologies and the potential strength of the crust during large-scale tectonic deformation (e.g., Karlstrom & Williams, 1998; Chapter 8). In 4500 Myr, one might imagine that differentiation processes could have produced a crust with a continuously graded chemical composition, which would produce a fairly simple velocity with depth relation and a standard continent-wide rheology profile. The crust has remained highly heterogeneous, a result of the counter-balancing forces of positive and negative buoyancy operating in Earth's convection cycle, and the segregation of the surface into oceanic and continental regions. On the continents, the result is large-scale lateral translation of detached packages of rocks, and infusions of heat and melt, but without the regularity of the oceanic system.

The importance of the horizontal translations, whereas not a surprise, cannot be overemphasized, as whole crustal detachments permeate even the cratonic, tectonic parts of the continents, and strike-slip boundaries pervade the orogens. The interlayering of metamorphic and igneous rocks thrust from lower crustal depths with recently deposited sedimentary rocks is a means of locally increasing heterogeneity. Another means of increasing heterogeneity is instanced by the localized melting processes resulting from injections of heat and melt into the crust that segregate the felsic from the mafic components of intermediate composition rocks. The felsic component may then either resolidify after only minor movement, or, if the physical conditions permit, migrate to shallower levels of the crust forming plutons.

2.10 Acknowledgements

We would like to thank many of our colleagues for valuable discussions related to the review we have presented in this chapter, particularly Gene Humphreys,

Ken Dueker, Randy Keller, Michael Williams, Sam Bowring, Louis Moresi, Walter Mooney, Cin-Ty Lee, Fenglin Niu, and Klaus Holliger. Support was provided by NSF grants EAR-0003539, EAR-0208020 to Levander, NSF Grant EAR-0001029 to Lenardic, and NSF grants EAR 9614787 and 0208473 to Karlstrom.

References

ANCORP Working Group (1999). Seismic reflection image revealing offset of Andean subduction-zone earthquake locations into oceanic mantle. *Nature*, **397**, 341–4.

Anderson, D. L. (1989). *Theory of the Earth*. Boston, MA: Blackwell Scientific.

Ansorge, J., Holliger, K., Valasek, P., Ye, S., Finckh, P., Freeman, R., Frei, W., Kissling, E., Lehner, P., Maurer, H., Mueller, S., Smithson, S. B. and Staeuble, M. (1991). Integrated analysis of normal incidence and wide-angle reflection measurements across the eastern Swiss Alps. In *Continental Lithosphere: Deep Seismic Reflections*, ed. R. O. Meissner, L. D. Brown, H. -J. Duerbaum, W. Franke, K. Fuchs and F. Seifert. Geodynamics Series, 22. Washington: American Geophysical Union, pp. 195–205.

Ashwal, L. D., Morgan, P., Kelley, S. A. and Percival, J. A. (1987). Heat production in an Archean crustal profile and implications for heat flow and mobilization of heat-producing elements. *Earth and Planetary Science Letters*, **85**, 439–50.

Ballard, S. and Pollack, H. N. (1987). Diversion of heat flow by Archean cratons: a model for southern Africa and Namibia. *Earth and Planetary Science Letters*, **85**, 253–64.

Birch, F., Roy, E. R. and Decker, E. R. (1968). Heat flow and thermal history of New England and New York. In *Studies of Appalachian Geology: Northern and Maritime*, ed. E. Zen, W. S. White, J. B. Hadley and J. B. Thompson. New York: Wiley Interscience, pp. 437–51.

Blackledge, J. M. (1989). *Quantitative Coherent Imaging: Theory Methods, and Application*. London: Academic Press.

Block, L. and Royden, L. H. (1990). Core complex geometries and regional scale flow in the lower crust. *Tectonics*, **9**, 557–67.

Bostock, M. G., Hyndman, R. D., Rondenay, S. and Peacock, S. M. (2002). An inverted continental Moho and serpentinization of the forearc mantle. *Nature*, **417**, 536–8.

Braile, L. W., Keller, G. R., Mueller, S. and Prodehl, C. (1995). Seismic techniques. In *Continental Rifts: Evolution, Structure and Tectonics*, ed. K. H. Olsen. Amsterdam: Elsevier, pp. 61–87.

Brown, L. D. (1998). Seismic stratigraphy of the continental lithosphere. *9th International Symposium on Deep Seismic Profiling of the Continents and their Margins*, Barcelona, Spain (abstract).

Brown, L. D., Zhao, W., Nelson, K. D., Hauck, M., Alsdof, D., Ross, A., Cogan, M., Liu X. and Che, J. (1996). Bright spots, structure and magmatism in southern Tibet from INDEPTH Seismic Reflection Profiling. *Science*, **274**, 1688–90.

Burke, K. (1977). Aulacogens and continental breakup. *Annual Reviews of Earth and Planetary Science Letters*, **5**, 371–96.

Calvert, A. J., Sawyer, E. W., Davis, W. J. and Ludden, J. N. (1995). Archean subduction inferred from seismic images of a mantle suture in the Superior Province. *Nature*, **375**, 670–4.

Chen, L., Booker, J. R., Jones, A. G., Wu, N., Unsworth, M. J., Wei, W. and Tan, H. (1996). Electrically Conductive Crust in Southern Tibet from INDEPTH Magnetotelluric Surveying. *Science*, **274**, 1694–6.

Christensen, N. I. and Mooney, W. D. (1995). Seismic velocity structure and composition of the continental crust: a global view. *Journal of Geophysical Research*, **100**, 9761–88.

Clark, M. K. and Royden, L. H. (2000). Topographic ooze; building the eastern margin of Tibet by lower crustal flow. *Geology*, **28**, 703–6.

Cook, F. A. (1995). Lithospheric processes and products in the southern Canadian Cordillera: a Lithoprobe perspective. *Canadian Journal of Earth Sciences*, **32**, 1803–24.

Cook, F. A., Brown, L. D, Kaufman, S., Oliver, J. E. and Petersen, T. A. (1981). COCORP seismic profiling of the Appalachian Orogen beneath the coastal plain of Georgia. *Geological Society of America Bulletin*, **92**, 738–48.

Cook, F. A., van der Velden, A. J., Hall, K. W. and Roberts, B. J. (1999). Frozen subduction in Canada's Northwest Territories: lithoprobe deep lithospheric reflection profiling of the western Canadian Shield. *Tectonics*, **18**, 1–24.

DeVoogd, B., Serpa, L. and Brown, L. (1988). Crustal extension and magmatic processes: COCORP profiles from Death Valley and the Rio Grande Rift. *Geological Society of America Bulletin*, **100**, 1550–67.

Dueker, K., Yuan, H. and Zurek, B. (2001). Thick-structured Proterozoic lithosphere of the Rocky Mountain region. *GSA Today*, **11**, 4–9.

Dunbar, J. A. and Sawyer, D. S. (1989). How preexisting weaknesses control the style of continental breakup. *Journal of Geophysical Research*, **94**, 7278–92.

Elder, J. W. (1967). Convective self-propulsion of continents. *Nature*, **214**, 657–60.

Ellis, S., Beaumont, C., Jamieson, R. A. & Quinlan, G. (1998). Continental collision including a weak zone: the vise model and its application to the Newfoundland Appalachians. *Canadian Journal of Earth Science*, **35**, 1323–46.

Evans, B. and Kohlstedt, D. L. (1995). Rheology of Rocks. In *Rock Physics and Phase Relations*, ed. T. J. Ahrens. Washington, DC, American Geophysical Union, pp. 148–165.

Fliedner, M. M. and Klemperer, S. L. (1999). Structure of an island-arc; wide-angle seismic studies in the eastern Aleutian Islands, Alaska. *Journal of Geophysical Research*, **104**, 10 667–94.

Fliedner, M. M., Ruppert, S. D., Malin, P. E., Park, S. K., Jiracek, G. R., Phinney, R. A., Saleeby, J. B., Wernicke, B. P., Clayton, R. W., Keller, G. R., Miller, K. C., Jones, C. H., Luetgert, J. H., Mooney, W. D., Oliver, H. L., Klemperer, S. L. and Thompson, G. A. (1996). Three-dimensional crustal structure of the southern Sierra Nevada from seismic fan profiles and gravity modelling, US, Southern Sierra Nevada Continental Dynamics Working Group, United States. *Geology*, **24**, 367–70.

Fountain, D. M., Salisbury, M. H. and Furlong, K. P. (1987). Heat production and thermal conductivity of rocks from the Pikwitonei–Sachigo continental cross section, central Manitoba: implications for the thermal structure of Archean crust. *Canadian Journal of Earth Science*, **24**, 1583–94.

Fuis, G. S., Levander, A., Lutter, W. J., Wissinger, E. S., Moore, T. E. and Christensen, N. I. (1995). Seismic images of the Brooks Range, Arctic Alaska, reveal crustal-scale duplexing. *Geology*, **23**, 65–8.

Fuis, G. S., Murphy, J. M., Lutter, W. J., Moore, T. E., Bird, K. J. and Christensen, N. I. (1997). Deep seismic structure and tectonics of Northern Alaska; crustal-scale duplexing with deformation extending into the upper mantle. *Journal of Geophysical Research*, **102**, 20 873–96.

Goff, J. A. and Levander, A. (1996). Incorporating 'sinuous connectivity' into stochastic models of crustal heterogeneity; examples from the Lewisian gneiss complex, Scotland, the Franciscan Formation, California, and the Hafafit gneiss complex, Egypt. *Journal of Geophysical Research, Solid Earth and Planets*, **101**, 8489–501.

Griffin, W. L. and O'Reilly, S. Y. (1987). Is the continental Moho the crust–mantle boundary? *Geology*, **15**, 241–4.

Guillou, L. and Jaupart, C. (1995). On the effect of continents on mantle convection. *Journal of Geophysical Research*, **100**, 24 217–38.

Guillou, L., Mareschal, J. C., Jaupart, C., Gariepy, C., Bienfait, G. and Lapointe, R. (1994). Heat flow, gravity and structure of the Abitibi belt, Superior Province, Canada: implications for mantle heat flow. *Earth and Planetary Science Letters*, **136**, 103–23.

Guillou-Frottier, L., Mareschal, J. C., Jaupart, C., Gariepy, C., Lapointe, R. and Bienfait, G. (1995). Heat flow variations in the Grenville Province, Canada. *Earth and Planetary Science Letters*, **136**, 447–60.

Guillou-Frottier, L., Jaupart, C., Mareschal, J. C., Gariepy, C., Bienfait, G., Cheng, L. Z. and Lapointe, R. (1996). High heat flow in the Trans-Hudson Orogen, central Canadian Shield. *Geophysical Research Letters*, **23**, 3027–30.

Gupta, M. L., Sundar, A. and Sharma, S. R. (1991). Heat flow and heat generation in the Archean Dharwar cratons and implications for the southern Indian shield geotherm and lithospheric thickness. *Tectonophysics*, **194**, 107–22.

Gurnis, M. (1988). Large-scale mantle convection and the aggregation and dispersal of supercontinents. *Nature*, **332**, 695–9.

Hansen, F. D. and Carter, N. L. (1982). Creep of selected crustal rocks at 1000 MPa. *Eos, Transactions of the American Geophysical Union*, **63**, 437.

Henstock, T. J. and The Deep Probe Working Group (1998). Probing the Archean and Proterozoic lithosphere of western North America. *GSA Today*, **8**, 1–5 & 16–17.

Hoffman, P. F. (1988). United plates of America, the birth of a craton: early Proterozoic assembly and growth of Laurentia. *Annual Review of Earth and Planetary Sciences*, **16**, 543–603.

Holbrook, W. S., Catchings, R. D. and Jarchow, C. M. (1991). Origin of deep crustal reflections; implications of coincident seismic refraction and reflection data in Nevada. *Geology*, **19**, 175–9.

Holbrook, W. S., Lizarralde, D., McGeary, S., Bangs, N. and Diebold, J. (1999). Structure and composition of the Aleutian island arc and implications for continental crustal growth. *Geology*, **27**, 31–4.

Holliger K. and Goff, J. A. (2003). A generic model for the 1/f-nature of seismic velocity fluctuations. In *Heterogeneity in the Crust and Upper Mantle*, ed. J. A. Goff and K. Holliger. New York: Kluwer Academic/Plenum Publishers, pp. 131–54.

Holliger, K. and Levander, A. R. (1992). A stochastic view of the lower crust based on the Ivrea Zone. *Geophysical Research Letters*, **19**, 1153–6.

Holliger K. and Levander, A. (1994a). Seismic structure of gneissic/granitic upper crust: geological and petrophysical evidence from the Strona–Ceneri zone (northern Italy) and implications for crustal seismic exploration. *Geophysical Journal International*, **119**, 497–510.

Holliger, K. and Levander, A. (1994b). Fine-scale structure and seismic response of extended continental crust: stochastic analysis of the Ivrea and Strona–Ceneri Zones. *Geology*, **22**, 79–82.

Holliger, K., Levander, A. and Goff, J. A. (1993). Stochastic modeling of the reflective lower crust: petrophysical and geological evidence from the Ivrea Zone (northern Italy). *Journal of Geophysical Research*, **98**, 11 967–80.

Humphreys, E. D. and Dueker, K. G. (1994a). Western U.S. upper mantle structure. *Journal of Geophysical Research*, **99**, 9615–34.

Humphreys, E. D. and Dueker, K. G. (1994b). Physical state of the Western U.S. upper mantle. *Journal of Geophysical Research*, **99**, 9635–50.

Hurich, C. A. (1996). Statistical description of seismic reflection wavefields: a step towards quantitative interpretation of deep seismic reflection profiles. *Geophysical Journal International*, **125**, 719–28.

Hurich, C. A. and Kocurko, M. J. (2000). Statistical approaches to interpretation of seismic reflection data. *Tectonophysics*, **329**, 251–67.

James, D. E., Fouch, M. J., VanDecar, J. C., van der Lee, S. and The Kaapvaal Seismic Group (2001). Tectospheric structure beneath southern Africa. *Geophysical Research Letters*, **28**, 2485–8.

Jarchow, C. M., Thompson, G. A., Catchings, R. D. and Mooney, W. D. (1993). Seismic evidence for active magmatic underplating beneath the Basin and Range Province, Western United States. *Journal of Geophysical Research*, **98**, 22 095–108.

Jaupart, C. (1983). Horizontal heat transfer due to radioactivity contrasts: causes and consequences of the linear heat flow relation. *Geophysical Journal of the Astronomical Society*, **75**, 411–35.

Jaupart, C., Mareschal, J. C., Guillou-Frottie, L. and Davaille, A. (1998). Heat flow and the thickness of the lithosphere in the Canadian Shield. *Journal of Geophysical Research*, **103**, 15 269–86.

Jiracek, G. R., Haak, V. and Olsen, K. H. (1995). Practical magnetotellurics in a continental rift environment. *Developments in Geotectonics*, **25**, 103–29.

Jordan, T. H. (1975). The continental tectosphere. *Reviews of Geophysical and Space Physics*, **13**, 1–12.

Jordan, T. H. (1988). Structure and formation of the continental tectosphere. *Journal of Petrology Special Lithosphere Issue* 1988, 11–37.

Karlstrom, K. E. and Williams, M. L. (1998). Heterogeneity of the middle crust; implications for strength of continental lithosphere. *Geology*, **26**, 815–18.

Karlstrom, K. and The CDROM Working Group (2002). Structure and evolution of the lithosphere beneath the Rocky Mountains. *GSA Today*, **12**, 4–10.

Ketcham, R. A. (1996). Distribution of heat-producing elements in the upper and middle crust of southern and west central Arizona: evidence from core complexes. *Journal of Geophysical Research*, **101**, 13 611–32.

Khazenehdari, J., Rutter, E. H. and Brodie, J. K. (2001). High-pressure–high-temperature seismic velocity structure of the midcrustal and lower crustal rocks of the Ivrea–Verbano Zone and Serie de Laghi. *Journal of Geophysical Research*, **105**, 13 843–58.

Kind, R., Ni, J., Zhao, W., Wu, J., Yuan, X., Zhao, L., Sandoval, E., Reese, C., Nabelek, J. and Hearn, T. (1996). Evidence from earthquake data for a partially molten crustal layer in southern Tibet. *Science*, **274**, 1692–4.

Klemperer, S. L. (1987). A relation between continental heat flow and the seismic reflectivity of the lower crust. *Journal of Geophysics*, **61**, 1–11.

Klemperer, S. L. and Hobbs, R. (1991). *The BIRPS Atlas: Deep Seismic Reflection Profiles around the British Isles*. Cambridge: Cambridge University Press.

Klemperer, S. L., Hauge, T A., Hauser, E. C., Oliver, J. E. and Potter, C. J. (1986). The Moho in the northern Basin and Range Province, Nevada, along the COCORP 40 degrees N seismic-reflection transect. *Geological Society of America Bulletin*, **97**, 603–18.

Kohlstedt, D. L., Evans, B. and Mackwell, S. J. (1995). Strength of the lithosphere. *Journal of Geophysical Research*, **100**, 17 587–602.

LaFlame, L. M. (1999). *Estimating Crustal Heterogeneity in the Northern Basin and Range from COCORP and PASSCAL Seismic Data*, M. A. thesis. Texas: Rice University.

Larkin, S. P. (1996). *Combining Deterministic and Stochastic Velocity Fields in the Analysis of Deep Crustal Seismic Data*, Ph.D. thesis. Texas: Rice University.

Larkin S. P., Levander, A., Henstock, T. and Pullammanappallil, S. (1997). Is the Moho flat? Seismic evidence for a rough crust–mantle interface beneath the northern Basin and Range. *Geology*, **25**, 451–4.

Lee, C-T. A. (2003). Compositional variation of density and seismic velocities in natural peridotites at STP conditions: implications for seismic imaging of compositional heterogeneities in the upper mantle. *Journal of Geophysical Research*, doi: 10.1029/2003JB002413.

Lenardic, A. (1997). On the heat flow variation from Archean cratons to Proterozoic mobile belts. *Journal of Geophysical Research*, **102**, 709–21.

Lenardic, A. (1998). On the partitioning of mantle heat loss below oceans and continents over time and its relationship to the Archean paradox. *Geophysical Journal International*, **134**, 706–20.

Lenardic, A. L., Moresi, N. and Muhlhaus, H.-B. (2000). The role of mobile belts for the longevity of deep cratonic lithosphere: the crumple zone model. *Geophysical Research Letters*, **27**, 1235–8.

Lenardic, A., Moresi, L.-N. and Muhlhaus, H. (2003). Longevity and stability of cratonic lithosphere: insights from numerical simulations of coupled mantle convection and continental tectonics. *Journal of Geophysical Research*, doi: 10.1029/2002JB001859.

Levander, A. (2003). US Array design implications for wavefield imaging in the lithosphere and upper mantle. *The Leading Edge*, **22**, 250–5.

Levander, A. R. and Holliger, K. (1992). Small-scale heterogeneity and large-scale velocity structure of the continental crust. *Journal of Geophysical Research*, 8797–804.

Levander, A., England, R. W., Smith, S. K., Hobbs, R. W., Goff, J. A. and Holliger, K. (1994a). Stochastic characterization and seismic response of upper and middle crustal rocks based on the Lewisian Gneiss Complex, Scotland. *Geophysical Journal International*, **119**, 243–59.

Levander, A., Smith, S. K., Hobbs, R. W., England, R. W., Snyder, D. B. and Holliger, K. (1994b). The crust as a heterogeneous "optical" medium, or "crocodiles in the mist." *Tectonophysics*, **232**, 281–97.

Levander, A., Fuis, G. S., Wissinger, E. S., Lutter, W. J., Oldow, J. S. and Moore, T. E. (1994c). Seismic images of the Brooks Range fold and thrust belt, Arctic Alaska, from an integrated seismic reflection/refraction experiment. *Tectonophysics*, **232**, 13–30.

Levander, A., Zelt, C. A. and Magnani, M. B. (2005). Crust and upper mantle velocity structure of the southern Rocky Mountains from the Jemez Lineament to the

Cheyenne Belt. In *Lithospheric Structure and Evolution of the Rocky Mountain Region*, ed. K. E. Karlstrom and G. R. Keller. American Geophysical Union monograph (in press).

Lowman, J. P. and Jarvis, G. T. (1993). Mantle convection flow reversals due to continental collisions. *Geophysical Research Letters*, **20**, 2087–90.

Maggi, A. J., Jackson, A., McKenzie, D. and Priestly, K. (2000a). Earthquake focal depths, effective elastic thickness, and the strength of the continental lithosphere. *Geology*, **28**. 495–8.

Maggi, A. J., Jackson, A., Priestley, K. and Baker, C. (2000b). A re-assessment of focal depth distributions in southern Iran, the Tien Shan and northern India; do earthquakes really occur in the continental mantle? *Geophysical Journal International*, **143**, 629–61.

Makovsky, Y. S. and Klemperer, L. (1999). Measuring the seismic properties of Tibetan bright spots; evidence for free aqueous fluids in the Tibetan middle crust. *Journal of Geophysical Research*, **104**, 10 795–25.

Makovsky, Y., Klemperer, S. L., Ratschbacher, L., Brown, L., Li, M., Zhao, W. and Meng, F. (1996). INDEPTH wide-angle reflection observation of P-wave-to-S-wave conversion from crustal bright spots in Tibet. *Science*, **274**, 1690–1.

Mareschal, J. C., Pinet, C., Gariepy, C., Jaupart, C., Bienfait, G., Dalla-Coletta, G., Jolivet, J. and Lapointe, R. (1989). New heat flow density and radiogenic heat production data in the Canadian Shield and the Quebec Appalachians. *Canadian Journal of Earth Science*, **26**, 845–52.

Mareschal, J. C., Guillou-Frottier, L., Cheng, L., Gariepy, C. and Jaupart, C. (1997). *Heat Flow in the Trans-Hudson Orogen*. Lithoprobe Report 62, pp. 106–14.

Mareschal, J. C., Guillou-Frottier, L., Cheng, L., Gariepy, C. and Jaupart, C. (1999). Heat flow in the trans-Hudson orogen of the Canadian Shield; implications for Proterozoic continental growth. *Journal of Geophysical Research*, **104**, 29 007–24.

Mechie, J., Sobolev, S. V., Ratschbacher, L., Babeyko, A. Yu., Bock, G., Jones, A. G., Nelson, K. D., Solon, K. D., Brown, L. D. and Zhao, W. (2004). Precise temperature estimation in the Tibetan crust from seismic detection of the α–β quartz transition. *Geology*, **32**, 601–4.

Mooney, W. D. (1989). Seismic methods for determining earthquake source parameters and lithospheric structure. *Memoir, Geological Society of America*, **172**, 11–34.

Mooney, W. D. and Brocher, T. M. (1987). Coincident seismic reflection/refraction studies of the continental lithosphere: a global review. *Reviews of Geophysics*, **25**, 723–42.

Morgan, P. (1985). Crustal radiogenic heat production and the selective survival of ancient continental crust. Proceedings of the 15th Lunar Planetary Science Conference, Part 2. *Journal of Geophysical Research*, Supplement C561–70.

Morgan, P. and Sass, J. H. (1984). Thermal regime of the continental lithosphere: a review. *Journal of Geodynamics*, **2**, 143–66.

Nelson, K. D., Zhao, S. W., Brown, L. D., Kuo, J., Che, J., Liu, X., Klemperer, S. L., Makovsky, Y., Meissner, R., Mechie, J., Kind, R., Wenzel, R., Ni, J., Nabelek, J., Leshou, C., Tan, H., Wei, W., Jones, A. G., Booker, J., Unsworth, M., Kidd, W. S. F., Hauck, M., Alsdorf, D., Ross, A., Cogan, M., Wu, E., Sandvol, E. and Edwards, M. (1996). Partially molten middle crust beneath southern Tibet: synthesis of Project INDEPTH Results. *Science*, **274**, 1684–7.

Nicolaysen, L. O., Hart, R. J. and Gale, N. H. (1981). The Vredefort radioelement profile extended to supracrustal strata at Carletonville, with implications for continental heat flow. *Journal of Geophysical Research*, **86**, 10 653–61.

Nyblade, A. A. and Pollack, H. N. (1993a). A global analysis of heat flow from Precambrian terrains: implications for the thermal structure of Archean and Proterozoic lithosphere. *Journal of Geophysical Research*, **98**, 12 207–18.

Nyblade, A. A. and Pollack, H.N. (1993b). A comparative study of parameterized and full thermal convection models in the interpretation of heat flow from cratons to mobile belts. *Geophysical Journal International*, 747–51.

Oldow, J. S., Bally, A. W. and Ave Lallemant, H. G. (1990). Transpression, orogenic float, and lithospheric balance. *Geology*, **18**, 991–4.

Olsen, K. H. and Morgan, P. (1995). Introduction: progress in understanding continental rifts. *Developments in Geotectonics*, **25**, 3–26.

Parsons, B. and McKenzie, D. P. (1978). Mantle convection and the thermal structure of the plates. *Journal of Geophysical Research*, **83**, 4485–96.

Parsons, B. and Sclater, J. G. (1977). An analysis of the variation of ocean floor bathymetry and heat flow with age. *Journal of Geophysical Research*, **82**, 803–27.

Pekeris, C. L. (1935). Thermal convection in the interior of the Earth, Monthly Notices. *Royal Astronomical Society Geophysical Supplement*, **3**, 343–67.

Pinet, C. and Jaupart, C. (1987). The vertical distribution of radiogenic heat production in the Precambrian crust of Norway and Sweden: geothermal implications. *Geophysical Research Letters*, **14**, 260–3.

Pinet, C., Jaupart, C., Mareschal, J. C., Gariepy, C., Bienfait, G. and Lapointe, R. (1991). Heat flow and structure of the lithosphere in the eastern Canadian Shield. *Journal of Geophysical Research*, **96**, 19 941–63.

Polet, J. and Anderson, D. L. (1995). Depth extent of crations as inferred from tomographic studies. *Geology*, **23**, 205–8.

Pollack, H. N. (1982). The heat flow from the continents. *Annual Review of Earth and Planetary Science Letters*, **10**, 459–81.

Pullammanappallil, S., Levander, A. and Larkin, S. P. (1997). Estimation of stochastic crustal parameters from seismic exploration data. *Journal of Geophysical Research*, **102**, 15 269–86.

Pysklywec, R. N., Beaumont, C. and Fullsack, P. (2000). Modeling the behavior of the continental mantle lithosphere during plate convergence. *Geology*, **28**, 655–9.

Rondenay, S., Bostock, M. G. and Shragge, J. (2001). Multiparameter two-dimensional inversion of scattered teleseismic body waves: 3, Application to the Cascadia 1993 data set. *Journal of Geophysical Research*, **106**, 30 795–807.

Rudnick, R. L. & Fountain, D. M. (1995). Nature and composition of the continental crust: a lower crustal perspective. *Reviews of Geophysics*, **33**, 267–309.

Rumpel, H. -M., Snelson, C. M., Prodehl, C. & Keller, G. R. (2005). Results of the CD-ROM project seismic refraction/wide-angle reflection experiment. In *Lithospheric Structure and Evolution of the Rocky Mountain Region*, ed. K. E. Karlstrom and G. R. Keller. American Geophysical Union monograph (in press).

Rutter, E. H., Khazanehdari, J., Brodie, K. H., Blundell, D. and Waltham, D. (1999). Synthetic seismic reflection profile through the Ivrea Zone. *Geology*, **27**, 79–82.

Ryberg, T., Tittgemeyer, M. & Wenzel, F. (2000). Finite difference modeling of P-wave scattering in the upper mantle. *Geophysics Journal International*, **141**, 787–800.

Schott, B., Yuen, D. A. and Schmeling, H. (2000). The diversity of tectonics from fluid-dynamical modeling of the lithosphere–mantle system. *Tectonophysics*, **321**, 35–51.

Sheehan, A. F., Abers, G. A., Jones, C. H. and Lerner-Lam, A. L. (1995). Crustal thickness variations across the Colorado Rocky Mountains from teleseismic receiver functions. *Journal of Geophysical Research*, 20 391–404.

Smith, M. and Mosley, P. (1993). Crustal heterogeneity and basement influence on the development of the Kenya rift, East Africa. *Tectonics*, **12**, 591–606.

Smithson, S. B. and Johnson, R. A. (2003). Petrological causes of seismic heterogeneity in the continental crust. In *Heterogeneity in the Crust and Upper Mantle*, ed. J. A. Goff and K. Holliger. New York: Kluwer Academic/Plenum Publishers, pp. 37–66.

Smithson, S. B., Johnson, R. A. and Wong, Y. K. (1981). Mean crustal velocity: a critical parameter for interpreting crustal structure and crustal growth. *Earth and Planetary Science Letters*, **53**, 323–32.

Snelson, C. M., Henstock, T. J., Keller, R. G., Miller, K. C. and Levander, A. (1998). Crustal and uppermost mantle structure along the Deep Probe seismic profile. *Rocky Mountain Geology*, **33**, 181–98.

Taylor, S. R. and McLennan, S. M. (1985). *The Continental Crust: Its Composition and Evolution*. Oxford: Blackwell Scientific.

Thybo, H. and Perchuc, E. (1997). The seismic 8° discontinuity and partial melting in continental mantle. *Science*, **275**, 1626–9.

Tommasi, A., Vauchez, A. and Daudre, B. (1995). Initiation and propagation of shear zones in a heterogeneous continental lithosphere. *Journal of Geophysical Research*, **100**, 22 083–101.

Turcotte, D. L. and Oxburgh, E. R. (1967). Finite amplitude convective cells and continental drift. *J. Fluid Mechanics*, **28**, 29–42.

Turcotte, D. L. and Schubert, G. (1982). *Geodynamics: Applications of Continuum Physics to Geological Problems*. New York: John Wiley & Sons, Inc.

Valasek, P, Mueller, S., Frei, W. and Holliger, K. (1991). Results of NFP 20 seismic reflection profiling along the Alpine section of the European Geotraverse (EGT). *Geophysical Journal International*, **105**, 85–102.

Van der Hilst, R. D., Kennett, B. L. N. and Shibutani, T. (1998). Upper mantle structure beneath Australia from portable array deployments. In *Structure and Evolution of the Australian Continent, Geodynamics Series*, ed. J. Braun, J. Dooley, B. Goleby, R. van der Hilst and C. Klootwijk. American Geophysical Union, Geodynamics Monograph 26, 39–58.

Warner, M. R., (1990). Modeling of synthetic seismic reflection data: CCSS workshop 1987, data set V. In *Studies of Laterally Heterogeneous Structures Using Seismic Refraction and Reflection Data*, ed. A. G. Green. Geological Survey of Canada Paper, pp. 219–24.

Weaver, B. L. and Tarney, J. (1984). Empirical approach to estimating the composition of the continental crust. *Nature*, **310**, 575–7.

Willet, S., Beaumont, C. and Fullsack, P. (1993). Mechanical model for the tectonics of doubly vergent compressional orogens. *Geology*, **21**, 371–4.

Wilson, D. and Aster, R. (2003). Imaging crust and upper mantle seismic structure in the southwestern United States using teleseismic receiver functions. *The Leading Edge*, **22**, 232–7.

Wissinger, E. S., Levander, A. and Christensen, N. I. (1997). Seismic imaging of crustal duplexing and continental subduction in the Brooks Range. *Journal of Geophysical Research*, **102**, 20 847–72.

Wissinger, E. S., Levander, A., Oldow, J. S., Fuis, G. S. and Lutter, W. J. (1998). Seismic profiling constraints on the evolution of the central Brooks Range, Arctic Alaska. In *Architecture of the Central Brooks Range Fold and Thrust Belt*, ed. J. S. Oldow and H. G. Ave Lallemant. Geological Society of America Special Paper 324, pp. 269–92.

Wollenberg, H. A. and Smith, A. R. (1987). Radiogenic heat production of crustal rocks, an assessment based on geochemical data. *Geophysical Research Letters*, **14**, 295–8.

Yuan, H. and Dueker, K. (2005). Upper mantle tomographic V_p and V_s images of the Middle Rocky Mountains in Wyoming, Colorado, and New Mexico: evidence for a thick heterogeneous chemical lithosphere. In *Lithospheric Structure and Evolution of the Rocky Mountain Region*, ed. K. E. Karlstrom and G. R. Keller. American Geophysical Union monograph (in press).

Yuan, X., Sobolev, S. V., Kind, R., Oncken, O., Bock, G., Asch, B., Schurr, F., Graeber, R., Hanka, W., Wylegalla, K., Tibi, R., Haberland, C., Rietbrock, A., Giese, P., Wigger, P., Rower, P., Zandt, G., Beck, S., Wallace, T., Pardo, M. and Comte, D. (2000). Subduction and collision processes in the Central Andes constrained by converted seismic phases. *Nature*, **408**, 958–61.

Zhong, S. and Gurnis, M. (1993). Dynamic feedback between a continent-like raft and thermal convection. *Journal of Geophysical Research*, **98**, 12 219–32.

3

Thermo-mechanical controls on heat production distributions and the long-term evolution of the continents

MIKE SANDIFORD AND SANDRA MCLAREN

3.1 Introduction

The heterogeneous nature of continental geology is testimony to the complex superposition of tectonic processes associated with the growth, differentiation, and reactivation of the continental lithosphere. Despite this heterogeneity, geophysical data sets provide compelling evidence for at least two levels of geochemical ordering, both of which impact on the long-term strength of the continents. The first-order layering is evident in the density contrast across the crust–mantle boundary, or Moho, which typically lies at depths of 30–40 km. The Moho is associated with a dramatic change in mineralogy. Consequently, the depth of the Moho exerts a profound control on the strength of the lithosphere (e.g., Brace & Kohlstedt, 1980; Sonder & England, 1986; Molnar, 1989; Ranalli, 1997).

A more subtle level of geochemical ordering is evident in terms of the distribution of *heat producing elements* (HPEs). On geological time scales there are only three elements, and in particular four isotopes, ^{238}U, ^{235}U, ^{232}Th and ^{40}K, that occur in sufficient abundance to contribute to the lithospheric thermal budget; these elements are referred to as the HPEs. The great proportion of the HPEs are contained within felsic igneous rocks such as granites and granodiorites, which have average volumetric heat production of $2.5\,\mu W\,m^{-3}$ and $1.5\,\mu W\,m^{-3}$, respectively (Haenel *et al.*, 1988). These values are significantly higher than mafic igneous rocks and most sedimentary rocks, and suggest that the distribution of HPEs is broadly related to the distribution of lithotypes. At the crustal scale, an ordering of the HPEs is evident from an analysis of surface heat flow and heat production data. These data sets show that in most stable continental regions the lithospheric complement of HPEs is largely confined to the upper 10–15 km of the crust and is responsible for about one-half to two-thirds of the characteristic surface heat

Evolution and Differentiation of the Continental Crust, ed. Michael Brown and Tracy Rushmer. Published by Cambridge University Press. © Cambridge University Press 2005.

flow of the continents (i.e., about 30–40 mW m^{-2}, e.g., McLennan & Taylor, 1996).

The distribution of HPEs exerts a crucial influence on continental thermal regimes. Because lithospheric rheology is extremely sensitive to the thermal regime (e.g., Brace & Kohlstedt, 1980; Ranalli, 1997), the distribution of HPEs must also impact on the strength of the lithosphere. Consequently, processes that redistribute the HPEs should be expected to affect the long-term strength of the lithosphere. Conventional wisdom holds that the absolute abundance of the HPEs in the continents and the concentration of HPEs in the upper half of the continental crust is a primary feature of magmatic-related crustal growth processes. In particular, the lithophile nature of the HPEs implies that they will be progressively differentiated during the partial melting processes attendant with primary crustal growth. However, there is only limited understanding of why such processes should lead to the characteristic abundance and distribution of HPEs inferred from surface heat flow and heat production measurements. As noted by Oxburgh (1980), " ... the problem of crustal heat production and its distribution and re-distribution by physical and chemical processes during crustal evolution is of fundamental importance and is at present little understood."

The primary objective of this chapter is to show that minor tectonic reworking of the continental lithosphere may modify the distribution of HPEs. In particular, we focus on the role of isostatic coupling between deformation and surface processes (i.e., erosion and sedimentation). The temperature sensitivity of lithospheric rheology requires that tectonic reworking of the continental lithosphere is sensitive to the thermal state of the lithosphere. By making the link between tectonic reworking and HPE distributions we will illustrate a potentially profound feedback relation that might serve as a long-term control on the distribution of HPEs in the continents. Although the ideas and models developed in this chapter contribute to our understanding of lithospheric evolution in a very general sense, we emphasize that our one-dimensional pure shear models are particularly relevant to the evolution of continental interiors, and the role played by intracontinental, rather than plate-margin deformation. We begin by outlining a new parameterization of crustal HPE distributions that permits a simple graphical scheme for evaluating some important thermal and mechanical consequences of HPE distributions.

3.2 Length scales and heat production distributions

Surface heat flow and heat production measurements imply that the upper 10–15 km of the continental crust is significantly enriched in HPEs relative to

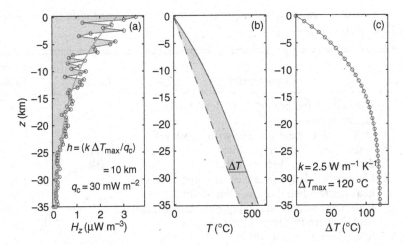

Fig. 3.1. (a) Hypothetical heat production distribution approximating an exponentially decreasing distribution of the form $H(z) = H_{sur} (\exp(-z/h_r))$ with added "noise" designed to simulate the heterogeneity of the continental lithosphere. The model exponential distribution is shown by the solid line. (b) The temperature field produced by such a distribution illustrating the component due to the reduced heat flow (the dashed line), and the component due to heat production in the lithosphere (the shaded area). (c) The temperature deviation, ΔT, is defined as the component of the temperature field due to heat production and corresponds to the shaded region in Fig. 3.1(b). Note that the maximum value, ΔT_{max}, is attained at the base of the heat-producing layer.

the deeper crust and mantle. The general form of the HPE distributions in the crust, and particularly the way HPE concentrations vary with depth, has been the subject of considerable interest. The heterogeneous composition of the continental crust implies that any distribution will be complex and discontinuous (Fig. 3.1a). However, for a number of purposes it is desirable to approximate this distribution with simple analytical models. Since the classic studies of Birch *et al.* (1968), Roy *et al.* (1968) and Lachenbruch (1968), heat-production distributions have been characterized in terms of both a characteristic volumetric heat production value (H, measured in units of $W\,m^{-3}$) and a characteristic length-scale (h, measured in kilometers). The most celebrated of these heat-production distributions is the *exponential* model (Lachenbruch, 1968, 1970) where the heat production at any depth z (defined to be negative downwards) is given by

$$H(z) = H_{sur} \exp\left(\frac{z}{h_r}\right). \qquad (3.1)$$

Table 3.1. *Parameters used in text and figures with default values applying to calculations unless otherwise stated.*

Symbol	Description	Units	Default value
z	Depth within the crust. z is defined to be negative downwards	km	
$H(z)$	Distribution of heat sources with depth (z)		
H_{sur}	Surface heat production value	$(\mu)W\,m^{-3}$	
H_{sed}	Sediment heat production value	$(\mu)W\,m^{-3}$	
h_r	Characteristic length scale of the exponential heat source distribution. Originally defined from the linear relationship between surface heat flow and heat production data (e.g., Lachenbruch, 1968)	km	15
h	General formulation for the characteristic length scale of the heat source distribution (this contribution). Defined in text in Eq. (3.7)	km	
k	Thermal conductivity	$W\,m^{-1}\,K^{-1}$	3
q_r	Reduced heat flow which applies at deep levels within the lithosphere, beneath all significant heat production	$(m)\,W\,m^{-2}$	
q_c	Depth-integrated heat production. q_c is always positive (i.e., heat flowing out of the crust)	$(m)\,W\,m^{-2}$	
T_{qc}	Temperature contribution due to crustal heat sources	°C	
T'_{qc}	Maximum temperature contribution due to crustal heat sources. T'_{qc} is reached at the depth at which heat production becomes negligible	°C	
z_c	Crustal thickness	km	35–40 km
z_s	Amount of subsidence during basin formation	km	
β	Ratio of the thickness of the crust prior to and following deformation; $\beta < 1$ for shortening, $\beta > 1$ for extension	dimensionless	
ρ_c	Density of the crust	$kg\,m^{-3}$	2780
ρ_m	Density of the mantle	$kg\,m^{-3}$	3300
ρ_{sed}	Density of the basin filling sediments	$kg\,m^{-3}$	2400

Table 3.1 lists definitions of all parameters, their units of measurement and default values used in calculations reported here. In Eq. (3.1), the characteristic length-scale h_r is defined as the depth at which heat production falls to $1/e$ of the characteristic surface value, H_{sur} (i.e., h_r is the e-fold length of the exponential distribution).

The use of such idealized, largely pedagogical models is motivated in part by the desire to obtain simple analytical expressions for steady-state temperature distributions. For example, assuming that there is no lateral variation in heat production, and that the thermal conductivity is constant throughout the crust (and independent of temperature), the steady-state temperature field for the exponential model is given by

$$T_z = -\frac{q_r z}{k} + \frac{H_{sur} h_r^2 (1 - \exp(-z/h_r))}{k} \tag{3.2}$$

In Eq. (3.2), q_r is the reduced heat flow that applies at deep levels within the lithosphere, beneath all significant heat production. One interpretation of q_r is that it represents the heat flux provided to the base of the lithosphere by convective processes in the deeper mantle. However, as noted by Jaupart (1983) and Vasseur and Singh (1986), the value of q_r deduced from regression of surface heat-flow–heat-production data will invariably overestimate the non-radiogenic component of surface heat-flow and underestimate the length scale of the crustal heat production. This is because the lateral transfer of heat is significant in regions where there are significant lateral contrasts in heat production. In such cases both the slope and intercept of any fit of surface heat-flow with surface heat production are dependent on the horizontal variation in heat-production parameters, as well as the vertical distribution.

An alternative model for the vertical distribution of heat production is a layer extending to depth h_r with constant heat production, H_{sed}, beneath which heat production is negligible. In this *homogeneous* model the temperature field in the heat-producing layer is

$$T_z = -\frac{q_r z}{k} + \frac{H_{sur} z (h_r - z/2)}{k} \tag{3.3}$$

Note that in Eqs. (3.2) and (3.3), the first term on the right-hand side of the equations represents the component of the temperature field due to the heat flow from beneath the heat-producing parts of the lithosphere (e.g., dashed line in Fig. 3.1(b)). The second term on the right-hand side of the equation represents the contribution due to heat sources in the crust (e.g., the shaded area in Fig. 3.1(b)). We use this second term to define the quantity T_{qc}, the temperature contribution due to crustal heat production. T_{qc} reaches its maximum value (T'_{q_c}) at the depth at which heat production becomes negligible (i.e., at the base of the heat-producing layer; Fig. 3.1(c)). The appropriate expressions for T'_{q_c} for the exponential and homogeneous models are, respectively,

$$T'_{qc} = \frac{H_{sur} h_r^2 (1 - \exp(-z_c/h_r))}{k} \tag{3.4}$$

and

$$T'_{qc} = \frac{H_{\mathrm{sur}} h_r^2}{2k} \tag{3.5}$$

In Eq. (3.4), z_c is the depth at which the exponential heat production distribution is terminated, which we approximate as the Moho. Note that provided that $z_c > h_r$, then Eq. (3.4) reduces to

$$T'_{qc} \approx \frac{H_{\mathrm{sur}} h_r^2}{k}. \tag{3.6}$$

One attribute of these two models is that the length scale h_r may be evaluated from the regression of surface heat-production and heat-flow data (with the proviso outlined above). Typically, such regressions yield a value of $\sim 10\,\mathrm{km}$ (e.g., Lachenbruch, 1968).

An alternative and more general formulation of the characteristic length-scale (h), that makes no explicit assumption about the form of the heat-production distribution, may be made in terms of the heat source distribution

$$h = \frac{1}{q_c} \int_0^{z_c} (H(z))\mathrm{d}z \tag{3.7}$$

where q_c is the depth integrated heat production

$$q_c = \int_0^{z_c} H(z)\mathrm{d}z. \tag{3.8}$$

One attribute of this definition is that it leads to a very simple relation between T'_{qc}, q_c and h, as follows

$$T'_{qc} = \frac{q_c h}{k}. \tag{3.9}$$

Equation (3.9) implies that very different HPE distributions characterized by similar values of h (cf. Fig. 3.2a–d) and similar integrated abundances of HPEs (i.e., similar q_c) will contribute the same temperature deviation at all depths beneath the heat-producing parts of the lithosphere. This parameterization also highlights the fact that the temperature deviation within the heat-producing parts of the lithosphere due to the crustal heat production, T_{qc}, is very sensitive to the depth at which the heat production is located (cf. Fig. 3.2a, b, e, f). We note that for both the *homogeneous* and *exponential* heat production models outlined above

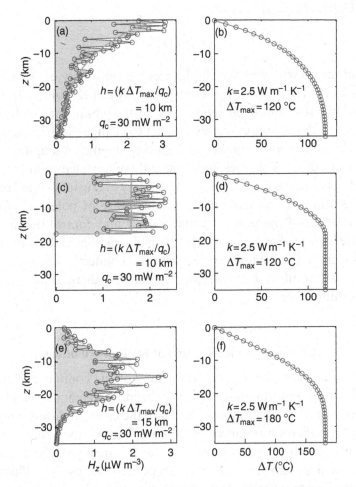

Fig. 3.2. Illustration of various heat-production models and their associated thermal effects, used in this chapter. Each model is based on an analytical expression with associated geological noise. The analytic expressions are respectively: (a and b) the exponential model; (c and d) the homogeneous model, and (e and f) show that, for a given complement of HPEs, an increase in the depth of burial (as represented by the parameter h) results in greater perturbation of lower crustal temperatures.

$$q_c = h_r H_{sur} \tag{3.10}$$

allowing h to be expressed in terms of h_r. Consequently, for the exponential model

$$h = h_r(1 - \exp(-z_c/h_r)) \tag{3.11}$$

that reduces to

$$h = h_r \text{ for } z_c \gg h_r \tag{3.12}$$

and, for the homogeneous model,

$$h = h_r/2 \tag{3.13}$$

The analysis presented above emphasizes the sensitivity of crustal thermal regimes not only to the total heat contributed by crustal sources (i.e., q_c) but also to its depth distribution as represented by the parameter h. An important consequence of this is that processes that act to change the distribution of the HPEs (i.e., q_c and/or h) are capable of inducing significant long-term changes in deep crustal thermal regimes, independently of any change in the heat flux from the deeper mantle. Moreover, the sensitivity of crustal thermal regimes to these distribution parameters raises the possibility of potentially profound feedback relations if the processes that modify and control the distribution of HPEs are themselves sensitive to the thermal structure of the lithosphere. In the remainder of this chapter we explore the connection between tectonic processes that modify HPE distributions and the thermal and mechanical state of the lithosphere.

3.3 Tectonic modification of HPE distributions

Tectonic processes associated with reworking of the continental lithosphere (e.g., magmatism, deformation, metamorphism, sedimentation and erosion) effect changes in the distribution of HPEs in the crust, and thus must impact on its long-term thermal and mechanical state. Because the long-term changes in the thermal structure of the lithosphere may be directly related to changes in the distribution parameters, the h–q_c plane may be used to illustrate the thermal consequences of such tectonic processes.

Figure 3.3 shows qualitatively how various processes attendant with crustal reworking are likely to change the HPE distribution parameters. For example, the generation of granites by crustal melting provides a mode for redistributing HPEs upwards in the crust with a corresponding reduction in h without reducing q_c (Fig. 3.3a). Expressed in terms of relative changes in the heat-production parameters, infracrustal magmatism is characterized by $\Delta q_c = 0$ and $\Delta h < 0$.

Deformation results in changes in both the amount and the depth of crustal heat sources. Crustal shortening ($\beta < 1$ where β is the ratio of the thickness of the crust prior to, and following, deformation) increases the total complement of heat sources within the crust, due to structural repetition, and at the same time buries it to deeper crustal levels (i.e., $\Delta q_c > 0$ and $\Delta h > 0$). In contrast, crustal extension (where $\beta > 1$) tends to attenuate the pre-existing

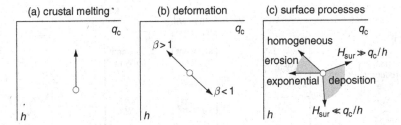

Fig. 3.3. Schematic illustration of the way tectonic processes associated with reworking of continental lithosphere modify the HPE distribution parameters, h and q_c.

heat production and move it to shallower levels (i.e., $\Delta q_c < 0$ and $\Delta h > 0$). Erosion reduces q_c ($\Delta q_c < 0$), particularly so when HPEs are already concentrated in the upper crust. The impact of erosion on the length scale depends on the general form of the HPE distribution in the crust (Fig. 3.3c). The length-scale for *exponential* distributions is essentially preserved during erosion, while for *homogeneous* distributions the length scale is reduced in direct proportion to the reduction in q_c. The effect of sedimentation depends on the heat production character of the sediments (H_{sed}) relative to the pre-existing crust (Fig. 3.3c). If H_{sed} is much greater than the mean heat-production of the pre-existing crust ($\sim q_c/h$), then sedimentation will result in a significant increase in q_c and may reduce h ($\Delta q_c \gg 0$ and $\Delta h > 0$). In contrast, if $H_{sed} \ll q_c/h$, then sedimentation will lead to a significant increase in h for a comparatively minor increase in q_c ($\Delta q_c > 0$ and $\Delta h > 0$).

3.4 Thermal modeling

In the long term, deformation, erosion, and sedimentation are linked through the isostatic response of the lithosphere, with the surface processes tending to restore the long-term elevation of the continents to near sea-level. Potentially, this coupling may serve to order the distribution of the HPEs. In this section we illustrate quantitatively the way in which deformation, erosion, and sedimentation modify various heat-source distributions (including the *exponential* and *homogeneous* models described in the previous section). The assumptions used in the development of these models are listed here.

First, we link deformation and surface processes through an isostatic response that tends, in the long term, to restore the surface elevation to some constant value (Fig. 3.4). Thus, when deformation results in crustal thickening, the isostatic response of surface uplift induces erosion, whereas crustal

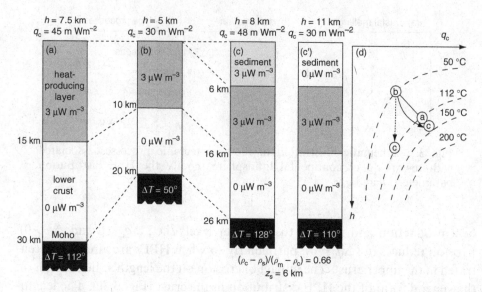

Fig. 3.4. Illustration of the role of deformation coupled with an isostatic response on heat-production distributions in the crust. Given an initial lithosphere (a), stretching (b) results in attenuation of pre-existing heat production, while the surface response of basin formation (c) moves it to deeper levels (i.e., an increase in h). The final heat-production distribution following restoration of the original surface elevation depends on the heat-production character of the basin-fillings sediments (with q_c either decreased or increased). The changes in h and q_c may be mapped on to the h–q_c plane (d) that may also be used to illustrate the steady-state contribution of crustal radiogenic sources to thermal regimes at deep crustal levels (i.e., ΔT_{max}).

thinning deformation is coupled with subsidence and sedimentation (i.e., basin formation). By long term we imply a sufficient time to dissipate all thermal transients accruing from the act of change itself (i.e., a period greater than the thermal time constant of the lithosphere). The treatment of the lithosphere as 1-D vertical column requires application of a local isostatic balance appropriate to the first-order dynamics of continental lithosphere for processes that operate on spatial scales greater than the thickness of the lithosphere (i.e., greater than several hundred kilometers).

Second, in order to preserve surface elevation, we make several assumptions about the densities of the various components of the crust. We assume that there is no significant density difference between the upper and lower crust. The important implication of this assumption, for all the models presented here, is that, following crustal thickening, the amount of erosion required to restore the original surface elevation will also, to a first approximation, restore the original crustal thickness. Further, we also assume that the basin-filling sediments are significantly less dense than the crust. We express this density in terms of the density of the basin filling sediments (ρ_{sed}), crust (ρ_c) and mantle (ρ_m), as follows

$$\rho' = \frac{\rho_c - \rho_{sed}}{\rho_m - \rho_c} \approx 0.66. \tag{3.14}$$

Consequently, the crust shows significant long-term thinning following extension. In terms of the changes in the compositional structure, the subsidence scales as

$$z_s = \frac{z_c(1 - 1/\beta)}{1 - \rho'}. \tag{3.15}$$

Third, in our models, deformation is assumed to operate homogeneously in the crustal column. Rather different changes to the HPE distribution may result when large-scale discontinuities in the deformation pattern are allowed, such as crustal-scale thrust faulting. Clearly, such discontinuities are important components of deformation in the continental crust when large strains accumulate. They are less likely to be important when finite strains are small, as is typical of intra-continental (or intra-plate) deformation associated with reworking of continental interiors. In such settings, deformation rarely results in the exposure of deep crustal rocks. As discussed in Section 3.6, we believe our analysis is particularly relevant to the role played by intracontinental deformation.

Lastly, in modeling the effects of sedimentation, we assume a typical upper crustal heat production of 1.5 $\mu W\ m^{-3}$ for the basin fill (e.g., McLennan & Taylor, 1996). The results of this modeling are summarized in Figs. 3.5 and 3.6. The initial HPE distributions used in the calculations correspond to the *exponential* (Fig. 3.5) and *homogeneous* models (Fig. 3.6) described in the previous section.

Figures 3.5a, b and 3.6a, b show the effect of a 10% change in crustal thickness (i.e., $\beta = 0.9$ for crustal thickening and $\beta = 1.1$ for crustal thinning). Figures 3.5c, d and 3.6c, d show the effects of an associated surface response (erosion or sedimentation) that, as noted above, restores the original surface elevation assuming local isostatic balance. Figures 3.5e, f and 3.6e, f show the combined effects of the deformation and the linked surface response. Figures 3.5 and 3.6 illustrate a number of important features concerning the way in which heat-production parameters evolve as a consequence of deformation and the surface processes, which are summarized below.

The coupling of crustal thickening (i.e., $\beta < 1$) and erosion provides an effective means for modifying HPE parameters, particularly where crustal

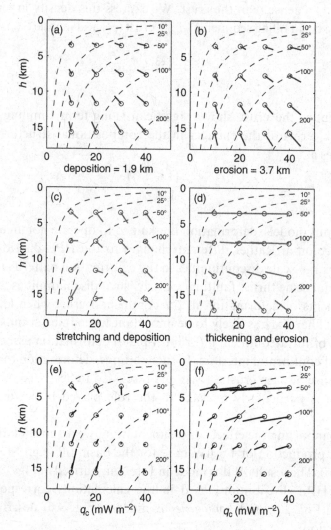

Fig. 3.5. (a), (b) illustrate the effects of a 10% attenuation ($\beta = 1.1$) and a 10% thickening ($\beta = 0.9$) on the distribution parameters h and q_c for initial *exponential* distributions of varying h and q_c (see Fig. 3.2a); (c), (d) show, respectively, the role of deposition and erosion sufficient to restore the original value surface elevation, and (e), (f) show, respectively, the combined effect of stretching and basin formation, and thickening and erosion. Each figure is contoured for ΔT_{max} (see Fig. 3.1).

heat-production is already significantly differentiated (e.g., 3.5f and 3.6f). Whereas crustal thickening tends to increase q_c, erosion will reduce it. If the existing heat production is already strongly concentrated near the surface (i.e., small h) the loss of heat production during erosion is much greater than

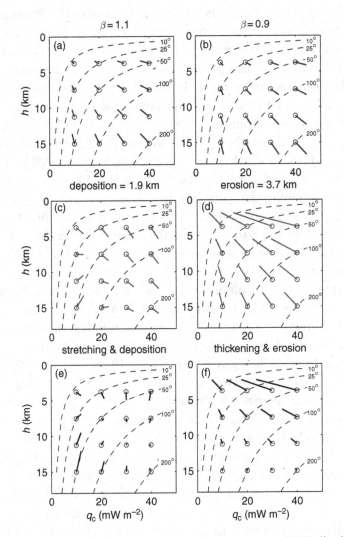

Fig. 3.6. As for Fig. 3.5, but with an initial *homogeneous* HPE distribution (i.e., Fig. 3.2c).

the increase in heat production due to thickening (see Fig. 3.4b), resulting in a dramatic decrease in q_c. This reflects the fact that the coupling of crustal thickening and erosion essentially removes upper crustal material and replaces it with lower crustal material. The consequences for the length scale, h, depend on the initial distribution of heat sources. For distributions approximating the homogeneous model, h will be reduced as a consequence of combined homogeneous thickening and erosion (Fig. 3.6).

For a distribution which is approximately exponential, then providing $h \ll z_c$, h will increase by the factor $1/\beta$, because erosion preserves the exponential length scale established by the deformation (Lachenbruch, 1968). Note that the coupling of crustal thickening and erosion is ineffective in changing the heat production parameters when the crust is largely undifferentiated (i.e., large h), because the heat-production character of the lower crust is essentially the same as the upper crust. Figures 3.7c, d show that the long-term changes in the thermal structure of the deep crust resulting from thickening and erosion may be significant. For typical crustal heat production parameters, a 10% thickening followed by erosion will lead to a reduction in deep crustal temperatures by 10–30 °C. Because the Moho will be returned to its original depth following erosion, it will be expected to experience all of this cooling.

For crustal thinning deformations ($\beta > 1$) the long-term thermal consequences are critically dependant on whether basin formation occurs and, if so, the heat production parameters of the basin-filling sediments. As noted above, in calculating Figs. 3.4 and 3.5 we have assumed that the basin fill has heat production of $1.5 \,\mu W \,m^{-3}$ (e.g., McLennan & Taylor, 1996). Crustal thinning without basin formation will result in a reduction in both q_c and h by the factor $1/\beta$ for all heat source models. Where the subsidence resulting from crustal thinning is accompanied by sedimentation sufficient to restore the original surface elevation, the resulting changes in h and q_c may lead to either an increase or decrease in T'_{qc} (e.g., Sandiford, 1999). As shown by Figs. 3.7a, b, increases in T'_{qc} are expected when h is less than about 10 km. For initial configurations characterized by high h, the combination of stretching and basin filling may lead to long-term reductions in T'_{qc} of up to ~ 10 °C for an initial 10% stretch. However, note that these values are sensitive to both the assumed density and heat-production character of the basin-filling sediments, with more dense and/or more radiogenic sediments leading to increased values of T'_{qc} for a given stretch (see more detailed discussion in Sandiford, 1999). Because the Moho will not be returned to its original depths in the long term for typical basin-fill densities, the estimates for T'_{qc} as shown in Figs. 3.7a, b cannot be directly equated with long-term changes in the Moho temperature. The amount of additional cooling of the Moho due to its long-term shallowing depends on the thermal gradient at the Moho level. For a thermal gradient of $10 \,°C \,km^{-1}$, this long-term shallowing of the Moho due to a stretch of 10% leads to Moho cooling (ΔT_{Moho}) of ~ 13 °C, with the change in Moho temperature given by $T'_{qc} - \Delta T_{Moho}$. For the parameters used in our calculations, only extremely high q_c, low h initial configurations lead to long-term Moho heating following crustal stretching (i.e., the shaded region in the upper right of Fig. 3.7a, b).

The long-term evolution of many continental interiors is punctuated by the accumulation of small to very small magnitude deformation events,

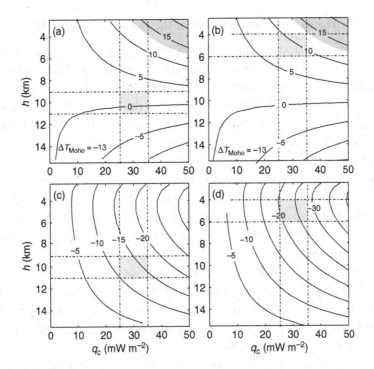

Fig. 3.7. The h–q_c plane contoured for changes in ΔT_{max} following a 10% stretch and an associated surface response that restores the initial surface elevation. (a) and (b) show the effects of a 10% stretch with basin formation for an initial *exponential* and *homogeneous* HPE distribution, respectively. (c) and (d) show the effects of a 10% thickening followed by erosion, for an initial *exponential* and *homogeneous* HPE distribution, respectively. In the case of stretching, the depth of the Moho is reduced in the long term because the density of sediments is less than that of the crust. Because of this reduction, the Moho is cooled (by ~13 °C for the parameters used here). The change in Moho temperature is given by $T'_{qc} - \Delta T_{Moho}$. The gray box shows the relatively narrow range in h–q_c parameters that characterize stable modern continental crust. This data is summarized from a compilation by Taylor and McLennan (1985) (modified from Vitorello and Pollack, 1980, and Morgan, 1984).

associated with minor reworking and reactivation. Such events involve both crustal shortening and extension, and are typically accompanied by an associated surface response that tends, in the long term, to keep the crustal thickness at the continental average of ~35 km. Figures 3.5–3.7 show that such small deformation events may effect significant redistribution of the HPEs, raising the possibility that tectonic reworking of continental interiors plays an important role in the geochemical ordering of the continental lithosphere.

3.5 Mechanical consequences of the redistribution of HPEs

In the previous section we have shown that significant long-term thermal consequences may accrue from the redistribution of crustal heat sources that accompanies deformation when coupled with associated surface processes, even for relatively minor deformation events that are likely to typify continental interiors. The strong temperature-dependence of lithospheric rheology suggests that deformation is likely to be localized in continental regions characterized by elevated thermal regimes, such as would be associated with high q_c and/or h values. If all other factors, including composition and strain rate, are equal, then continental crust with low q_c and h values would be expected to be significantly stronger than crust with higher q_c and h values. Therefore, such crust would be less likely to experience deformation in response to an applied tectonic load. In this section we attempt to quantify how variations in q_c and h affect the strength of the lithosphere. We use a lithospheric rheology in which the lithosphere is assumed to deform by a combination of frictional sliding and ductile creep (e.g., Brace & Kohlstedt, 1980; Table 3.2). Throughout the remainder of this section we refer to this model as the *Brace–Goetze* rheology.

The application of the *strength-envelope* approach implicit in this model has become widespread in addressing geodynamic problems. Nevertheless, it is important to note that any calculations of lithospheric strength are subject to very large uncertainties. Basic uncertainty stems from a number of sources (e.g., Paterson, 1987), including: (1) the extrapolation of rheological flow laws from laboratory conditions and time scales to the geological realm; (2) an imprecise understanding of the compositional and mineralogical structure of the lithosphere, which results in uncertainties in knowing what particular flow laws to apply to which part of the lithosphere, and (3) uncertainties in the absolute thermal structure of the lithosphere due to imprecise knowledge of its thermal property structure, particularly thermal conductivity. These problems make any calculation of absolute strength subject to very large uncertainties, and in the following section we emphasize the role of temperature changes accompanying the various tectonic processes to *relative* changes in lithospheric strength. Whereas the absolute magnitude of these relative changes will vary with the composition and thermal structure of the reference frame, the relative changes are likely to be far more robust than any absolute measure of strength, because many of the uncertainties alluded to above will cancel.

Recognizing that the h–q_c plane may be contoured for thermal structure (e.g., Figs. 3.5–3.7) also allows it to be contoured for lithospheric strength

Table 3.2. *Rheological parameters used in thermo-mechanical models*

Model rock type	Flow parameter, A (MPa^{-n} s^{-1})	Power index, n	Activation energy, Q (kJ mol^{-1})	Reference
Crust	Adirondack granulite	7.93×10^{-3}	3.1	243	Wilks & Carter (1990)
Upper mantle	Aheim Dunite	398.11	4.5	535	Chopra & Paterson (1981)

(Fig. 3.8). Figure 3.8a shows variations in the strength of a *Brace–Goetze* lithosphere forced to deform at a specified strain rate, whereas Fig. 3.8b shows variations in the effective strain rate for the lithosphere when subject to an imposed tectonic load (the rheological parameters assumed in the calculations are summarized in Table 3.2). For the *Brace–Goetze* lithosphere, varying either h or q_c by a factor of 2 leads to a variation in effective strength by a factor of 2–3, and effective strain rates by 1–2 orders of magnitude. The sensitivity of effective strain rates to the heat-production parameters highlights the inherent temperature sensitivity of the *Brace–Goetze* lithosphere and raises the possibility, to be discussed in the next section, of a feedback relation between deformation processes and HPE distributions in the crust. One way to view Fig. 3.8b is that it expresses the likelihood that a lithosphere will experience significant deformation when subject to an imposed tectonic load.

Whereas Fig. 3.8 illustrates the way in which variations in HPE parameters may alter the strength of the lithosphere, it is important to realize that other factors also influence lithospheric strength. In particular, long-term changes in the depth of the Moho, as may be expected following basin formation (see earlier discussion), will also contribute to variations in lithospheric strength. In the case of basin formation, the cooling associated with shallowing of the Moho tends to increase the strength of the lithosphere, while the changes in HPE parameters usually tend to decrease its strength (e.g., Fig. 3.7a, b). Depending on the initial HPE distribution parameters, basin formation may either result in mild long-term weakening or strengthening of the lithosphere (a detailed discussion is given in Sandiford, 1999).

Figure 3.8 suggests that the mechanical response of the lithosphere to imposed tectonic loads is extremely sensitive to the distribution of HPEs. In order to explore how the distribution of HPEs impacts on the way the lithosphere responds, and the way in which this response might change HPE

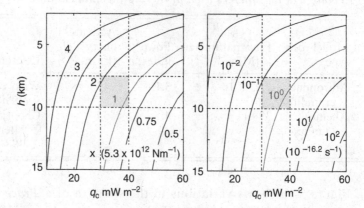

Fig. 3.8. (a) h–q_c plane contoured for integrated lithospheric strength. (b) h–q_c plane contoured for rate of deformation subject to an imposed tectonic load. The assumed rheology is that of the *Brace–Goetze* lithosphere in which deformation is governed by a combination of frictional sliding and temperature-dependent creep. The strength parameters are shown normalized against a configuration characterized by $q_c = 45\,\mathrm{mW\ m^{-2}}$ and $h = 7\,\mathrm{km}$. The shaded area illustrates that, all other factors being equal, a 25% reduction in h and q_c will result in an increase in strength by a factor of 2 or a decrease in strain rate in response to an imposed load by an order of magnitude. Note that the calculations are sensitive to a large number of assumed parameters including thermal conductivity ($k = 3\,\mathrm{W\ m^{-1}\ K^{-1}}$) and mantle heat flux ($q_m = 30\,\mathrm{mW\ m^{-2}}$), as well as the material parameters constraining creep and frictional sliding of crustal and mantle rocks.

distributions, we have carried out a set of numerical simulations in which a series of model 1-D *Brace–Goetze* lithospheric columns, differing only in terms of their HPE parameters, are subject to an identical loading history (Figs. 3.9 and 3.10). The load history, as shown in Figs. 3.9g and 3.10g, is scaled so as to provide a mildly compressional stress regime, as is suggested by the potential energy differences between mid-ocean ridges and normal continental crust (e.g., Coblentz *et al.*, 1994), with a hierarchy of fluctuations of up to $\sim 10^{13}$ $\mathrm{Nm^{-1}}$ over a range of timescales up to $\sim 500\,\mathrm{Ma}$. Deformation rates are relatively insignificant ($\varepsilon_{zz} < 10^{-16}\,\mathrm{s^{-1}}$, Figs. 3.9h and 3.10h) especially when compared to modern plate-margin deformation rates, with the result that strain increments during any given episode are only small (individual stretch increments are $< 15\%$, Figs. 3.9e and 3.10e). The isostatic response to changes in crustal thickness (Figs. 3.9d and 3.10d) either elevates or depresses the surface (e.g., Figs. 3.9a and 3.10a), thereby inducing a surface response that tends to restore the surface to sea level (e.g., Figs. 3.9f and 3.10f) on a characteristic time scale. The three simulations shown in Fig. 3.9 are

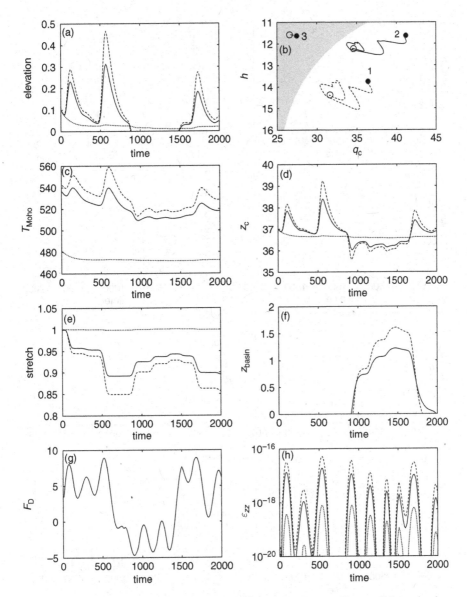

Fig. 3.9. Numerical simulations in which 1-D *Brace–Goetze* lithospheric columns are subject to a tectonic load (g). The response of three columns is shown, these columns being identical in all respects apart from the initial HPE distribution parameters (b), which translates to variations in the initial thermal regime (c). The initial HPE distribution is exponential. The tectonic response is measured in terms of vertical strain rate (h), as well as incremental stretching history (e). The interaction of deformation and surface response leads to long-term changes in surface elevation (a) and crustal thickness (d). Accumulation of sediments in basins, and its subsequent removal during basin inversion, is shown in (f). The change in HPE distribution parameters is shown in (b). Time is measured in Ma, with the thermal response of the lithosphere (c) calculated using a finite element algorithm. Closed circles show initial HPE parameters while open circles show final HPE parameters. The shaded region represents parameter range that shows stable (or cratonic) behavior throughout the imposed loading history. See text for discussion.

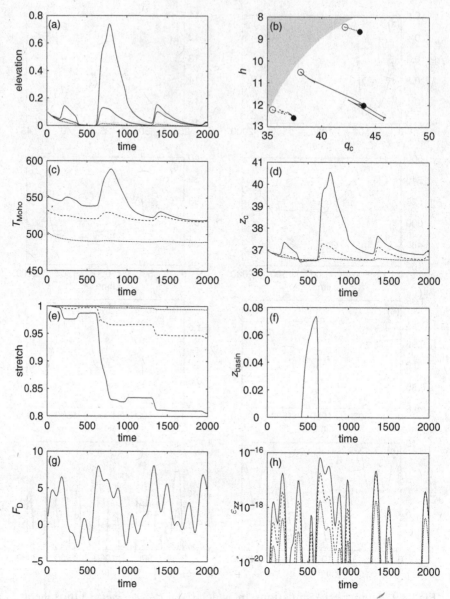

Fig. 3.10. As for Fig. 3.9, but with an initial HPE distribution approximating the homogeneous model (i.e., Fig. 3.2c).

characterized by an initially *exponential* heat source distribution with HPE parameters, $q_c = 37$, 42 and 27 mW m^{-2} and $h = 13.5$, 11.8, and 11.8 km, respectively (Fig. 3.9b), whereas Fig. 3.10 is characterized by an initially *homogeneous* distribution.

In accord with our earlier discussion, these simulations show that the response of the lithosphere is sensitive to its thermal state, and that HPE distributions are modified significantly by repeated small-scale tectonic reworking of the lithosphere. The simulations show that initial configurations characterized by high q_c ($> 35\,\text{mW m}^{-2}$) experience significant reductions in q_c during repeated tectonic reworking, even though the scale of the reworking is very minor (note that the amplitude in variations in crustal surface elevation is generally less than 500 m, e.g., Fig. 3.9a). For an initial exponential distribution, the erosion-dominated reduction in q_c tends largely to preserve the initial value of h, in accord with the results in Fig. 3.5. Figure 3.10b shows that initial homogenous distributions may show significant h reduction, as well as q_c reduction, on reworking. The reductions in q_c of up to 20% equate to long-term Moho cooling of $\sim$20–40 °C. In contrast, the initial configurations with low q_c and low h (point 3 in Fig. 3.9b) show no significant response to the imposed loads by virtue of their inherent mechanical strength, a minor reduction in q_c is induced as a consequence of the long-term erosional denudation of the initial topography.

While these models are clearly very sensitive to the input parameters, they do highlight a potentially profound role that tectonic reworking may play in modulating the HPE distributions in the continental crust. Crust initially configured with elevated q_c and/or h will, through the isostatic coupling of deformation and surface processes, tend to remove HPEs from the crust until it ceases to respond to the normal fluctuation in intraplate stress levels. In terms of the HPE distribution parameters, this is expressed as a reduction in q_c and, depending on the form of the initial distribution, a reduction in h.

3.6 Discussion

The origin of the characteristic HPE distribution inferred from the analysis of surface-heat flow and heat-production data is central to the differentiation of the continental crust, not in the least because the distribution of HPEs provides a fundamental control on the thermal and mechanical state of the lithosphere. The available data imply that HPEs are concentrated in the upper 10–15 km and contribute on average $\sim$30–40 mW m^{-2} to the observed surface-heat flow. An inevitable consequence of the lithophile nature of the HPEs is that primary magma-dominated, crustal growth processes will lead to length scales for the heat source distribution that are significantly less than the characteristic crustal thickness. Nevertheless, no compelling reason has emerged as to why such processes would naturally lead to a characteristic length scale or to a characteristic abundance of HPEs that contributes typically between one-half and two-thirds of the surface-heat flow (i.e., q_c = 30–40 mW m^{-2}).

It is important to realize that the gross chemical and structural architecture of the crust is not only a function of primary crustal accretion, but also reflects complex tectonic reworking in zones of continental deformation, both at plate boundaries (e.g., collisional orogens) and in intraplate settings. One of our main purposes in this chapter has been to show that, in intraplate settings in particular, HPEs are redistributed as a direct consequence of the deformation as well as through related processes such as erosion and sedimentation that follow as the isostatic consequence of the deformation. Such tectonic reworking provides an additional mechanism for ordering HPEs in the continental lithosphere.

The continents are subject to temporal fluctuations in tectonic loads due to changing plate configurations (e.g., Sandiford & Coblentz, 1994) as well as interactions with the convective mantle beneath. Hot, weak continental lithosphere will respond to relatively small loads by deforming. Under any prevailing load regime, the deformation may be limited through the accumulation of potential energy in the deformed lithosphere. Changes in the load regime will lead to further changes in the deformation with the isostatic response modulating the nature and efficiency of the associated surface processes. In contrast, cold strong continental lithosphere will be able to withstand much greater tectonic loads without appreciable deformation. In the previous sections we have introduced a simple parameterization of crustal HPE distributions that provides a framework for understanding how crustal-scale deformation associated with tectonic loading may effect long-term changes in thermal regimes through the redistribution of HPEs. This framework shows that deformation, when linked with a surface process through an isostatic response, is capable of effecting significant change in heat production parameters. The most significant response to the coupled deformation-surface response regime outlined here, when measured as a function of strain increments, occurs for crust already strongly differentiated. As noted above, this concurs with the notion that the primary differentiation of the HPEs within the crust is due to magmatic processes attendant with primary crustal growth.

The new insight provided by our analysis comes with the recognition that the likelihood of deformation affecting the continental lithosphere is dependant on the thermal regime and therefore on the abundance and distribution of HPEs. This is best exemplified by Fig. 3.8b, which shows the way in which the HPE distribution influences the rate of deformation of a *Brace–Goetze* lithosphere in response to an imposed tectonic load. This diagram, which may be considered as representing the probability of deformation in response to an imposed tectonic load, shows that for a given complement of HPEs (i.e., constant q_c), relatively undifferentiated crust (i.e., high h) is very much weaker than differentiated crust (i.e., low h).

An important consequence of the inherent weakness of high h–q_c configurations is their susceptibility to tectonic reworking. Even though high h–q_c distributions are expected to show only relatively modest changes in h–q_c parameters per unit increment of deformation, this susceptibility implies that tectonic loading will result in very much greater strains and thus may, in the long-term, effect significant redistribution of HPEs providing there is some primary differentiation. Indeed, only once reworking has effectively removed much of the heat production from the crust, or moved it to shallow levels, will such lithosphere be stabilized.

The correlation between the distribution of HPEs and the strength of the continental lithosphere implies that the differentiation of HPEs is essential for the long-term stability of the continental lithosphere. Our understanding of the stabilization of continental lithosphere is intimately intertwined with the notion of craton development. Whereas craton formation is undoubtedly a complex process, most probably involving long-term changes in crust–mantle interaction, the analysis of tectonic reworking presented in this chapter suggests that crustal-scale differentiation of HPEs is a necessary precursor to cratonization. Whereas cratons are manifest by long-term tectonic stability, cratonization is usually presaged by a lengthy "early" history involving not only crustal growth but also extensive crustal reworking during episodic tectonism spanning many hundreds of millions of years. In a thermal and mechanical sense, such reworking may be viewed in terms of the way it redistributes the HPEs (e.g., McLaren & Sandiford, 2001). Indeed, such a view may be consistent with the observation that Archean cratons are characterized not only by lower surface heat flow ($\sim$40 mW m^{-2}), but somewhat lower characteristic length scales than younger geological provinces.

The data compilation in Taylor and McLennan (1985), largely based on Vitorello and Pollack (1980) and Morgan (1984), suggests that the mean (and standard deviation) length scale for heat production in Archean terranes is 6.9 ± 1.7 km, compared to 10.1 ± 3.6 km in Proterozoic terranes. In part, the low surface heat flow in Archean cratons is due to the low present-day abundance of HPEs, but also to the presence of very thick mantle lithosphere, which has the effect of diminishing the mantle contribution. For example, McLennan and Taylor (1996) have suggested that the mantle heat flow beneath Archean cratons is typically $\sim$14 mW m^{-2}, implying that the HPEs contribute $\sim$25 mW m^{-2}. Due to secular decay heat production at 3 Ga was twice modern-day rates, and so the characteristic Archean q_c could conceivably have been as high as $50 \, mW \, m^{-2}$ (i.e., some 50–66% greater than typical of the modern continents). One way to stabilize lithosphere mechanically with such high values of crustal radiogenic heat production is to concentrate the HPEs at shallow levels in the crust (i.e., low h). We note with interest that in many Archean cratons the crustal HPE complement is carried in granites that form the cores of large "diapiric" domes (e.g., Choukroune *et al.*, 1997). Indeed, the origin of the characteristic "dome and basin" geometry of these Archean granite–greenstone terranes has been the subject of much discussion, and we speculate that such geometries may reflect a distinctive "high-q_c" mode of stabilizing continental crust by redistributing the HPEs into the shallow crust.

3.7 Acknowledgements

The ideas presented in this chapter have benefited from discussions with Martin Hand. We are grateful to Becky Jamieson, Chris Beaumont and Keith Klepeis for their helpful reviews of the manuscript.

References

Birch, F., Roy, R. F. and Decker, E. R. (1968). Heat flow and thermal history of New England and New York. In *Studies of Appalachian Geology: Northern and Maritime*, ed. E. Zen, W. S. White, J. B. Hadley and J. B. Thompson. New York: Wiley Interscience, pp. 437–51.

Brace, W. F. and Kohlstedt, D. L. (1980). Limits on lithospheric strength imposed by laboratory experiments. *Journal of Geophysical Research*, **85**, 6248–52.

Chopra, P. N. and Paterson, M. S. (1981). The experimental deformation of dunite. *Tectonophysics*, **78**, 453–73.

Choukroune, P., Ludden, J. N., Chardon, D., Calvert, A. J. and Bouhallier, H. (1997). Archean crustal growth and tectonic processes: a comparison of the superior province, Canada and the Dharwar Craton, India. In *Orogeny Through Time*, ed. J.-P. Burg and M. Ford. Geological Society of London Special Publications 121, pp. 63–98.

Coblentz, D., Richardson, R. M. and Sandiford, M. (1994). On the gravitational potential of the Earth's lithosphere. *Tectonics*, **13**, 929–45.

Haenel, R., Rybach, L. and Stegena, L. (1988). *Handbook of Terrestrial Heat-flow Density Determination; with Guidelines and Recommendations of the International Heat Flow Commission*. Dordrecht: Kluwer Academic Publishers.

Jaupart, C. (1983). Horizontal heat transfer due to radioactivity contrasts: causes and consequences of the linear heat flow relation. *Geophysical Journal of the Royal Astronomical Society*, **75**, 411–35.

Lachenbruch, A. H. (1968). Preliminary geothermal model of the Sierra Nevada. *Journal of Geophysical Research*, **73**, 6977–89.

Lachenbruch, A. H. (1970). Crustal temperature and heat production: implications of the linear heat-flow relation. *Journal of Geophysical Research*, **75**, 3291–300.

McLaren, S. and Sandiford, M. (2001). Long-term thermal consequences of tectonic activity at Mt Isa: implications for polyphase tectonism in the Proterozoic. In *Continental Reactivation and Reworking*, ed. J. Miller, I. Buick, M. Hand and R. Holdsworth. Geological Society of London, Special Publication pp. 184, 219–36.

McLennan, S. M. and Taylor, S. R. (1996). Heat flow and the chemical composition of continental crust. *Journal of Geology*, **104**, 369–77.

Molnar, P. (1989). Brace–Goetze strength profiles, the partitioning of strike-slip and thrust faulting at zones of oblique convergence and the stress-heat flow paradox of the San Andreas Fault. In *Fault Mechanics and Transport Properties of Rocks*, ed. B. Evans and T.-F. Wong. London: Academic Press, pp. 435–60.

Morgan, P. (1984). The thermal structure and thermal evolution of the continental lithosphere. *Physics and Chemistry of the Earth*, **15**, 107–85.

Oxburgh, E. R. (1980). Heat flow and magma genesis: In *Physics of Magmatic Processes*, ed. R. B. Hargraves. Princeton: Princeton University Press, pp. 161–200.

Paterson, M. S. (1987). Problems in the extrapolation of laboratory rheological data. *Tectonophysics*, **133**, 33–43.

Ranalli, G. (1997). Rheology of the lithosphere in space and time. In *Orogeny Through Time*, ed. J.-P. Burg and M. Ford. Geological Society of London Special Publication, 121, pp. 63–98.

Roy, R. F., Blackwell, D. D. and Birch, F. (1968). Heat generation of plutonic rocks and continental heat flow provinces. *Earth and Planetary Science Letters*, **5**, 1–12.

Sandiford, M. and Coblentz, D. (1994). Plate-scale potential energy distributions and the fragmentation of ageing plates. *Earth and Planetary Science Letters*, **126**, 143–59.

Sandiford, M. (1999). Mechanics of basin inversion. *Tectonophysics*, **305**, 109–20.

Sonder, L. and England, P. (1986). Vertical averages of rheology of the continental lithosphere; relation to thin sheet parameters. *Earth and Planetary Science Letters*, **77**, 81–90.

Taylor, S. R. and McLennan, S. M. (1985). *The Continental Crust: Its Composition and Evolution*. Oxford: Blackwell Scientific Publications.

Vasseur, G. and Singh, R. N. (1986). The effects of random horizontal variations in the radiogenic heat source distribution and its relationship with heat flow. *Journal of Geophysical Research*, **91**, 10 397–404.

Vitorello, I. and Pollock, H. N. (1980). On the variation of continental heat flow with age and the thermal evolution of continents. *Journal of Geophysical Research*, **85**, 983–95.

Wilks, K. R. and Carter, N. L. (1990). Rheology of some continental lower crustal rocks. *Tectonophysics*, **182**, 57–77.

4

Composition, differentiation, and evolution of continental crust: constraints from sedimentary rocks and heat flow

SCOTT M. MCLENNAN, STUART ROSS TAYLOR AND
SIDNEY R. HEMMING

4.1 Introduction

How the continental crust evolves may be viewed on a variety of time scales. At the most basic level, the mineralogical and chemical nature of many rocks at or near the surface of the crust do not resemble the expected products of mantle melting that generate the continental crust. For example, the chemical composition of near-surface crustal rocks, on average, is greatly enriched in the heat-producing elements (K, Th, U) far above that possible for reasonable average crustal compositions and measured heat flow. Thus, it is clear that the crust must undergo substantial chemical and mineralogical differentiation between the time of mantle extraction and that of final stabilization, typically a period on the order of from 10^7 to 10^8 yr. On much longer time scales, measured on the order of 10^9 yr, there is evidence from the compositions of sedimentary rocks and possibly from heat-flow data that the upper continental crust and bulk continental crust have changed their composition over geological time.

Knowing the composition of the continental crust in some detail is of great importance. Although the crust is about 0.5% of the mass of Earth, it is a major reservoir for the incompatible elements (e.g., Taylor, 1964, 1967). Accordingly, detailed understanding of the composition and inter-element relationships of the bulk continental crust may be used to constrain models of crust formation and of crust–mantle evolution (e.g., Taylor & McLennan, 1985, 1995; Rudnick & Fountain, 1995; Hofmann, 1997; Collerson & Kamber, 1999; Rudnick et al., 2000). Similarly, it is generally accepted that major chemical differentiation takes place within the continental crust to form a chemically and petrologically distinct upper crust, and detailed understanding of upper crustal compositions provides insight into the processes by which this is accomplished (e.g., Taylor & McLennan, 1985, 1995; Condie, 1993; Rudnick & Fountain, 1995; Rudnick & Gao, 2003).

Evolution and Differentiation of the Continental Crust, ed. Michael Brown and Tracy Rushmer. Published by Cambridge University Press. © Cambridge University Press 2005.

The chemical composition of sedimentary rocks provides fundamental constraints on the composition and evolution of the continental crust. There are a variety of approaches that have been taken in estimating upper crustal abundances, including grid sampling of large regions of upper crust where huge numbers of samples are combined to make manageable numbers of representative composite samples (e.g., Fahrig & Eade, 1968; Shaw *et al.*, 1976, 1986; Eade & Fahrig, 1971, 1973; Gao *et al.*, 1998), and mapping of exposed lithological proportions and appropriately "mixing" typical or average igneous compositions (e.g., Wedepohl, 1995; Condie & Selverstone, 1999). For a number of elements, these arduous and statistically difficult approaches may be short-circuited by using the natural sampling of the exposed continental crust provided by erosion, transport, and formation of well-mixed clastic sedimentary rocks (Taylor & McLennan, 1981, 1985; McLennan *et al.*, 1980; Condie, 1993; Plank & Langmuir, 1998; McLennan, 2001).

These approaches have been described in detail by Taylor and McLennan (1985, 1995) and the reader is referred to these publications for details. Estimating upper crustal abundances from sedimentary rocks provides unique insight unavailable from any other approach. There is a sedimentary record dating back to at least 3.8 Ga, so that, in principle, it allows for the composition of upper crust to be evaluated over geological time, although in practice there are a variety of complexities (as discussed below).

In this chapter, we review some current ideas on the relationships between sedimentary compositions and the continental crust, and consider the implications this has for understanding the composition and differentiation of the crust as a whole and the crust–mantle system over geological time. In particular, we will focus on three main issues where there has been recent controversy, including: (1) estimating upper crustal abundances from the sedimentary data base, and providing the most recent values; (2) constraining the composition of the bulk continental crust and continental crust differentiation processes using upper crustal chemistry and heat flow data, and (3) reviewing the evidence, and in particular the recent Pb isotope and heat flow evidence, that the upper crust and bulk continental crust have changed composition over geological time, with a major shift during the Archean–Proterozoic transition.

4.2 Sedimentary rocks and the upper continental crust

4.2.1 Upper crustal abundance estimates

The composition of the upper continental crust, which has long been considered as generally well-established, is close to the igneous rock-type granodiorite.

For many elements, estimated abundances come from the large-scale sampling programs carried out in the Canadian Shield by two groups (Shaw *et al.*, 1967, 1976, 1986; Fahrig & Eade, 1968; Eade & Fahrig, 1971, 1973), and with a few notable exceptions, these two sets of estimates agree reasonably well. However, as the number of regional sampling programs have grown (e.g., Condie & Brookins, 1980; Gao *et al.*, 1992, 1998; Condie & Selverstone, 1999), it is becoming clear that there is greater variability than previously thought. This may be seen in Fig. 4.1, where several of the regional averages are compared to the upper crustal estimate of Taylor and McLennan (1985) for selected trace elements. The explanation for such variation may lie in the intrinsic uncertainty of such approaches, analytical difficulties for individual elements, real regional differences in composition, or some combination of these. A number of other estimates have been made using a variety of approaches, including igneous rock balances, sedimentary compositions and "hybrid" models where different elements are derived from difference sources (e.g., Condie, 1993; Wedepohl, 1995; Plank & Langmuir, 1998; Barth *et al.*, 2000; McLennan, 2001; Rudnick & Gao, 2003).

4.2.2 General approaches from sedimentary compositions

The rare earth element (REE) distribution in average shale has long been widely accepted to be parallel to the REE pattern of the upper continental crust (e.g., Taylor, 1964). This is due to the remarkable uniformity of REE patterns in most shales and to simple mass balance considerations. Although erosion of the upper crust contributes REE to all sediment types, shales completely dominate the mass balance both because they are the most abundant sediment type (estimates of up to 70% of the sedimentary mass; Garrels & Mackenzie, 1971) and REE abundances are the highest in shales compared to any other common sedimentary lithology. Although a number of estimates of average shale REE have been published (e.g., North American Shale Composite (NASC), Post-Archean Average Australian Shale (PAAS), European Shale Composite (ES)), they differ only slightly. Among the most cited values are those for PAAS, derived from Nance and Taylor (1976), revised slightly by McLennan (1989), and these values are adopted as a starting-place (Fig. 4.2). McLennan *et al.* (1980; also Taylor and McLennan, 1985) provided a general approach for extending REE estimates to estimate abundances of other elements, using correlations with elements that behave similarly to REE during sedimentary processes (Th, Sc) or that may be estimated from reasonably well-established crustal ratios (U from $Th/U = 3.8$; K from $K/U = 10\ 000$; Rb from $K/Rb = 250$). The similarity between these estimates

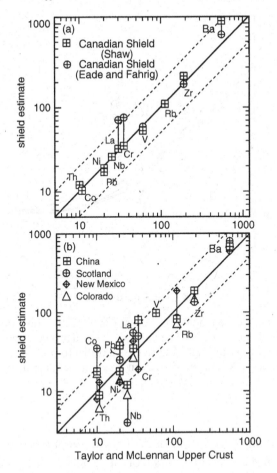

Fig. 4.1. Comparison of various estimates of regions of exposed crust with the estimate of the average upper crust from Taylor and McLennan (1985). Data sources are as follows: Canadian Shield – Shaw (Shaw *et al.*, 1986); Canadian Shield – Eade & Fahrig (Fahrig & Eade, 1968; Eade & Fahrig, 1971, 1973); China (Gao *et al.*, 1998); Scotland (Bowes, 1972); New Mexico (Condie & Brookins, 1980). Note that in the Canadian Shield comparison in Fig. 4.1a, results for Ni, Th, and Pb by Eade and Fahrig are so close to the results of Shaw that the symbols are obscured.

and various sampling programs in shield areas are compared in Table 4.1 (more extensive comparisons may be found in McLennan, 2001).

4.2.3 *Extending upper crustal abundance estimates from sedimentary data*

Upper crustal abundance estimates of Taylor and McLennan (1985) for Cs, Ti, Ta, and Nb were questioned by Plank and Langmuir (1998; also see Barth

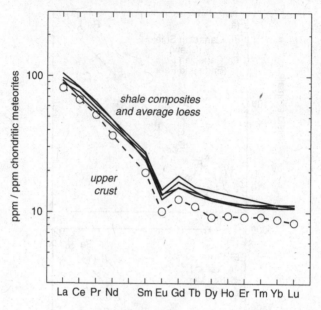

Fig. 4.2. Rare earth elements patterns for various shale averages and composites and average loess. The values for the average upper continental crust are derived by assuming the pattern is parallel to post-Archean Average Australian Shale (PAAS) but lower in absolute concentrations by 20% (adapted from McLennan, 1989).

et al., 2000) based on sedimentary data. They noted a correlation between Cs and Rb in modern marine sediments from a variety of tectonic and sedimentological regimes. Using this correlation and accepting a Rb upper crustal abundance of 112 ppm (McLennan *et al.*, 1980; Taylor & McLennan, 1985), they derived a new Cs estimate of 7.3 ppm, contrasting with the previous value of 3.7 ppm (based on a Rb content of 112 ppm and an assumed Rb/Cs ratio of 30). In addition, they noted correlations between Nb and Al_2O_3, Ti and Al_2O_3 and Nb and Ta in marine sediments. From these relationships and accepting the Al_2O_3 upper crustal estimate of Taylor and McLennan (1985), they further estimated TiO_2 at 0.76 wt.%, Nb at 13.7 ppm and Ta at 0.96 ppm. These values contrast with those of Taylor and McLennan (1985) of $TiO_2 = 0.5$ wt.%; $Nb = 25$ ppm and $Ta = 2.2$ ppm, based on large-scale sampling of the Canadian Shield for Ti and Nb (Shaw *et al.*, 1976) and a Nb/Ta ratio of 11.6 (Taylor & McLennan, 1981).

One difficulty with these approaches is that the geochemistry of Cs, Ti, Nb, and Ta in the sedimentary environment is complex. For example, the Rb/Cs ratios of weathering profiles appear to change systematically as a function of

Table 4.1. *Selected trace elements for estimates of various shield areas and the average upper continental crust*

Element (ppm)	Canadian Shield[1]	Canadian Shield[2]	East China[3]	Scotland[4]	New Mexico[5]	Colorado[6]	Upper crust[7]	Upper crust[8]	Upper crust[9]	Upper crust[10]	Upper crust[11]	Upper crust[12]
Sc	7.0	12	15			16	13.3		11		14	13.6
Ti	3120	3180	3900	2400	2700	4200	3300		3000	4555	3840	4100
V	53	59	98				86		60		97	107
Cr	35	76	80	<50	19	82	104		35		92	83
Co	12		17	35	8	17	18		10		17.3	17
Ni	19	19	38	25	13	43	56		20		47	44
Rb	110		82	85	187	72	83		112		84	(112)
Zr	237	190	188	135	180	148	160		190		193	(190)
Nb	26		12	4		9.1	9.8		25	13.7	12	12
Cs			3.55					5.8	3.7	7.3	4.9	4.6
Ba	1070	730	678	795	590	749	633	668	550		624	(550)
La	32.3	71	34.8	55	43	27	28.4		30		31	(30)
Hf	5.8		5.12			4.4	4.3		5.8		5.3	(5.8)
Ta	5.7		0.74			0.67	0.79	1.5	2.2	0.96	0.9	1.0
Pb	17	18	18			14	17		20		17	17
Th	10.3	10.8	8.95		13	6.2	8.6		10.7		10.5	(10.7)

Data sources: (1) Average Canadian shield (Shaw *et al.*, 1986); (2) Average Canadian Shield (Fahrig & Eade, 1971, 1973); (3) Average Central East China calculated on carbonate-free basis (Gao *et al.*, 1998); Average of crystalline basement N.W. Scotland Highlands (Bowes, 1972); (5) Average Precambrian surface terrane, New Mexico (Condie & Brookins, 1980); (6) Average of Colorado Plateau Upper Crust derived from equal proportions of northwest and southeast sections (Condie & Selverstone, 1999); (7) Average upper continental crust from "Map Model" (Condie, 1993); (8) Average upper crust for elements not taken from Shaw *et al*, 1986 (Wedepohl, 1995); (9) Average upper crust (Taylor & McLennan, 1985, 1995); (10) Upper continental crust derived from marine sedimentary record (Plank & Langmuir, 1998); (11) Average upper continental crust derived from a combination of approaches (Rudnick & Gao, 2003); (12) Average upper continental crust derived from sedimentary record (McLennan, 2001). Elements that are unchanged from Taylor & McLennan (1985) are shown in parentheses.

Rb content in both basaltic and granitic terranes (Price *et al.*, 1991; Nesbitt & Markovics, 1997; McLennan, 2001). Thus, although Rb and Cs are carried dominantly in clastic sediments, it is not obvious that the Rb/Cs ratio of marine sediment, dominated by very fine-grained clays, is necessarily representative of the upper crust. Titanium, Nb and Ta may be concentrated in certain heavy mineral suites (e.g., rutile, ilmenite, etc.) and thus affected by sedimentary transport processes, although likely to a lesser degree than are Zr and Hf (McLennan, 2001). In spite of these caveats, the proposed differences are so large that they require re-evaluation of upper crustal abundance estimates.

McLennan (2001; see also McLennan & Xiao, 1998) attempted to evaluate further the upper crustal trace element abundances from sedimentary data for a variety of elements in addition to those reported by McLennan *et al.* (1980) and Taylor and McLennan (1985). There are numerous other trace elements that are transferred from the upper crust primarily to the clastic sedimentary mass, such as Ti, Zr, Hf, Nb, Ta, Rb, Cs, Pb, Cr, V, Ni, and Co. However, until recently these elements were basically neglected, because of perceived problems such as fractionation during mineral sorting so that shales may not dominate the sedimentary mass balance (e.g., Ti, Zr, Hf, Nb, Ta, Pb), and possible redistribution during weathering, diagenesis or other sedimentary processes (e.g., Ba, Rb, Cs, Pb, Cr, V, Ni, Co).

The approach used by McLennan (2001) was to examine the ratios of REE (using La) to various trace elements for clastic sedimentary rock averages. Each average is based on a relatively large number of samples (from $n = 19$ to $> 33\,000$) for a variety of clastic lithologies (shales, loess, sandstones, tillites) from a variety of settings (active margins, passive margins, cratonic, deep marine). Several examples of such plots are shown in Fig. 4.3 (see McLennan (2001) for data sources) to illustrate the general approach. Since the elements under consideration have relatively low abundances in non-clastic sedimentary reservoirs (i.e., carbonates, evaporites), mass balance demands that upper crustal abundances must lie within the range of such ratios. The upper crustal values derived from this procedure are given in Table 4.1 and compared with other upper crustal estimates. The major differences include factor of 2 decreases for Nb and Ta (in agreement with Plank & Langmuir, 1998 and Barth *et al.*, 2000), factor of 2 increases for Cr, V, Ni, and Co (similar to values proposed by Condie, 1993), minor changes (from about 20 to 30%) for Ti, Cs, Sc, and Pb and no change for REE, Y, Th, Ba, Zr, Hf, and Rb.

An important, but not readily quantified, issue is the inherent uncertainty in such estimates. The uncertainties (both for upper crustal estimates and for

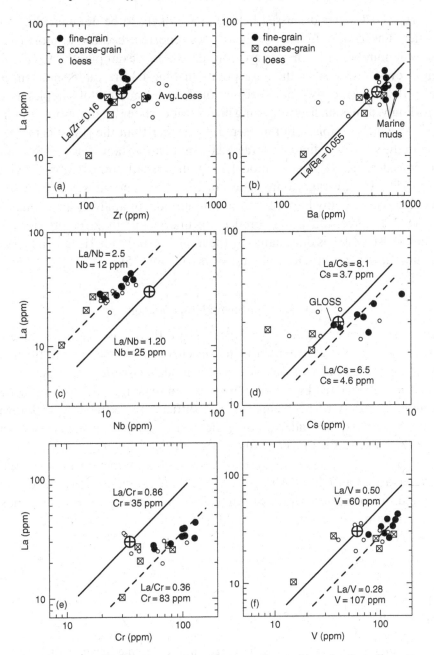

Fig. 4.3. Plots of La versus various elements for various clastic sedimentary rocks (adapted from McLennan, 2001; data from various sources). Each of the "fine-grain" and "coarse-grain" data points represent the average of a large number of samples (anywhere from 19 to > 30 000) from a variety of tectonic and sedimentological regimes (e.g., average river suspended sediment, average loess, GLOSS, NASC, PAAS, etc.; see McLennan, 2001 for details).

bulk crustal estimates discussed below) are unlikely to be dominated by the statistical uncertainty in the average values reported, because in most cases very large numbers of samples are considered. For example, the 95% confidence levels on most of the data points in Fig. 4.3 are quite small (in all cases < 10% and mostly < 5%) because of the large number of samples used. Accordingly, issues such as weighting factors, underlying assumptions, and the degree to which samples are representative, rather than the statistical uncertainty in the various sediment averages, are more important. A useful example is to consider the "Andesite" model for bulk crustal composition (Taylor, 1967, 1977). The average composition of island arc andesites is known very precisely because many hundreds (if not thousands) of analyses are available. However, that precision does not reflect the uncertainty in the composition of the bulk crust, which is dominated by the model assumption (is the bulk crust andesite?), and thus difficult to quantify (see Chapter 5).

4.3 Bulk continental crust models

In contrast to upper crustal abundances where there has long been general consensus (in spite of above discussion), the chemical composition of the bulk crust is far more controversial with recent models covering the range from basalt through dacite (Fig. 4.4). However, compositions at both extremes encounter a variety of problems that are difficult to reconcile with known crustal characteristics. Understanding the overall average chemical composition of the bulk continental crust and whether or not that may differ from the composition of recent major crustal additions is of fundamental importance to geochemistry because it leads to constraints on the basic processes of crustal growth, differentiation, and evolution (e.g., delamination, crust–mantle mass balance relationships, intra-crustal differentiation, crust–mantle recycling, and so forth).

Fig. 4.3. (cont.)

The small circles are regional loess averages (e.g., New Zealand, Argentina, China, etc.). Shown for reference is the average abundances (large cross-filled circle) and La/element ratio (solid diagonal line) for the upper crust from Taylor & McLennan (1985). Estimated La/element ratio of the clastic sedimentary mass is shown as a dashed line where the ratio differs significantly from the upper crust value of Taylor & McLennan (1985). The average upper crustal ratios for these elements are thought to fall within the range of clastic sedimentary rocks since these elements have very low abundances in non-clastics.

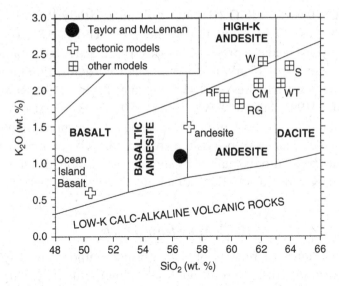

Fig. 4.4. Plot of K$_2$O versus SiO$_2$ showing the compositions of selected models of bulk crustal composition superimposed on the fields of calc-alkaline volcanic rock classification. Note that apart from the model of Taylor and McLennan (1985, 1995; McLennan & Taylor, 1996) most models indicate relatively felsic compositions with SiO$_2$ higher than average andesite. Compositional models include: S = Shaw *et al.* (1986); WT = Weaver and Tarney (1984); CM = Christensen and Mooney (1995); W = Wedepohl (1995); RF = Rudnick and Fountain (1995); RG = Rudnick and Gao (2003).

4.3.1 Plume models

One class of crustal models proposes that the continental crust grows predominantly by mantle plume related basaltic volcanism (e.g., Abbott & Mooney, 1995; Abbott *et al.*, 1997; Albarède, 1998; also see Stein & Hofmann, 1994). Basaltic continents, even those that are relatively enriched in incompatible elements such as ocean island basalts, encounter at least two difficulties. The most fundamental difficulty is that in order to produce a granodioritic upper crust that comprises between about one-quarter and one-third of the entire crust (Rudnick & Fountain, 1995), such crustal models require an ultramafic average composition for the lower crust, a feature that is inconsistent with a variety of xenolith and granulite studies and seismology (Rudnick & Fountain, 1995). A second problem centers on the predicted level of continental heat flow. Although average model compositions have not been published, it is unlikely that continents built from ocean island-like material would have K$_2$O contents much in excess of from 0.5 wt.% to 0.7 wt.% (for example, average Hawaiian glass averages 0.6 wt.% K_2O (Basaltic Volcanism Study Project (1981) and Albarède 1998) used 0.3 wt.% in his modeling). For Th/U = 3.8 and K/U = 10^4, such compositions predict a present-day crustal radioactive contribution to heat flow of less than about 15 mW m^{-2} (McLennan & Taylor, 1996), which is far below any previous estimates (e.g., McLennan & Taylor, 1996; Rudnick *et al.*, 1998; Nyblade, 1999; Jaupart & Mareschal, 1999, 2003). These difficulties are generally recognized by most authors and disposed of by proposing additional (but unspecified) sources of potassium (Albarède, 1998) and/or loss of ultra-mafic residues back into the mantle by subduction erosion, delamination, and so forth (e.g., Kay & Kay, 1991; Rudnick, 1995; Albarède, 1998).

4.3.2 More felsic models

The so-called "andesite model" of bulk crustal composition (Taylor, 1967, 1977) has fallen into disfavor (although see below). This is due largely to the observation that typical andesites are not unfractionated mantle-derived mag-mas and that, on petrological grounds, calc-alkaline basalt was the likely dominant mantle-derived product in arcs (e.g., Arculus, 1981; Gill, 1981; Ellam & Hawkesworth, 1988; Pearcy *et al.*, 1990; see Chapter 5). Although perhaps not as severe as ocean island basalt models, a continental crust consisting of island arc basalt encounters essentially the same problems discussed above.

Most crustal composition models use a variety of constraints from seismic velocity profiles and geochemistry of deep crustal cross-sections and lower crustal xenoliths (Pakiser & Robinson, 1966; Weaver & Tarney, 1984; Shaw *et al.*, 1986; Christensen & Mooney, 1995; Rudnick & Fountain, 1995; Wedepohl, 1995; Gao *et al.*, 1998; Condie & Selverstone, 1999; Rudnick & Gao, 2003). Although there is considerable variability among these models in detail (e.g., Fig. 4.4), they are all similar in that they predict compositions significantly more felsic than average andesite.

At the extreme, some of these models predict a crustal radiogenic component of heat flow that is in excess of overall average continental heat flow (Nyblade & Pollack, 1993; McLennan & Taylor, 1996; Jaupart & Mareschal, 2003) and at best may only be representative of local sections of continental crust. (Further discussion of heat flow constraints will be given below.) Felsic models en-counter additional difficulties. Although there are established mechanisms for extracting more felsic melts from the mantle (e.g., Rapp & Watson, 1995), these probably only operate for relatively young and hot subducted litho-sphere, and most workers appear to agree that mantle-derived melts are dominated by basaltic compositions. Accordingly, felsic models typically predict that an ultra-mafic–mafic component must be lost subsequent to crust forma-tion by processes such as delamination of the lower crust (e.g., Kay & Kay, 1991) or subduction erosion (Albarède, 1998). Such residues have not been unambigu-ously identified in the subcontinental lithospheric mantle (Arculus, 1999).

4.3.3 The "andesite model" revisited

Detailed seismic studies of the Izu–Bonin Island arc indicate an overall compo-sition that is more likely basaltic andesite in composition (Suyehiro *et al.*, 1996; Taira *et al.*, 1998; but see Holbrook *et al.*, 1999). Accordingly, Arculus (1999; Chapter 5) has resurrected the andesite model as a viable candidate for the generation of bulk continental crust during post-Archean time. In this case, ultra-mafic cumulates are lost during the crust-forming (subduction) process itself, by being incorporated into the mantle wedge beneath the arc and advect-ing into the deeper mantle. This greatly alleviates the requirement of significant amounts of ultra-mafic cumulate being present in the sub-continental litho-spheric mantle and the necessity of delaminating the lower crust long after its formation. Some support for an andesite model comes from a study of heat production in a vertical section of exposed arc in Japan (Furukawa & Shinjoe, 1997). Here, the estimated crustal radiogenic component of heat flow totaled $25\,mW\,m^{-2}$ over 30 km of arc crust. Normalized to average crustal thickness of 41 km, this translates into $34\,mW\,m^{-2}$, which is virtually identical to that pre-dicted by the andesite model of crustal composition (see average island arc composition tabulated in Taylor & McLennan, 1985).

4.3.4 Heat flow constraints

In spite of a variety of complications in interpreting heat-flow–surface-heat production relationships (e.g., Jaupart & Mareschal, 1999), the average heat flow from the continents remains as one of the few independent constraints on all models of crustal composition, because of the limits it provides on the abundances of the heat producing elements (HPEs). Survey of the global heat flow data set indicates that the average heat flow from tectonically stabilized regions of Precambrian continental crust is now reasonably well established at $48 \pm 1\,mW\,m^{-2}$ (Nyblade & Pollack, 1993). Although the number of heat flow measurements has about doubled in the last decade, the average values for the Precambrian have not changed significantly (Jaupart & Mareschal, 2003). McLennan and Taylor (1996) used this heat flow estimate to argue that crustal composition models with K_2O greater than about 1.6 wt.% (and appropriate corresponding Th and U abundances) could not be reconciled with the current state of knowledge of average continental heat flow because they predicted continental heat flow in excess of $48 \, mW \, m^{-2}$.

On the other hand, there is considerable debate as to whether or not heat flow data may be used to further refine crustal composition models (McLennan & Taylor, 1996; Rudnick *et al.*, 1998; Jaupart & Mareschal, 1999, 2003). The critical issue centers on whether or not the crustal and mantle contributions to total continental heat flow may be adequately distinguished. Continental heat flow may be divided into three major components (see Jaupart & Mareschal, 2003 for a more detailed subdivision and recent discussion).

(1) Heat derived from radioactive decay within the crust (*crustal heat*), which provides a measure of the average levels of heat producing elements (K, Th, U) by assuming $K/U = 10^4$ and $Th/U = 3.8$. The crustal component of heat may vary with crustal age if the composition of mantle additions to the crust has changed over time.
(2) Heat derived from the mantle beneath the continents (roughly equating to the so-called *reduced heat flow*). This component of heat may also vary as a function of crustal age and be influenced by mantle convection regimes.
(3) Heat derived from various processes associated with orogenic activity (the so-called *tectonothermal heat*). This component decays on time scales of several hundred million years after the last tectonothermal event.

Is it possible to unravel these contributions to a sufficient degree to further constrain bulk crustal compositions? In order to exclude tectonothermal heat some workers concentrate on crustal provinces where the last tectonothermal event was at least 500–600 Myr ago (e.g., Nyblade & Pollack, 1993). One concern here is that, if there has been a secular change in bulk crustal composition, then relatively young crust may be preferentially excluded from the average. Large regions of the continents (>35%) have Phanerozoic igneous crystallization ages and metamorphic ages. However, isotopic studies indicate that most of these regions were actually derived from the mantle during the Precambrian and have been largely reworked or recycled during later orogenic activity (e.g., Allègre & Ben Othman, 1980; McCulloch & Chappell, 1982). Virtually all models of crustal growth predict that from <5 to 10% of the continents were actually added during the Phanerozoic (see reviews in Taylor & McLennan, 1985, 1995) and accordingly, concentrating on stabilized Precambrian terranes is unlikely to introduce any significant uncertainty. Jaupart and Mareschal (2003) included Phanerozoic terranes (older than 200 Ma) in their evaluation and derived steady-state continental heat flow of $51 \, mW \, m^{-2}$, slightly higher than the value for stabilized Precambrian terranes of $48 \, mW \, m^{-2}$. However, since most of this Phanerozoic crust was derived from the mantle during the Precambrian, their analysis implies large increases in the K, Th, and U abundances of crustal additions at about the Precambrian–Phanerozoic transition for which there is no evidence in either arc-related rocks or in sedimentary rocks (e.g., Condie, 1993).

The major source of uncertainty comes from unraveling the mantle versus crustal contributions. As noted above, the average heat flow from stabilized continental crust is $48 \pm 1\,\mathrm{mW\,m^{-2}}$. How much of this is derived from the mantle? Early studies (e.g., Vitorello & Pollack, 1980; Morgan, 1984) assumed 1-D heat flow models and used the common linear relationship between heat flow and surface heat production in "Heat Flow Provinces" to estimate the mantle contribution (taken to be equivalent to reduced heat flow). However, most workers now accept that such models are overly simplistic and that some of the variation of continental heat flow is due to lateral variations in heat-producing elements within the crust itself (see below). The upshot is that classical estimates of reduced heat are likely to significantly overestimate mantle contributions to continental heat flow.

Accordingly, many studies of continental heat flow have attempted to constrain the relative contributions of crust and mantle heat by independently evaluating the distribution of chemical elements through the crust, mainly by examining deeply exposed crustal sections. Results are now available for numerous continental regions of a wide variety of age, including the Superior Province (Fountain *et al.*, 1987; Pinet *et al.*, 1991; Jaupart *et al.*, 1998; Jaupart & Maraschal, 1999), Indian Shield (Gupta *et al.*, 1991), Grenville Province (Guillou-Frottier *et al.*, 1995; Jaupart & Maraschal, 1999), Trans-Hudson orogen (Guillou-Frottier *et al.*, 1996; Mareschal *et al.*, 1999), North Atlantic craton (Weaver & Tarney, 1984; Pinet & Jaupart, 1987), Kaapvaal craton (Jones, 1988), Kalahari craton (Rudnick & Nyblade, 1999), Slave Craton (Russell *et al.*, 2001), Fennoscandian Shield (Kukkonen & Peltonen, 1999) and Yilgarn craton (Lambert & Heier, 1968; Sass & Lachenbruch, 1979). Without exception, the best estimates of mantle contribution of overall heat flow are in the range of $10–18\,\mathrm{mW\,m^{-2}}$, with an average of the ten regions being $14 \pm 3\,\mathrm{mW\,m^{-2}}$.

These studies have preferentially been conducted in Precambrian (especially Archean) shield areas and Archean terrains on average have lower total heat flow (see below) that in part may be due to a reduced mantle contribution of heat (Nyblade & Pollack, 1993), although this is controversial. Accordingly, these independent estimates of mantle contribution to continental heat flow probably represent minimum estimates by up to about $7\,\mathrm{mW\,m^{-2}}$. If this evaluation is correct, the mantle contribution to continental heat flow must average somewhere between 11 and $24\,\mathrm{mW\,m^{-2}}$.

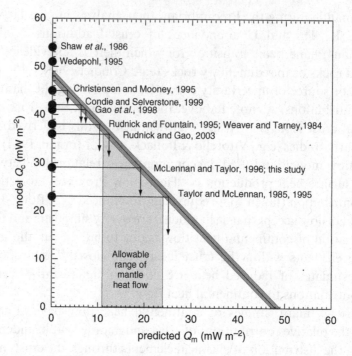

Fig. 4.5. Plot of the calculated component of global heat flow derived from the continental crust (Q_c) versus the predicted mantle component of heat flow (Q_m) for various models of crustal composition. Diagonal line and dark gray region represents the constraint on total crustal heat flow of $48 \pm 1\,\mathrm{mW\,m^{-2}}$. The light gray region represents the most likely range of average heat flow from the mantle based on a variety of studies of deep crustal sections (see text for discussion).

In either case, bulk crustal composition models that predict crustal radiogenic contributions to heat flow of $\geq 38\,\mathrm{mW\,m^{-2}}$ (48 ± 1 minus $14 \pm 3\,\mathrm{mW\,m^{-2}}$) appear to be inconsistent with a conservative interpretation of the global heat flow data base (Fig. 4.5). Accordingly, all of the relatively felsic crustal models (Condie & Brookins, 1980; Weaver & Tarney, 1984; Shaw et al., 1986; Christensen & Mooney, 1995; Rudnick & Fountain, 1995; Wedepohl, 1995; Gao et al., 1998; Condie & Selverstone, 1999) encounter difficulties, and at best may only be considered to be representative of relatively localized regions of the crust. The recent estimate of Rudnick and Gao (2003) falls within these constraints but predicts relatively low mantle contributions ($< 13\,\mathrm{mW\,m^{-2}}$).

Rudnick et al. (1998) argued that estimates of crustal contributions to heat flow for individual crustal section are highly variable, ranging from about 20 to $45\,\mathrm{mW\,m^{-2}}$. They concluded from this observation that there was too much uncertainty to use heat flow to constrain crustal composition models further.

We consider this to be an overly pessimistic evaluation of the heat flow data because it essentially neglects the very well-constrained value of average heat flow from tectonically stabilized Precambrian continents of $48 \pm 1\,\mathrm{mW\,m^{-2}}$ based on more than 1600 locations (Nyblade & Pollack, 1993, Jaupart & Mareschal, 2003). Regional variations in heat flow are very well documented (e.g., Jaupart & Mareschal, 1999, 2003), are likely to reflect variations in the mantle heat flux as well as variations in crustal composition, and do not seem to us to be particularly surprising.

The recent evaluation of global heat flow data by Jaupart and Mareschal (2003) deserves further comment. They adopted a value of $51\,\mathrm{mW\,m^{-2}}$ (from Stein, 1995) for continental heat flow from stabilized regions. This value is higher than that of Nyblade and Pollack (1993) because it also includes Phanerozoic regions older than about 200 Ma, which have relatively high heat flow that Jaupart and Mareschal (2003) interpret to be due to high crustal heat production (i.e., high abundances of K, Th, U). We are unconvinced by this interpretation because it implies a significant change in crustal composition (e.g., in arc magmatism) across the Precambrian–Phanerozoic transition. Most Phanerozoic crust was, in fact, derived from the mantle during the Precambrian (see discussion above), and thus their model also implies even greater differences in the composition of crustal additions from the mantle. There is no independent geochemical evidence in support of either of these predictions.

In the Jaupart and Mareschal (2003) analysis, the values for the mantle contributions to continental heat flow are essentially identical to the values presented here ($10\text{–}18\,\mathrm{mW\,m^{-2}}$), both being derived from studies in Precambrian (mainly Archean) regions. However, Jaupart and Mareschal (2003) adopt this value for crust of all ages, whereas we accept the possibility that part of the change in heat flow across the Archean–Proterozoic boundary is due to a change in mantle contribution (Nyblade & Pollack, 1993) rather than all being due to crustal contributions. The implication of these differing interpretations is that Jaupart and Mareschal (2003) predict a much greater change in the composition of crustal additions over geological time. In any case, the model of continental heat flow presented by Jaupart and Mareschal (2003) would suggest that the contribution from crustal heat production amounts to between 33 and $40\,\mathrm{mW\,m^{-2}}$, which is higher than the crustal composition presented here.

4.3.5 Taylor and McLennan (1985, 1995) revisited

The bulk crust K, Th, and U estimates of Taylor and McLennan (1985, 1995) were based on a model whereby the continental crust was composed of 75% Archean crust and 25% average andesite, and provided for $23\,\mathrm{mW\,m^{-2}}$ of average crustal heat flow, values that were consistent with the heat flow constraints as understood at the time. McLennan and Taylor (1996) revised these values upwards to be more in line with current views of heat flow constraints for Archean and total crust, and more recent views that the continental crust may have only grown to about 50–70% of its present-day mass by 2.5 Ga. The revised values were $K = 1.1$ wt.%, $Th = 4.2$ ppm and $U = 1.1$ ppm, leading to a crustal contribution to heat flow of 29 mW m^{-2} and predicting an average mantle contribution of about 18–20 mW m^{-2}. The revised method for estimating bulk crustal abundances has a significant effect only on the most incompatible elements Cs, Rb, K, Th, and U. To maintain a K/Rb of 300, Rb is revised upwards to 37 ppm (from 32 ppm) and Cs is estimated from the average island arc volcanic Rb/Cs ratio of 25 (there are insufficient high-quality Archean data for Cs) at 1.5 ppm.

Along with these changes, revised upper crustal abundances presented above have additional consequences for the crustal composition models of Taylor and McLennan (1985, 1995) because estimates of the lower crust are derived from simple mass balance by assuming that the upper crust on average represents 25% of the total crust. In Table 4.2, the revised crustal

Table 4.2. *Estimates of the chemical composition of the continental crust from Taylor & McLennan (1985) with revisions shown in underlined bold italic*

Element	Upper continental crust	Bulk continental crust	Lower continental crust	Archean upper crust	Archean bulk crust
Li (ppm)	20	13	11	–	–
Be (ppm)	3.0	1.5	1.0	–	–
B (ppm)	15	10	8.3	–	–
Na (wt.%)	2.89	2.30	2.08	2.45	2.23
Mg (wt.%)	1.33	3.20	3.80	2.83	3.56
Al(wt.%)	8.04	8.41	8.52	8.10	8.04
Si (wt.%)	30.8	26.8	25.4	28.1	26.6
P (ppm)	700	–	–	–	–
K (wt.%)	2.80	*1.1*	*0.53*	1.5	*1.0*
Ca (wt.%)	3.00	5.29	6.07	4.43	5.22
Sc (ppm)	*13.6*	30	*35*	14	30
Ti (wt.%)	*0.41*	0.54	*0.58*	0.50	0.60
V (ppm)	*107*	230	*271*	195	245
Cr (ppm)	*83*	185	*219*	180	230
Mn (ppm)	600	1400	1700	1400	1500
Fe (wt.%)	3.50	7.07	8.24	6.22	7.46
Co (ppm)	*17*	29	*33*	25	30
Ni (ppm)	*44*	128	*156*	105	130
Cu (ppm)	25	75	90	–	80
Zn (ppm)	71	80	83	–	–

Table 4.2. (cont.)

Element	Upper continental crust	Bulk continental crust	Lower continental crust	Archean upper crust	Archean bulk crust
Ga (ppm)	17	18	18	–	–
Ge (ppm)	1.6	1.6	1.6	–	–
As (ppm)	1.5	1.0	0.8	–	–
Se (ppb)	50	50	50	–	–
Rb (ppm)	112	*37*	*12*	50	28
Sr (ppm)	350	260	230	240	215
Y (ppm)	22	20	19	18	19
Zr (ppm)	190	100	70	125	100
Nb (ppm)	*12*	8.0	*6.7*	–	–
Mo (ppm)	1.5	1.0	0.8	–	–
Pd (ppb)	0.5	1	1	–	–
Ag (ppb)	50	80	90	–	–
Cd (ppb)	98	98	98	–	–
In (ppb)	50	50	50	–	–
Sn (ppm)	5.5	2.5	1.5	–	–
Sb (ppm)	0.2	0.2	0.2	–	–
Cs (ppm)	*4.6*	*1.5*	*0.47*	–	–
Ba (ppm)	550	250	150	265	220
La (ppm)	30	16	11	20	15
Ce (ppm)	64	33	23	42	31
Pr (ppm)	7.1	3.9	.8	4.9	3.7
Nd (ppm)	26	16	12.7	20	16
Sm (ppm)	4.5	3.5	3.17	4.0	3.4
Eu (ppm)	0.88	1.1	1.17	1.2	1.1
Gd (ppm)	3.8	3.3	3.13	3.4	3.2
Tb (ppm)	0.64	0.60	0.59	0.57	0.59
Dy (ppm)	3.5	3.7	3.6	3.4	3.6
Ho (ppm)	0.80	0.78	0.77	0.74	0.77
Er (ppm)	2.3	2.2	2.2	2.1	2.2
Tm (ppm)	0.33	0.32	0.32	0.30	0.32
Yb (ppm)	2.2	2.2	2.2	2.0	2.2
Lu (ppm)	0.32	0.30	0.29	0.31	0.33
Hf (ppm)	5.8	3.0	2.1	3	3
Ta (ppm)	*1.0*	*0.8*	*0.73*	–	–
W (ppm)	2.0	1.0	0.7	–	–
Re (ppb)	0.4	0.4	0.4	–	–
Os (ppm)	0.05	0.05	0.05	–	–
Ir (ppb)	0.02	0.1	0.13	–	–
Au (ppb)	1.8	3.0	3.4	–	–
Tl (ppb)	750	360	230	–	–
Pb (ppm)	*17*	8.0	*5.0*	–	–
Bi (ppb)	127	60	38	–	–
Th (ppm)	10.7	*4.2*	*2.0*	5.7	*3.8*
U (ppm)	2.8	*1.1*	*0.53*	1.5	*1.0*

abundances estimates are listed for the total continental crust and Archean continental crust and revisions from Taylor and McLennan (1985, 1995) are highlighted.

4.4 Secular variations in crustal composition

Earth has cooled substantially over the past 4.5 billion years due to whole body cooling, associated with accretion and differentiation into core, mantle, and crust, and diminished radioactivity. Consequently, there may be little doubt that the thermal regime of continental crust formation and crustal differentiation has changed over Earth history, but there is little consensus about the exact nature and overall effects of such changes. One major reason for the lack of agreement is that it is not clear whether the excess heat resulted in increased geothermal gradients or whether heat was more efficiently lost from the mantle during rapid sea floor spreading at a greater number of plate boundaries, or both (see review in Pollack, 1997).

The Archean–Proterozoic transition is widely recognized as a major benchmark in the chemical evolution of the crust–mantle system and among the differences seen for the Archean, compared to the post-Archean, are the following.

(1) During the Archean, the crust appears to be dominated by the so-called "bimodal suite" of basalt and tonalite–trondhjemite–granodiorite (TTG), contrasting with the range of mafic to intermediate rocks found in modern arcs (Gill, 1981; Chapter 5). Among the suggestions are that this results from the subduction of relatively young, warm oceanic lithosphere, compared to the modern regime, resulting in increased partial melting of the subducted lithosphere or direct melting of an enriched Archean upper mantle (e.g., Martin, 1986, 1993; Defant & Drummond, 1990; Rapp & Watson, 1995; Rapp, 1997; Evans & Hanson, 1997; Chapter 6).

(2) The relatively low $^{87}Sr/^{86}Sr$ and high $^{143}Nd/^{144}Nd$ ratios of Archean seawater, reflected in unaltered marine carbonates and banded iron formation, are thought to partially reflect the much higher contribution of mantle-derived components during the Archean, due to increased hydrothermal activity at mid-ocean ridges (e.g., Veizer, 1989; Derry & Jacobsen, 1988; Jacobsen & Pimentel-Klose, 1988).

(3) Archean terranes are characterized by significantly lower heat flow that may be a reflection of differing bulk compositions or related to greater thickness of subcontinental lithosphere beneath Archean terranes or both (e.g., Nyblade & Pollack, 1993).

(4) The upper continental crust, as reflected in the sedimentary record, appears to have undergone a fundamental change in overall composition and critical trace element characteristics (e.g., Eu/Eu^*, La/Sm, Gd/Yb, Th/Sc, U/Pb) at the

Archean–Proterozoic transition, due to an increasing role of intracrustal differentiation resulting, among other things, in the post-Archean upper crust being more felsic and incompatible element-enriched in composition (e.g., Taylor & McLennan, 1985, 1995; McLennan *et al.*, 2003; Veizer & Mackenzie, 2003).

4.4.1 Seismology, heat flow, and secular variations in bulk crust composition

Has the bulk composition of the continental crust changed over geological time? Although this question is of fundamental importance for understanding the nature of the crust-forming process and crust–mantle evolution, it is extremely difficult to constrain because there is no unambiguous way to sample bulk crust in a representative fashion over geological time.

In the most comprehensive study to date, Rudnick and Fountain (1995) found no systematic differences in V_p and V_s seismic velocity structure for Archean and younger crust. In contrast, Zandt and Ammon (1995) suggested a possible secular change in Poisson's ratio (V_p/V_s), with older crust having higher ratios (i.e., more mafic compositions) although all values overlapped at the one standard deviation level.

In spite of, what is at best, ambiguous evidence for any difference in seismic velocity structure between Archean and post-Archean crust, models of Archean bulk crustal composition derived from seismic structure result in a major difference in composition compared to the overall crust. Thus Rudnick and Fountain (1995) calculated that Archean crust on average is depleted in the most incompatible elements (Rb, Cs) by up to a factor of 3–4 and by a factor of about 2 for the HPE compared to the overall bulk crust. In their evaluation of global continental heat flow, Jaupart and Mareschal (2003) suggested post-Archean crust was enriched in the HPE by a factor of about 1.4 over Archean crust. In contrast, the models of Taylor and McLennan (1985, 1995) and McLennan and Taylor (1996), which are constrained by a variety of other factors (crustal growth models, bimodal suite compositions, heat flow), predict far more modest differences between Archean and post-Archean bulk crust with the most incompatible elements being depleted by a factor of about 1.3–1.4, and a factor of about 1.1 for the HPE.

The contrast becomes even greater when one considers a comparison of the implied compositions of average Archean crustal additions to that of average post-Archean crustal additions. The relative proportions of crustal additions during the Archean and post-Archean is a matter of some uncertainty, but few models predict that less than about half of the crust was formed during the Archean. In Fig. 4.6, estimates of the composition of post-Archean crustal additions are calculated for the compositional models of Rudnick and

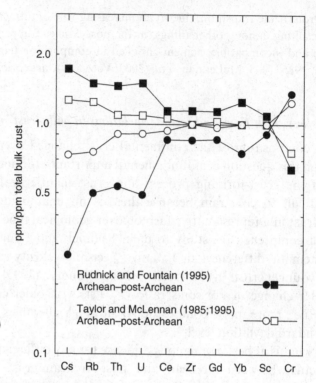

Fig. 4.6. Plots of Archean and post-Archean bulk crustal additions, for selected elements, normalized to the average crustal composition for models of Rudnick and Fountain (1995) and Taylor and McLennan (1985, 1995); McLennan and Taylor (1996; this chapter). Post-Archean crustal additions are estimated assuming that 50% of the continental crust was added during the Archean.

Fountain (1995) and Taylor and McLennan (1985, 1995, and McLennan and Taylor (1996), assuming that Archean and post-Archean crustal additions to the bulk crust are equal. If the continental crust is more than 50% Archean in age, the contrasts are even greater. The models of Taylor and McLennan (1985, 1995) predict relatively modest changes in composition with the most incompatible elements (Rb, Cs) differing by a factor of about 1.6–1.7 (about 1.2 for the HPE), whereas Rudnick and Fountain (1995) predict a major change with the most incompatible elements differing by a factor of about 3.0–6.5 (about 2.0–2.5 for the HPE). In comparison, Jaupart and Mareschal (2003) predict an increase for the HPE by a factor of 1.7.

Heat flow measurements represent a second means of evaluating any changes in bulk crustal composition over geological time, but again the data are somewhat ambiguous. Stabilized Archean-aged continental crust on

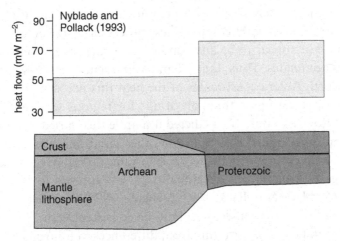

Fig. 4.7. Diagram schematically illustrating the jump in crustal heat flow associated with the transition to thinner subcontinental lithosphere at the boundary between Archean and Proterozoic terrains (adapted from Nyblade & Pollack, 1993). On average, total heat flow increases by about $14 \pm 3 \, \mathrm{mW \, m^{-2}}$.

average produces $41 \pm 1 \, \mathrm{mW \, m^{-2}}$ of heat flow, whereas stabilized continental crust on average produces a heat flow of $48 \pm 1 \, \mathrm{mW \, m^{-2}}$. It is more difficult to evaluate the intrinsic heat production of post-Archean crustal additions because most continental crust is an intimate mixture of crustal additions of various ages. However, Nyblade and Pollack (1993) noted that there was a step change in crustal heat flow from dominantly Archean terrains ($41 \pm 1 \, \mathrm{mW \, m^{-2}}$) to Proterozoic terrains far removed from Archean cratons ($55 \pm 2 \, \mathrm{mW \, m^{-2}}$). This is shown schematically in Fig. 4.7. This higher value of heat flow may be as good an estimate as possible of the average heat flow of stabilized crust added during the post Archean.

At least two factors may contribute to this difference. One factor is that the difference in heat flow is a consequence of differing heat production within the continental crust (Morgan, 1984; Taylor & McLennan, 1985). If so, then the maximum difference in the heat production between Archean and post-Archean crustal additions would be about a factor of 2 ($55 \pm 2 - 14 \pm 3 \, \mathrm{mW \, m^{-2}}$ for post-Archean, versus 41 ± 1 minus $14 \pm 3 \, \mathrm{mW \, m^{-2}}$ for Archean). However, it is well established that the subcontinental lithosphere is thickest under old Archean cratons (see reviews in Fei *et al.*, 1999). It has been suggested that this thick lithosphere may act to divert heat, derived from deeper in the mantle, away from the thick cratonic lithosphere into the surrounding suboceanic convecting mantle where it is lost through the oceanic crust (e.g., Ballard & Pollack, 1987; Nyblade & Pollack, 1993; Nyblade, 1999). In this scenario, the differing heat flow from

Archean and younger terrains does not require any change in crustal compositions. Of course, some combination of these processes also may be possible.

There does not appear to be any consensus about whether one or other of these factors dominates. Thus, Jaupart and Mareschal (1999) have argued that in eastern North America, estimates of the heat flux across the crust–mantle boundary do not vary as a function of age for regions with similar overall surface heat flow, as would be predicted if mantle flux varied as a function of age (i.e., lithospheric thickness). On the other hand, the offset in heat flow between Archean cratons and Proterozoic terrains far removed from Archean cratons (Fig. 4.7), based on the global data set, appears to be robust and compelling (Nyblade & Pollack, 1993; Nyblade, 1999). Geothermobarometry of upper mantle xenoliths also appears to be consistent with at least some of the difference in heat flow resulting from differences in lithospheric thickness (Boyd *et al.*, 1985; Rudnick *et al.*, 1998; Rudnick & Nyblade, 1999).

Thermal modeling similarly offers contrasting results. For example, modeling where continents sit above mantle upwelling predicts a strong relationship between mantle heat flux and lithospheric thickness (Ballard & Pollack, 1987; Nyblade & Pollack, 1993; Lenardic, 1997), whereas models that have continents sitting above mantle downwelling do not predict a strong correspondence between continental lithospheric thickness and continental heat flow (Lenardic, 1997). Nyblade (1999) has pointed out that these models represent the extreme configurations of the continents with respect to the convecting mantle and thus a combination of lithospheric thickness and crustal heat production may best explain the heat flow offset.

In summary, the heat flow data suggest a significant offset between Archean cratons and Proterozoic terrains that are far removed from Archean cratons, and consequently may approximate the heat flow from post-Archean crustal additions. Since there is no seismic evidence for a difference in Archean and post-Archean crustal thickness, this heat flow offset may be explained by an increase in heat production (i.e., K, Th, U abundances) or by increased heat flux across the crust–mantle boundary (related to lithospheric thickness), or both. Accordingly, models that predict that the continental crust has become less enriched in incompatible elements over time (DePaolo, 1988) are not consistent with the heat flow data. If the offset is due entirely to a change in crustal heat production, then an upper limit on the increase of the HPE across the Archean Proterozoic boundary is set at a factor of about 2.0. If some fraction of the offset is due to differing mantle heat flux, then any change in composition would be correspondingly less. For example, using the average heat flow values and assuming that half of the offset in heat flow is due to differing heat production, then a 20–30% difference in the HPE abundances is predicted for Archean and post-Archean crustal additions (the difference between average Archean and overall average crust would be about half this value).

4.4.2 Secular variations in upper crustal compositions

It has long been recognized that a variety of sedimentary rock types (carbonates, shales, average clastic sedimentary rock) change their composition over geological time, with a major shift at about the Archean–Proterozoic boundary (e.g., Garrels & Mackenzie, 1971; Ronov, 1972; van Moort, 1973; Engel *et al.*, 1974; Veizer, 1973, 1979; Nance & Taylor, 1976, 1977; McLennan & Taylor, 1980; McLennan, 1982; Taylor & McLennan, 1985; Condie, 1993). On the other hand, there is less agreement regarding the interpretation of these various secular trends. Both the presence and absence of secular trends in sedimentary rocks may be difficult to interpret in terms of changing compositions of the upper continental crust. Among the possible complications are the following.

(1) Due to cannibalistic sedimentary recycling processes, the composition of the sedimentary mass is strongly buffered, such that any change in the composition in the upper crust is only very slowly and inefficiently monitored by the sedimentary record (Veizer & Jansen, 1979; McLennan, 1988).
(2) Because sedimentary rocks of different tectonic regimes are preserved in the geological record with fundamentally differing efficiency, there is a substantial bias in the preserved sedimentary record (Veizer & Jansen, 1985). Although complex in detail, the over-riding bias is that sediments deposited in volcanically active tectonic settings are preferentially lost from the geological record compared to those deposited in stable tectonic settings (Veizer & Mackenzie, 2003).
(3) The nature of secondary chemical processes that affect the composition of sediments may have changed over geological time in response to factors such as changing oxygen content of the atmosphere and hydrosphere (affecting redox reactions), and evolution of animals and plants (the latter, for example, affecting weathering processes).

Consequently, among the various models that have been proposed to explain one or more of the secular trends seen in chemical composition of sedimentary rocks are those that:

(1) Reflect an evolution in the composition of the upper continental crust (Engel *et al.*, 1974; Taylor & McLennan, 1985).
(2) Reflect the influences of sedimentary processes (e.g., weathering, diagenesis) that have a cumulative effect due to cannibalistic recycling processes (Garrels & Mackenzie, 1971, 1974; Cox & Lowe, 1995).

(3) Reflect some combination of evolution and recycling (Veizer, 1973; McLennan & Taylor, 1980; Taylor & McLennan, 1985; McLennan, 1988; McLennan & Hemming, 1992; Veizer & Mackenzie, 2003).

(4) Reflect a sampling bias related to tectonic setting (Gibbs *et al.*, 1986; Condie, 1993).

(5) Have trends that are statistically insignificant and do not exist (e.g., Dia *et al.*, 1990), or are the opposite of that suggested by the sedimentary data (DePaolo, 1988).

Below we review some of the more recent data that bear on the question of possible secular changes in the composition of the upper continental crust.

Rare earth elements

It has long been suggested that there has been a secular change in the REE distribution of sedimentary rocks marked by an abrupt change in character- istics at the Archean–Proterozoic transition (Jakes & Taylor, 1974; Nance & Taylor, 1976, 1977; McLennan *et al.*, 1979). These changes are interpreted to reflect a change in upper crustal abundances during the late Archean and the implications for these differences have been discussed at length (Taylor & McLennan, 1985, 1995). There are three aspects of the sedimentary REE patterns that are thought to vary with time (e.g., Fig. 4.8).

(1) Post-Archean sedimentary rocks are characterized by a negative Eu-anomaly (average $Eu/Eu^* = 0.65 \pm 0.05$), whereas Archean sedimentary rocks are typically more variable but on average have only a negligible negative Eu-anomaly (Fig. 4.8). Negative Eu-anomalies in the upper crust result from intra-crustal differentiation processes such that Eu is held back in plagioclase feldspar in the lower crust, mainly during partial melting, to form the granodioritic upper crust.

(2) Post-Archean sedimentary rocks are characterized by relatively flat heavy rare earth element (HREE) distributions ($Gd_N/Yb_N = 1.0$–2.0), whereas Archean sedimentary rocks are characterized by highly variable HREE patterns, with Gd_N/Yb_N commonly well in excess of 2.0 (Fig. 4.8). Steep HREE distributions are characteristic of the TTG end-member of the Archean bimodal suite and typically thought to reflect crust formation by melting of mafic rocks within the garnet stability field such that HREE-enriched garnet is a residual phase and leading HREE-depletion in the melts.

(3) On average, post-Archean sedimentary rocks are characterized by steeper LREE patterns (i.e., higher La/Sm, lower Sm/Nd) compared to the Archean (which are highly variable) reflecting a more differentiated (i.e., felsic) composition. More differentiated (more felsic) post-Archean upper crustal compositions have also been suggested from major element data and trace element data; this is discussed in greater detail below.

Although the interpretations of the above features appear to be generally agreed upon, the actual existence of significant changes in sedimentary REE

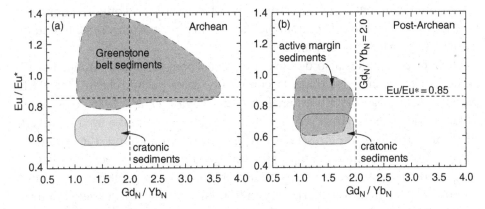

Fig. 4.8. Plots of Eu/Eu* versus Gd$_N$/Yb$_N$ for sedimentary rocks of Archean and post-Archean age. The field for "cratonic" sedimentary rocks of all ages is shown on both diagrams. This is compared to the fields for sedimentary rocks in greenstone belts during the Archean (a) and deposited at volcanically active tectonic settings during the post-Archean (b). Each field encompasses about 90% of the available data. Archean active margin settings differ from the post-Archean in having less relative depletion in Eu and more commonly displaying HREE depletion (high Gd/Yb). The extent of Archean cratonic regions was considerably less than during the post-Archean, whereas volcanically active terrains were far more common. Accordingly, on average, Archean upper crustal REE differed from average post-Archean upper crustal REE patterns. (From Taylor & McLennan, 1995, The geochemical evolution of the continental crust. *Reviews of Geophysics*, **33**, 241–65. © 1995 American Geophysical Union. Reproduced by permission of American Geophysical Union.)

patterns across the Archean–Proterozoic boundary is more controversial. Alternatives include suggestions that the data are statistically insignificant or that changes reflect a bias in sampling, with cratonic settings being under-represented in the Archean data base and volcanically active settings being preferentially sampled (e.g., Gibbs *et al.*, 1986; Goldstein & Jacobsen, 1988; DePaolo, 1988; Dia *et al.*, 1990; Condie, 1993; Gao & Wedepohl, 1995; Cox & Lowe, 1995; Jahn & Condie, 1995).

These apparent concerns fail on several accounts. Early studies of Archean sedimentary rocks indeed relied almost exclusively on samples from relatively low-grade greenstone belts and some (but by no means all) of these certainly represent volcanically active tectonic settings. However, it is now understood that Archean greenstone belts represent a variety of tectonic settings, including back arcs, continental arcs, oceanic island arcs, and perhaps continental rifts. Although modern sediments found in purely oceanic island arcs may lack Eu-anomalies and have other features superficially similar to Archean greenstone sediments, most tectonically active settings contain abundant material with negative Eu-anomalies and other post-Archean sedimentary features (e.g., McLennan *et al.*, 1990).

More recent studies also documented that where stabilized Archean crust existed, the associated sediments commonly have post-Archean-type REE patterns with negative Eu-anomalies (e.g., Taylor *et al.*, 1986; Condie & Wronkiewicz, 1990). However, these terrains appear to make up a relatively small proportion of the preserved Archean crust. In addition, the record of preservation of tectonic setting with geological age unequivocally indicates that cratonic terranes are preferentially preserved over geological time and volcanically active tectonic terrains preferentially lost due to a variety of intra-crustal and crust–mantle recycling processes (Veizer & Jansen, 1985; McLennan, 1988; Veizer & Mackenzie, 2003). This is the exact opposite of that required to explain the difference by preservation bias. Accordingly, the distinctions in Eu-anomalies and HREE distributions shown in Fig. 4.8 are considered to reflect the differences in average upper crustal REE patterns for the Archean and post-Archean.

Incompatible elements

Studies of major elements in sediments over time generally indicated an increase in K_2O/Na_2O ratios at the Archean–Proterozoic transition (Engel *et al.*, 1974, Veizer, 1979; McLennan, 1982). This change is consistent with a general increase in the level of incompatible elements and a more felsic overall composition for the post-Archean upper crust. An alternative interpretation of these data is that the change recorded the cumulative effects of weathering and diagenesis, as sediments were repeatedly recycled (Garrels & Mackenzie, 1974). However, detailed examination of the sedimentary records suggested that while sedimentary processes associated with cannibalistic sedimentary recycling certainly have the potential of changing sedimentary compositions (e.g., Cox & Lowe, 1995), the major change in sedimentary compositions takes place during a fairly brief interval at the Archean–Proterozoic boundary and thus was not dominated by the more progressive changes that would be expected for a recycling model (Veizer & Jansen, 1979; McLennan, 1988).

The trace element pair Th and Sc is considered to be especially sensitive for evaluating changes in the overall composition of the upper continental crust. Both elements are transferred almost exclusively to the clastic sedimentary record during weathering, erosion and transport, and are not redistributed widely by secondary processes, such as diagenesis. Thorium is a highly incompatible element ($D \ll 1$) in most magmatic systems, whereas Sc is typically more compatible ($D \geq 1$). Accordingly, the ratio Th/Sc is an especially sensitive and reliable

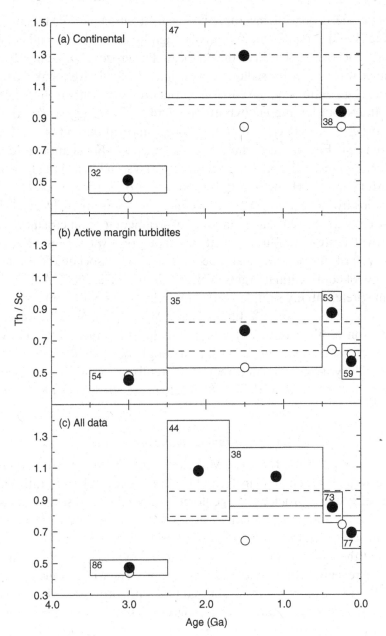

Fig. 4.9. Plots of Th/Sc ratio versus stratigraphic age for sedimentary rocks deposited in stable continental regions (shales) and tectonically active tectonic regions (shales/greywackes). Sample numbers for each time interval are shown in or near the boxes. Solid symbols and boxes are arithmetic means and 95% confidence intervals over time span considered, and open symbols are geometric means. Dashed lines represent 95% confidence intervals for average of all post-Archean samples. Adapted from McLennan and Hemming (1992).

measure of the degree of incompatible element enrichment. McLennan and Hemming (1992) compiled more than 300 high-quality analyses and confirmed a dramatic increase in Th/Sc at the Archean–Proterozoic boundary (Fig. 4.9). The change was greater for sediments associated with relatively stable continental setting (e.g., cratons, passive margins) than for sediments from tectonically active settings, but was present in both cases and strongly suggests that the upper continental crust underwent a change in composition at the end of the Archean.

The data on Fig. 4.9 also show a small but probably significant decrease in Th/Sc for young sediments. A similar reversal is seen for sedimentary Sm/Nd ratios (McLennan & Hemming, 1992) and was also seen for sedimentary K_2O/Na_2O ratios by Engel *et al.* (1974). One interpretation for such reversals that is consistent with current understanding of sedimentary and crustal recycling, is that they represent transient features that are always present in young geological terrains but are lost due to recycling associated with tectonic maturation of continental margins (Veizer & Mackenzie, 2003). At the present, sediments preferentially sample young and relatively undifferentiated crust at volcanically active tectonic settings (Veizer & Jansen, 1985). However, such sediment is most likely to be lost from the sedimentary record due to a variety of processes including metamorphism, melting, and sediment subduction, or masked as the young sediment is recycled and mixed with the overall sedimentary mass (Veizer & Jansen, 1979, 1985; McLennan, 1988; Veizer & Mackenzie, 2003).

Thorium, uranium, and lead isotopes

On the basis of relatively few data, McLennan and Taylor (1980) examined the evidence for possible changes in Th and U abundances and Th/U ratios in fine-grained sedimentary rocks over geological time. They observed an abrupt increase in Th and U abundances at the Archean–Proterozoic boundary, consistent with a change towards more felsic compositions. They also suggested that the Th/U ratio may show a gradual increase over time, especially since the Archean–Proterozoic boundary, consistent with progressive oxidation of U^{4+} to the more readily mobilized U^{6+}, and progressive loss of U from the clastic sedimentary record during progressive sedimentary recycling. Since relatively oxidizing surficial conditions appear to have evolved on Earth by about 2.3 Ga, such processes would be restricted to the post-Archean sedimentary record. These findings were generally confirmed and refined by Taylor and McLennan (1985) from a somewhat enlarged database ($n = 230$; Fig. 4.10). Collerson and Kamber (1999) have also observed the expected reverse trend of Th/U (decreasing Th/U ratio with decreasing age) in samples representing the upper mantle.

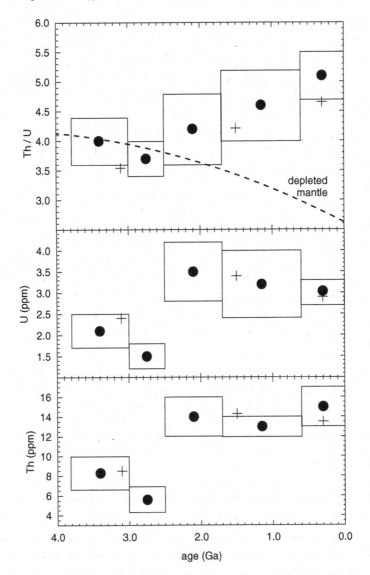

Fig. 4.10. Plots of Th, U, and Th/U ratios versus age for fine-grained sedimentary rocks (adapted from Taylor & McLennan, 1985). Solid symbols and boxes represent arithmetic means and 95% confidence intervals for the timespan considered. Crosses are the average Archean, Proterozoic, and Phanerozoic shales from the compilation of Condie (1993). The dashed line represents the estimate for the evolution of Th/U ratio in the upper mantle by Collerson and Kamber (1999).

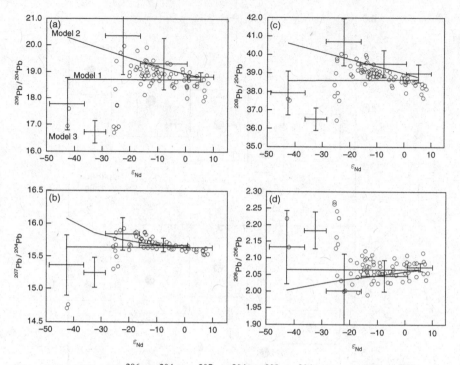

Fig. 4.11. Plots of $^{206}Pb/^{204}Pb$, $^{207}Pb/^{204}Pb$, $^{208}Pb/^{204}Pb$, and $^{208}Pb/^{206}Pb$ versus ε_{Nd} for modern turbidites, river-suspended sediment and deep-sea terrigenous sediment. Depleted mantle model ages of 3.0, 2.0, and 1.0 Ga for average upper crust correspond approximately to ε_{Nd} values of -36, -19, and -2, respectively. Three models of Pb isotope evolution are shown (labeled in (a)), and discussed in text. Adapted from Hemming and McLennan (2001).

The patterns of abrupt increases in Th and U abundances have recently been confirmed by Hemming and McLennan (2001), who evaluated the Pb–Nd isotope characteristics of young marine and fluvial sediments from a variety of tectonic settings. Neodymium isotopes provide a rough measure of the average age of mantle extraction of the various crustal components and provide a measure of provenance age. Figure 4.11 plots various Pb isotope ratios against ε_{Nd}. The remarkable observation is that there is a marked discontinuity in Pb isotopic characteristics for sediments with ε_{Nd} values more negative than -20 (or Nd-model ages in excess of about 2.0 Ga, and thus dominated by an Archean provenance) with older samples showing significantly lower $^{206}Pb/^{204}Pb$, $^{207}Pb/^{204}Pb$, $^{208}Pb/^{204}Pb$, and higher $^{208}Pb/^{206}Pb$ ratios. For $^{206}Pb/^{204}Pb$, $^{207}Pb/^{204}Pb$, and $^{208}Pb/^{204}Pb$ there also appears to be an inverse correlation between Pb and Nd isotopic compositions for samples with ε_{Nd} values that are less negative than -20 (i.e., with Nd model ages less than about 2.0 Ga).

The simplest explanation for these trends is that Archean upper crustal sources have substantially lower U/Pb and Th/Pb ratios compared to the post-Archean upper continental crust. Three models of upper crustal Pb isotope evolution are also shown in an attempt to understand these trends (Hemming & McLennan, 2001). The first (Model 1) assumes μ (^{238}U/^{204}Pb) and κ (^{232}Th/^{238}U) for the upper continental crust equivalent to that given in the Stacey and Kramers Pb isotope evolution model, thus assuming no U/Pb or Th/Pb fractionation between crust and mantle (Stacey & Kramers, 1975). This model results in horizontal trends on all diagrams and is clearly not consistent with the sedimentary Pb isotope data. A second model (Model 2) assumes mantle evolution according to Stacey and Kramers but with an increase in U/Pb, equivalent to $\mu = 12$ and $\kappa = 4$, at the Nd-model age, thus assuming a significant increase in U/Pb and Th/Pb during formation of upper continental crust. Such a model predicts the negative correlation between Pb-isotope ratios and ϵ_{Nd}, but does not predict the sharp drop in Pb isotope ratios for samples with Archean provenance ages.

The third model (Model 3) also adopts a Stacey and Kramers mantle evolution but assumes that the sedimentary sources evolve with Th, U, and Th/U given by the sedimentary data shown in Taylor and McLennan (1985; Fig. 4.10), normalized to $\mu = 12.0$ for the interval 0.6–0.0 Ga. Accordingly, uncertainty in upper crustal common lead (i.e., ^{204}Pb) may allow the model values to shift up and down on Fig. 4.11 but the relative trends remain unchanged. This model predicts both the negative correlation between Pb isotopic compositions and ϵ_{Nd} for samples with young model ages as well as the direction and magnitude of the offset for samples with Archean model ages. Accordingly, the Pb isotope data provide convincing evidence for a substantial increase in Th and U abundances in the upper continental crust at the Archean–Proterozoic boundary. Since all of these samples are "zero" age, some of the questions of possible preservation bias that are introduced by looking at stratigraphic trends are eliminated.

The stratigraphic trends (Fig. 4.10) also suggest a possible increase in Th/U ratios in fine-grained sedimentary rocks over geological time, related to a decrease in U, notably for samples of post-Archean age. If this trend is present it is thought to be due to progressive loss of U from the fine-grained sedimentary record associated with repeated recycling of sedimentary material with ultimate loss of U to the mantle during hydrothermal alteration of ocean crust and to continental ore deposits (McLennan & Taylor, 1980; see also Collerson & Kamber, 1999). Model 3 incorporates these trends and the Pb isotope data are thus consistent with such secular changes in Th/U. However, the scatter in the data is so large that the Pb isotopes provide no firm constraint on such trends. For example, the trends predicted for $^{208}Pb/^{206}Pb$ during the post-Archean are similar for each of Models 1, 2, and 3.

4.4.3 Upper crustal evolution and sedimentary recycling

In summary, the sedimentary data appear to provide convincing evidence for significant difference in the composition of the upper continental crust between the Archean and post-Archean. The major differences are as follows.

(1) Higher levels of incompatible elements in the post-Archean (e.g., K/Na, Th/Sc, Th, U), suggesting an overall more felsic composition.

(2) More pronounced negative Eu-anomalies in the post-Archean suggesting that intracrustal differentiation (partial melting, fractional crystallization) is a far more important and widespread process in the late Archean and post-Archean compared to the earlier Archean.

(3) More common occurrence of HREE-depletion (high Gd/Yb) ratios in Archean upper crustal rocks compared to post-Archean upper crustal rocks. It is not clear that on average the Archean and post-Archean upper crusts have differing Gd/Yb, because the Archean TTG igneous suite is balanced by the presence of more mafic rocks (McLennan & Taylor, 1984; Taylor & McLennan, 1985, Condie, 1993); however, the highly variable Gd/Yb ratios in Archean upper crustal sources is good evidence that the processes of crust formation may have differed.

Condie (1993) attempted to track the composition of the upper continental crust over time by evaluating the composition of mapped regions and stratigraphic sections of various ages and attempting to "restore" material that has been lost from erosion. Such an approach is fraught with uncertainties related to, among other things, sampling statistics, estimating lithological proportions, assumptions about the nature of eroded materials, and preferential recycling of certain tectonic regimes. Nevertheless, it is notable that this approach recognizes most of the trends observed from the sedimentary data, although the magnitudes of change are commonly different. One notable difference is that Condie (1993), while observing a shift in Eu/Eu* at the Archean–Proterozoic boundary, suggested that the magnitude of the change is only about half that indicated in Fig. 4.8.

Changes in upper crustal composition are clearly not the only factor controlling secular trends in sedimentary compositions. In the case of possible increases in Th/U ratio during the post-Archean, the progressive influence of cannibalistic sedimentary recycling coupled with the introduction of a relatively oxidizing atmosphere/hydrosphere during the early Proterozoic are likely to be playing the dominant roles. However, in spite of this control by exogenic processes, such secular changes may still influence crust–mantle evolution, for example, by influencing the Th/U ratio of the upper mantle (Collerson & Kamber, 1999). For a number of trends (e.g., Sm/Nd, Th/Sc, K_2O/Na_2O; Fig. 4.9) there is an apparent reversal during the post-Paleozoic. Such reversals appear to be statistically significant but, rather than reflecting long-term chemical evolution of the upper continental crust, may be a transient feature that is always present in the youngest part of the sedimentary mass but effectively lost as a variety of recycling processes take place over time (Engel *et al.*, 1974; Veizer & Jansen, 1979; McLennan & Hemming, 1992; Veizer & Mackenzie, 2003).

Caution is also warranted in interpreting the apparent lack of change in composition during the post-Archean for most elements. The sedimentary record is intrinsically cannibalistic and, on average, 70% of most sedimentary rocks are derived from the erosion of pre-existing sedimentary rocks. This recycling has a strong buffering effect on the ability of the sedimentary record to reflect changes in upper crustal composition (Veizer & Jansen, 1979; Taylor & McLennan, 1985; McLennan, 1988; Veizer & Mackenzie, 2003), and only major changes are likely to be observed. Accordingly, the sedimentary data are consistent with no significant changes in upper crustal composition during the post-Archean, but only within the constraints imposed by the buffering capacity of sedimentary recycling.

4.5 Epilogue

In contrast to the transient surface history of the oceanic crust that is everywhere less than about 200 Myr, the continental crust has grown slowly and episodically through geological time. On average, the continental crust is more than 10 times the age but only about twice the volume of the oceanic crust. The continental crust, unlike the oceanic crust, is buoyant and is subject to widespread surficial erosion. The upper crust composition may thus be arrived at by using the abundances of immobile elements in sedimentary rocks, whereas heat flow data may be used to constrain bulk crustal composition.

The Archean crust was formed from mixtures of the two dominant igneous lithologies, Na-rich igneous rocks such as tonalites, trondhjemites and granodiorites (the TTG suite and their volcanic equivalents) and basalts. An increase in the growth rate of the continental crust occurred over an extended period between 3.2 and 2.6 Ga, with the exact age differing for individual cratonic regions (e.g., Taylor & McLennan, 1985). During this time, the number of plates became fewer as global heat flow diminished in the late Archean and modern-style plate tectonics became the dominant tectonic theme (e.g., Pollack, 1997). As a consequence, oceanic crust is now both older and colder by the time it reaches the subduction zone, compared to the Archean. This older oceanic crust returns to the deep mantle without being remelted, but fluids from dehydration of the slab rise into the overlying mantle wedge, where they induce melting. This results in the production of the present subduction-zone calc-alkaline suite and in the ultimate addition of this material to the slowly growing crust (Gill, 1981).

Isotopic evidence shows that some continental sediments have been recycled into the mantle, but geological, geochemical, and isotopic constraints limit the amount of subducted sediments to no more than a few percent of arc sources (e.g., McLennan, 1988). Those models, which propose massive recycling of the continental crust through the mantle, encounter various difficulties. Data from Pb isotopes, although sometimes cited in support, provide no independent constraints on this problem. The K–Ar isotopic systematics limit the mass of continental crust recycled through the mantle to no more than about 0.3 crustal masses over geological time (Coltice *et al.*, 2000), which is at least a factor of 5–10 too little to support steady-state crustal evolution models (Armstrong, 1981; 1991). In summary, the growth of the continental crust has proceeded in an episodic fashion throughout geological time with a major increase in the growth rate in the Late Archean. The crust continues to grow at a reduced rate at present, dominantly by island-arc volcanism and related magmatism. Continental crust generated during the Archean may have been somewhat more mafic in overall composition compared to more recent crustal additions, but any difference was likely to be fairly modest.

Following extraction of the crustal material from the mantle, subsequent melting within the deep crust has produced K-rich granodiorites and granites, rich in elements such as K, U, Th, Ba, Rb, and LREE, but depleted in Eu. These dominate the composition of the upper crust after the late Archean. This process is responsible for the changes observed at the Archean–Proterozoic boundary, as reflected in REE patterns observed in the post-Archean clastic sediments, which typically display a significant depletion in Eu. The lower crust is depleted in incompatible elements and is more basic in overall composition (e.g., Rudnick & Fountain, 1995). Such intracrustal melting typically occurs within 50–100 Myr of the derivation of new crust from the mantle (e.g., Moorbath, 1978). Mantle plume activity beneath the crust is considered to be a prime cause of crustal melting (e.g., Harley, 1989; but also see Barboza *et al.*, 1999). The end-result of about 4 billion years of this activity is that the continental crust now contains up to about 50% of the total Earth budget of the most incompatible of the elements.

The continental crust is unique compared to the dominantly basaltic crusts on other planets in the inner solar system (Taylor, 1992). Basalt is the primary magma derived by partial melting of the silicate mantles of the terrestrial planets (although Mercury is a probable exception). Along with the operation of the unique plate tectonic cycle on Earth, this is a consequence of the presence of free water on Earth (Campbell & Taylor, 1983) that enables recycling and remelting of the primary basaltic crust.

4.6 Acknowledgements

We are grateful to Terry Plank and Paul Sylvester for helpful reviews of this chapter.

References

Abbott, D. and Mooney, W. (1995). The structural and geochemical evolution of the continental crust: support for the oceanic plateau model of continental growth. In *Reviews in Geophysics* (Supplement) US National Report IUGG 1991–1994, pp. 231–42.

Abbott, D., Drury, R. and Mooney, W. D. (1997). Continents as lithological icebergs: the importance of buoyant lithospheric roots. *Earth and Planetary Science Letters*, **149**, 15–27.

Albarède, F. (1998). The growth of continental crust. *Tectonophysics*, **296**, 1–14.

Allègre, C. J. and Ben Othman, D. (1980). Nd–Sr isotopic relationship in granitoid rocks and continental crust development: a chemical approach to orogenesis. *Nature*, **286**, 335–42.

Arculus, R. J. (1981). Island arc magmatism in relation to the evolution of the crust and mantle. *Tectonophysics*, **75**, 113–33.

Arculus, R. J. (1999). Origins of the continental crust. *Royal Society of New South Wales. Journal and Proceedings*, **132**, 83–110.

Armstrong, R. L. (1981). Radiogenic isotopes: the case for crustal recycling on a near-steady-state no-continental-growth Earth. *Philosophical Transactions of the Royal Society of London*, **A301**, 443–72.

Armstrong, R. L. (1991). The persistent myth of crustal growth. *Australian Journal of Earth Science*, **38**, 613–30.

Ballard, S. and Pollack, H. N. (1987). Diversion of heat by Archean cratons: a model for southern Africa. *Earth and Planetary Science Letters*, **85**, 253–64.

Barboza, S. A., Bergantz, G. W. and Brown, M. (1999). Regional granulite facies metamorphism in the Ivrea zone: is the Mafic Complex the smoking gun or a red herring? *Geology*, **27**, 447–50.

Barth, M. G., McDonough, W. F. and Rudnick, R. L. (2000). Tracking the budget of Nb and Ta in the continental crust. *Chemical Geology*, **165**, 197–214.

Basaltic Volcanism Study Project, (1981). *Basaltic Volcanism on the Terrestrial Planets*. Pergamon Press: New York.

Bowes, D. R. (1972). *Geochemistry of Precambrian Crystalline Basement Rocks, North-west Highlands of Scotland*. 24th International Geological Congress, vol. 1, pp. 97–103.

Boyd, F. R., Gurney, J. J. and Richardson, S. H. (1985). Evidence for a 150–200 km thick Archean lithosphere from diamond inclusion thermobarometry. *Nature*, **315**, 387–9.

Campbell, I. H. and Taylor, S. R. (1983). No water, no granites – no oceans, no continents. *Geophysical Research Letters*, **10**, 1061–4.

Christensen, N. I. and Mooney, W. D. (1995). Seismic velocity structure and composition of the continental crust: a global view. *Journal of Geophysical Research*, **100**, 9761–88.

Collerson, K. D. and Kamber, B. S. (1999). Evolution of the continents and the atmosphere inferred from Th–U–Nb systematics of the depleted mantle. *Science*, **283**, 1519–22.

Coltice, N., Albarède, F. and Gillet, P. (2000). ^{40}K–^{40}Ar constraints on recycling continental crust into the mantle. *Science*, **288**, 845–7.

Condie, K. C. (1993). Chemical composition and evolution of the upper continental crust: contrasting results from surface samples and shales. *Chemical Geology*, **104**, 1–37.

Condie, K. C. and Brookins, D. G. (1980). Composition and heat generation of the Precambrian crust in New Mexico. *Geochemical Journal*, **14**, 95–9.

Condie, K. C. and Selverstone, J. (1999). The crust of the Colorado Plateau: new views of an old arc. *Journal of Geology*, **107**, 387–97.

Condie, K. C. and Wronkiewicz, D. A. (1990). A new look at the Archean–Proterozoic boundary: sediments and the tectonic setting constraint. In *The Precambrian Continental Crust and its Economic Resources*, ed. S. M. Naqvi. Amsterdam: Elsevier, pp. 61–89.

Cox, R. and Lowe, D. R. (1995). A conceptual review of regional-scale controls on the composition of clastic sediment and the coevolution of continental blocks and their sedimentary cover. *Journal of Sedimentary Research*, **65**, 1–12.

Defant, M. J. and Drummond, M. S. (1990). Derivation of some modern arc magmas by melting of young subducted lithosphere. *Nature*, **347**, 662–5.

DePaolo, D. J. (1988). Age dependence of the composition of continental crust: evidence from Nd isotopic variations in granitic rocks. *Earth and Planetary Science Letters*, **90**, 263–71.

Derry, L. A. and Jacobsen, S. B. (1988). The Nd and Sr isotopic evolution of Proterozoic seawater. *Geophysical Research Letters*, **15**, 397–400.

Dia, A., Dupré, B., Gariépy, C. and Allègre, C. J. (1990). Sm–Nd and trace-element characterization of shales from the Abitibi Belt, Labrador Trough, and Appalachian Belt: consequences for crustal evolution through time. *Canadian Journal of Earth Science*, **27**, 758–66.

Eade, K. E. and Fahrig, W. F. (1971). Geochemical evolutionary trends of continental plates – a preliminary study of the Canadian shield. *Bulletin of the Geological Survey of Canada Bulletin*, **179**.

Eade, K. E. and Fahrig, W. F. (1973). Regional, lithological, and temporal variation in the abundances of some trace elements in the Canadian shield. *Geological Survey of Canada Paper*, 72–46.

Ellam, R. M. and Hawkesworth, C. J. (1988). Is average continental crust generated at subduction zones? *Geology*, **16**, 314–17.

Engel, A. E. J., Itson, S. P., Engel, C. G., Stickney, D. M. and Cray, E. J. (1974). Crustal evolution and global tectonics: a petrogenetic view. *Geological Society of America Bulletin*, **85**, 843–58.

Evans, O. C. and Hanson, G. N. (1997). Late- to post-kinematic Archean granitoids of the S.W. Superior Province: derivation through direct mantle melting. In *Greenstone Belts*, ed. M. deWit and L. D. Ashwal. Oxford: Oxford University Press, pp. 280–95.

Fahrig, W. F. and Eade, K. E. (1968). The chemical evolution of the Canadian shield. *Canadian Journal of Earth Science*, **7**, 1247–51.

Fei, Y., Bertka, C. M. and Mysen, B. O., eds (1999). *Mantle Petrology: Field Observations and High Pressure Experimentation*. Geochemical Society Special Publication 6. The Geochemical Society.

Fountain, D. M., Salisbury, M. H. and Furlong, K. P. (1987). Heat production and thermal conductivity of rocks from the Pikwitonei–Sachigo continental cross-section, central Manitoba: implications for the thermal structure of Archean crust. *Canadian Journal of Earth Science*, **24**, 1583–94.

Furukawa, Y. and Shinjoe, H. (1997). Distribution of radiogenic heat generation in the arc's crust of the Hokkaido Island, Japan. *Geophysical Research Letters*, **24**, 1279–82.

Gao, S. and Wedepohl, K. H. (1995). The negative Eu anomaly in Archean sedimentary rocks – implications for decomposition, age and importance of their granitic sources. *Earth and Planetary Science Letters*, **133**, 81–94.

Gao, S., Luo, T.-C., Zhang, B.-R., Zhang, B.-R., Zhang, H.-F., Han, Y.-W., Zhao, Z.-D. and Hu, Y.-K. (1992). Chemical composition of the continental crust in the Qinling orogenic belt and its adjacent North China and Yangtze cratons. *Geochimica et Cosmochimica Acta*, **56**, 3933–50.

Gao, S., Luo, T.-C. and 5 others (1998). Chemical composition of the continental crust as revealed by studies in East China. *Geochimica et Cosmochimica Acta*, **62**, 1959–75.

Garrels, R. M. and Mackenzie, F. T. (1971). *Evolution of Sedimentary Rocks*. New York: W.W. Norton.

Garrels, R. M. and Mackenzie, F. T. (1974). Chemical history of the oceans deduced from post-depositional changes in sedimentary rocks. In *Studies in Paleoceanography*, ed. W. W. Hay. Special Publication Society of Economic Paleontologists and Mineralogists, 20, pp. 193–204.

Gibbs, A. K., Montgomery, C. W., O'Day, P. A. and Erslev, E. A. (1986). The Archean–Proterozoic transition: evidence from the geochemistry of metasedimentary rocks of Guyana and Montana. *Geochimica et Cosmochimica Acta*, **50**, 2125–41.

Gill, J. B., (1981). *Orogenic Andesites and Plate Tectonics*. New York: Springer-Verlag.

Goldstein, S. J. and Jacobsen, S. B. (1988). Nd and Sr isotopic systematics of river-water suspended material: implications for crustal evolution. *Earth and Planetary Science Letters*, **87**, 249–65.

Guillou-Frottier, L., Mareschal, J. C., Jaupart, C., Gariepy, C., Lapointe, R. and Bienfait, G. (1995). Heat flow in the Grenville Province, Canada. *Earth and Planetary Science Letters*, **136**, 447–60.

Guillou-Frottier, L., Jaupart, C., Mareschal, J. C., Gariepy, C., Bienfait, G., Cheng, L. Z. and Lapointe, R. (1996). High heat flow in the Trans-Hudson orogen, central Canadian Shield. *Geophysical Research Letters*, **23**, 3027–30.

Gupta, M. L., Sundar, A. and Sharma, S. R. (1991). Heat flow and heat generation in the Archean Dharwar cratons and implications for the Southern Indian Shield geotherm and lithospheric thickness. *Tectonophysics*, **194**, 107–22.

Harley, S. L. (1989). The origins of granulites: a metamorphic perspective. *Geological Magazine*, **126**, 215–331.

Hemming, S. R. and McLennan, S. M. (2001). Pb isotope composition of modern deep sea turbidites. *Earth and Planetary Science Letters*, **184**, 489–503.

Hofmann, A. W. (1997). Mantle geochemistry: the message from oceanic volcanism. *Nature*, **385**, 219–29.

Holbrook, W. S., Lizarralde, D., McGeary, S., Bangs, N. and Diebold, J. (1999). Structure and composition of the Aleutian island arc and implications for continental crustal growth. *Geology*, **27**, 31–4.

Jacobsen, S. B. and Pimentel-Klose, M. R. (1988). Nd isotopic variations in Precambrian banded iron formations. *Geophysical Research Letters*, **15**, 393–6.

Jahn, B. M. and Condie, K. C. (1995). Evolution of the Kaapvaal craton as viewed from geochemical and Sm–Nd isotopic analyses of intracratonic pelites. *Geochimica et Cosmochimica Acta*, **59**, 2239–58.

Jakes, P. and Taylor, S. R. (1974). Excess europium content in Precambrian sedimentary rocks and continental evolution. *Geochimica et Cosmochimica Acta*, **38**, 739–45.

Jaupart, C. and Mareschal, J. C. (1999). The thermal structure and thickness of continental roots. *Lithos*, **48**, 93–114.

Jaupart, C. and Mareschal, J.-C. (2003). Constraints on crustal heat production from heat flow data. In *The Crust*, vol. 3, ed. R. L. Rudnick. Amsterdam and Oxford: Elsevier, pp. 65–84.

Jaupart, C., Mareschal, J. C., Gulliou-Frottier, L. and Davaille, A. (1998). Heat flow and thickness of the lithosphere in the Canadian Shield. *Journal of Geophysical Research*, **103**, 15 269–86.

Jones, M. Q. W. (1988). Heat flow in the Witwatersrand Basin and environs and its significance for the South African shield geotherm and lithosphere thickness. *Journal of Geophysical Research*, **93**, 3243–60.

Kay, R. W. and Kay, S. M. (1991). Creation and destruction of lower continental crust. *Geologische Rundschau*, **80**, 259–78.

Kukkonen, I. T. and Peltonen, P. (1999). Xenolith-controlled geotherm for the central Fennoscandian Shield: implications for lithosphere–asthenosphere relations. *Tectonophysics*, **304**, 301–15.

Lambert, I. B. and Heier, K. S. (1968). Estimates of the crustal abundances of thorium, uranium and potassium. *Chemical Geology*, **3**, 233–8.

Lenardic, A. (1997). On the heat flow variation from Archean cratons to Proterozoic mobile belts. *Journal of Geophysical Research*, **102**, 709–21.

Mareschal, J. C., Jaupart, C., Cheng, L. Z., Rolandone, F., Gariépy, C., Bienfait, G., Guillou-Frottier, L. and Lapointe, R. (1999). Heat flow in the Trans-Hudson orogen of the Canadian Shield: implications for Proterozoic continental growth. *Journal of Geophysical Research*, **104**, 29 007–24.

Martin, H. (1986). Effects of steeper Archean thermal gradient on geochemistry of subduction-zone magmas. *Geology*, **14**, 753–6.

Martin, H. (1993). The mechanisms of petrogenesis of the Archaean continental crust – comparison with modern processes. *Lithos*, **30**, 373–88.

McCulloch, M. T. and Chappell, B. W. (1982). Nd isotopic characteristics of S- and I-type granites. *Earth and Planetary Science Letters*, **58**, 51–64.

McLennan, S. M. (1982). On the geochemical evolution of sedimentary rocks. *Chemical Geology*, **37**, 335–50.

McLennan, S. M. (1988). Recycling of the continental crust. *Pure and Applied Geophysics*, **128**, 683–724.

McLennan, S. M. (1989). Rare earth elements in sedimentary rocks: influence of provenance and sedimentary processes. *Reviews in Mineralogy*, **21**, 169–200.

McLennan, S. M. (2001). Relationships between the trace element composition of sedimentary rocks and upper continental crust. *Geochemistry, Geophysics, Geosystems*, **2**(4), doi: 10.1029/2000 GC000109, 20d.

McLennan, S. M. and Hemming, S. (1992). Samarium/neodymium elemental and isotopic systematics in sedimentary rocks. *Geochimica et Cosmochimica Acta*, **56**, 887–98.

McLennan, S. M. and Taylor, S. R. (1980). Th and U in sedimentary rocks: crustal evolution and sedimentary recycling. *Nature*, **285**, 621–4.

McLennan, S. M., and Taylor, S. R. (1984). Archaean sedimentary rocks and their relation to the composition of the Archaean continental crust. In *Archaean Geochemistry*, ed. A. Kröner, *et al.* New York: Springer–Verlag, pp. 47–72.

McLennan, S. M. and Taylor, S. R. (1996). Heat flow and the chemical composition of continental crust. *Journal of Geology*, **104**, 377–96.

McLennan, S. M. and Xiao, G. (1998). Composition of the upper continental crust revisited: insights from sedimentary rocks (abstract). *Mineralogical Magazine*, **62A**, 983–4.

McLennan, S. M., Fryer, B. J. and Young, G. M. (1979). Rare earth elements in Huronian (Lower Proterozoic) sedimentary rocks: composition and evolution of the post-Kenoran upper crust. *Geochimica et Cosmochimica Acta*, **43**, 375–88.

McLennan, S. M., Nance, W. B. and Taylor, S. R. (1980). Rare earth element–thorium correlations in sedimentary rocks, and the composition of the continental crust. *Geochimica et Cosmochimica Acta*, **44**, 1833–9.

McLennan, S. M., Taylor, S. R., McCulloch, M. T. and Maynard, J. B. (1990). Geochemical and Nd–Sr isotopic composition of deep sea turbidites: crustal evolution and plate tectonic associations. *Geochimica et Cosmochimica Acta*, **54**, 2015–50.

McLennan, S. M., Bock, B., Hemming, S. R., Hurowitz, J. A., Lev, S. M. and McDaniel, D. K. (2003). The roles of provenance and sedimentary processes in the geochemistry of sedimentary rocks. In *Geochemistry of Sediments and Sedimentary Rocks: Evolutionary Considerations to Mineral Deposit-Forming Environments*, ed. D. R. Lenz. Geological Association of Canada GEOtext 5, pp. 1–31.

Moorbath, S. (1978). Age and isotope evidence for the evolution of continental crust. *Philosophical Transactions of the Royal Society of London*, **A288**, 401–13.

Morgan, P. (1984). The thermal structure and thermal evolution of the continental lithosphere. *Physics and Chemistry of the Earth*, **15**, 107–93.

Nance, W. B. and Taylor, S. R. (1976). Rare earth element patterns and crustal evolution – I. Australian post-Archean sedimentary rocks. *Geochimica et Cosmochimica Acta*, **40**, 1539–51.

Nance, W. B. and Taylor, S. R. (1977). Rare earth element patterns and crustal evolution – II. Archean sedimentary rocks from Kalgoorlie, Australia. *Geochimica et Cosmochimica Acta*, **41**, 225–31.

Nesbitt, H. W. and Markovics, G. (1997). Weathering of granodioritic crust, long-term storage of elements in weathering profiles, and petrogenesis of siliciclastic sediments. *Geochimica et Cosmochimica Acta*, **61**, 1653–70.

Nyblade, A. A. (1999). Heat flow and the structure of Precambrian lithosphere. *Lithos*, **48**, 81–91.

Nyblade, A. A. and Pollack, H. N. (1993). A global analysis of heat flow from Precambrian terrains: implications for the thermal structure of Archean and Proterozoic lithosphere. *Journal of Geophysical Research*, **98**, 12 207–18.

Pakiser, L. C. and Robinson, R. (1966). Composition and evolution of the continental crust as suggested by seismic observations. *Tectonophysics*, **3**, 547–57.

Pearcy, L. G., DeBari, S. M. and Sleep, N. H. (1990). Mass balance calculations for two sections of island arc crust and implications for the formation of continents. *Earth and Planetary Science Letters*, **96**, 427–42.

Pinet, C. and Jaupart, C. (1987). The vertical distribution of radiogenic heat production in the Precambrian crust of Norway and Sweden: geothermal implications. *Geophysical Research Letters*, **14**, 260–3.

Pinet, C., Jaupart, C., Mareschal, J. C., Gariepy, C., Bienfait, G. and Lapointe, R. (1991). Heat flow and structure of the lithosphere in the eastern Canadian Shield. *Journal of Geophysical Research*, **96**, 19 941–63.

Plank, T. and Langmuir, C. H. (1998). The chemical composition of subducting sediment and its consequences for the crust and mantle. *Chemical Geology*, **145**, 325–94.

Pollack, H. N. (1997). Thermal characteristics of the Archaean. In *Greenstone Belts*, ed. M. deWit and L. D. Ashwal. Oxford: Oxford University Press, pp. 223–32.

Price, R. C., Gray, C. M., Wilson, R. E., Frey, F. A. and Taylor, S. R. (1991). The effects of weathering on rare-earth element, Y and Ba abundances in Tertiary basalts from southeastern Australia. *Chemical Geology*, **93**, 245–65.

Rapp, R. P. (1997). Heterogeneous source regions for Archaean granitoids: experimental and geochemical evidence. In *Greenstone Belts*, ed. M. deWit and L. D. Ashwal. Oxford: Oxford University Press, pp. 267–79.

Rapp, R. P. and Watson, E. B. (1995). Dehydration melting of metabasalt at 8–32 kbar: implications for continental growth and crust–mantle recycling. *Journal of Petrology*, **36**, 891–931.

Ronov, A. B. (1972). Evolution of rock composition and geochemical processes in the sedimentary shell of the Earth. *Sedimentology*, **19**, 157–72.

Rudnick, R. L. (1995). Making continental crust. *Nature*, **378**, 571–8.

Rudnick, R. L. and Fountain, D. M. (1995). Nature and composition of the continental crust: a lower crustal perspective. *Reviews in Geophysics*, **33**, 267–309.

Rudnick, R. L. and Gao, S. (2003). Composition of the continental crust. In *The Crust*, ed. R. L. Rudnick. Treatise on Geochemistry. Amsterdam: Elsevier, 3, pp. 1–64.

Rudnick, R. L. and Nyblade, A. A. (1999). The thickness and heat production of Archean lithosphere: constraints from xenolith thermobarometry and surface heat flow. In *Mantle Petrology: Field Observations and High Pressure Experimentation*, ed. Y. Fei, C. M. Bertka and B. O. Mysen. Geochemical Society Special Publication 6, pp. 3–12.

Rudnick, R. L., McDonough, W. F. and O'Connell, R. J. (1998). Thermal structure, thickness and composition of continental lithosphere. *Chemical Geology*, **145**, 395–411.

Rudnick, R. L., Barth, M., Horn, I. and McDonough, W. F. (2000). Rutile-bearing refractory eclogites: missing link between continents and depleted mantle. *Science*, **287**, 278–81.

Russell, J. K, Dipple, G. M. and Kopylova, M. G. (2001). Heat production and heat flow in the mantle lithosphere, Slave craton, Canada. *Physics of The Earth and Planetary Interiors*, **123**, 27–44.

Sass, J. H. and Lachenbruch, A. H. (1979). Thermal regime of the Australian continental crust. In *The Earth: Its Origin and Evolution*, ed. M. W. McElhinny. Academic Press: New York, pp. 301–51.

Sclater, J. G., Jaupart, C. and Galson, D. (1980). The heat flow through oceanic and continental crust and the heat loss of the Earth. *Reviews of Geophysics and Space Physics*, **18**, 269–311.

Shaw, D. M., Reilly, G. A., Muysson, J. R., Patterden, G. E. and Campbell, F. E. (1967). An estimate of the chemical composition of the Canadian Precambrian shield. *Canadian Journal of Earth Sciences*, **4**, 829–53.

Shaw, D. M., Dostal, J. and Keays, R. R. (1976). Additional estimates of continental surface Precambrian shield composition in Canada. *Geochimica et Cosmochimica Acta*, **40**, 73–83.

Shaw, D. M., Cramer, J. J., Higgins, M. D. and Truscott, M. G. (1986). Composition of the Canadian Precambrian shield and the continental crust of the earth. In *The Nature of the Lower Continental Crust*, ed. J. B. Dawson, D. A. Carswell, J. Hall and K. H. Wedepohl. Geological Society Special Publication 24, pp. 275–82.

Stacey, J. S. and Kramers, J. D. (1975). Approximation of terrestrial lead isotope evolution by a two-stage model. *Earth and Planetary Science Letters*, **26**, 207–21.

Stein, C. A. (1995). Heat flow of the Earth. In *Global Earth Physics. A Handbook of Physical Constants*, ed. T. J. Ahrens. American Geophysical Union Reference Shelf 1, pp. 144–58.

Stein, M. and Hofmann,. A. W. (1994). Mantle plumes and episodic crustal growth. *Nature*, **372**, 63–8.

Suyehiro, K., Takahashi, N. and Ariie, Y. (1996). Continental crust, crustal underplating, and low-Q upper mantle beneath an oceanic island arc. *Science*, **271**, 390–2.

Taira, A., Saito, S.; Aoike, K., Morita, S., Tokuyama, H., Suyehiro, K., Takahashi, N., Shinohara, M., Kiyokawa, S., Naka, J. and Klaus, A. (1998). Nature and growth rate of the northern Izu-Bonin (Ogasawara) arc crust and their implications for continental crust formation. *The Island Arc*, **7**, 395–407.

Taylor, S. R. (1964). Abundance of chemical elements in the continental crust: a new table. *Geochimica et Cosmochimica Acta*, **28**, 1273–85.

Taylor, S. R. (1967). The origin and growth of continents. *Tectonophysics*, **4**, 17–34.

Taylor, S. R. (1977). Island arc models and the composition of the continental crust. *American Geophysical Union Maurice Ewing Series*, **1**, 325–35.

Taylor, S. R. (1992). *Solar System Evolution: A New Perspective*. New York: Cambridge University Press.

Taylor, S. R. and McLennan, S. M. (1981). The composition and evolution of the continental crust: rare earth element evidence from sedimentary rocks. *Philosophical Transactions of the Royal Society of London*, **A301**, 381–99.

Taylor, S. R. and McLennan, S. M. (1985). *The Continental Crust: Its Composition and Evolution*. Oxford: Blackwell.

Taylor, S. R. and McLennan, S. M. (1995). The geochemical evolution of the continental crust. *Reviews of Geophysics*, **33**, 241–65.

Taylor, S. R., Rudnick, R. L., McLennan, S. M. and Eriksson, K. A. (1986). Rare earth element patterns in Archean high-grade metasediments and their tectonic significance. *Geochimica et Cosmochimica Acta*, **50**, 2267–79.

van Moort, J. C. (1973). The magnesium and calcium contents of sediments, especially pelites, as a function of age and degree of metamorphism. *Chemical Geology*, **12**, 1–37.

Veizer, J. (1973). Sedimentation in geologic history: recycling vs. evolution or recycling with evolution. *Contributions to Mineralogy and Petrology*, **38**, 261–78.

Veizer, J. (1979). Secular variations in chemical composition of sediments: a review. *Physics and Chemistry of the Earth*, **11**, 269–78.

Veizer, J. (1989). Strontium isotopes in seawater through time. *Annual Review of Earth and Planetary Science Letters*, **17**, 141–67.

Veizer, J. and Jansen, S. L. (1979). Basement and sedimentary recycling and continental evolution. *Journal of Geology*, **87**, 341–70.

Veizer, J. and Jansen, S. L. (1985). Basement and sedimentary recycling – 2: Time dimension to global tectonics. *Journal of Geology*, **93**, 625–43.

Veizer, J. and Mackenzie, F. T. (2003). Evolution of sedimentary rocks. In *Treatise on Geochemistry*, vol. 7, ed. F. T. Mackenzie. Amsterdam: Elsevier, pp. 369–407.

Vitorello, I. and Pollack, H. N. (1980). On the variation of continental heat flow with age and the thermal evolution of the continents. *Journal of Geophysical Research*, **85**, 983–95.

Weaver, B. L. and Tarney, J. (1984). Empirical approach to estimating the composition of the continental crust. *Nature*, **310**, 575–7.

Wedepohl, K. H. (1995). The composition of the continental crust. *Geochimica et Cosmochimica Acta*, **59**, 1217–32.

Zandt, G. and Ammon, C. J. (1995). Continental crust composition constrained by measurements of crustal Poisson's ratio. *Nature*, **374**, 152–4.

5

The significance of Phanerozoic arc magmatism in generating continental crust

JON P. DAVIDSON AND RICHARD J. ARCULUS

5.1 Introduction: characterizing the continental crust

An important first step in addressing the significance of Phanerozoic arc magmatism in the generation of continental crust is to ensure that we have a common understanding of: "what is the continental crust?" It may seem unnecessary to ask for such a definition, given the voluminous literature (including this book) relating to the subject. But it is important from the perspective of geochemistry and petrology to remind ourselves that the continental crust is defined as that portion of the continental lithosphere that lies above the Mohorovičić discontinuity. Therefore, the continental crust is defined seismically as material with P-wave velocities (V_p) less than $8 \, km \, s^{-1}$. The seismic velocity of a material is determined by the density and modulus, which in turn are a function of pressure, temperature, and composition, so it is perhaps not surprising that there is a relationship between mineralogy and seismic velocity.

We have some understanding of the bulk composition of the crust as defined in this way, after decades of effort and a variety of chemical and physical approaches. Recent reviews that include discussions of the various geological, geochemical, and geophysical approaches that may be deployed to derive compositional estimates have been presented by Rudnick (1995), Rudnick and Fountain (1995), Rudnick and Gao (2003), Taylor and McLennan (1985, 1995), Wedepohl (1995) and Kemp and Hawkesworth (2003). A granodioritic bulk upper (<15 km depth) continental crustal composition (~66 wt.% SiO_2) has been relatively straightforwardly established (for summaries, see Table 5.1, and the Geochemical Earth Reference Model web site at http://www.earthref.org/germ/) through a combination of brute force grid sampling, deep continental drilling, or by analysis of crustal-derived sediments formed via large-scale natural processes such as those deposited in glacial moraines and periglacial (e.g., loess) activity (Chapter 4).

Evolution and Differentiation of the Continental Crust, ed. Michael Brown and Tracy Rushmer. Published by Cambridge University Press. © Cambridge University Press 2005.

Table 5.1. *Estimated composition of the continental crust. Data from Rudnick and Fountain (1995).*

	Lower	Middle	Upper	Bulk
SiO_2	52.3	60.6	66.0	59.1
TiO_2	0.8	0.7	0.5	0.7
Al_2O_3	16.6	15.5	15.2	15.8
FeO_T	8.4	6.4	4.5	6.6
MnO	0.1	0.1	0.08	0.11
MgO	7.1	3.4	2.2	4.4
CaO	9.4	5.1	4.2	6.4
Na_2O	2.6	3.2	3.9	3.2
K_2O	0.6	2.01	3.4	1.88
P_2O_5	0.1	0.1	0.4	0.2
Li	6	7	20	11
Sc	31	22	11	22
V	196	118	60	131
Cr	215	83	35	119
Co	38	25	10	25
Ni	88	33	20	51
Cu	26	20	25	24
Zn	78	70	71	73
Ga	13	17	17	16
Rb	11	62	112	58
Sr	348	281	350	325
Y	16	22	22	20
Zr	68	125	190	123
Nb	5	8	25	12
Cs	0.3	2.4	5.6	2.6
Ba	259	402	550	390
La	8	17	30	18
Ce	20	45	64	42
Pr	2.6	5.8	7.1	5
Nd	11	24	26	20
Sm	2.8	4.4	4.5	3.9
Eu	1.1	1.5	0.9	1.2
Gd	3.1	4	3.8	3.6
Tb	0.48	0.58	0.64	0.56
Dy	3.1	3.8	3.5	3.5
Ho	0.68	0.82	0.8	0.76
Er	1.9	2.3	2.3	2.2
Yb	1.5	2.3	2.2	2
Lu	0.25	0.41	0.32	0.33
Hf	1.9	4	5.8	3.7
Ta	0.6	0.6	2.2	1.1
Pb	4.2	15.3	20	12.6
Th	1.2	6.1	10.7	5.6
U	0.2	1.6	2.8	1.42

In the case of the middle and lower crust, we share the general problem of lack of exposure that students of the uppermost mantle face even more acutely. For the middle-to-lower crust, we may locate areally extensive terranes, now exposed at Earth's surface, that acquired their mineralogical and chemical characteristics through tectonic cycling at pressures matching those normally characteristic of the middle to lower crust in the 0.5–1 GPa range. For the lowermost crust, reliance has been placed on a few areally extensive exposures of continental crust inferred to be exhumed from close to the Moho (Salisbury & Fountain, 1988), such as the Ivrea Zone in northern Italy, or the Horoman Complex in Hokkaido, coupled with suites of xenoliths that have been brought to the surface in explosive volcanic eruptions. Critical additional constraints are velocity–density profiles acquired through analysis of seismic waves transmitted through the crust (Mooney & Brocher, 1987; Holbrook *et al.*,1992), and measurements of (near) surface heat flow (e.g., Nyblade & Pollack, 1993). Inevitably, these latter techniques are integrative over much larger volumes of crust than by the geochemical analysis of individual rock samples.

Given that we know that crustal compositions are heterogeneous on kilometer to tens or hundreds of kilometer scales, the concept of a bulk crustal average is necessarily synthetic and serves only in petrological and geochemical modeling, such as presented here. Furthermore, the seismic definition of the crust is predicated on the basis of major element composition (reflected in the major mineralogy), whereas many of the methods for estimating composition employ principally trace element approaches. A consensus (Rudnick & Fountain, 1995; Rudnick & Gao, 2003) seems to have been reached, despite these caveats, that overall, the continental crust is chemically and lithologically stratified below a felsic uppermost crust (compressional seismic wave velocity $(V_p) = {\sim}6.2$ km s^{-1}). The middle crust (${\sim}15$–25 km depth) is lithologically heterogeneous, of amphibolite facies mineralogy, and intermediate (${\sim}60$ wt.% SiO$_2$, 3.5 wt.% MgO) in terms of bulk composition ($V_p = 6.2$–6.5 km s^{-1}). The lowermost crust (${\sim}25$–40 km depth), is also lithologically heterogeneous, overall of mafic granulite facies mineralogy consisting of predominantly aluminous pyroxenes and plagioclase feldspar, and broadly of basaltic (${\sim}52$ wt.% SiO$_2$, 7 wt.%. MgO) composition ($V_p = 6.9$ to 7.2 km s^{-1}) (Table 5.1).

More specifically, we note that, regardless of the method of estimate, the *bulk* continental crust (Fig. 5.1) has the following chemical characteristics. It is relatively high in SiO$_2$ and low in MgO (compared with mantle-derived basalts), and relatively enriched (50- to one 100-fold compared with best estimates of primitive silicate Earth abundances) in incompatible trace elements (e.g., Cs, Rb, Ba, Th, and the light rare earth elements (LREEs)). Specific trace element characteristics such as high large ion lithophile

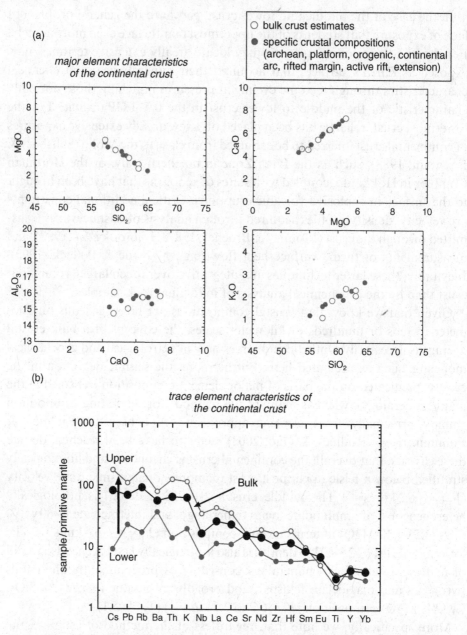

Fig. 5.1. (a) Bulk crustal averages together with specific crust averages from different tectonic environments (data from Rudnick & Fountain, 1995) demonstrating the rather restricted compositional range. (b) Trace element characteristics of the continental crust (data from Rudnick & Fountain, 1995; normalizing values from Sun & McDonough, 1989).

(LILE)/high field strength element (HFSE), and in particular, the low Ce/Pb ($\sim$4) and Nb/U ($\sim$10), are consistent with generation of the continental crust (at least in part) in an arc setting (e.g., Hofmann *et al.*, 1986). The continental crust also has relatively high $^{87}Sr/^{86}Sr$ coupled with low $^{143}Nd/^{144}Nd$, indicating that specific trace element characteristics, such as high Rb/Sr and LREE enrichment, are long-term features.

It is important to note that the trace element abundance pattern of the continental crust requires at least three fundamental influences. First, a low degree of mantle melting, or a two-stage process in which a low volume percent partial melt is important in one of the stages in order to generate high overall incompatible element enrichment. The second requirement is the ubiquitous involvement of a phase, most likely garnet, capable of fractionating the REE either by crystallizing from the magma or as a residual mineral in the source. Lastly, a fluid/solid elemental fractionation is required to account for the marked enrichment of Pb and LILE compared with other trace elements of similar silicate melt/solid incompatibility.

There are several attractive features of the geochemical composition of Phanerozoic arc magmas that render them popular candidates as key components of continental crustal genesis. The apparent equivalence of average andesite with a plausible bulk continental crust composition inspired Taylor (1967) to propose the "Andesite" model of continent genesis. More recently, the striking fact that only supra-subduction zone magmas possess the low Nb/U and Ce/Pb (compared with mid-ocean-ridge basalt (MORB) and ocean-island basalt (OIB); Nb/U $= 47 \pm 10$ and Ce/Pb $= 25 \pm 5$ (Hofmann *et al.*, 1986)) characteristic of the continental crust, has seemed to be a powerful argument that arc systems must be involved to some significant extent in the formation of continents. The critical point is that fractionation of Nb/U and Ce/Pb appears to require the involvement of fluid (i.e., super-critical hydrous) in addition to a silicate melt/residue fractionation event, and this is most likely to occur in the subduction zone rather than MORB or OIB environment (e.g., Arculus, 1994).

Clearly a flux of material is erupted or emplaced at present-day convergent plate margins. These materials do not have the short residence time at Earth's surface that is experienced by oceanic lithosphere, which is destroyed by subduction. In other words, the volcanic and plutonic rocks formed at island and continental arcs are not subducted *en masse* within $\sim$100 Myr. Accordingly, it is sensible to consider what the magnitudes of mass fluxes at Tertiary–Quaternary supra-subduction zones are, and what impact these might have on continental crustal growth. In the extreme, we might discover that the volumes are so trivial as to be insignificant.

5.2 Where is crust made today?

Let us define crustal additions as those masses that cross the oceanic and continental Moho. The largest proportion of material is added at divergent plate boundaries, followed by convergent plate margins and then plumes (Fisher and Schmincke (1984) estimate roughly 62 vol.% at divergent margins, 24 vol.% at convergent margins, and 12 vol.% at hotspots above plumes). Nevertheless, nearly all the mass that forms the oceanic lithosphere is recycled via subduction over relatively short time periods (within ~200 Myr), so that we may conclude that long-term addition of material to crust today is concentrated at convergent plate margins.

Although many geochemists and petrologists regard convergent margins as the primary factory for continental crustal production, it is worth noting that a growing band advocate a prominent role for oceanic plateau accretion (e.g., Hill *et al.*, 1992; Abbott *et al.*, 1997; Albarède, 1998). Oceanic plateaus are generally believed to be the products of melting in mantle plume heads, with volumetrically large basaltic piles formed in a restricted time frame (e.g., Coffin & Eldholm, 1994). Such events are episodic and probably under-represented in a present-day inventory of mass fluxes from the mantle. Although supra-subduction zone magmas are the only basalt type to share the prominent and distinctive continental crustal characteristic of low Nb/U and Ce/Pb (as discussed above), it may be argued that this reflects a minor addition of arc magmas to a volumetrically more abundant oceanic plateau lithology.

Several features of oceanic plateaus are attractive from the point of view of involvement in continental crust genesis. For example, clear Phanerozoic examples exist for the accretion of extensive oceanic plateau terranes in western North America (e.g., Wrangellia (Howell, 1989)) and Ecuador (Arculus *et al.*, 1999). It appears that basaltic crust is unsubductable when thicknesses exceed ~25 km (Abbott *et al.*, 1997). In the case of the world's largest extant oceanic plateau (Ontong Java Plateau east of Papua New Guinea, the Solomon Islands and Vanuatu), the crustal thickness averages ~35 km and subduction ceased when the Plateau encountered the trench along those islands. By the same token, it seems likely that the Caribbean Plate, ringed by suprasubduction zone activity, is the thick basaltic crust formed by melting of the Galapagos plume head (Kerr *et al.*, 1997). Preservation of oceanic plateau material as a basement for arc magmatic products is therefore possible in these cases.

The episodic pattern of continental crust formation recognized in Arabia and West Africa (Boher *et al.*, 1992) seems to be more consistent with the style of mantle plume episodicity than the supposed continuum of subduction zone activity (Albarède, 1998). A significant plume component in continental crust genesis provides a clear element transfer mechanism from the relatively fertile (in terms of incompatible elements such as Cs, Rb, Th) deeper mantle. This particular aspect is an attractive feature of the plateau model over a subduction zone origin for the crust. Continent construction via the tapping of a MORB-like mantle wedge in supra-subduction zone environments, even with the addition of components from subducted oceanic lithosphere, is challenged by the strongly depleted (in terms of incompatible trace elements) character of these upper mantle sources.

However, both plateau and subduction zone models are afflicted with a common problem. There is unequivocal agreement that the bulk composition (supra-Moho) of any oceanic plateau is predominantly basaltic. As we shall see later, the mass flux from the mantle at present day (and probably Phanerozoic in general) arcs is similarly basaltic. Accordingly, the fundamental problem is the requirement for reprocessing a basaltic building block into a granodioritic upper crust without retaining the complementary more mafic residue at depth in the crust. While we should not lose sight of the potential role of plumes to Phanerozoic crust, the remainder of this contribution will focus explicitly on the role of arcs.

5.3 How much crust is made at arcs?

As a starting-point for considering growth of the continental crust, the average growth rate required to produce all of the continental crustal volume in $\sim 4 \times 10^9$ yr (the age of the oldest crust) is $\sim 1.6\,\mathrm{km^3\,yr^{-1}}$. This really should be considered as an incremental linear survival rate because we have neglected erosion and subduction of continental materials. However, from this point onwards we encounter considerably greater difficulties in our mass flux calculations for convergent margins than is the case, for example, with computation of the rate of oceanic crust production. In this latter situation, we may straightforwardly take the average rate of ocean floor spreading (e.g., centimeters per year), the total ridge length ($\sim 5.5 \times 10^4$ km), and the average crustal thickness (7 km) to calculate a present-day production rate of $\sim 20\,\mathrm{km^3\,yr^{-1}}$, equivalent to $\sim 360\,\mathrm{km^3\,km^{-1}}$ of ridge length/Myr of spreading activity. By way of comparison, Reymer and Schubert (1984) estimate average production rates of oceanic crust production from the Mesozoic to Cenozoic of $25\,\mathrm{km^3\,yr^{-1}}$, equivalent to $450\,\mathrm{km^3\,km^{-1}}$ ridge length/Myr.

In the case of island and continental arcs, the situation is complicated in a number of ways. For example, very poor constraints exist in general concerning the nature (proportion and age) of the basement of any arc.

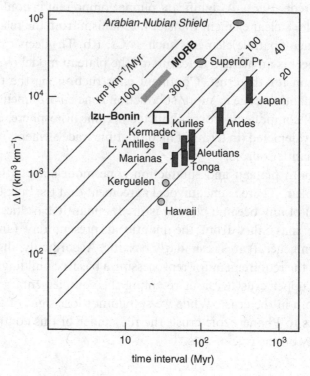

Fig. 5.2. Growth rates of: 1. Mid-ocean ridges (labeled MORB), representative hot spots (in circles), and individual arc systems (after Reymer & Schubert (1984)); 2. Izu–Bonin arc system (after Arculus (1996) and Taira *et al.* (1998)).

Generally, the time interval over which any given volume of arc magmas has accumulated is also not well known. In addition, the nature and depth of the Moho are typically poorly constrained. Accordingly, it is difficult to determine what the crustal production rate is at present-day convergent margins. Nevertheless, Reymer and Schubert (1984) presented the results of a valiant attempt to surmount these difficulties. For intra-oceanic arcs, they calculated arc volumes from seismically derived crustal profiles, subtracted the equivalent of 6 km of "oceanic basement" upon which the arcs were assumed to have been built, and assumed durations of activity from estimates of dates of arc inception (Fig. 5.2).

The average arc crustal addition rate resulting from Reymer and Schubert's (1984) estimate is $30 \, \text{km}^3 \, \text{km}^{-1}$ arc strike length/Myr. Given the total present-day, active arc length ($\sim 3.7 * 10^4 \, \text{km}$), a global production rate of $1.1 \, \text{km}^3 \, \text{yr}^{-1}$ is derived. Despite the nature of the assumptions built into this estimate, it is important to realize that $1.1 \, \text{km}^3 \, \text{yr}^{-1}$ is within the range ($1.6 \, \text{km}^3 \, \text{yr}^{-1}$) required of an (conservative) average growth rate for the continental crust.

However, the notion of an average crustal growth rate is rather arbitrary. The rates of crustal growth through time are still poorly constrained (Chapter 4), although there is general consensus that, if anything, crustal production has decreased through time after a burst of growth in the Archean. In fact, many current calculations suggest that present-day net crustal growth is close to zero. If this is the case, then the arc flux calculated above must be balanced by an equal volume of arc materials removed from the continents each year.

While there is good evidence that the continental crust may have reached ~50–70% of its mass by the Archean–Proterozoic boundary (Chapter 4), extensive continental crustal growth has also taken place in the Phanerozoic. For example, a considerable fraction of recently mantle-derived components are recognized to be significant in the volumetrically extensive Phanerozoic crustal additions of much of eastern Australia east of the Rodinia break-up margin, and the Altaids of central Asia (Sengör & Natal'in, 1996; Jahn *et al.*, 2000).

5.3.1 Mass fluxes at island arcs – initial constraints from the Izu–Bonin system

In the past few years, the results of a major controlled-source, two-ship seismic reflection–refraction survey of the intra-oceanic Izu–Bonin arc between latitudes of 31–33 °N have become available (Suyehiro *et al.*, 1996; Takahashi *et al.*, 1998). Allied with our comprehensive understanding of the tectonic and magmatic evolution of this arc system, stemming primarily from decades of on-land and particularly deep-sea drilling programs (e.g., Taylor, 1992), we have the opportunity for the first time of attempting a well-constrained, crustal growth calculation for an arc. The Izu–Bonin crustal profile is reproduced in Fig. 5.3. The important features of the Izu–Bonin system to the current discussion are as follows.

(1) The difference in thickness of ~10 km between the remnant arc, represented by the Palau–Kyushu Ridge (~0 km thick; Li *et al.*, 1997) and the active arc (~20 km thick).
(2) During the opening of the Shikoku backarc basin and abandonment of the Palau–Kyushu remnant arc, there was minimal (if any – see Taylor, 1992; Arculus *et al.*, 1995) arc activity.
(3) The incremental (~10 km) thickness of the Izu–Bonin active arc was created in the 15 Myr following cessation of backarc spreading and resumption of arc activity.
(4) The width of the active arc is 300 km.

From these observations, we calculate an arc crustal growth rate of ~200 km^3 km^{-1} arc strike length/Myr (Arculus, 1996 – open box on Fig. 5.2).

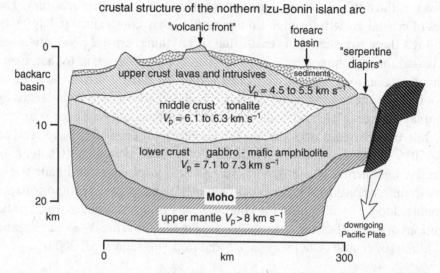

Fig. 5.3. The crustal structure of the Izu–Bonin island arc between the latitudes of 31–33 °N (after Suyehiro *et al.* (1996)).

This is equivalent to a global arc production rate of $7.4 \, km^3 \, yr^{-1}$, but might only be representative for the past 15 Myr in the Izu–Bonin system. Taira *et al.* (1998) made a similar calculation assuming the Izu–Bonin arc was constructed on 6-km thick oceanic crust, and was constructed over 45 Myr (assumed age of inception now known to be at least 49 Ma (Cosca *et al.*, 1998)) resulting in a flux of $80 \, km^3 \, km^{-1}$ arc strike length/Myr or global production rate of $\sim 3 \, km^3 \, yr^{-1}$. If the proto-arc was formed in an extensional (but still supra-subduction zone) environment, and not built on any pre-existing oceanic basement (as argued by Taylor, 1992), then the crustal growth rate to produce the entire 300 km wide × 20 km thick arc crust in ~ 50 Myr is $120 \, km^3 \, km^{-1}$ arc strike length/Myr. DeBari *et al.* (1999) have identified some fragments of the Philippine Sea Plate trapped in the Izu–Bonin forearc, possibly forming a basement to the arc system, so that $120 \, km^3 \, km^{-1}$ arc strike length/Myr would have to be an upper limit over 50 Myr. The main point to note about these estimates is that they are a factor of from 3 to 6 times greater than the global average estimate of Reymer and Schubert (1984).

Of course, there are a number of difficulties and problems with these estimates as well as dangers inherent in extrapolation of the estimates. For example, the estimates are for a comparatively limited period of time for an intra-oceanic arc with episodic growth that has undergone strong rotation since inception (Hall *et al.*, 1995). The magmatism (intra- and underplated) is assumed to have occurred over the full arc width of 300 km; in general, however, the major eruptive flux of magmas in arcs appears to be limited to the volcanic front – the specific locus marking the nearest-to-trench volcanic activity. Nevertheless, we know from intersection through drilling in the Mariana forearc of a Pleistocene sill (Marlow *et al.*, 1992) that the extent of forearc construction through magmatism has likely been underestimated.

Although the Izu–Bonin eruptive products possess the characteristic low Nb/U and Ce/Pb of island arc magmas generally, a striking feature is the generally flat (relative to a chondritic reference) REE abundance patterns (e.g., Gill *et al.*, 1994) unlike that of the bulk continental crust. Accordingly, whereas Phanerozoic intra-oceanic arc magmas of the Izu–Bonin type may contribute a significant fraction of long-lived additions to the continental crust, it is clear that other components (with elevated LREEs/heavy rare earth elements (HREEs)) are also required. These clearly include older, recycled, continental crust (Chapter 4) but an additional source for direct Phanerozoic additions are the continental margin arcs such as the Andes (the type locality of andesite; von Buch, 1836).

5.3.2 Mass fluxes at continental arcs – an Andean solution?

It has been known for some time that rocks from the Andes arc are comparable with crustal averages in terms of major and trace element compositions *and* isotopic ratios (Fig.5.4). It is therefore tempting to suggest that these rocks are the solidified products of magmas derived from the mantle, and thus providing an elegant and simple solution to the problem of where crust is produced, if not exactly how it is produced. Unfortunately, a large number of petrogenetic studies have concluded that Andean magmas are highly differentiated – the products of extensive fractionation *and* contamination by pre-existing crust. It seems that Andean magmas look like crust simply because they have interacted extensively with it (e.g., Hildreth & Moorbath, 1988; Davidson *et al.*, 1991).

This raises an important caveat – the distinctive trace element geochemical signature of the crust is an enduring feature. Once acquired it cannot be completely removed, even allowing for a considerable dilution effect. Any magma contaminated by crust will look as though it was generated in a subduction zone (Arculus, 1987). This is particularly important if we are to concede that plume-derived magmas may form a significant contribution to the crust. If polluted by interaction with the crust, they will resemble subduction zone magmas in the rock record.

It is really not possible at present to constrain the Moho-crossing mass flux in the Andean arcs because we have no unambiguous way of knowing, unlike the Izu–Bonin case, the incremental mass addition per unit of time.

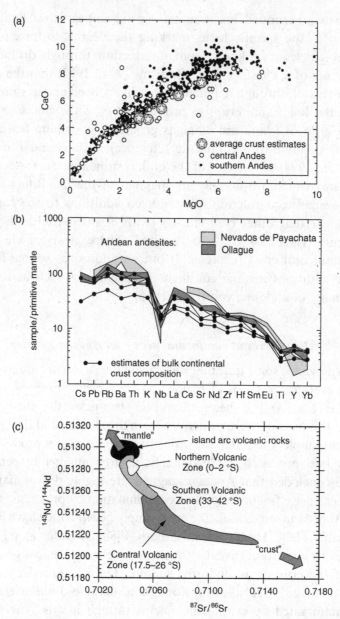

Fig. 5.4. Compositions of Andean magmatic rocks (Harmon *et al.*, 1984; Davidson *et al.*, 1988; 1990; Feeley & Davidson, 1994) compared with estimates of bulk continental crust (data compiled in Rudnick & Fountain, 1995). (a) Major elements as represented by CaO versus MgO; (b) incompatible trace element "spidergram", and (c) isotope compositions.

Of course, there is no a priori reason to assume that the magmas leaving the mantle beneath the Andes are significantly different from those generated in supra-subduction zone systems elsewhere, including the oceans. If we accept the standard model whereby primary intra-oceanic arc magmas are derived from hydration-induced melting of an advecting (i.e., asthenospheric) mantle wedge, then these magmas are not subjected to a predisposed influence in terms of the crust type they will encounter during ascent. A selection of rare primary (or near-primary), high-MgO arc magmas, which have been erupted through thin oceanic crust, shows them to be high-MgO, -Ni and -Cr basalts (Fig. 5.5; Table 5.2). Despite high LILE/REE and high LILE/HFSE, overall concentrations of incompatible trace elements in the primitive or primary magmas are all low as emphasized in the case of the Izu–Bonin system. We will return to the significance of the Andean magmas shortly.

5.3.3 Mass fluxes at arcs – constraints from cross-sections (exposed and seismic)

Arculus (1981) and Gill (1981) suggested that the overall Moho-crossing flux from the mantle in Phanerozoic island arcs is basalt. Mass balancing of lithologies from well-exposed sections of some arcs has reached similar conclusions (e.g., Pearcy et al., 1990). Clearly, this is in conflict and inconsistent with production of an andesitic bulk continental crust directly from the mantle. However, these observations might be reconciled if crustal compositions may be generated by crystal fractionation and the fractionated cumulates returned to the mantle. In this regard, there are other aspects of the Izu–Bonin seismically determined crustal profile (Fig. 5.3) that merit further consideration. The most striking of these is the presence of a layer between 5 and 15 km depth with a V_p of 6.1–6.3 km s^{-1}, consistent with a granitic (*sensu lato*) composition (Suyehiro et al., 1996; Takahashi et al., 1998). Taira et al. (1998) suggest that this layer corresponds to the tonalitic (~66 wt.% SiO$_2$) plutons (Kawate & Arima, 1998) outcropping in the Tanzawa Mountains of the Izu–Bonin arc crust that has been accreted to Honshu. Depending on the assumptions made concerning the compositions of the other layers, Taira et al. (1998) estimate a bulk overall andesitic (54 wt.% SiO$_2$) composition for the Izu–Bonin arc crust, and question whether the assumption by Arculus (1981) of a bulk basaltic composition for intra-oceanic arcs is correct.

However, there are a number of alternative conclusions that may be drawn from the evidence concerning the structure of the Izu–Bonin crust. In our view, there is unequivocal evidence both from the widespread examples of basaltic

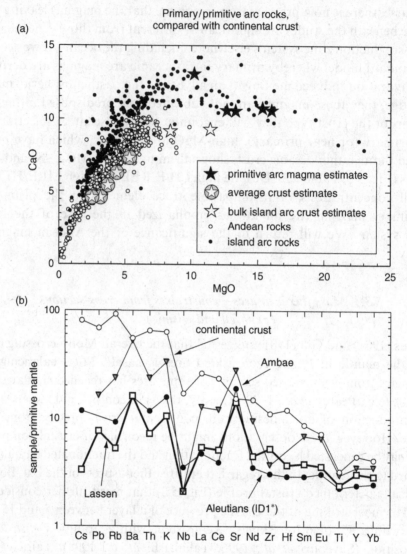

Fig. 5.5. (a) CaO–MgO relations of Andean rocks as in Fig. 5.4a, and island arcs (sources in Davidson, 1996) compared with the compositions of primitive arc magmas (Bacon *et al.*, 1997; Eggins; 1993; Nye & Reid, 1986) and of the bulk compositions of island arc crust (Pearcy *et al.*, 1990). (b) Incompatible trace element compositions of primitive arc basalts (sources as in (a)) compared with continental crust. These plots underscore the large differences between primitive arc magmas and continental crust.

Table 5.2. *Selected primitive arc magma compositions*[1]

Locality	Mt Adams	Lassen	Okmok	Ambae
Sample Number	MA953	LC88–1398	ID1*	68638
Reference[2]	1	1	2	3
SiO_2	49.13	49.15	48.19	48.90
TiO_2	1.35	0.84	0.49	0.65
Al_2O_3	15.48	16.28	13.32	13.24
$Fe_2O_{3(TOT)}$	9.50	10.00	9.19	11.09
MnO	0.14	0.17	0.17	0.18
MgO	9.55	10.23	15.85	10.43
CaO	9.74	10.43	10.48	12.13
Na_2O	3.15	2.39	1.80	2.20
K_2O	1.47	0.40	0.47	1.00
P_2O_5	0.52	0.11	0.06	0.18
Cs	0.225	0.319	0.377	–
Pb	4.24	2.32	–	3
Rb	23.5	6.2	5.8	15
Ba	547	268	110	366
Th	5.09	2.5	1	2
Nb	15	2	2.5	1.5
La	38.45	7.94	3.2	8.7
Ce	81.32	17.93	6.9	21.1
Sr	967	237	404	585
Nd	45.5	10.69	4.7	12.9
Zr	204	76	34	41
Hf	4.19	1.74	0.9	–
Sm	8.59	2.78	1.3	3.05
Eu	2.42	0.96	0.52	1.08
Y	24	21	11	16
Yb	1.9	2.58	1.2	1.73

[1] N.B. Major elements are compared with all iron as Fe_2O_3, and normalized to 100% volatile free (– = not determined/reported).

[2] 1 = Bacon *et al.*, (1997); 2 = Nye and Reid (1986); 3 = Eggins (1993).

magmas erupted in arcs together with extensive experimental evidence, that basalt is the primary (wet) partial melting product of the peridotitic mantle wedge. There is also considerable observational and theoretical evidence for the prevalence of fractional crystallization processes in arc systems (e.g., Gill, 1981). The crystalline phases that dominate the earliest cumulates in basaltic arc magmas are spinel–olivine–clinopyroxene (given the suppression of plagioclase crystallization by dissolved H_2O). Accepting the fact that the bulk composition of the supra-Moho Izu–Bonin crust is a mafic andesite, we suggest that the bulk of the early-formed olivine–clinopyroxene-rich cumulate phases are, in fact, located below the Moho. Furthermore, in the longer term, it is likely that these materials are incorporated in the advecting mantle wedge, or in other words, swept away from the base of the arc crust and recycled into the mantle.

A comparable seismic study of the Aleutian island arc finds a significantly different crustal structure from the Izu–Bonin, suggesting that perhaps the magmatic processes responsible for the construction of arcs vary considerably among locations. Holbrook *et al.* (1999) conclude that the bulk composition of the Aleutian crust is basaltic, despite a typical continent-like thickness of 30 km. They argue that if arc building is an important element in the development of the continental crust, then such crust must be subsequently modified. Specifically, some process (or processes) is needed to remove a significant mass of mafic to ultra-mafic material and leave a more felsic complement with a continent-like composition.

It is an interesting fact that exposed examples of olivine–clinopyroxene-dominated plutonic roots of island arcs are rare (cf. Pearcy *et al.*, 1990). We might expect that deep crustal sections of island and continental arcs would more commonly expose the type of dunite–clinopyroxenite–websterite cumulate sequences characteristic of the Jijal Complex of Kohistan (Jan & Windley, 1981). However, in the case of this deeply exhumed Himalayan example, garnet is an important phase of the more evolved mafic cumulate pile that was developed at the base of a Mesozoic arc. There is no support in general for the presence of garnet in the intra-oceanic arc sequences of the Izu–Bonin system. A much closer analog in terms of predicted phase sequence and calculated melt compositions to the early fractional crystallization products of tholeiitic arc basalts of the Izu–Bonin system, is the layered Permo-Triassic Greenhills Complex of the South Island of New Zealand (Mossman, 1970; Spandler *et al.*, 2000), where dunite, olivine clinopyroxenite and gabbro form the most mafic layers. Cumulate blocks of appropriate phase proportions and composition are among the volcanic ejecta of a number of arcs (e.g., Arculus & Wills, 1980; Conrad & Kay, 1984). But the point remains that volumetrically extensive, ultra-mafic–mafic sub-arc cumulates are not prominent components of the continental crust. It appears that advective sweeping into the mantle wedge of these materials generally, and recycling into the mantle, is a plausible model that fits the present observational and theoretical facts.

A corollary of the suggestion that a cumulate fraction of the basaltic flux from the mantle wedge may be located sub-Moho for the Izu–Bonin system, is that the total supra-subduction zone flux in this instance is greater than the

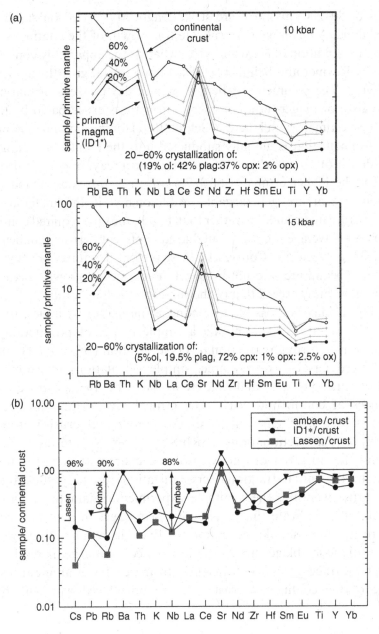

Fig. 5.6. (a) Crystal fractionation models using experimentally determined phase proportions (Draper & Johnson, 1992) and starting with a nominal primitive arc basalt (ID1* from Okmok, Aleutians; Nye & Reid, 1986). Simple fractional crystallization cannot produce the overall enriched incompatible trace element pattern of the continental crust unless: (1) the starting composition is considerably less depleted, or (2) the fractionating assemblage/distribution coefficients are drastically different from those published and used here.

estimates of $80-200\,km^3\,km^{-1}$ arc strike length/Myr based on supra-Moho volumes alone. An average survival rate for growth of the continental crust requires a net addition of $\sim 1.6\,km^3\,yr^{-1}$, whereas the apparently conservative estimate by Reymer and Schubert (1984) of arc crustal growth supplies only $1.1\,km^3\,yr^{-1}$. The volumetric deficit, if present-day convergent margins are representative of longer-term continental growth, is exacerbated by the fact that our best estimate (von Huene & Scholl, 1991) for the amount of continent-derived sediment that is globally subducted into the mantle is $\sim 1\,km^3\,yr^{-1}$. The net deficit of $\sim 1.5\,km^3\,yr^{-1}$ of crustal growth may, of course, be readily supplied if the higher estimates of $3-7.4\,km^3\,yr^{-1}$ fluxes estimated for the Izu–Bonin system are generalizable. Additionally, if the material supra-Moho fluxes are mafic andesite rather than basalt, then the required continental growth rate ($=$ average arc flux $-$ subducted continent-derived sediment) may be balanced satisfactorily. Conversely, these present-day fluxes are embarrassingly large if island arcs are not involved in continental growth, because the volumes significantly exceed that of the subducted sediment flux.

Although it may be possible to reconcile the mass flux issues, there are geochemical problems in accepting a simple model of production of an andesitic crust through fractional crystallization of primary, mantle wedge-derived basalt. For example, simple calculations to examine this possibility (Fig. 5.6) are unable to reproduce the average trace element abundances of continental crust. The critical fact is that the degree of incompatible element enrichment in the continental crust is too high. Furthermore, the fractionations in Rb/Sr and Sm/Nd are inadequate for generating the observed distinctive isotopic characteristics over time. In addition to the initial fractional crystallization (possibly sub-Moho) of spinel-bearing wehrlite, the available experimental data (e.g., Draper & Johnston, 1992) indicate that cumulate assemblages in neutrally buoyant mafic arc magmas should be gabbroic. Even in combination, wehrlitic and gabbroic assemblages are not able to effect the large increases in the incompatible trace elements sufficiently to match those abundances characteristic of the continental crust. We are now faced with an apparent

Fig. 5.6. (cont.)

(b) An alternative way to consider the implications of (a) – an illustration of the enrichments that would be required to produce crust from primary arc magmas (taken from references in Fig. 5.5a) through fractional crystallization, and the minimum amount of fractionation determined by the elements with the maximum enrichment (assumed $D_i = 0$).

dilemma – the main focus of mantle-to-crust mass flux appears to be along convergent plate margins, yet the composition of the flux is not the same as the time-integrated flux represented by the continental crust.

5.4 Models for the origin of the continental crust

Accounting for the origin of the continental crust is a challenging task for a number of reasons. These include the complexity of the petrogenetic processes required for fractionation of quartzofeldspathic compositions from a primary ultramafic (mantle) lithology. Our relatively poor knowledge of the lowermost continental crust is a distinct hindrance in attempts to achieve mantle–crust geochemical mass balances. Furthermore, from a geochemical point of view, it is not clear what significance may be attached to the seismically recognized boundary between crust and mantle (the Moho) and its global equivalence. Further complicating the task is the strong likelihood of secular changes in crustal production processes, and our poor grasp of present-day mass fluxes (both in and out of the mantle) at convergent plate margins, where current continental crust production is concentrated. Finally, there is a distinct possibility that production of continental crust is significantly non-linear through time, with major pulses of activity super-imposed on an overall peak production (or survival) interval straddling the Archean–Proterozoic boundary.

It has long been recognized (e.g., Taylor & McLennan, 1985) that a minimum of two stages is required for creation of a granodiorite-dominated continental crust from a peridotitic (upper mantle) lithology; a first stage of basalt or high-Mg andesite production followed by an intracrustal melting stage, possibly accompanied by other non-igneous processes (e.g., Arculus & Ruff, 1990). It has also been clear that understanding the origins of granite (*sensu lato*) is fundamental to unraveling the process of continental crust formation (e.g., Tuttle & Bowen, 1958; White & Chappell, 1977; Campbell & Taylor, 1985; see also Chapters 9, 10, and 12). In contrast to formation of the predominantly basaltic oceanic crust, which may theoretically be produced in a single-stage melting of the mantle combined with upward melt transport, manufacture of the continental crust clearly involves a more complex genesis (e.g., Borg & Clynne, 1998; Barnes *et al.*, 1996). However, for the purposes of this chapter, the critical point is that the granitic upper continental crust is fundamentally generated, at least through the late Proterozoic and Phanerozoic, not by simple fractional crystallization of basalt alone, but predominantly by intra-crustal processing of one-or-more parental lithologies. These lithologies might include a supra-Moho flux of bulk basaltic andesite composition coupled with recycled continental crustal components. Two important questions relating to this proposition are, first, "what are the parental lithologies, and how are they made?", and second, "can we reconcile our observations of the overall chemical stratification of the continental crust with models of intra-crustal processing?"

A number of potential explanations may be proposed to resolve the dilemma that the composition of the major supra-Moho magmatic flux at convergent margins is not the same as the time-integrated flux represented by the continental crust. Many of these models have been discussed in some detail in earlier contributions (e.g., Rudnick, 1995). We will present these briefly, identifying potential problems, and then return to the compelling similarity between Andean arc rocks and the continental crust in general, from which we suggest a general model for crust growth and evolution, at least through the post-Archean.

5.4.1 Model I

The crust has actually become more mafic through time and the bulk time-integrated crust composition is therefore not the same as the composition of the crust being added today (e.g., DePaolo, 1988).

Based on the sedimentary record, McLennan *et al.* (Chapter 4) argue that significant compositional differences developed in the upper continental crust between the Archean and post-Archean. However, they argue further that extensive recycling of sediments through intra-crustal granite-forming processes, and inherited buffering of trace element characteristics such as elevated LREE/HREE renders the detection of secular changes in Proterozoic and Phanerozoic crustal additions a difficult task.

Problems

Estimates of Archean and post Archean crust composition are similar in terms of major elements (e.g., Taylor & McLennan, 1985; Chapter 4). Furthermore, decreasing mantle temperatures since Earth formation make primary magmas likely to be *less* mafic through time, as seems to be suggested by the occurrence of high-MgO komatiites only in the Archean. Further evidence comes from a global comparison of the nature of the sub-Moho continental lithosphere. For example, Griffin *et al.* (1998) have documented secular changes in the nature of the sub-continental lithospheric mantle. They note that sub-calcic harzburgites are present only in mantle beneath Archean terrains, and mildly sub-calcic harzburgites are common beneath Archean terrains, less abundant beneath Proterozoic terrains, and essentially absent beneath Neoproterozoic and Phanerozoic terrains. Griffin *et al.* (1998) comment that the degree of depletion (in basaltic components) has gradually decreased in peridotite preserved in subcontinental lithospheric mantle, observing:

The Archean–Proterozoic boundary represents a major change in the processes that form continental lithospheric mantle; since 2.5 Ga there has been a pronounced, but more gradual, secular change in the nature of these processes. Actualistic models of lithospheric formation based on modern processes may be inadequate, even for Proterozoic time. The correlation between mantle type and crustal age indicates that the continental crust and underlying lithospheric mantle are formed together, and generally stay together for periods of eons.

5.4.2 Model II

Crust-forming magmatism (and bulk crust composition) is actually high-Mg andesite (Kelemen, 1995; Kelemen et al., 1998, 2003) – so that the high-Mg basalt flux observed/inferred for most Phanerozoic arcs is inappropriate.

Problems

Experiments suggest that high-Mg andesite may only be made directly from peridotite mantle at relatively low pressures with high water contents (e.g., "boninites") or through extensive re-equilibration of peridotite-derived basalt with (refractory) harzburgite during ascent (e.g., Kelemen, 1995). High-Mg andesite (of *adakitic* type – see Defant & Drummond, 1990) derived by partial melting of the mafic layers of subducted lithosphere, while volumetrically trivial in the Phanerozoic, might have been more significant in the Archean (e.g., Martin, 1986). However, Smithies (2000) has shown that distinct geochemical differences exist between Phanerozoic adakites and the Archean tonalite–trondhjemite–granodiorite (TTG) series, and that melting of hydrous basaltic material at the base of a thickened crust rather than subducted lithosphere is more appropriate for the Archean.

5.4.3 Model III

The incompatible trace element enrichment that characterizes continental crust is provided by the addition and recycling of small-degree, mantle-derived partial melts such as lamprophyres or strongly alkaline basalts – perhaps during rifting (e.g., O'Nions & McKenzie, 1988).

Problems

This cannot explain bulk (major element) compositions, as these small-degree melts would still be basaltic, and, by virtue of being small-degree, would not be a large mass fraction. Furthermore, the distinctive negative Eu anomaly characteristic of post-Archean upper continental crust is unlikely to be a mantle-derived signature.

5.4.4 Model IV

Crustal differentiation takes place largely through surficial (weathering) processes, which preferentially dissolve "mafic" components [e.g., Mg] *leaving SiO$_2$-rich residua* [e.g., Albarède & Michard, 1986].

Problems

Weathering does not appear to fractionate REE adequately to produce "crust" from primitive arc basalt compositions. Bulk sediment compositions would need to be "ultramafic" but they are not (they are closer to "bulk crust" – although estimates of bulk crust compositions based on sediments are therefore necessarily circular!)

5.4.5 Model V

An additional process is needed to fractionate arc crust. This may be achieved by delamination of cumulate layers back into the mantle after orogenic thickening, through thermal erosion by convection of the sub arc wedge (Fig. 5.7), or by incorporation into the advecting wedge as advocated by Arculus (1999) for the Izu–Bonin arc. Alternatively, the cumulate complement to the evolved crust composition may be present in deeper levels of the lithosphere. As in the case of the oceanic lithosphere, the continental Moho is simply a seismological distinction and much of the material that is genetically integral to the formation of the continental crust is then actually sub-Moho. For example, Ducea and Saleeby (1998) have shown that the ~30-km thick column of Sierra Nevada granitoids is underlain by ~70-km thickness of genetically related, partly sub-Moho, mafic/ultramafic residues and cumulates.

Problems

As shown above, simple removal of cumulates, either physically by delamination or by virtue of the Moho separating them from superjacent differentiates,

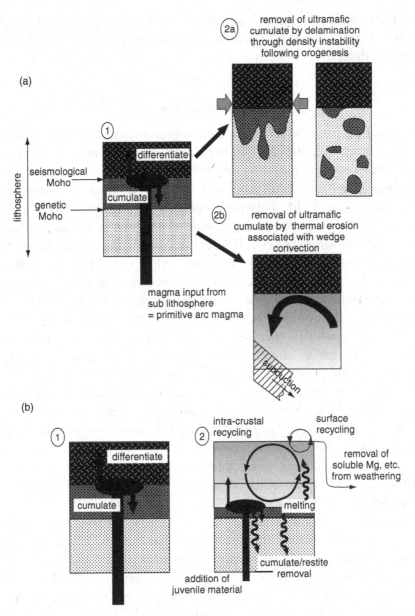

Fig. 5.7. (a) Possible models for generating continental crust (see text) – the key feature is that the complement to primitive arc magma derived from the mantle must be either left at the Moho or physically recycled to deeper levels. (b) A compromise evolution model requires the cumulate removal as in (a) but is accompanied by processes such as contamination, intracrustal melting and weathering.

while preserving the overall arc signature cannot easily produce crustal compositions. Many lithospheric mantle xenoliths do not appear to represent simple cumulates.

5.4.6 Discussion

All five models for producing the continental crust, as defined geochemically, have appealing aspects, yet are simultaneously apparently seriously flawed. We propose a potential compromise, in which crust generation is a progressive hybridization process, perhaps combined with suggestions III–V above (Fig. 5.7b). This progressive hybridization, which includes many of the "alphabet" processes such as melting, assimilation, storage and homogenization (MASH); (Hildreth & Moorbath, 1988), assimilation–fractional crystallization (AFC), (DePaolo, 1981), intra-crustal partial melting and so on, may enrich incompatible trace element compositions and produce crust-like compositions, as we will illustrate below. Vertical compositional stratification will be achieved through time by many of these processes, along with physical transfer mechanisms such as the late-stage sinking of the mafic complements of granitoid plutons proposed by Glazner and Miller (1997). However, it is important that regardless of the exact mechanism(s) by which the continental crust evolves, the residues and/or cumulates from the original mantle flux must be either located beneath the Moho or have been recycled into the mantle (Jull & Kelemen, 2001). As suggested above, they may be physically transferred to deeper levels by density sinking/delamination, or they may simply serve to define the Moho, in which case it is the Moho itself that changes position (cf. Arndt & Goldstein, 1989).

The concept of delamination of the lower parts of the continental lithosphere, which may include the lower crust, has been widely accepted as a dynamic consequence of plate tectonic processes (e.g., Kay & Kay, 1993). The geochemical effects of such processes have been speculated on. Recent work in the Sierra Nevada of the western US appears to have recognized the products of crustal recycling in mantle-derived xenoliths (Ducea & Saleeby, 1998). The timing of recycling of mafic/ultra-mafic material in the overall development of continental crust remains obscure. On the one hand, it may be contemporaneous with, or closely post-date, local subduction. In such cases recycling may reflect delamination of density-unstable material produced by crustal thickening episodes in response to changing subduction zone architecture. Recycling might also be achieved by tectonic erosion of the lower leading edge of the lithosphere by a shallowing slab (e.g., von Huene & Scholl, 1991). On the other hand, recycling may be completely separate in time and space from the subduction that produced the initial mass flux forming a given crustal column. Resolving these issues requires further research into the history of the continental crust in the vertical dimension.

We note further, that to a first approximation, the marked enrichment of a number of alkali elements (around 100-fold) in the continental crust compared with estimates of primitive mantle abundances, is matched by complementary depletions in the global mantle source regions tapped by mid-ocean ridge basalts (see Hofmann, 1988). In other words, the "depleted mantle" source of the most abundant crustal rock type (oceanic) represents a residue from continental crust extraction. In fact, depending on the primitive mantle abundances chosen, it is probable that ~60 vol.% at least of the mantle has been involved in formation of the continental crust. Consequently, there is a strong geochemical argument that any depleted ("refractory") mantle component of the continental lithosphere may only represent a small fraction of the total mantle involved in continental crust formation. There may be no enduring retention of *all* associated residual mantle in any vertical, continental crust–mantle differentiation process, and any juxtaposition may simply represent a fortuitous linkage of buoyant crust and mantle (i.e., a "life raft" model of lithospheric survival).

5.5 Magma genesis in the Andes – mass flux and maturation of the continental crust

If Andean arc rocks compositionally resemble "continental crust," as claimed earlier in this chapter, then the processes that modify Andean magmas subsequent to separation from the sub-arc mantle may represent those responsible for the characteristics of the continental crust in general. The following broad observations are pertinent to the discussion.

The crustal thickness along the Andes varies from ~35 km in the southern volcanic zone (SVZ) to ~70 km in the central volcanic zone (CVZ). The ages of the crust vary from Phanerozoic in the SVZ, much of it post-Paleozoic, to as old as 2 Ga in the CVZ (Fig. 5.8). Where the crust is thickest and oldest, the isotopic characteristics of young volcanic rocks are the most "crust-like" (high $^{87}Sr/^{86}Sr$, low $^{143}Nd/^{144}Nd$). Whereas this might be regarded as a coincidence, there is most likely to be a fundamental connection between the

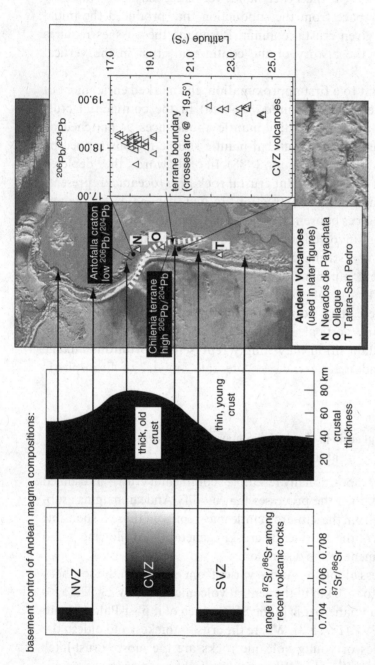

Fig. 5.8. Relationship between the isotope compositions of Andean arc rocks and the crust with which they are associated. Variations in isotope composition of recent volcanic rocks relative to crustal thicknesses along strike of the arc. SVZ, CVZ, NVZ = southern, central, and northern volcanic zones, respectively (after Harmon *et al.*, 1984; Davidson *et al.*, 1991). Dashed line represents terrane boundary between Arequipa block (Antofalla craton) and Chilenia terrane basements (after Ramos, 1988). Note that significant differences in Pb isotope composition occur among the volcanic rocks in crossing this terrane boundary, underscoring the likely strong control of basement character on magma composition.

isotopic compositions of the magmatic rocks and those of the basement through which they have ascended (Davidson *et al.*, 1991). This association is particularly clear from Pb isotope compositions of the volcanic rocks that bear a striking connection to those of the basement (Fig. 5.8; Wörner *et al.*, 1992).

Clear differences in major element trends are seen where magmas pass through crust of different thickness (Fig. 5.9). The CVZ magmas are consistent with both mixing and higher-pressure fractionation (high clinopyroxene/olivine ratio). In this respect, we note the relatively high Mg# and Ni of continental crust and central Andean magmatic rocks relative to simple island arc basalt differentiates. This feature has been suggested by Kelemen (1995) to support a model of high-Mg andesite flux from the mantle, but we argue here that it is a simple consequence of contamination and mixing in the crust. The differences in major element characteristics appear, therefore, to be related to crustal thickness. Plank and Langmuir (1988) have suggested this type of global correlation reflects the degree of melting in the sub-arc mantle wedge – a function of the vertical distance over which melting occurs (inversely related to crustal thickness). However, this cannot satisfactorily explain isotopic differences. The alternative preferred here is that thicker crust impedes ascent leading to more and deeper contamination. Higher pressure leads to higher clinopyroxene/olivine in the fractionating assemblage, and consequently lower CaO/MgO in the differentiates (Fig. 5.9b). Among CVZ magmas, relative enrichment in incompatible trace elements (e.g., LREE-enrichment) occurs with differentiation, probably as a combination of AFC and high-pressure fractionation where amphibole and garnet are able to rotate REE patterns (Fig. 5.10). Protracted feldspar differentiation, perhaps at shallower levels, increases Rb/Sr. With time, the higher Rb/Sr ratios and lower Sm/Nd ratios generate the relatively high ^{87}Sr/^{86}Sr and low ^{143}Nd/^{144}Nd that are characteristic of the crust.

Isotope compositions are modified during differentiation by AFC/MASH. Where the crust is thickest, no primary magma isotope characteristics are preserved (Figs. 5.8 and 5.11). In the central Andes, even the most mafic magmas (barely basaltic = "baseline"; Davidson *et al.*, 1991) are isotopically modified by ubiquitous deep-level interaction with crust (Fig. 5.11a). Elevation of δ^{18}O beyond the range for mantle and high-T differentiates (from +5.5 to +6.5%) implicates interaction with crust (Fig. 5.11b).

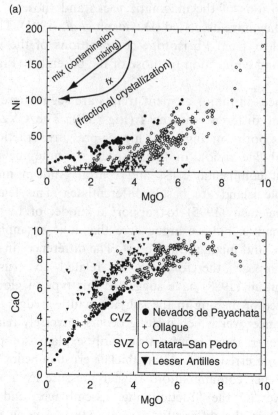

Fig. 5.9. (a) Compositions of suites from southern and central Andean volcanoes respectively. The Lesser Antilles data are shown for comparison as an example of an arc developed on oceanic crust. Differences in trends appear to be related to differences in the thickness of the crustal column traversed. The cartoon in (b) shows two different ways in which the CaO–MgO trends may be related to crustal thickness. The observed trends, as illustrated in the diagram above, are for higher CaO at a given degree of differentiation (6 wt.% MgO) to correspond to thinner crust. This may be explained by (on the left – as proposed by Plank & Langmuir, 1988) higher degrees of melting, which transfers more of the Cpx (CaO) component into the melt, or (on the right) fractionation at lower pressure, with a lower proportion of clinopyroxene in the fractionating assemblage (e.g., Grove & Kinzler, 1986), which therefore is less effective in reducing CaO. In the latter case, fractionation in the thick crust occurs deeper and is probably associated with more contamination/mixing, accentuating the low CaO trend.

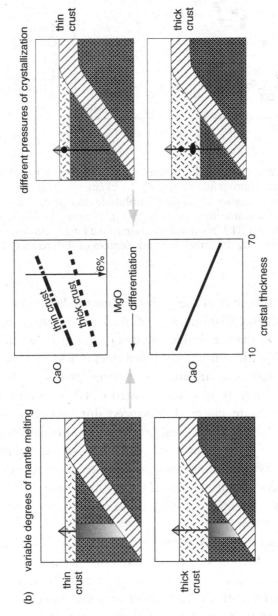

Fig. 5.9. (cont.)

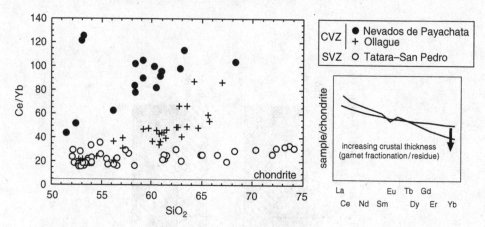

Fig. 5.10. Trace element characteristics of Andean rocks related to crustal thickness (data sources as in Fig. 5.4). Thicker crust promotes crystallization at deeper levels (higher P) so that amphibole and/or garnet are more likely to be in the fractionating assemblage. The propensity of these minerals to concentrate heavy REE results in rotation of the REE patterns, as expressed by elevated Ce/Yb – again correlated with degrees of differentiation (SiO_2).

We argue that the sum of these observations indicates that the continental crust-like geochemical signatures of Andean magmas are generated by differentiation of mantle-derived magmas at the base and within the arc crust. The differentiation must involve interaction with pre-existing crust, some of which has been affected by surficial processes (low-T interaction with H_2O is required to develop elevated $\delta^{18}O$). In addition, separation is required of cumulate material produced during crystallization, perhaps accompanied by residual material produced by partial melting, particularly partial melting of juvenile material such as mafic underplates. This process of progressive hybridization and its chemical consequences is preserved in rare instances of deep crustal exposures, such as the Ivrea Zone, Northern Italy (Voshage *et al.*, 1990), and Fiordland, New Zealand (Klepeis *et al.*, 2003).

Given an initial mass flux of material with an arc-trace element signature, the ultimate production of continental crust as defined compositionally, is therefore dependent on a second stage of maturation, in which thickening is a fundamental factor. A thick crust impedes magma ascent, promoting stalling, differentiation, and contamination. Furthermore, a thick crust stabilizes mineral phases capable of fractionating the REE during contamination, partial melting or fractional crystallization as magmas differentiate.

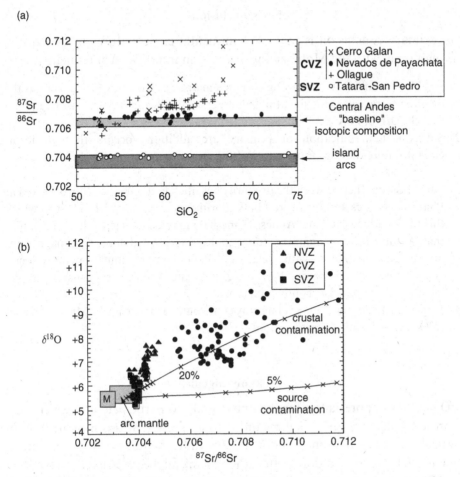

Fig. 5.11. Isotopic characteristics of Andean magmas (compare with Fig. 5.4c). (a) $^{87}Sr/^{86}Sr$ ratios are distinct between CVZ rocks (erupted through thick crust – see Fig. 5.8) and SVZ rocks. The isotope ratios may (e.g., Ollague, Cerro Galan) or may not (e.g., Nevados de Payachata) correlate with indices of differentiation. (b) $\delta^{18}O - {}^{87}Sr/^{86}Sr$ relations (after Davidson *et al.*, 1991) – the strong increases in $\delta^{18}O$ are consistent with crustal contamination. Addition of the same crustal component via subduction (i.e., from the slab – "source contamination") would be expected to be ineffective in significantly changing $\delta^{18}O$ based on the arguments of James (1981) and Taylor (1986). "M" = unmodified oceanic mantle.

5.6 Conclusions

Our primary conclusion is that the continental crust matures by processing through time, with the Moho acting as an open interface. Consequently:

(1) Initial mass additions are through arc magmatism as basalts, which generates the distinctive trace element signatures that persist in all continental crust (e.g., high Ba/Nb, U/Nb, Pb/Ce).

(2) Subsequent modification of primary arc additions occurs through intra-crustal differentiation, especially melting or contamination by the deep mafic lower crust.

(3) Part of the conflict between the primarily basaltic mantle-derived flux observed at Phanerozoic arcs and the overall bulk dioritic composition of the crust to which this flux is added (i.e., an "Andesite" model") may be reconciled if we recognize that a sub-Moho assignment (on geophysical grounds) of ferromagnesian-dominated cumulates with relatively high compressional wave velocity ($V_p > 8\,km^{-1}$), accompanied by residue, dislocates the necessary mafic complement from a more felsic continental crust.

(4) Upper crustal sedimentary cycling may contribute to increasing Si/Mg, to elevated $\delta^{18}O$ and to chemical stratification.

5.7 Acknowledgements

JPD expresses appreciation for the opportunity to participate in the Verbania Penrose Conference on crustal melting – to all present at this meeting, and in particular to Tracy Rushmer and Mike Brown for inviting us to participate in the current book. In addition, both authors thank all the participants of the 1999 IUGG symposium on "Building Continents From Arc Lithosphere," which stimulated our thinking on the issue. JPD thanks NSF and UCLA for support. RJA similarly thanks the Australian Research Council for ongoing support. We appreciate constructively critical reviews by Allen Glazner and Susan DeBari.

References

Abbott, D. H., Drury, R. and Mooney, W. D. (1997). Continents as lithological icebergs: the importance of buoyant lithospheric roots. *Earth and Planetary Science Letters*, **149**, 15–27.

Albarède, F. (1998). The growth of continental crust. *Tectonophysics*, **296**, 1–14.

Albarède, F. and Michard, A. (1986). Transfer of continental Mg, S, O and U to the mantle through hydrothermal alteration of oceanic crust. *Chemical Geology*, **57**, 1–15.

Arculus, R. J. (1981). Island arc magmatism in relation to the evolution of the crust and mantle. *Tectonophysics*, **75**, 113–33.

Arculus, R. J. (1987). The significance of source versus process in the tectonic controls of magma genesis. *Journal of Volcanology and Geothermal Research*, **32**, 1–12.

Arculus, R. J. (1994). Aspects of magma genesis in arcs. *Lithos*, **33**, 189–208.

Arculus, R. J. (1996). Source variability and magma fluxes in arcs. Subduction zone structure, dynamics, and magmatism. Abstract, Joint Association for Geophysics and Royal Astronomical Society Meeting, London, pp. 13–14.

Arculus, R. J. (1999). Origins of the continental crust. *Journal and Proceedings of the Royal Society of New South Wales*, **132**, 83–110.

Arculus, R. J. and Ruff, L. J. (1990). Genesis of continental crust: evidence from island arcs, granulites and exospheric processes. In *Granulites and Crustal Differentiation*, ed. D. Vielzeuf and Ph. Vidal. NATO AS1 Series. London: Kluwer Academic Publishers, pp. 1–18.

Arculus, R. J. and Wills, K. J. A. (1980). The petrology of plutonic blocks and inclusions from the Lesser Antilles Island arc. *Journal of Petrology*, **21**, 743–99.

Arculus, R. J., Gill, J. B., Cambray, H., Chen, W. and Stern, R. J. (1995). Geochemical evolution of arc systems in the western Pacific: the ash and turbidite record recovered by drilling. In *Active Margins and Marginal Basins of the Western Pacific*, ed. B. Taylor and J. Natland. American Geophysical Union, Geophysical Monograph Series, 88, pp. 78–101.

Arculus, R. J., Lapierre, H. and Jaillard, E. (1999). Geochemical window into subduction and accretion processes; Raspas metamorphic complex, Ecuador. *Geological Society of America*, **6**, 547–50.

Arndt, N. T. and Goldstein, S. L. (1989). An open boundary between lower continental crust and mantle; its role in crust formation and crustal recycling. *Tectonophysics*, **161**, 201–12.

Bacon, C. R., Bruggman, P. E., Christiansen, R. L., Clynne, M. A., Donnelly-Nolan, J. M. and Hildreth, W. (1997). Primitive magmas at five Cascades volcanic fields: melts from hot, heterogeneous sub-arc mantle. *Canadian Mineralogist*, **35**, 397–424.

Barnes, C. G., Petersen, S. W., Kistler, R. W., Murray, R. and Kays, M. A. (1996). Source and tectonic implications of tonalite–trondhjemite magmatism in the Klamath Mountains. *Contributions to Mineralogy and Petrology*, **123**, 40–60.

Boher, M., Abouchami, W., Michard, A., Albarède, F. and Arndt, N. T. (1992). Crustal growth in West Africa at 2.1 Ga. *Journal of Geophysical Research*, **97**, 345–69.

Borg, L. E. and Clynne, M. A. (1998). The petrogenesis of felsic calc-alkaline magmas from the southernmost Cascades, California: origin by partial melting of basaltic lower crust. *Journal of Petrology*, **39**, 1197–222.

Campbell, I. H. and Taylor, S. R. (1985). No water, no granites – No oceans, no continents. *Geophysical Research Letters*, **10**, 1061–4.

Coffin, M. F. and Eldholm, O. (1994). Large igneous provinces: crustal structure, dimensions, and external consequences. *Reviews of Geophysics*, **32**, 1–36.

Conrad, W. K. and Kay, R. W. (1984). Ultramafic and mafic inclusions from Adak Island: crystallization history and implications for the nature of primary magmas and crustal evolution in the Aleutian arc. *Journal of Petrology*, **25**, 88–125.

Cosca, M. A., Arculus, R. J., Pearce, J. A. and Mitchell, J. G. (1998). ^{40}Ar/^{39}Ar and K–Ar geochronological age constraints for the inception and early evolution of the Izu–Bonin–Mariana arc system. *The Island Arc*, **7**, 579–95.

Davidson, J. P. (1996). Deciphering mantle and crustal signatures in subduction zone magmatism. In *Subduction: Top to Bottom*, ed. G. E. Bebout, D. W. Scholl, S. H. Kirby and J. P. Platt. American Geophysical Union, Geophysical Monograph Series, 96, pp. 251–62.

Davidson, J. P., Ferguson, K. M., Colucci, M. T. and Dungan, M. A. (1988). The origin and evolution of magmas from the San Pedro–Pellado Volcanic Complex, S. Chile: multicomponent sources and open system evolution. *Contributions to Mineralogy and Petrology*, **100**, 429–45.

Davidson, J. P., McMillan, N. J., Moorbath, S., Wörner, G., Harmon, R. S. and Lopez-Escobar, L. (1990). The Nevados de Payachata volcanic region (18 °S, 69 °W, N. Chile) II. Evidence for widespread crustal involvement in Andean magmatism. *Contributions to Mineralogy and Petrology*, **105**, 412–32.

Davidson, J. P., Harmon, R. S. and Wörner, G. (1991). The source of central Andean magmas: some considerations. In *Andean Magmatism and its Tectonic Setting*, ed. R. S. Harmon and C. W. Rapella. Geological Society of America Special Paper, 265, pp. 233–44.

DeBari, S. M., Taylor, B., Spencer, K. and Fujioka, K. (1999). A trapped Philippine Sea plate origin for MORB from the inner slope of the Izu–Bonin trench. *Earth and Planetary Science Letters*, **174**, 183–97.

Defant, M. J. and Drummond, M. S. (1990). Derivation of some modern arc magmas by melting of young subducted lithosphere. *Nature*, **347**, 662–5.

DePaolo, D. J. (1981). Trace element and isotopic effects of combined wallrock assimilation and fractional crystallization. *Earth and Planetary Science Letters*, **53**, 189–202.

DePaolo, D. J. (1988). Age dependence of the composition of the continental crust: evidence from Nd isotopic variations in granitic rocks. *Earth and Planetary Science Letters*, **90**, 263–71.

Draper, D. S. and Johnston, A. D. (1992). Anhydrous PT phase relations of an Aleutian high-MgO basalt: an investigation of the role of olivine–liquid reaction in the generation of arc high alumina basalts. *Contributions to Mineralogy and Petrology*, **112**, 501–19.

Ducea, M. N. and Saleeby, J. B. (1998). The age and origin of a thick mafic–ultramafic keel from beneath the Sierra Nevada batholith. *Contributions to Mineralogy and Petrology*, **133**, 169–85.

Eggins, S. M. (1993). Origin and evolution of picritic arc magmas, Ambae (Aoba), Vanuatu. *Contributions to Mineralogy and Petrology*, **114**, 79–90.

Feeley, T. C. and Davidson, J. P. (1994). Petrology of calc-alkaline lavas at Volcan Ollague and the origin of compositional diversity at central Andean stratovolcanoes. *Journal of Petrology*, **35**, 1295–340.

Fisher, R. V. and Schmincke, H. U. (1984). *Pyroclastic Rocks*. Berlin: Springer-Verlag.

Gill, J. B. (1981). *Orogenic Andesites and Plate Tectonics*. Berlin: Springer-Verlag.

Gill, J. B., Hiscott, R. N. and Vidal, Ph. (1994). Turbidite geochemistry and evolution of the Izu–Bonin arc and continents. *Lithos*, **33**, 135–68.

Glazner, A. F. and Miller, D. M. (1997). Late stage sinking of plutons. *Geology*, **25**, 1099–102.

Griffin, W. L., O'Reilly, S. Y., Ryan, C. G., Gaul, O. and Ionov, D. A. (1998). Secular variation in the composition of subcontinental lithospheric mantle: geophysical and geodynamic implications. *Geodynamics*, **26**, 1–26.

Grove, T. L. and Kinzler, R. J. (1986). Petrogenesis of andesites. *Annual Review of Earth and Planetary Science Letters*, **14**, 417–54.

Hall, R., Fuller, M., Ali, J. R. and Anderson, C. D. (1995). The Philippine Sea Plate: magnetism and reconstructions. In *Active Margins and Marginal Basins of the Western Pacific*, ed. B. Taylor and J. Natland. American Geophysical Union, Geophysical Monograph Series, **88**, pp. 371–404.

Harmon, R. S., Barreiro, B. A., Moorbath, S., Hoefs, J., Francis, P. W., Thorpe, R. S., Deruelle, B., McHugh, J. and Viglino, J. A. (1984). Regional O- Sr- and Pb isotope relationships in late Cenozoic calc-alkaline lavas of the Andean Cordillera. *Geological Society of London Journal*, **141**, 803–22.

Hildreth, W. and Moorbath, S. (1988). Crustal contributions to arc magmatism in the Andes of central Chile. *Contributions to Mineralogy and Petrology*, **98**, 455–99.

Hill, R. I., Campbell, I. H., Davies, G. F. and Griffiths, R. W. (1992). Mantle plumes and continental tectonics. *Science*, **256**, 186–93.

Hofmann, A. W. (1988). Chemical differentiation of the Earth: the relationship between mantle, continental crust, and oceanic crust. *Earth and Planetary Science Letters*, **90**, 297–314.

Hofmann, A. W., Jochum, K. P., Seufert, M. and White, W. M. (1986). Nb and Pb in oceanic basalts: new constraints on mantle evolution. *Earth and Planetary Science Letters*, **79**, 33–45.

Holbrook, W. S., Mooney, W. D. and Christensen, N. I. (1992). *The Seismic Velocity Structure of the Deep Continental Crust*, ed. D. M. Fountain, R. J. Arculus and R. W. Kay. Amsterdam: Elsevier, pp. 1–43.

Holbrook, W. S., Lizarralde, D., McGeary, S., Bangs, N. and Diebold, J. (1999). Structure and composition of the Aleutian island arc and implications for continental crustal growth. *Geology*, **27**, 31–4.

Howell, D. G. (1989). *Tectonics of Suspect Terranes: Mountain Building and Continental Growth*. New York: Chapman and Hall.

Jahn, B., Wu, F. and Chen, B. (2000). Massive granitoid generation in Central Asia: Nd isotope evidence and implication for continental growth in the Phanerozoic. *Episodes*, **23**, 82–92.

James, D. E. (1981). The combined use of oxygen and radiogenic isotopes as indicators of crustal contamination. *Annual Review of Earth and Planetary Science*, **9**, 311–44.

Jan, M. Q. and Windley, B. F. (1981). The mineralogy and geochemistry of the metamorphosed basic and ultrabasic rocks of the Jijal Complex, Kohistan, NW Pakistan. *Journal of Petrology*, **22**, 85–126.

Jull, M. and Kelemen, P. B. (2001). On the conditions for lower crustal convective stability. *Journal of Geophysical Research*, **106**, 6423–46.

Kawate, S. and Arima, M. (1998). Petrogenesis of the Tanzawa plutonic complex, central Japan: exposed felsic middle crust of the Izu–Bonin–Mariana arc. *The Island Arc*, **7**, 342–58.

Kay, R. W. and Kay, S. M. (1993). Delamination and delamination magmatism. *Tectonophysics*, **219**, 177–89.

Kelemen, P. B. (1995). Genesis of high Mg# andesites and the continental crust. *Contributions to Mineralogy and Petrology*, **120**, 1–19.

Kelemen, P. B., Hart, S. R. and Bernstein, S. (1998). Silica enrichment in the continental upper mantle via melt/rock reaction. *Earth and Planetary Science Letters*, **164**, 387–406.

Kelemen, P. B., Yogodzinski, G. M. and Scholl, D. W. (2003). Along strike variation in lavas of the Aleutian island arc: implications for the genesis of high Mg# andesite and the continental crust. In *Inside the Subduction Factory*, ed. J. Eiler. Geophysical Monograph Series 138, American Geophysical Union.

Kemp, A. I. S. and Hawkesworth, C. J. (2003). Granitic perspectives on the generation and secular evolution of the continental crust. In *The Continental Crust*, ed. R. L. Rudnick. Treatise of Geochemistry 12. Oxford: Elsevier Science.

Kerr, A., Tarney, J., Marriner, G. F., Nivia, A. and Saunders, A. D. (1997). The Caribbean–Colombian Cretaceous Igneous Province: the internal anatomy of an oceanic province. In *Large Igneous Provinces: Continental, Oceanic, and Planetary Flood Volcanism*, ed. J. Mahoney and M. F. Coffin. *Geophysical Monograph Series* **100**, American Geophysical Union, pp. 123–44.

Klepeis, K. A., Clarke, G. L. and Rushmer, T. (2003). Magma transport and coupling between deformation and magmatism in the continental lithosphere. *GSA Today*, **13**, 4–11.

Li, K., Shinohara, M., Suyehiro, K., Kurashimo, E., Miura, S. and Nishisaka, H. (1997). Crustal structure of north Kyushu–Palau Ridge by ocean bottom seismographic observation. *Proceedings of the Seismological Society of Japan*, **2**, 38.

Marlow, M. S., Johnson, L. E., Pearce, J. A., Fryer, P. B., Ledabeth, B., Pickthorn, L. G. and Murton, B. J. (1992). Late Cenozoic volcanism in the Mariana Forearc revealed from drilling at ODP site 781. *Proceedings of the Ocean Drilling Program, Scientific Results*, **125**, 293–310.

Martin, H. (1986). Effect of steeper Archean thermal gradient on geochemistry of subduction-zone magmas. *Geology*, **14**, 753–6.

Mooney, W. D. and Brocher, T. M. (1987). Coincident seismic reflection/refraction studies of the continental lithosphere: a global review. *Reviews of Geophysics*, **25**, 723–42.

Mossman, D. J. (1970). The geology of the Greenhills Ultramafic Complex, Bluff Peninsula, Southland, New Zealand. *Geological Society of America Bulletin*, **81**, 3753–6.

Nyblade, A. A. and Pollack, H. (1993). A global analysis of heat flow from Precambrian terrains: implications for the thermal structure of Archean and Proterozoic lithosphere. *Journal of Geophysical Research*, **98**, 12 207–18.

Nye, C. J. and Reid, M. R. (1986). Geochemistry of primary and least-fractionated lavas from Okmok Volcano, central Aleutians: implications for arc magma genesis. *Journal of Geophysical Research*, **91**, 10 271–87.

O'Nions, R. K. and McKenzie, D. P. (1988). Melting and continent generation. *Earth and Planetary Science Letters*, **90**, 449–56.

Pearcy, L. G., DeBari, S. M. and Sleep, N. H. (1990). Mass balance calculations for two sections of island arc crust. *Earth and Planetary Science Letters*, **96**, 427–42.

Plank, T. and Langmuir, C. H. (1988). An evaluation of the global variations in the major element chemistry of arc basalts. *Earth and Planetary Science Letters*, **90**, 349–70.

Ramos, V. A. (1988). Late Proterozoic–Early Paleozoic of South America – a collisional history. *Episodes*, **11**, 168–74.

Reymer, A. and Schubert, G. (1984). Phanerozoic addition rates to the continental crust and crustal growth. *Tectonics*, **3**, 63–77.

Rudnick, R. L. (1995). Making continental crust. *Nature*, **378**, 571–8.

Rudnick, R. L. and Fountain, D. M. (1995). Nature and composition of the continental crust: a lower crustal perspective. *Reviews of Geophysics*, **33**, 267–309 (and references therein).

Rudnick, R. L. and Gao, S. (2003). The composition of the continental crust. In *Treatise on Geochemistry*, Vol. 3, ed. H. D. Holland and K. K. Turekian. Oxford: Elsevier, pp. 1–64.

Salisbury, M. H. and Fountain, D. M. (1988). *Exposed Cross-sections of the Continental Crust*. NATO ASI Series, **317**, Dordrecht: Kluwer.

Sengör, A. M. C. and Natal'in, B. A. (1996). Turkic-type orogeny and its role in the making of the continental crust. *Annual Review of Earth and Planetary Sciences*, **24**, 263–337.

Smithies, R. H. (2000). The Archean tonalite–trondhjemite–granodiorite (TTG) series is not an analogue of Cenozoic adakite. *Earth and Planetary Science Letters*, **182**, 115–25.

Spandler, C., Eggins, S. M., Arculus, R. J. and Mavrogenes, J. A. (2000). Using melt inclusions to determine parent-magma compositions of layered intrusions: application to the Greenhills Complex (New Zealand), a platinum group minerals-bearing, island-arc intrusion. *Geology*, **28**, 991–4.

Sun, S. S. and McDonough, W. M. (1989). Chemical and isotopic systematics of oceanic basalts: implications for mantle composition and processes. In *Magmatism in the Ocean Basins*, ed. A. J. Saunders and M. J. Norry. Geological Society of London Special Publication, **42**, pp. 313–45.

Suyehiro, K., Takahashi, N. and Ariie, Y. (1996). Continental crust, crustal underplating, and low-Q upper mantle beneath an oceanic island arc. *Science*, **271**, 390–2.

Taira, A., Saito, S., Aoike, K., Morita, S., Tokuyama, H., Suyehiro, K., Takahashi, N., Shinohara, M., Kiyokawa, S., Naka, J. and Klaus, A. (1998). Nature and growth rate of the northern Izu–Bonin (Ogasawara) arc crust and their implications for continental crust formation. *The Island Arc*, **7**, 395–407.

Takahashi, N., Suyehiro, K. and Shinohara, M. (1998). Implications from the seismic crustal structure of the northern Izu–Bonin arc. *The Island Arc*, **7**, 383–94.

Taylor, B. (1992). Rifting and the volcanic–tectonic evolution of the Izu–Bonin–Mariana arc. *Proceedings of the Ocean Drilling Program, Scientific Results*, **126**, 625–51.

Taylor, H. (1986). Igneous rocks II. Isotopic case studies of circumpacific magmatism. In *Stable Isotopes in High Temperature Geological Processes*, ed. J. W. Valley, H. P. Taylor, Jr., and J. R. O'Neill. Reviews in Mineralogy, **16**, pp. 273–317.

Taylor, S. R. (1967). The origin and growth of continents. *Tectonophysics*, **4**, 17–34.

Taylor, S. R. and McLennan, S. M. (1985). *The Continental Crust: Its Composition and Evolution*. Oxford: Blackwell.

Taylor, S. R. and McLennan, S. M. (1995). The geochemical evolution of the continental crust. *Reviews of Geophysics*, **33**, 241–65.

Tuttle, O. F. and Bowen, N. L. (1958). Origin of granite in the light of experimental studies in the system $NaAlSi_3O_8$–$KAlSi_3O_8$–SiO_2–H_2O. *The Geological Society of America Memoir*, **74**.

von Buch, L. (1836). Ueber Erhebungscratere und Vulkane. In *Sächsische Akademie der Wissenschaften Biographisch-Literarisches Handwörter der Exakten Naturwissenschaften*, ed. J. C. Poggendorf, **37**, pp. 169–90.

von Huene, R. and Scholl, D. W. (1991). Observations at convergent margins concerning sediment subduction, subduction erosion, and the growth of continental crust. *Reviews of Geophysics*, **29**, 279–316.

Voshage, H., Hofmann, A. W., Mazzucchelli, M., Rivalenti, G., Sinigoi, S., Raczek, I. and Demarchi, G. (1990). Isotopic evidence from the Ivrea Zone for a hybrid lower crust formed by magmatic underplating. *Nature*, **347**, 731–6.

Wedepohl, K. H. (1995). The composition of the continental crust. *Geochimica et Cosmochimica Acta*, **59**, 1217–32.

White, A. J. R. and Chappell, B. W. (1977). Ultrametamorphism and granitoid genesis. *Tectonophysics*, **43**, 7–22.

Wörner G., Moorbath, S. and Harmon, R. S. (1992). Andean Cenozoic volcanic centres reflect basement isotopic domains. *Geology*, **20**, 1103–6.

6

Crustal generation in the Archean

HUGH ROLLINSON

6.1 Introduction

Earth is unique among the rocky planets of the Solar System in possessing a chemically evolved continental crust. It is the presence of such a crust that has ultimately permitted the appearance of human life. This chapter uses the insights of geochemistry to explore the first-order question of how Earth came to be so different from its planetary neighbors in possessing an evolved continental crust. It will be argued that a major portion of Earth's continental crust formed early in Earth history. This chapter is primarily about the processes of crustal genesis, applied especially to the early history of our planet; it complements chapters that are about intra-crustal processes, in particular crustal differentiation or "crustal ripening."

When Earth is viewed alongside the other planets of our Solar System it becomes apparent that it has many similarities in composition with its near neighbors Mercury, Venus, and Mars. These are the rocky planets, with densities of 3.9–5.4 g cm^{-3}, and are known as the "terrestrial planets" because of their Earth-like features. In contrast, the larger outer planets Jupiter, Saturn, Uranus, and Neptune are gaseous planets, made up largely of a mixture of hydrogen and helium that have densities of 0.7–1.8 g cm^{-3}. However, what is surprising is that, despite the similarities between Earth and other rocky planets, only Earth has a significant mass of continental crust: 41.2% of the surface area and 0.35% of the mass (or 0.5% of the mass of the silicate Earth).

There are two principal observations that are necessary to constrain models for the origin and growth of Earth's continental crust. First is knowledge of the average *composition* of the continental crust, for if its composition is known, then a mechanism for deriving this composition from Earth's mantle may be deduced. Second is the *age structure* of Earth's continental crust. From this information, the growth history of the continental crust may be understood,

Evolution and Differentiation of the Continental Crust, ed. Michael Brown and Tracy Rushmer. Published by Cambridge University Press. © Cambridge University Press 2005.

and rates of continental growth may be calculated. A comparison between former crustal growth rates and modern crustal growth rates sets further constraints on the mechanism whereby the continental crust formed.

6.2 Constraints on the origin of the continental crust I: Estimating the average composition of the continental crust

Given the geological complexity of Earth's continental crust, making an estimate of its average composition is not trivial. Two different types of evidence are required in order to estimate the average composition of the continental crust. First, geochemical compositional data are needed for each of the different components of the crust. In addition, it is important to have information on the layered structure of the crust and the relative proportions of the different compositional layers. These latter data are best provided by geophysical studies and come mainly from seismic surveys. Further constraints on crustal compositions come from heat flow studies, for these data may be used to estimate the proportions of heat-producing elements (K, Th, U) in the continental crust.

One of the earliest attempts to estimate the composition of Earth's continental crust was reported by Taylor (1977) and Taylor and McLennan (1981). The approach adopted by these authors was principally based upon geochemical evidence and made two important assumptions. First, that Earth's crust was made up of two geochemically distinct portions, an upper and lower crust that are present in the proportions 1 : 2. Second, it was assumed that the bulk composition of the crust approximated to the composition of an andesite – a rock-type intermediate in composition between granite and basalt, and dominant at sites of modern crust generation. This second assumption gave rise to the term "the andesite model" for this particular model of crustal genesis – a model that was to become the subject of intense debate. The approach adopted by Taylor and McLennan was very straightforward. They had assumed a composition for the whole crust. They then estimated the average composition of the upper continental crust, by sampling fine-grained clastic sedimentary rocks. From these two compositions they calculated a composition for the lower continental crust and found it to be basaltic.

The work of Taylor and McLennan (1981) was criticized by Weaver and Tarney (1982) who showed that observed lower continental crust had a different geochemical signature from that calculated by Taylor and McLennan (1981). Using measured lower crustal compositions from the Lewisian gneisses of northwest Scotland, Weaver and Tarney (1984) produced an alternative and empirical estimate of the composition of the continental crust, which differed in detail from the Taylor and McLennan (1981) model (see Table 6.1).

Table 6.1. *Average compositions of the continental crust, Archean TTGs and sanukitoids*

	Continental crust – average			Archean TTGs – average				Archean sanukitoid
	1	2	3	4	5	6	7	8
SiO_2	59.10	63.20	57.30	59.70	61.61	69.79	69.66	55.24
TiO_2	0.70	0.60	0.90	0.68	0.67	0.34	0.35	0.77
Al_2O_3	15.80	16.10	15.90	15.70	15.04	15.56	15.86	14.58
FeO(tot)	6.60	4.90	9.10	6.50	5.56	2.78	2.81	7.73
MnO	0.11	0.08	0.18	0.09	0.09	0.05	0.04	0.13
MgO	4.40	2.80	5.30	4.30	3.65	1.18	1.14	7.35
CaO	6.40	4.70	7.40	6.00	5.39	3.19	3.38	8.14
Na_2O	3.20	4.20	3.10	3.10	3.18	4.88	4.70	3.13
K_2O	1.90	2.10	1.33	1.80	2.58	1.76	1.59	1.54
P_2O_5	0.20	0.19	–	0.11	0.17	0.13	0.11	0.33
	98.41	98.87	100.51	97.98	97.94	99.66	99.64	98.93
Mg#	54.3	50.5	50.9	54.1	53.9	43.1	41.9	62.9
Cs	2.6	–	1.5	–	3.4	–	–	–
Rb	58.0	61.0	37.0	53.0	78.0	55	65	–
Ba	390.0	707.0	250.0	429.0	584.0	690	660	–
Th	5.6	5.7	4.2	5.5	8.5	6.9	8	–
U	1.4	1.3	1.1	1.4	1.7	1.6	2	–
K	15771.9	17432.1	10999.4	14941.8	21398.7	14609.8	13198.6	12742.04
Nb	8.0	13.0	11.0	8.0	19.0	6.4	7.5	8.10
Ta	0.7				1.1	0.71		
La	18.0	28.0	16.0	18.0	30.0	32	30	
Ce	42.0	57.0	33.0	42.0	60.0	56	56	68.70
Pb	12.6	15.0	8.0	13.0	14.8	–		
Pr	5.0		3.9		6.7			
Sr	325.0	503.0	260.0	299.0	333.0	454	435	638.50
Nd	20.0	23.0	16.0	–	27.0	21.4	22	34.05
Sm	3.9	4.1	3.5	4.0	5.3	3.3	3.4	6.30

Table 6.1. (cont.)

	Continental crust – average					Archean TTGs – average		Archean sanukitoid
Zr	123.0	210.0	100.0	118.0	203.0	152	160	152.50
Hf	3.7	4.7	3.0	3.4	4.9	4.5	4	–
Eu	1.2	1.1	1.1	1.2	1.3	0.92	1	1.71
Ti	4196.5	3597.0	5395.5	4076.6	4010.0	2038.3	2098.3	4586.18
Y	20.0	14.0	20.0	21.0	24.0	7.5	13	–
Ho	0.8	–	0.8	–	0.8	–	–	–
Yb	2.0	1.5	2.2	1.9	2.0	0.55	1	1.96

Sources:

1. Rudnick & Fountain (1995); Nb & Ta from Barth *et al.* (2000)

2. Weaver & Tarney (1984)

3. Taylor & McLennan (1985, 1995; Chapter 6)

4. Condie (1997)

5. Wedepohl (1995)

6. Martin (1994)

7. Condie (1981)

8. Shirey & Hanson (1984)

(– = not determined/reported)

A further criticism of the work of Taylor and McLennan (1981) was that they failed to recognize that much crustal growth took place during the early history of Earth and may therefore have a composition different from that assumed in the andesite model – the model underpinning their average crustal composition.

This led Taylor and McLennan (1985) to propose a modified andesite model in which an andesitic composition is assumed only for post-Archean crustal growth (post-2500 Ma). The composition of older, Archean crust was calculated to be a 2 : 1 mixture of the mafic and felsic lithologies found in areas of Archean crust. The revised bulk crustal composition was based upon a 75% Archean crust, 25% andesitic crust, mixture (Taylor and McLennan, 1985). This latter model gives a crust that is richer in Fe, Mg, Ni, and Cr than the andesite model and lower in the heat-producing elements K, Th, and U. Other aspects of the Taylor and McLennan model that have been subject to scrutiny are their upper crustal averages, which were based on the composition of Australian shales and which are not now regarded as representative of Earth's crust as a whole, and the average crustal thickness used in their calculations. It had been thought that average thickness of the continental crust was 36 km, whereas Christensen and Mooney (1995) showed that it is 41 km. Thus, the compositional averages of Taylor and McLennan need to be adjusted to take account of the larger than realized crustal volume.

A particularly difficult problem in estimating crustal compositions is that of the nature of the *lower* continental crust. This is a problem that has been investigated by geochemists and geophysicists over the past 20 years and over which there is a real divergence of opinion. Deep seismic reflection experiments show that P-wave velocities increase with depth in the continental crust, and that in many cases there is a lower crustal layer several kilometers thick that has a P-wave velocity $>6.9 \, \mathrm{km \, s^{-1}}$ (Durrheim & Mooney, 1991; Rudnick & Fountain, 1995). Many studies have equated this high-velocity layer with basaltic compositions and thus have inferred that the lower continental crust is basaltic in composition (e.g., Zandt & Ammon, 1995). However, a recent study in China has shown that a basaltic composition for the lower crust may be an over-simplification, for much of the lower crust beneath that continent may be divided into two layers – an upper lower crust with velocities of 6.7–$6.8 \, \mathrm{km \, s^{-1}}$ and a lowermost crust with velocities of 7.0–$7.2 \, \mathrm{km \, s^{-1}}$ (Gao *et al.*, 1998). In this model, the bulk lower continental crust is a mixture of 20–40% mafic rocks and 80–60% rocks of a more felsic composition.

Geochemical opinion on the nature of Earth's lower continental crust is based upon three main types of evidence. First, there are deep crustal xenoliths brought to the surface during volcanism. Of course, xenoliths present only a partial view of the lower continental crust. However, generally they are mafic in composition (Rudnick & Taylor, 1987) and support the geophysical evidence for a mafic composition for the lower continental crust. Second, in contrast, deep crustal rocks found in metamorphic belts – granulites – are much more varied in composition and include both felsic and mafic examples (Rollinson & Blenkinsop, 1995), as suggested by Gao *et al.* (1998) for the lower crust beneath China. Third, the composition of the deep crust may be estimated from granites, for the trace element and isotopic character of granite may be used to "image" its source region (Wyborn *et al.*, 1992). Such compositions imply a significant felsic component to the lower continental crust.

The debate over the composition of the lower continental crust raises a secondary question, for it is important to consider whether or not lower continental crust compositions have changed with time. Durrheim and Mooney (1991, 1994) suggested that this is the case, and argued that the lower continental crust has become more mafic with time. Their study was based upon data from seismic sections through Archean and Proterozoic cratons and suggests that Archean continental crust is 27–40 km thick and has a thin layer with a *P*-wave velocity >7.0 km s^{-1} at its base. This high-velocity layer makes up 5–10% of the total crustal thickness. In contrast, Proterozoic crust is 40–55 km thick and has a lower mafic layer that makes up 20–30% of the crustal thickness. In contrast, more recent reviews have argued against this finding (Wever, 1992; Christensen & Mooney, 1995; Rudnick & Fountain, 1995; Zandt & Ammon, 1995), and Rudnick and Fountain (1995) state: "we find no significant difference between Archean and Proterozoic shields in either crustal thickness or velocity structure." More recent studies of crustal structure have focused upon the differences that arise as a consequence of the tectonic setting, rather than the age of crust formation. For example, there is good evidence to show that an average crustal section in an active rift is different from the average crustal section of an ancient shield or platform. Thus, it is argued that differences in crustal structure that arise as a consequence of differences in tectonic setting are more important than those that do so from temporal differences.

A related debate has focused on heat flow data from different regions of the continental crust. Nyblade and Pollack (1993) showed that average heat flow measurements in Archean cratons (41 ± 11 mW m^{-2}) are lower than those for Proterozoic cratons (55 ± 17 mW m^{-2}). Two quite different inferences have been drawn from this observation. First, it has been argued there is a difference in crustal heat production (i.e., in U, Th, K concentrations) between Archean and Proterozoic cratons. This implies that there are differences in bulk composition between the two. Second, it has been suggested that the observed differences in heat flow are not a function of crustal composition but reflect a different mantle heat-flow contribution to the two cratons. This is explained in terms of the observed difference in lithospheric thicknesses between Proterozoic and Archean cratons. Thicker lithosphere beneath Archean cratons is thought to divert heat from the convecting mantle away from the base of the craton, resulting in the lower observed crustal heat flow (Nyblade, 1999; Rudnick *et al.*, 1998).

Jaupart and Mareschal (1999) have recently argued that "average" heat flow values for Archean and Proterozoic cratons are meaningless because the variation within each group of values is so large. These authors emphasize the importance of the differences in crustal heat-flow values within Archean cratons and within Proterozoic cratons. It is thought that these within-craton differences in heat flow demonstrate that there are different types of Archean crust and different types of Proterozoic crust. This observation strongly suggests that the observed variations in heat flow arise through differences in crustal composition, but cannot be used to argue for a difference in bulk composition between Archean and Proterozoic cratons. Despite the failure of these largely geophysical arguments to demonstrate a difference in bulk crustal composition with time, there are a number of geochemical arguments that do support a difference, and these are reviewed in Section 6.5.

Probably the most rigorous estimate of the bulk composition of the continental crust has been made by Rudnick and Fountain (1995; Fig. 6.1). This model combines the insights of both geochemistry and geophysics in order to constrain first the composition of the lower continental crust and then the composition of the whole crust. Thus, seismic reflection measurements of the lower crust were used to estimate its broad composition and the results refined using heat flow data; from that, information on the heat-producing element content was obtained. The results were then combined with the measured compositions of known lower crustal rocks, largely xenolith data, and from that Rudnick and Fountain (1995) were able to calculate, in a systematic way the probable composition of the lower continental crust. Second, these authors addressed the other major difficulty in calculating bulk crustal compositions (that of the relative proportions of upper to lower crust). Using the observation that crustal thickness and structure vary with tectonic setting and with crustal age, Rudnick and Fountain (1995) used a range of different types of measured crustal section where the proportions of upper to lower crust are known, either from seismic evidence or from observation. Combining crustal sections of differing age and from different tectonic settings, together with estimates of their areal extent, they computed an average crustal composition.

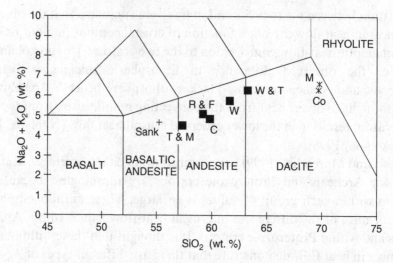

Fig. 6.1. Total alkalis versus silica classification for the average crustal compositions listed in Table 6.1. Solid squares – average continental crust after Taylor and McLennan (1985) [T & M], Condie (1997) [C], Rudnick and Fountain (1995) [R], Weaver and Tarney (1984) [W & T], Wedepohl (1995) [W]; stars – average TTG compositions after Martin (1994) [M], Condie (1981) [Co]; cross average sanukitoid composition (after Shirey & Hanson, 1984, samples 5–80 and 53–80) [Sank].

Interestingly, the differences between the various recent estimates of crustal composition are not great (Rudnick, 1995). Most models indicate an average continental crustal composition with between 57–64 wt.% SiO2, molar Mg/(Mg + Fe) between 0.50 and 0.55, indicating bulk compositions in the andesite range. Average crustal compositions (Fig. 6.2) are also marked by a strong enrichment of the trace elements Cs, Rb, Ba, Th, U, K, and Pb and the light rare earth elements (REEs), relative to the composition of Earth's primitive mantle.

In the following sections of this chapter, average (time integrated) crustal compositions are used to review and evaluate models of crustal evolution. Particularly useful are trace element concentrations and trace element ratios, for these may be combined with values calculated for Earth's primitive mantle to set limits on what may or may not be complementary residues to the process of crust formation. In addition, trace element ratios may be used to constrain competing magmatic processes in crustal evolution models. For example, Barth *et al.* (2000) used La/Nb ratios to estimate the relative contributions of plume-related and subduction-related magmatism to the average crustal composition.

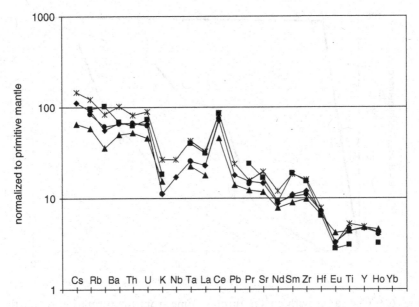

Fig. 6.2 Trace element concentrations in the average crustal compositions listed in Table 6.1, normalized to the composition of the primitive mantle (after Sun & McDonough, 1989). Triangles – Taylor and McLennan (1985); circles – Condie (1997); diamonds – Rudnick and Fountain (1995); squares – Weaver and Tarney (1984); stars – Wedepohl (1995).

6.3 Constraints on the origin of the continental crust II: crustal growth through time

The rate at which Earth's continental crust has grown with time is the subject of some considerable discussion. There are three main views (Fig. 6.3). Initially, it was thought that the continental crust had grown at a linear rate through geological time from formation at about 3.1 Ga to the present (Hurley & Rand, 1969). This model was based on the geochronological data that was available in the late 1960s (mainly K–Ar, Rb–Sr, U–Pb age determinations) for the North American continent. These age determinations were used to determine the "crystallization ages" of individual samples from which "age provinces" were mapped, and it was the analysis of these age data that provided evidence for a relatively late start to the formation of the continental crust and its subsequent growth at a linear rate.

Second, at the other end of the spectrum there is a group of models that argue for no-growth of the continental crust through time (Armstrong, 1981, 1991). In other words, these authors believe that the continental crust formed very early in the history of the planet and that there has been minimal net growth in

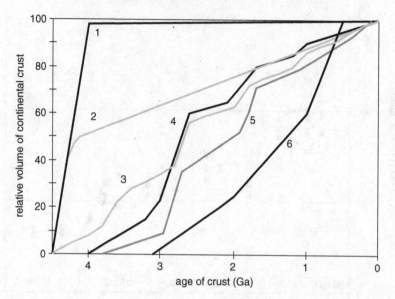

Fig. 6.3. Crustal growth curves through time from the formation of Earth to the present. No-growth models illustrated by curve 1 (Armstrong, 1981); two-stage progressive growth – curve 2 (after Reymer & Schubert, 1984); multi-stage progressive growth – curves 3 (after McCulloch & Bennett, 1994), 4 (after Taylor & McLennan, 1985) and 5 (after Condie, 1990); the linear growth curve for North America of Hurley and Rand (1969) is shown as curve 6.

the continental mass through time. An essential part of this argument is that early-formed crust has been either reworked during Earth history through subsequent geological activity, or that it has been recycled through Earth's mantle at a rate equivalent to that of new continental growth.

A third group of models emphasizes that the crust has grown progressively with time, but in a non-linear fashion (Condie, 1998). These models suggest that the crust has grown in a stepwise manner and that a significant part of the continental crust (at least 50%) was formed early in the history of the planet – before the end of the Archean at 2.5 Ga ago.

These competing models may be evaluated on the basis of a number of critical parameters that include the following.

(1) An estimate of the relative volumes of continental crust of different age presently exposed. Even this apparently straightforward measurement might be a gross over-simplification, for if continents grow to any extent from below (i.e., by accretion through underplating) then age provinces exposed at the surface will be biased in their true age distribution (Corfu, 1987). Equally, any remelting of existing crust will make the observed rock ages younger than the crustal material from which

they formed. This point was recognized even in the early work of Hurley and Rand (1969) although it has only been more recently that geochronological methods have been developed that could more reliably identify the time of crust formation. Over the past two decades, Nd-isotope measurements have permitted the calculation of "crust formation" ages for differing volumes of the continental crust (i.e., the time at which the continental crust was first extracted from the mantle).

(2) The trace element evolution of the mantle. Certain trace elements are strongly fractionated from Earth's mantle into the continental crust, leaving a proxy record of crustal evolution in the mantle. Of particular importance are the elements Nb–Th–U and recent discussions have focused on the way in which crust and mantle Nb/U ratios have changed through time.

(3) The forward modeling of radiogenic isotopes. Isotopic ratios for the main terrestrial reservoirs at the time of their formation combined with their present-day values provide a means of investigating how they have evolved with time. Such modeling has been used to constrain both the volume of the continental crust through time and the amount of crustal recycling that took place (Kramers & Tolstikhin, 1997).

(4) The time at which continental crust might have first formed. This is constrained by: (a) the time of core mantle separation, thought to have taken place between 4.45 and 4.4 Ga (Halliday, 2000; Kramers, 1998); (b) the age of the oldest terrestrial zircons, recovered from quartzites in the Jack Hills in Western Australia and are dated at 4.4 Ga (Wilde *et al.*, 2001), and (c) the age of the oldest terrestrial rocks – the volumetrically small Acasta gneisses of northern Canada, the oldest known components of which were dated by Bowring and Williams (1999) at 4.03 Ga.

6.3.1 No growth models

Freeboard arguments

It has been argued (e.g., Kasting & Holm, 1992) that the continental freeboard, the position of sea level relative to the continental masses, has remained approximately constant (within 1 km of the present level) for at least 2000 Ma. This implies that the continental volume has remained essentially constant, relative to the oceans, throughout the last 2000 Ma. However, Reymer and Schubert (1984) pointed out that the freeboard argument may be used to argue *for, rather than against*, crustal growth because the secular decline in terrestrial heat flow will mean that the volume of the ocean basins will expand with time and that continental growth is necessary to maintain an approximately constant freeboard. A recent discussion by Hynes (2001) showed that if Archean crustal volumes were similar to modern ones, then continental crustal thicknesses in the Archean would have been greater, and that Archean terrains today should be deeply eroded. This is inconsistent with the geological record.

Trace element ratios in basalts

Hofmann *et al.* (1986) and Sylvester *et al.* (1997) showed that the Nb/U ratios of basalts could be used to constrain the timing of continental growth. They used the observation that, during the partial melting of the mantle, Nb and U are partitioned into a basaltic melt such that the Nb/U ratio of the basalt reflects that of its mantle source. Given that primitive mantle has an Nb/U of ~30 and modern ocean island basalts (OIBs) and mid-ocean ridge basalts (MORBs) have a ratio of ~47, the change in Nb/U ratio through time is attributed to the removal of the continental crust (Nb/U ~6) from a mantle source. Sylvester *et al.* (1997) present evidence to show that some 2.7 Ga basalts have Nb/U ratios equivalent to that of modern OIBs and MORBs. One possible interpretation of these data is that by 2.7 Ga, Earth's mantle had experienced depletion equivalent to that of the modern depleted upper mantle. If it may be assumed that depletion of the upper mantle is due to continental crust extraction, then it may be argued that at 2.7 Ga there was an extensive continental crust. This argument was subsequently extended to include basalts as old as 3.5 Ga (Green *et al.*, 2000), suggesting a very early time of formation of the continental crust (Fig. 6.3).

However, there are problems with this argument. The first is that in an oxidizing environment U is mobile and that the relative mobility of U will have changed with time as Earth's atmosphere has become progressively more oxidizing (Collerson & Kamber, 1999). Second, as a consequence of its increased mobility in an oxidizing atmosphere and hydrosphere, U will be more readily recycled into the mantle (Staudigel *et al.*, 1995). The final difficulty arises from recent studies that show that the two-reservoir model, assumed by Sylvester *et al.* (1997), is inadequate for explaining some trace element ratios in Earth's crust and depleted mantle. In particular, trace element ratios involving Nb face this difficulty and require a third silicate reservoir (Barth *et al.*, 2000). This problem is discussed more fully below.

Crustal recycling

An argument used to support zero net growth of the continents with time depends upon a comparison of the rates of crust generation with those for the rate of crust destruction. This is a very inexact science, but at least one estimate of the flux of crustal material returned to the mantle (von Huene & Scholl, 1991) is similar to the crustal growth rate of Reymer and Schubert (1984), implying that the two are equally balanced. However, it should be recognized that time-averaged rates of sediment subduction are very difficult to obtain from the single "snapshot" available at the present day.

We would also expect to find evidence for crustal recycling in the geo-chemistry of the mantle. In particular, we would expect to find a "crustal" trace element signature in the source for OIB basalts, and yet the available evidence suggests that it is not there. The similarity of Pb/Ce and Nb/U ratios in OIB and basalt (White, 1989) and the oxygen and Nd isotope signature of OIBs (Woodhead *et al.*, 1993) imply that there is only a small (maximum 9%) crustal contribution to the OIB source. McLennan *et al.* (1990) are of the view that in the modern period, crustal recycling is pre-ferentially constrained to arc environments where newly formed crust is recycled into subduction zones.

Isotopic evidence

There is no record in the history of our planet of large volumes of early felsic crust. Indeed, Tarney and Jones (1994) argue that it is difficult to generate such a large volume of crust in such a short time. Early Archean terrains only rarely preserve very old zircons (Nutman, 2001), strongly suggesting that very early-formed felsic crust was not abundant during the early Archean. This was also demonstrated by Stevenson and Patchett (1990) using the Hf-isotope system-atics of zircon. These authors used Hf-isotope model ages from zircons in clastic sediments as a proxy for the volume of the continental crust at the time of sediment deposition. They argued that if there were large volumes of felsic crust in existence, then clastic sediments would record Hf-isotope model ages that would be older than the depositional age of the sediments. If, on the other hand, there was little felsic crust, then the Hf model ages would be similar to the age of sediment deposition. In a study of clastic sediments from Archean and Proterozoic basins, they showed that in most late Archean sedimentary sequences (2.6–3.0 Ga) zircon Hf model ages are within 10% of the depositional age, implying a small volume of felsic crust. Proterozoic sediments in contrast show a 400 Ma disparity between the zircon Hf model age and their depositional age, signifying a large volume of felsic crust. These results imply that a sub-stantial volume of felsic crust was created in the time interval 3.0–2.5 Ga.

Recently, Kramers and Tolstikhin (1997) explored the U–Th–Pb isotopic constraints on the growth of the continental crust through time, in the context of a U–Th–Pb transport model for the entire crust–mantle–core system. In a similar study, Nagler and Kramers (1998) explored an equivalent Nd model. An important outcome of these calculations for an integrated crust–mantle–core system is that no more than 10% of the present-day crustal mass could have existed at 4.4 Ga.

Summary

Taken together, these different lines of reasoning provide a significant argument against "No-Growth models". However, they do allow that some of Earth's continental crust might be extremely old.

6.3.2 *Progressive growth models*

Progressive (step-wise) growth models have now replaced the linear growth model of Hurley and Rand (1969) as more and better geochronological data have become available. The geochronological data collected in the past three decades show two important features: first that the continents are older than previously recognized, and second that growth was probably episodic. Two types of progressive growth model have been proposed. There are models that require growth in two stages, and those that require many stages (Fig. 6.3). An example of the two-stage model is that of Reymer and Schubert (1984) who estimated, from modern island arc volcanic rocks, a net crustal growth rate of about $1 \, km^3 \, yr^{-1}$. They argued that this is insufficient to produce the present mass of the continents (which they estimated to be $7.6 \times 10^9 \, km^3$) and therefore the growth rate must have been greater in the past. They proposed an early growth history of the continents, with more than 50% of crust formed before 4.0 Ga. This model has many features in common with the No-Growth models outlined in the section above and therefore faces similar difficulties. In addition it is possible that the estimate of crustal growth rates may be incorrect; Holbrook *et al.* (1999), for example, recently showed that growth rates in some geologically recent arcs are twice that calculated by Reymer and Schubert (1984).

Many authors now favor a progressive growth model for the continental crust in which there were several episodes of crustal growth. In these models the late Archean and the Proterozoic appear to represent important intervals of rapid crustal growth (Tarney & Jones, 1994; Rudnick, 1995). Typical of these models is that of Taylor and McLennan (1995), who calculated a step-wise progressive growth curve for the continental crust based on isotopic age distribution patterns in crustal rocks. They proposed that about 20% of the continental crust had formed by 3000 Ma, but that 60–70% had formed by 2500 Ma, implying a period of rapid crustal growth in the late Archean (Fig. 6.3, curve 4). A similar curve for the growth of Laurentia (the North American continent with Greenland and Scandinavia) is given by Condie (1990), although this latter curve implies a less rapid episode of crustal growth during the late Archean (Fig. 6.3, curve 5). McCulloch and Bennett (1994) calculated a stepwise crustal growth curve based upon crust formation ages determined using Nd geochronology and U–Pb zircon geochronology for the present-day distribution of the continental crust in the Australian, North American, and Scandinavian cratons. This model assumes no mantle recycling and shows episodic crustal growth, with periods of rapid crust-formation at 3.5–3.6 Ga, 2.5–2.7 Ga, and 1.8–2.0 Ga (Fig. 6.3, curve 3). A unique feature of this model is that it was the first time that the episodic growth of the continental crust had been calculated to mirror simultaneously the growth in volume of Earth's depleted mantle reservoir.

The principal argument in favor of a progressive growth model is the age distribution of crustal rocks, as deduced from isotopic measurements. The Nd isotope system is particularly useful in this respect (Nelson & DePaulo, 1985; Patchett & Arndt, 1986), although the cautions of Arndt and Goldstein (1987) and Chavagnac (1999) must be borne in mind when using these methods. In a particularly elegant study, Patchett and Arndt (1986) calculated the proportion of newly formed crust for a large province of Proterozoic rocks (1.9–1.7 Ga old), from Europe, Greenland, and North America (Fig. 6.4). Their method uses Nd isotope measurements coupled with precise age determinations of the rocks studied (usually through U–Pb in Zr). The age determination is used to estimate the crust-formation age, and the Nd isotopes are used to estimate the extent of crustal recycling. Calculated ε_{Nd} values are plotted on a time–ε_{Nd} diagram that shows the evolution curves for depleted mantle and for older (Archean) continental crust. Most data plot between the two curves, suggesting that they are the product of mixing. Calculations based upon estimates of Nd concentrations in mantle melts and crustal melts are used to contour the diagram for percent mixing and thereby infer the extent to which 1.7–1.9 Ga crust is contaminated with older Archean crust. The results of this study showed that 34% of the crust in the North American–European continent formed in the time interval 1.7–1.9 Ga, implying a growth rate of 1.25 km^3 yr^{-1}.

Neodymium isotope studies on sediments (see, for example, Miller *et al.*, 1986) have also been used to estimate an average age for the continental crust, thereby further constraining the rate and timing of continental growth. However, there are problems with this method. McCulloch and Bennett (1994) showed that measured ε_{Nd} values for the continental crust do not match those predicted by their model for upper mantle evolution through time. They found that sediments have a less negative ε_{Nd} than that expected for the continental crust at any particular moment. Thus, integrated age measurements obtained from sediments only provide minimum estimates of the crustal growth rate. The likely explanation is that a significant volume of

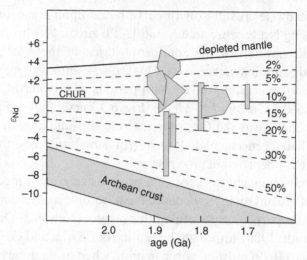

Fig. 6.4. A plot of ε_{Nd} versus time for samples of Proterozoic crust from Europe, Greenland, and North America (grey boxes) showing how the samples have compositions between that of the depleted mantle and Archean crust. The percent curves indicate the proportion of Archean crust (data from Patchett & Arndt, 1986).

sediment is recycled within the crust, although determining a reliable "recycling factor" is also fraught with difficulty, for this is dependent upon tectonic setting. Typically, sediment ages are biased towards younger sediments, since often the major contribution to the sedimentary mass is from orogenic belts that tend to be dominated by younger sediments.

The transport balance model of Kramers and Tolstikhin (1997) for the U–Th–Pb isotope system is particularly sensitive to the extent to which crustal material may be recycled into the mantle and offers some constraints on the mechanism of crustal recycling. Kramers and Tolstikhin (1997) concluded that the rate of recycling of crustal material back into Earth's mantle has increased with time, particularly since 2.0 Ga. The results of their model also indicate that recycling through erosion is the dominant process, rather than the process of lower crustal delamination – a process that will be discussed more fully below.

Progressive, stepwise growth curves for the continental crust tend to have two main features. First, they show that a major portion of the crust formed during a relatively short span of time. For example, the Taylor and McLennan (1985) model indicates that at least 45% of the crust was made in only 13% of Earth history (Fig. 6.3, curve 4). Second, whereas very little of Earth's crust had formed by 3.5 Ga, a major part had formed by the end of the Archean at 2.5 Ga, suggesting that a significant percentage by volume of the continental crust formed in the mid- to late-Archean. Estimates vary from 50–60% (McCulloch & Bennett, 1994) to 60–70% (Taylor & McLennan, 1995), although this is a much greater volume of Archean crust than is evident from its present areal extent of 14% (Goodwin, 1991). It is this episode of rapid, early crustal growth that is particularly the focus of the later part of this chapter.

One of the significant findings of the studies by Patchett and Arndt (1986) and Nelson and DePaulo (1985) is that there was a second interval of rapid crustal growth during the Proterozoic. It appears that between 20–30% of the continental crust in Europe and North America formed at this time. The likely mechanism of crust formation is discussed in some detail by Patchett and Arndt (1986), who argue that the available evidence strongly supports an island-arc accretion model for crustal growth in the Proterozoic.

Why crustal growth might be episodic is the subject of some debate. As a first-order observation it is generally agreed that the rate of continent formation has decreased since the Archean and that the rate of crustal recycling has increased (Kramers & Tolstikhin, 1997). However, the evidence for episodicity needs to be looked at with care, for many crustal growth curves are regional rather than global, and the time resolution of Nd-model ages is broad – commonly more than 100 Ma. Further, it may be argued that the crust preserved is a product of its preservation potential rather than a product of the processes of crust formation (Morgan, 1985) and that the episodicity is apparent rather than real.

6.4 Continental crust–mantle reservoirs and fluxes

It is generally agreed that the ultimate source of Earth's continental crust is from the mantle. Thus, the growth history of the continental crust through time is also the history of its separation from the upper mantle. This section briefly reviews current ideas on how the crust and mantle reservoirs may be linked and how they might have changed through time. It then considers the evidence for present-day fluxes from upper mantle to continental crust.

There is good trace-element and isotopic evidence to support the view that the continental crust and the depleted upper mantle are complementary geochemical reservoirs relative to the composition of Earth's primitive mantle. McCulloch and Bennett (1994) show that there are a number of ways in which this relationship may be modeled. The simplest approach is the three-reservoir model, in which an initial primitive mantle composition is progressively differentiated through time into a depleted mantle reservoir and the continental crust. This view is supported by evidence from many isotopic systems and may

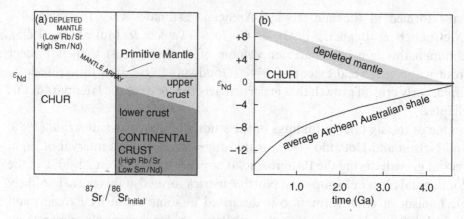

Fig. 6.5. (a) ε_{Nd} versus initial Sr plot showing the complementary compositions of the depleted mantle and the continental crust relative to that of the primitive mantle; (b) ε_{Nd} versus time diagram for the depleted mantle (a number of different evolution curves are combined within the depleted mantle array) and average Archean Australian shale – a proxy for the composition of the continental crust, showing the complementary evolution of each with time, relative to the chondritic reservoir (CHUR) (data from Rollinson, 1993).

be seen clearly on bivariate Nd–Sr isotope plots (Fig. 6.5a) and on an Nd isotope versus time diagram (Fig. 6.5b). A more problematic variant of this theme is the static three-reservoir model in which the differentiation takes place very early in Earth history (the no-growth model for the continental crust). In this case, the reservoirs are maintained in volume by matching new continental growth with the recycling of an equivalent mass into the mantle.

It should be stressed that the differentiation of the primitive mantle into crust and depleted mantle may not be a single-stage process. Although O'Nions and McKenzie (1988) showed that Earth's continental crust is the geochemical equivalent to a 1% melt of the upper mantle, McKenzie (1989) argued that it is not possible to extract this melt, and that such a process is inconsistent with the findings of experimental petrology (O'Hara, 1968). The extraction of the continental crust from the mantle therefore has to be a multi-stage process.

Whereas three-reservoir models work well for some isotopic systems, mass balance arguments based on trace elements indicate that additional reservoirs must also be present. Galer and Goldstein (1988, 1991) proposed a four-reservoir model in order to satisfy the evidence for a very early (4.0–3.7 Ga) and extensive depletion of the upper mantle (Bennett *et al.*, 1993; Bowring & Housh, 1995; Collerson *et al.*, 1991), without having to produce abundant, very old felsic crust, for which there is little evidence in the geological record. In models of this type, it is argued that basaltic, ultra-mafic, or alkalic crust formed very early in the history of the planet and was subsequently stored in the mantle for in excess of 1.0 Ga. It should be noted that this model has been challenged recently by Moorbath *et al.* (1997), who have argued that the isotopic evidence for a very old depleted mantle and extensive early continental crust is unreliable, and is the product of later, open-system behavior and the resetting of the Sm–Nd system in the samples studied. Such a model is also inconsistent with the higher temperatures inferred for the early Archean mantle and the consequent rapid homogenization of the mantle. Nor is there support for such a view either from the geochemistry of early Archean basalts that might be expected to preserve the evidence of such an event in their source, or from the felsic melts of the tonalite–trondhjemite–granodiorite (TTG) suite, derived from such basalts, which might also be expected to preserve a geochemical signal of such an event.

A different four-reservoir model was proposed by McDonough (1991) who argued that, whereas the depleted mantle and the continental crust are complementary for many trace elements, this is not the case for the elemental ratios Nb/La and Ti/Zr. Both these elemental ratios are sub-chondritic and therefore cannot be regarded as complementary. A further reservoir is required with supra-chondritic Nb/La and Ti/Zr ratios. McDonough (1991) suggested that this enriched reservoir was eclogite, the refractory residue of a sinking slab, which became isolated in the deep mantle. A more recent study (Rudnick *et al.*, 2000) shows that eclogite xenoliths from kimberlites from regions of Archean continental crust have supra-chondritic Nb/La, Nb/Ta, and Ti/Zr ratios and that the Nb and Ta are located in rutile – an accessory phase in the eclogite (Fig. 6.6). Mass balance calculations suggest that this eclogitic reservoir is between 0.5–6.0% of the mass of the silicate Earth and that the eclogite is located deep in the mantle.

A potentially important reservoir, and one that is not frequently discussed in the context of continent formation, is the sub-continental lithospheric mantle. The sub-continental lithosphere is made up of Mg-rich peridotite, depleted in the elements Fe, Ca, and Al, which has equilibrated at relatively low temperatures (850–1100 °C) and appears to form a thick mantle "keel" to the continental crust. These relationships first became apparent from geochronological and thermobarometric studies of silicate inclusions from diamonds from beneath the >3.5 Ga Kaapvaal craton in South Africa (Richardson *et al.*, 1985) and is now well established from the study of Os isotopes in mantle xenoliths (Pearson, 1999; Nagler *et al.*, 1997). In detail, the mass of the sub-continental lithosphere is ill-constrained and is at most about

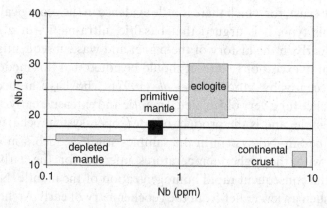

Fig. 6.6. Neodymium versus Nb/Ta plot showing the compositions of the primitive mantle, the depleted mantle and the continental crust. The relationships illustrate that with respect to Nb–Nb/Ta, the continental crust and depleted mantle cannot be complementary reservoirs derived from the primitive mantle and require an additional reservoir with a higher Nb/Ta than that of the primitive mantle. It is likely that rutile-bearing eclogites represent this missing reservoir (after Rudnick *et al.*, 2000).

2% of the mass of the mantle (McDonough, 1991), it is compositionally heterogeneous, typically has a multistage history, and is insufficiently enriched in any important major or trace element to shift the mass balance in global models. However, within the context of a discussion of the origin of the continents, the sub-continental lithosphere is important, for there is now a well-established link between the timing of crust formation and that of the underlying sub-continental lithosphere. In addition, there is also a close relationship between the thickness of the sub-continental lithosphere and the age of the overlying continental crust, such that Archean cratons are underlain by thick, old, (cold) sub-continental lithosphere, whereas younger continental crust is underlain by a thinner lithospheric keel (Sun, 1989; Menzies, 1989).

   The significance of the close relationship between the age of the early Archean continental crust and its underlying lithospheric mantle is currently the subject of much discussion. It is likely that the Archean sub-continental lithospheric mantle stabilizes and preserves Archean continental crust. In addition, as proposed by Kramers (1987, 1988), there may be a genetic link between the two, although recent Os isotopic investigations suggest that such a complementary relationship is not possible for any reasonable sizes of the crustal and sub-continental lithosphere reservoirs (Nagler *et al.*, 1997). Hawkesworth *et al.* (1990) argued that the composition of the Archean sub-continental lithosphere is best explained by the extraction of a komatiitic melt, triggered maybe by the interaction with a mantle plume, although this model requires very large volumes of komatiite production. Pearson (1999) has shown recently that the lack of a simple age–depth structure in the subcontinental lithosphere beneath the Archean Kaapvaal craton implies relatively rapid growth, suggesting either rapid arc accretion or plume-related differentiation.

6.4.1 Geologically recent crustal growth and mantle fluxes

Geologically recent crustal growth is thought to take place principally at destructive plate boundaries (Chapter 5). These zones of plate convergence are regarded as the main tectonic setting for the return of volatiles (especially water) to Earth's mantle. Crustal growth takes place as a consequence of mantle melting triggered by the introduction of fluids released by the dehydration of minerals in the subducting slab. For many years it was believed that the primary melt products of mantle melting during plate convergence were andesites – hence the "andesite crustal growth model" of Taylor and McLennan (1981), described above (see also Chapter 5). However, it is now known that the primary melt is basaltic and that the frequently observed andesite is a derivative of basalt through fractionation in sub-surface magma chambers (Arculus, 1994; Holbrook *et al.*, 1999). Basaltic melts formed in this way are enriched in a number of trace elements carried by the fluids from the subducted slab. In addition, some subduction zone magmas are also thought to contain a chemical component derived from the melting of sediments, introduced into the mantle during the subduction of the oceanic slab. However, this is a minor component compared with the mantle contribution and is restricted to a few percent of the total mass of new crust. Crustal growth through island arc formation is thought to take place through the eventual accretion of island arcs to each other and to older continental masses, through collision or following the reorientation of plate motions. Reymer and Schubert (1984), in a study of seventeen different island arcs, estimated a worldwide crustal addition rate of $1.65 \, \text{km}^3 \, \text{yr}^{-1}$. For a fuller discussion of this subject see Chapter 5.

Crustal growth associated with plate convergence also takes place at the margins of continents. At the present time, this may be seen along convergent continental margins such as the Andes and the western Cordillera of the USA. These are regions where the continental crust is particularly thick, in places up to 70 km, from the addition of new material to the crust through andesitic volcanism and magmas of equivalent composition emplaced deeper in the crust as plutons. However, in these regions the relative volume of new crust is more difficult to estimate because of the inevitable mixing between old and new crust (Hildreth & Moorbath, 1988).

Although the dominant site of contemporary crustal growth is at convergent plate boundaries, there are two other tectonic settings in which crustal growth also takes place. The rapid eruption of very large volumes of basaltic magma within continental areas to form continental flood basalts of the type found in the Deccan Plateau in India or the Karoo Province of southern Africa, not only add huge volumes of basaltic lava at Earth's surface but frequently underplate large volumes of magmatic rocks to the base of the continental crust (Voshage *et al.*, 1990). A variant of this model is where voluminous magmatism takes place in ocean basins leading to the formation of oceanic plateaus. These are areas of thickened oceanic crust, where the crust may be as much as 40 km thick. Oceanic plateaus are not easily subducted and it has been proposed that their partial disruption and melting and accretion to existing continental crust may be a further form of crustal growth, and one which was particularly important early in Earth history (Saunders *et al.*, 1996; White *et al.*, 1999). Recently, this model of continental growth has had a number of advocates (Abouchami *et al.*, 1990; Stein & Hofmann, 1994; Stein & Goldstein, 1996), although Barth *et al.* (2000) estimate that this mode of crustal growth accounts for, at most, between 5 and 20% of the total crustal mass.

A further site of crustal growth has recently been recognized from seismic reflection studies of volcanic rifted continental margins of the type found adjacent to the North Atlantic. Here, large volumes of basalt have accreted to the continental margin during continental break-up (Holbrook, 1998). Their relative contribution to crustal growth is, as yet, undetermined.

In each of the three tectonic settings discussed above – convergent margins, within-plate (oceanic and continental crust), and at rifted margins – the dominant flux from the mantle to the continental crust is basaltic. This poses a major problem for the generation of the continental crust, which is inter-mediate and felsic in composition (broadly andesitic), from the mantle, for the main flux from the mantle to the continental crust in the geologically recent past appears to have been basaltic.

The disparity between the calculated average intermediate composition of the continental crust and the observed dominantly basaltic flux from Earth's mantle (see Fig. 6.1) cannot immediately be resolved. One possible explanation lies in a better understanding of the composition of the lower continental crust. If, as some authors have suggested on geophysical, xenolith, and field-based evidence, the lower continental crust is basaltic in composition (e.g., Rudnick & Taylor, 1987), this could weight the calculated crustal composition more closely towards the composition predicted from the study of modern mantle fluxes. If this is the case, why is it that a basaltic lower crust is not more commonly observed?

One possible explanation is that basaltic crust under lower crustal conditions converts to eclogite and undergoes a density inversion such that it detaches from the base of the continental crust and sinks into the mantle (England & Houseman, 1989). In this way, by its very nature, a basaltic lower crust is never seen. Indirect evidence for lower crustal delamination comes from an examination of the density structure and thickness of crustal sections. Gao *et al.* (1998) argued that in eastern China the crust is thinner (34 km), has a lower P wave velocity (by from 0.2 to 0.45 km s^{-1}) and a more evolved average composition than the global average of Rudnick and Fountain (1995) and suggest that a part of the lower crust is missing due to lower crustal delamination. They show from mass balance modeling, that the eclogite in the Dabie–Sulu ultra-high pressure metamorphic belt of China is the likely delaminated material. Similarly, Ducea and Saleeby (1998) proposed that the thin granite crust (30–40 km) of the Sierra Nevada batholith, underlain by peridotitic upper mantle, could be explained by the delamination of a thick eclogitic root. However, Rudnick (1995) has argued that, if delamination is an important process in crustal evolution, then it is most likely to operate in regions of thickened crust. If this observation is true, then delamination is probably not a major process in the mechanism of crust formation but rather relates to processes that significantly post-date its formation. This is consistent with the findings of Kramers and Tolstikhin (1997) who showed, from their Pb-isotope forward transport model, that lower crustal delamination is a minor process. If therefore, lower crustal delamination is not an important process for modifying crustal compositions, then the disparity between the measured crustal compositions and those predicted from the present-day flux from the upper mantle remains, and an alternative mechanism has to be found.

6.5 The composition of Earth's early continental crust

In reviewing the evidence for the average bulk composition of the continental crust and in assessing its relative growth through time, two central features emerge that are crucial to an understanding of the origin of the continental crust. First, it is apparent that the continental crust is intermediate and felsic in composition and is different from the basaltic composition predicted from modern mantle fluxes. Second, there is evidence to suggest that Earth's continental crust grew at a greater rate in the past than it does at present and that a significant portion of Earth's continental crust formed early in Earth history, between 3.5 and 2.5 Ga. Taken together, these two observations strongly suggest that the mechanism of continent formation may have changed with time and that it was different early in the history of Earth, when a significant portion of the crust was formed.

An important question that has not yet been explored in this chapter is to enquire whether rocks that formed during the Archean are any different from rocks forming today. The answer is not difficult to find, for rocks of Archean age are found on all the major continents and in many cases form extensive continental shields hundreds of kilometers across. These regions are dominated by two contrasting lithological associations. These are the Archean greenstone belts, linear belts of folded mafic volcanic rocks, and clastic and chemical sediments (e.g., Rollinson, 1999), and felsic magmas, predominantly of the TTG suite. Archean greenstone belts preserve rocks formed in a diverse range of former tectonic settings including continental rifts, island arcs, and oceanic plateaus. Their origin and evolution has recently been well reviewed in a comprehensive volume by DeWit and Ashwal (1997). However, it is the magmas of the TTG suite that dominate areas of ancient continental crust (Rollinson & Windley 1980; Rollinson & Fowler, 1987; Rollinson, 1996; Rollinson & Blenkinsop, 1995; Berger & Rollinson, 1997).

Magmas of the TTG type form today, although they are not common. In contrast, in the Archean, they were *the* dominant felsic magmatic suite. Furthermore, compositionally they are similar throughout the Archean, such that the magmatic protolith to the 3.65 Ga Amitsoq gneisses of west Greenland is very similar in composition to TTG magmas formed in the late Archean (Kamber & Moorbath, 1998). In some respects, TTG magmas are similar to the tonalites found at modern active continental margins, but in many ways they are quite different. Modern (post-Archean) granitoids are typically granodiorite to granitic in composition with K_2O/Na_2O ratios close to 1.0, whereas Archean TTG granitoids are more sodic with K_2O/Na_2O ratios lower than 0.5. There are also important trace element differences (Martin, 1986; Ellam & Hawkesworth, 1988). These are discussed more fully below.

In order to look more closely at the possibility of different mechanisms of crust production operating at different times in Earth history it is instructive to examine the way in which continental crustal compositions might have changed with time. The focus here is upon specific groups of trace elements and illustrated by three important studies.

First, there is good evidence to show that the trace element composition of some types of sedimentary rocks have changed over time. For example, using fine-grained sediments as a proxy for the composition of the upper continental crust, Taylor and McLennan (1985) showed that Archean fine-grained

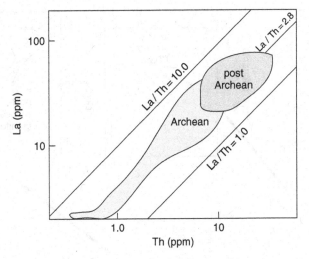

Fig. 6.7. La–Th concentrations in Archean and post-Archean shales (after Taylor & McLennan, 1985).

sedimentary rocks have compositions that are different from their post-Archean equivalents (see Chapter 4). In particular, Archean shales have less steep REE patterns, with no Eu anomaly, a lower Th/Sc ratio and lower concentrations of U, La, and Th (Fig. 6.7) than is found in post-Archean shales. Similarly, Veizer (1983) showed that the Sr isotopic composition of carbonate rocks increases significantly at the end of the Archean. Strontium isotopes in carbonates are thought to reflect the composition of the sea water from which they formed and thus are a proxy for the change in the composition of the continental mass being eroded. Thus, a change in Sr isotope composition of the sea water with time reflects a change in the bulk composition of the continents with time. These interpretations of sedimentary geochemistry were challenged by Gibbs *et al.* (1986) who argued that the observed differences were the products of differences in tectonic setting rather than that of a secular evolutionary trend.

A second and significant study of the temporal difference in crustal compositions was made by Martin (1986). Martin plotted the REE composition of granitoids formed throughout Earth history and showed that they fall into two distinct groups according to their age (Fig. 6.8). On a plot of $(La/Yb)_n$ versus Yb_n (the chondrite normalized ratio of light to heavy members of the REE group plotted against the normalized concentration of the HREE Yb) Martin showed that granitoids that are older than 2.5 Ga tend to plot with high $(La/Yb)_n$ ratios and low Yb_n concentrations, whereas post-Archean granitoids plot with lower $(La/Yb)_n$ ratios and higher Yb_n concentrations.

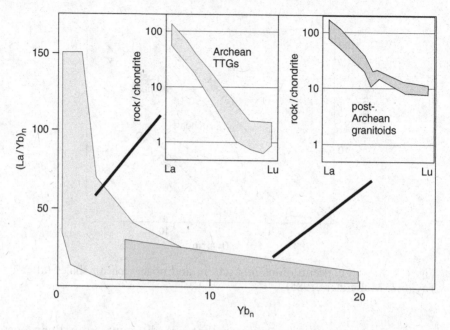

Fig. 6.8. Chondrite normalized plot of La/Yb versus Yb showing the field of Archean TTGs (high $(La/Yb)_n$, low Yb_n) and the field for post-Archean granitoids (low $(La/Yb)_n$, high Yb_n). The inset diagrams show typical REE patterns for Archean TTGs and post-Archean granitoids (data from Martin, 1994).

Martin drew two important conclusions from his observation: (a) that Archean granitoids are of a different type from post-Archean granites and belong to the TTG magma suite, and (b) the reason there is a geochemical difference between the two types of granitoid is that the mechanism of granitoid genesis was different. In the case of Archean granitoids, the low Yb_n concentrations and high $(La/Yb)_n$ ratios indicate a significant propor-tion of present garnet during melt generation. This is not the case during the formation of most post-Archean granitoids. The abundance of garnet during granite melt generation has important implications for the melting process, for it sets limits on the composition of the material that melted and on the depth of melting. Martin (1986) used these observations to make a number of important points about the way in which the subduction process may have changed with time. These ideas are discussed more fully in a later section of this chapter.

A third study, by Ellam and Hawkesworth (1988), identified further evi-dence for a different mechanism of crust generation early in Earth history. Ellam and Hawkesworth (1988) examined the trace element pair Rb–Sr in

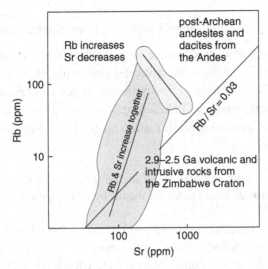

Fig. 6.9. Rubidium–Strontium concentrations in: (a) Archean felsic igneous rocks from Zimbabwe, and (b) post-Archean felsic igneous rocks from the central Andes (after Ellam & Hawkesworth, 1988).

ancient and modern crustal rocks in order to resolve a paradox in crustal evolution. They showed that, if the dominant flux from mantle to crust at sites of modern crust generation in island arcs is basaltic, then the Rb/Sr ratio of that flux would be low. This conflicts with the observed Rb/Sr ratio of the crust, which is high. Exploring in some detail the way in which Rb/Sr ratios may change during magma genesis, Ellam and Hawkesworth (1988) found that there is an important difference between Archean magmatic rocks and those formed more recently in a destructive margin setting. In Archean volcanic and plutonic rocks from Zimbabwe, the elements Rb and Sr show a positive correlation, whereas in andesites and dacites from the Andes, Rb and Sr show a negative correlation (Fig. 6.9). The difference is thought to reflect the role of plagioclase in the melting process, for the mineral plagioclase is the principal control on the behavior of the element Sr in crustal magmatic systems. Ellam and Hawkesworth (1988) proposed that in the Archean, crustal rocks were produced through the deep melting of a mafic source in which plagioclase was absent, either because it had entered the melt or because melting was at a higher pressure and it was replaced by garnet. In contrast, geologically recent crustal rocks as illustrated by the Andean example are produced in the presence of plagioclase, either through shallow melting or through intra-crustal fractionation of the melt. The importance of garnet, inferred here from Rb/Sr ratios, strongly supports the findings of Martin (1986) discussed above, based on La/Yb ratios.

6.6 The formation of Earth's early continental crust

The three examples discussed above demonstrate that Archean continental crust is subtly different in composition from crust that has formed since Archean times. This has important implications for the *process* of crust generation early in Earth history. Taken together with observations made earlier in this paper about crustal growth rates, this leads to three key propositions. They are as follows.

(1) The *mechanism* whereby Earth's continental crust formed during the Archean (>2.5 Ga ago) was different from the means by which continental crust is formed today.
(2) The *rate* at which Earth's continental crust formed during the Archean was greater than the rate at which continental crust forms today.
(3) Because Archean crust is dominated by TTG compositions, understanding the process of continental crust formation in the early part of Earth history requires an understanding of *the process of TTG genesis*.

Whereas the importance of the TTG magma suite for the process of Archean crust generation has already been outlined, here the emphasis is on their precise geochemical characteristics as a basis for a more detailed discussion of their origin. Archean TTGs are preserved, albeit as gneisses, in all the Archean provinces of the world and are found in all the continents (Martin, 1994). They are defined geochemically by their CIPW normative feldspar content, as plotted on an O'Connor diagram (see Rollinson, 1993).

Average compositions plot close to the tonalite–trondhjemite–granodiorite triple point on the classification diagram (Fig. 6.10a). Most compositions are silica and alumina-rich, have low $FeO(total) + MgO + MnO + TiO_2$, and have low K_2O/Na_2O ratios (average 0.36), (see Fig. 6.1). Most Archean trondhjemites are of the high-Al_2O_3 group of trondhjemites as classified by Barker and Arth (1976), and have $Al_2O_3 > 15\%$ at $SiO_2 = 70\%$. The trace element character of Archean TTGs is Sr-rich (average 450 ppm), low Rb/Sr (average 0.12), low to moderate K/Rb (mostly <550), a marked negative Nb–Ta anomaly (relative to La) and a negative Ti anomaly (relative to Sm). Nickel, Cr and V concentrations are all low (Ni average 14 ppm, Cr 29 ppm and V 35 ppm), and the HREEs are also always low ($Yb_n \sim 2.6$) (Table 6.1.) The average TTG REE pattern is strongly depleted in HREE and shows a strongly concave-up pattern at the HREE end (Fig. 6.8) (Martin, 1994).

In a very general way, the process of TTG genesis is now understood (Defant & Drummond, 1990; Martin, 1994). The constraints of isotope geochemistry carried out in the 1970s and 1980s showed that TTG melts were

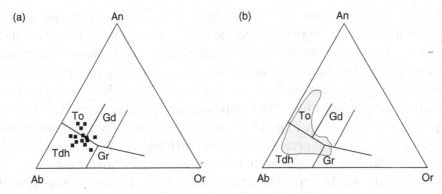

Fig. 6.10. O'Connor Ab-An-Or diagram showing the fields of tonalites (To), trondhjemites (Tdh), granodiorites (Gd) and granites (Gr): (a) typical melt compositions produced during the dehydration melting of amphibolites, and (b) typical TTG compositions (after Martin, 1994).

derived from Earth's mantle, either in a single process, or in several stages. The trace element studies of Martin (1986) and Ellam and Hawkesworth (1988) showed that the likely source was basaltic. Thus, by the mid-1980s, a working model had been accepted whereby the partial melting of a basaltic source was thought to be the principal means whereby TTG melts are produced. The presence of garnet, signified by low HREE concentrations, and absence of plagioclase indicated by high Sr-concentrations, constrained the depth of partial melting to the garnet stability field (>1.0 GPa). A need for water to lower the melting temperature of the parent basalt highlighted the importance of hornblende as an essential mineral in the source. Evidence for the presence of hornblende also comes from the shape of TTG REE patterns, for the concave-up form of the heavy REE strongly indicates the presence of hornblende in the melting residue.

In addition to the partial melting model, a number of other mechanisms have been proposed for the formation of TTGs from a basaltic parent. These include the crystal fractionation of basaltic melts and the partial melting of crustal rocks (see the reviews of Drummond & Defant, 1990; Martin, 1994). Whereas examples of these other processes are well documented in the literature, it is argued here that the partial melting of a basaltic source is likely to be the most efficient process for producing the large volumes of magma required for generating new continental crust.

It is important to note that the model proposed for continental crust formation in the Archean is significantly different from that proposed for modern crust generation (e.g., Gill, 1981). On modern Earth, basaltic oceanic lithosphere is subducted, water bound in hydrous minerals is released, and partial melting is initiated in the overlying mantle. Thus, the principal source of modern island arc magmas lies in the mantle overlying a subducting slab and the melts that are produced are basaltic. In contrast, the current model suggests that in the Archean, basaltic oceanic lithosphere is subducted and undergoes partial melting. In this case, the source of the melt is basaltic and the melts that are produced have the composition of TTG magmas. This fundamental difference in mode of magma production is normally explained by the higher thermal gradients operating early in Earth history. These require that a subducting slab will melt, rather than dehydrate during subduction.

Thus far, the model for TTG genesis and, hence, for Archean continent formation, has been sketched only in outline. Now six lines of evidence are presented that support the basalt-melting model proposed above. These expand upon and develop the arguments presented above.

6.6.1 *The field evidence*

About 14% of the rocks exposed in the continents are Archean (>2.5 Ga) in age. There are two main compositional types: basaltic and felsic. The felsic rocks are commonly deformed into gneisses but compositionally are TTGs, and careful fieldwork reveals that they are magmatic in origin. Basaltic rocks are dominant in Archean greenstone belts, but there is also a separate group of Archean basalts that are found as a minor component of the TTG gneiss regions. These basalts, which are generally now amphibolite or granulite, are commonly cut by veins of TTG magmas and form fragments of amphibolite or mafic granulite from a few meters to several kilometers in length enclosed within the TTG suite rocks (Rollinson, 1987; Rollinson & Lowry, 1992), indicating that they are slightly older than the TTGs. Some of the amphibolites and mafic granulites described from Archean TTG terrains are associated with metamorphosed sediments, such as iron formation, suggesting by association that the mafic rocks were originally formed in a near-surface environment, possibly as volcanic rocks. The close association between older basalt and younger TTGs as an integral part of Archean TTG gneiss complexes world wide, is supportive of a genetic model whereby the basalt is parental to the TTG.

6.6.2 *Experimental evidence*

The critical evidence for the origin of TTG magmas by the partial melting of a basaltic parent comes from an experimental test of the hypothesis. There is now a large body of experimental evidence (Rushmer, 1991; Rapp *et al.*, 1991; Wyllie & Wolf, 1993; Wolf & Wyllie, 1994; Winther & Newton, 1991;

Sen & Dunn, 1994; Rapp & Watson, 1995) in which basalts of various types have been melted under a range of experimental conditions. The chemical analysis of the resulting melts has confirmed the broad hypothesis (Fig. 6.10b), but has also set some constraints on the melting process. The intricacies of these studies have recently been reviewed by Wyllie and Wolf (1993), Wyllie *et al.* (1997) and Rapp (1997).

In brief, it is now known that there are a number of factors that are critical to producing TTG melts from a basaltic starting composition. These include the degree of melting of the basaltic parent, which is a non-linear function of the temperature and pressure of melting and strongly influences the mineralogy of the unmelted residue, and the composition of the starting materials and the water content. In individual melting experiments it may be difficult to unravel the separate effects of these different variables, for over some temperature intervals there may be a rapid increase in melt production over a small temperature range.

Of foremost importance is the presence of water. Water is necessary to lower the solidus of the basalt into the temperature range where melting may take place. Dry melting requires unreasonably high melting temperatures. For example, the dry melting of mid-ocean ridge basalt commences at 1425 °C at 3 GPa (Yasuda *et al.*, 1994). Since the presence of free water is unrealistic, it is necessary that water bound in hydrate minerals fluxes partial melting. This is the process of dehydration melting, and amphibole, typically hornblende, is thought to be the likely source of this water during melting of amphibolite or mafic granulite.

It is the presence of water that is the major difference between Earth and the other terrestrial planets. The presence of liquid surface-water throughout much of Earth history has had a profound influence on the interior workings of the planet and the introduction of water back into the mantle, through subduction, at the end of the Archean may explain the contrasting tectonic patterns in the Archean and post-Archean Earth (McCulloch, 1993). It has been suggested that water is *the* vital ingredient in the formation of a felsic crust ("No water, no granites – no oceans, no continents" – Campbell & Taylor, 1983). Hence, it may be no coincidence that Earth, the only planet with abundant liquid water, is also the only planet with a substantial felsic continental crust. The link between liquid surface-water and the presence of a stable continental crust becomes more apparent when looking at the history of Mars and Venus. Mars lacks any evidence of recent tectonic activity, and yet was once the home to violent outpourings of water. This may be inferred from its deeply channeled surface. Venus, on the other hand, has been geologically active in its recent history, but is very hot and cannot sustain liquid water.

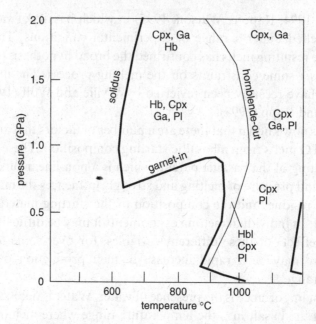

Fig. 6.11. Pressure–temperature phase diagram showing the amphibolite solidus and the residual minerals in equilibrium with TTG melts at different pressures and temperatures (after Wyllie *et al.*, 1997).

Since neither planet appears to have a continental crust of the type found on Earth, this comparison shows that it is not thermal energy alone, as on Venus, that gives rise to a continental crust, nor is it the presence of water alone, as on Mars. It is both.

During the dehydration melting of an amphibolite source, the mineralogy of the residue in equilibrium with a TTG melt is controlled by the pressure and temperature of melting and by the water content of the starting material. The critical reactions are the location of the solidus, the garnet-in reaction and the hornblende-out curve (Figs. 6.11, 6.12). These reactions subdivide the region between the solidus and liquidus into six distinctive assemblages of residual minerals coexisting with TTG melts (Wyllie *et al.*, 1997). The location and shape of the solidus is discussed in detail by Wyllie and Wolf (1993), who show that at pressures below about 1.0 GPa melting does not commence until about 900 °C, whereas at higher pressures melting takes place between 600–700 °C.

The garnet-in curve is pressure-sensitive, with a slight positive slope. The hornblende-out curve is temperature-sensitive at low pressures as hornblende dehydrates to clinopyroxene; at higher pressures it has a negative slope as hornblende breaks down to jadeitic clinopyroxene + garnet. Experiments by Winther and Newton (1991) and Winther (1996) show that as the water content

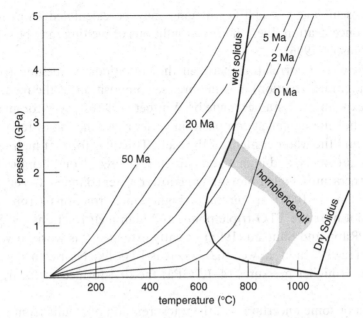

Fig. 6.12. Pressure–temperature phase diagram showing the results of the thermal modeling experiment of Peacock *et al.* (1994) for the subduction of ocean crust at a rate of 1 cm yr^{-1}. The curves labeled 0–50 Ma show the *P–T* path for subducted ocean floor 0–50 Ma old. Also shown are the wet and dry solidi for basalt melting and the hornblende-out curve. It may be seen that only the *P–T* trajectories of young ocean floor intersect the wet basalt solidus.

of the starting material is increased, the amphibole phase volume increases, the plagioclase phase volume decreases, and the garnet-in curve shifts to higher pressure. Sen and Dunn (1994) have described this process in terms of melting reactions. They identify two stages in the dehydration melting of basalt. The first represents the breakdown of amphibole + plagioclase ± quartz to produce melt + garnet + clinopyroxene. The second stage is the breakdown of clinopyroxene + garnet to produce melt. As already discussed, trace element studies set some limits on the phases present in equilibrium with typical TTG melts. These include the presence of garnet and the absence (or small percentage) of plagioclase. A further constraint is the presence of hornblende.

Wyllie *et al.* (1997) summarize from the available experimental data the pressure–temperature range of the different residual mineral assemblages. Of particular importance is the observation that some TTGs evolved in equilibrium with a garnet–clinopyroxene (eclogitic) residue. A map of residual mineral assemblages plotted in *P–T* space (Fig. 6.11) means that if the residual mineral assemblage for a TTG magma may be calculated from its trace element concentration, then the *P–T* conditions of melting may be estimated (e.g., Rollinson, 1996).

There is also a correlation between the conditions of melting (pressure, temperature, melt fraction) and the precise composition of the melt. A synthesis of experimental studies made by Winther (1996) showed for an average Archean tholeiite how melt composition varies as a function of temperature, pressure, and the water content of the melt. Tonalites form at high temperatures, low pressures, and high water contents, whereas trondhjemites form at lower temperatures, higher pressure, and lower water contents. In general, low melt fractions ($\sim$10%) are granitic, higher melt fractions (from >10 to <25%) plot near the TTG triple point, and high melt fractions (>25%) are tonalitic. Rapp and Shimizu (1998) recently extended this work to very high pressures (12 GPa) and showed that, whereas TTG-like Na-rich melts form up to 4 GPa, at higher pressures (4–12 GPa) the first-formed melts are K-rich granitoids.

An area of some uncertainty in the interpretation of results from dehydration melting experiments is the composition of the basalt that is being melted. For example, Wolf and Wyllie (1994) used very Ca-rich basalt in their experiments, which yielded highly calcic melts, not at all representative of typical TTGs. It is difficult to decide which type of basalt is most appropriate to model as a source of Archean TTGs, for authors start with different assumptions about the nature of the processes that they are modeling. This, in turn, becomes one of the major sources of uncertainty in using the results of such experimental studies, for it may be difficult to combine results from a number of different experimental studies, when there are significant differences in the composition of the starting materials.

6.6.3 *Evidence from trace element modeling*

A further test of the basalt–TTG link is to calculate the trace element composition of a likely melt, given a specific basaltic composition. If this is done for a suite of rocks where there is thought to be a parent–daughter relationship between basalt and TTG, then the degree to which the calculated trace element composition of the TTG matches the measured TTG composition is a measure of the reliability of the method and a further test of the hypothesis. This approach, of course, makes a number of assumptions. These include the type of melting process, the degree of melting, and the minerals present in the unmelted residue. The method also assumes knowledge of trace element partition coefficients for the minerals present in the residue, which for the

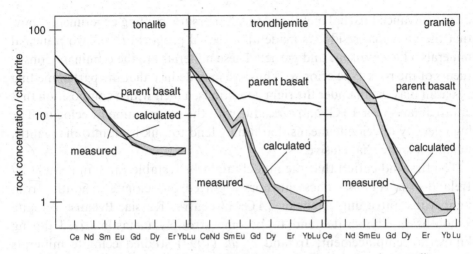

Fig. 6.13. Calculated values for selected rare earth elements in tonalites, trondhjemites and granites from the Lewisian Complex at Gruinard Bay (heavy lines), compared with measured values (shaded areas). The calculations are based upon the starting composition of amphibolite 104 (parent basalt), using 10–30% batch melting and clinopyroxene, hornblende and garnet in the residue (after Rollinson & Fowler, 1987).

REE, are now well known (Rollinson, 1993). Recent experimental studies have determined partition coefficients appropriate to the high pressures encountered during slab melting (Rapp & Shimizu, 1995; Klein *et al.*, 2000) and these new partition coefficients will permit an even greater degree of confidence to be placed in modeling of this type.

As a first order solution the method works well, Rollinson and Fowler (1987) showed that basaltic rocks enclosed in TTG gneisses from the Gruinard Bay area of the Lewisian gneisses in Northwest Scotland have the character of enriched Archean tholeiites and form an appropriate source for the gneisses. Trace element modeling for the REE, assuming batch melting, showed that the Gruinard Bay tonalites formed by 30% melting in equilibrium with a residue of 70% clinopyroxene, 23% hornblende, and 7% garnet. Trondhjemite REE patterns were modeled assuming 20% batch melting and a residue comprising 70% clinopyroxene, 10% hornblende and 20% garnet (Fig. 6.13).

6.6.4 *Evidence from eclogite chemistry*

An important outcome of the experimental study of dehydration melting of basaltic compositions is the information provided about the minerals present during advanced partial melting. At high pressure, melting experiments show that the unmelted residue is made up of equal proportions of two principal minerals: clinopyroxene and garnet. These minerals are the dominant constituents of the rock-type eclogite. Thus, we may deduce that the partial melting of basalt may yield, under the right conditions, a melt with a composition that approximates to a TTG, and a residue with the composition of eclogite. The high density of eclogite means that it may tend to sink back into the mantle, making it somewhat elusive.

The first indication that the model might be testable came in a study by Ireland *et al.* (1994). These authors reported on eclogite xenoliths from kimberlites intruding Archean TTGs in central Russia. Because eclogite xenoliths in kimberlite tend to become strongly metasomatized during kimberlite emplacement, Ireland *et al.* (1994) studied eclogite minerals found as inclusions within diamonds. Using ion microprobe analyses of trace elements in the garnet and clinopyroxene, they were able to show that the garnet and clinopyroxene had formed in equilibrium with TTG melts, which had subsequently been removed. These data suggested that the eclogite samples were residues from basalt melting during TTG genesis (Rollinson, 1995). This hypothesis was confirmed by a comparison between the composition of TTGs, which now make up the Archean West African Craton, and eclogites (now found as xenoliths in kimberlites) from more than 150 km beneath that craton (Rollinson, 1997). This major element geochemical study confirmed the complementarities between TTGs and eclogites as is required by the melting experiments, further establishing the importance of a basaltic parent (Fig. 6.14). More recent studies have sought further confirmation of this hypothesis using trace elements (Rudnick *et al.*, 2000) and demonstrate the importance of the eclogite residue as a significant mantle reservoir that is complementary to the continental crust (Fig. 6.6). The argument has been further extended by Jacob and Foley (1999) who argued that Siberian eclogite xenoliths preserve features that are indicative of hydrothermal alteration in the basaltic protolith to the eclogite. This in turn is taken as evidence for an ocean floor origin for the basaltic protolith.

6.6.5 *Thermal modeling of subducting systems*

A further outcome of the experimental investigation of the dehydration melting of basalt was a study that combined the results of these experiments with those from thermal modeling of partial melting during subduction. Peacock *et al.* (1994) investigated the *P–T* path followed by the top of subducting ocean

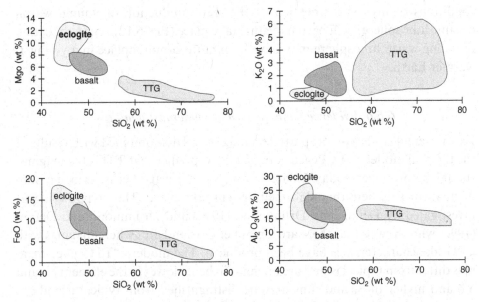

Fig. 6.14. Major element silica variation diagrams for TTGs, eclogites, and basalts for the Archean of Sierra Leone. The trends shown strongly support the complementarities of the TTGs and eclogites as melt and residue, respectively, of a basaltic parent (after Rollinson, 1997).

crust during subduction in a subduction zone with a 27° angle of dip. The experiments investigated the extent to which the upper part of a descending slab heats up through the conduction of heat from the overlying mantle wedge. In addition, the experiment investigated the relative contribution to slab melting of the following variables: shear heating along the top of the subduction shear zone; the subduction of young, warm lithosphere; subduction beneath a hot mantle wedge, and the effects of transient thermal conditions. Each variable was investigated for a range of convergence rates between 1 and 10 cm yr^{-1}.

The results of the experiments showed that transient thermal conditions and a hot mantle wedge are unlikely to give rise to partial melting. Shear heating may give rise to melting when convergent rates are high and shear stresses large. However, the most important conclusion of the experiment was that the age of the subducting slab exerts a strong influence upon whether or not the slab melts. Hot, young, oceanic lithosphere more readily undergoes partial melting than cooler, older, oceanic lithosphere. Oceanic lithosphere being subducted must be less than 2 Ma old at subduction rates of 10 cm yr^{-1} and less than 5 Ma old for subduction rates of 1 cm yr^{-1} if slab melting is to take place. These results imply that only under rather special conditions (i.e., under conditions of high shear heating or the slow subduction of young, warm, oceanic lithosphere), will partial melting take place (Fig. 6.12). The subduction of young warm lithosphere may have been more commonplace in a younger, warmer Earth.

6.6.6 *Evidence from modern arc magmas: adakites*

Experimental studies on the partial melting of basalts combined with results of the thermal modeling by Peacock *et al.* (1994) predict that TTG-like magmas should be produced today at sites where young, warm, crust is being subducted, for it is here that slab melting will take place. This hypothesis was investigated by Defant and Drummond (1990) and Drummond and Defant (1990) who showed that there are a number of island arcs where magmas with TTG-like characteristics have been produced. The modern "TTG-like" magmas differ from typical island arc magmas in being lower in the elements Y and Yb and higher in Sr and thus may be distinguished from typical island arc magmas on a Sr/Y versus Y trace element diagram. Modern TTG-like magmas are known as adakites, because they were first documented in Adak Island in the Aleutian Islands, Alaska (Kay, 1978). An important feature in their genesis is that they are associated with subducting oceanic lithosphere that is younger than 25 Ma old.

Geochemically, adakites are characterized by $SiO_2 \geq 56$ wt.%, $Al_2O_3 \geq 15$ wt.%, $MgO < 3$ wt.% (rarely > 6 wt.%), low Y and HREE ($Y \leq = 18$ ppm, $Yb \leq 1.9$ ppm), high Sr (normally > 400 ppm), a positive or absent Eu anomaly in the REE pattern, low HFSEs and $^{87}Sr/^{86}Sr$ ratios <0.7040 (Defant and Drummond, 1990). Petrographically, adakites contain hornblende and plagioclase $\pm$ clinopyroxene, orthopyroxene, biotite, and opaque minerals. On a plot of Sr/Y ratio versus Y concentration they overlap with the compositional field of Archean TTGs (Fig. 6.15). A number of adakite localities have now been described in detail. A good example is that described by Defant *et al.* (1992) in which high heat flow ocean floor of the Cocos Plate and the Cocos Ridge are subducted beneath Panama and Costa Rica in Central America. Magmatic arc rocks, associated with this subduction system include rare adakites. Peacock *et al.* (1994) review some of these occurrences and show that for some they fulfill both the geochemical and thermal criteria for being slab melts.

The similarity in composition between modern adakites and Archean TTGs is generally taken to confirm the view that Archean TTGs were produced by slab melting (see, for example, the review by Martin, 1999). Further support for this argument has come from a study by Schiano *et al.* (1995),

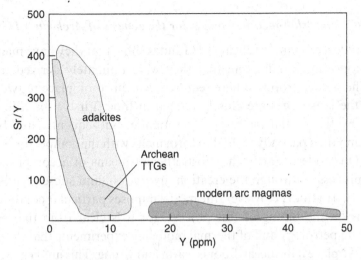

Fig. 6.15. Plot of Sr/Y versus Y showing the fields for adakites, Archean TTGs and modern arc magmas (after Drummond & Defant, 1990).

who showed that melt inclusions in olivine from the mantle wedge above a subducting slab are adakitic in composition. Schiano *et al.* (1995) examined the major and trace element composition of *c.* 20 μm glass inclusions in olivine from harzburgite xenoliths in lavas from the Philippine arc. They showed that the inclusions represent potassic (up to 4 wt.% K_2O), hydrous (up to 5 wt.% H_2O) and silica-rich (up to 68 wt.% SiO_2) melts with the trace element character of adakites. These were interpreted as small fraction melts of subducted basaltic crust, which have percolated upwards into the overlying mantle wedge.

A contrary view has recently been expressed by Smithies (2000), who has argued that there are significant differences in composition between modern adakites and Archean TTGs. Smithies (2000) argues that there are different types of adakite (see also Defant & Kepezhinskas, 2001), some which have formed as slab melts, and others as melts of basaltic lower crust. On a Mg# (i.e., $(Mg/Mg + Fe)_{atomic}$) versus SiO_2 plot, those adakites that are slab melts may be recognized by their elevated Mg#– the product of their interaction with the overlying mantle wedge. In contrast, adakites that are melts of basaltic lower crust do not show elevated Mg#. Smithies (2000) argues that Archean TTGs tend to have lower Mgs and so look more like "lower crustal melt" adakites than slab-melt adakites" on a Mg#–SiO_2 plot. From this observation, he infers that TTGs are not slab melts, but rather melts of basaltic lower crust. This proposal needs to be tested further.

6.6.7 *The "slab-melting" model for the genesis of Archean TTGs*

Isotopic studies of many Archean TTG suites show that they have mantle-like initial isotope ratios indicating that they were ultimately derived from the mantle. The model proposed here requires that this took place in two stages, although the two stages are closely related in time. First, a basaltic melt is produced by the partial melting of the mantle. Subsequently, the basalt is hydrated and then partially melted to form melts with the composition of TTG magmas. On modern Earth, the most efficient means of accomplishing this two-step process is through the creation, hydration, and subsequent subduction of the ocean floor. However, the model imposes particular conditions and restrictions on the timing of this process, for it is clear from the results of experimental petrology and of thermal modeling experiments that this process will only take place if the ocean floor is warm and young. This implies a relatively short time interval between the creation and destruction of the oceanic lithosphere, consistent with the radiogenic isotope evidence. The model is supported by the observation that in a few locations, where young, warm, crust is being subducted today, melts with a TTG-like character are being formed.

In the Archean, when the heat flux from Earth's mantle was greater as a result of its higher concentration of heat-producing elements, it is thought that this process may have been more widespread. Calculations based upon the decay of radioactive elements suggest that heat production was twice its present level at 2.6 Ga and 3 times present levels at 3.6 Ga. It was on this basis that Bickle (1978) suggested that the rate of ocean-floor creation was greater in the Archean, for this process provided an efficient heat-loss mechanism for the planet. If the creation of ocean crust was greater in the Archean, then the possibility of the subduction of young, warm, oceanic crust was also much greater in the Archean than today.

It should be noted that whereas the process of TTG-formation is normally discussed within the context of slab melting, this is not a requirement of the model. In essence, TTG genesis requires the dehydration melting of hydrated basalt at depth. This process may take place very efficiently during the subduction of a basaltic slab, but other mechanisms have been proposed. For example, it has been suggested that the dehydration melting of basalt at depth may also take place where basalts, underplated onto continental crust, are melted through interaction with a mantle plume, or where basalts deep in an oceanic plateau are melted during plateau collapse. However, these models require water to be present, most probably in the form of amphibole, for melting to take place; introducing water into the deep crust may be the limiting step in this alternative model.

6.7 Problems with the "standard" TTG model for Archean crust generation

The model described above, may be regarded as the "standard model" for Archean crust generation. It was widely accepted as a working hypothesis in the late 1980s, but during the 1990s, as more detailed geochemical evidence has been collected on regions of Archean continental crust, a number of weaknesses have become apparent in the model. This section outlines four areas of recent research that pose problems for the standard slab melting model for Archean crust formation.

6.7.1 *The mismatch between experimental and natural TTG compositions*

Two recent reviews have compared the compositions of Archean TTGs with the experimental products of dehydration melting of amphibolite (Rudnick, 1995; Rapp, 1997). Both authors show that, whereas in broad terms experimental melt compositions conform to TTG compositions, in detail this is not the case. On the standard O'Connor normative *An-Ab-Or* feldspar plots, there is a close agreement between experimental melt compositions and those of Archean TTGs (Fig. 6.10). However, on Harker diagrams, showing individual oxides plotted against silica content, there are some discrepancies. Rudnick (1995) showed that the Al_2O_3 and TiO_2 contents are higher and the *mg* number lower in the experimental melts than in TTGs (Fig. 6.16; see also Martin (1999)). In addition, Rapp (1997) showed that Na_2O contents are also lower in the experimental melts than in TTGs.

These discrepancies, between what is found and what is expected, may be explained, at least in part, by considering the precise basaltic composition used as starting materials in the melting experiments. Indeed, Arndt *et al.* (1997) have shown that Archean tholeiitic basalts have a number of distinctive compositional features and relative to modern mid-ocean ridge basalts are enriched in Fe and Ni, and depleted in Al and Ti. In addition, it is likely that Archean basalt undergoing subduction will be chemically altered by interaction with sea water and this may give rise to a starting material with a higher K_2O content than is typical in unaltered Archean tholeiite (Rudnick, 1995). Thus, the disagreement between experimental and actual TTG melt compositions may arise from uncertainty over the precise composition of the basalt that was melted. The problem is complex, for not only does it require knowledge of the composition of the Archean oceanic crust (if subduction is assumed to be the principal process) but also of the effects of submarine hydration/alteration on that oceanic crust.

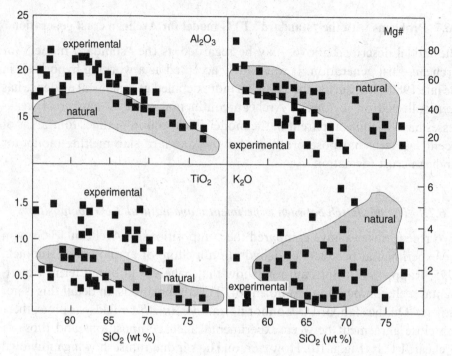

Fig. 6.16. Silica variation diagrams showing the compositions of experimental TTG melts and natural TTG melts (shaded areas) (after Rudnick, 1995).

However, there is another possible explanation for the mismatch between the melt compositions formed in amphibolite melting experiments and those of TTGs. That is, that the process of slab melting, generally regarded as the principal means of TTG production, may be inadequate. The three sections that follow explore additional processes that may be involved in TTG genesis, which may contribute to the geochemical diversity of TTG melts.

6.7.2 *Sanukitoids and mantle melting*

During the 1980s, the principal focus of studies of Archean crust generation was on the genesis of TTG magmas, and the basalt slab-melting model quickly became accepted as a working hypothesis. However, at the same time, another group of Archean magmas was being investigated that seemed to tell a different story. These magmas are less siliceous and more magnesian than TTGs and were classified as diorites, monzodiorites, and trachyandesites. They have subsequently become known as "sanukitoids." In many Archean cratons they appear to be of relatively small volume compared with TTGs, but in terms of process they are extremely important.

Sanukitoids were first described from the Archean of Ontario, Canada, by Shirey and Hanson (1984). The chief chemical characteristics of these rocks are that they have initial Pb, Sr, and Nd isotope compositions close to those of the mantle, and the more primitive magmas have high Mg# (>0.6), high Ni and Cr, REE 6–12 times that of chondrite with strong LREE enrichment, and relatively high K, Sr, Zr, and Nb concentrations. Shirey and Hanson (1984) compared the chemistry of these rocks to that of Miocene high-Mg andesites from the Setouchi Volcanic Belt, Japan, rocks known as "sanukites". The composition of sanukites had been explained as primitive magmas, produced by the direct melting of the mantle. This was in agreement with experimental studies that show that melts of this composition may be produced by the shallow melting of the mantle under both hydrous and anhydrous conditions. The mantle-like character of the Archean sanukitoids (mantle initial isotope ratios, high Mg#, high Ni, Cr) combined with their enriched character (high LREE, K, Sr, Zr, Nb) led Shirey and Hanson (1984) to postulate that they were produced by the direct melting of an enriched mantle source. This model directly conflicts with the two-step model for the derivation of TTGs from the mantle.

More recent studies (see, for example, Stern & Hanson, 1991; the review by Evans & Hanson, 1997) have shown that sanukitoids and related rocks are relatively well known in northern Canada and are present in a number of other Archean terrains (Martin, 1994; Moyen *et al.*, 1997). Stern and Hanson (1991) showed that high-Mg granodiorites from the Superior Province, Canada, were derived by fractional crystallization from a dioritic, sanukitoid parent and introduced the term "sanukitoid suite" to include rocks of intermediate to felsic composition, produced by fractionation from the primary mantle melts. However, the isotopic studies of Stern and Hanson (1991), and more recently by Stevenson *et al.* (1999), highlight a paradox. For, whereas the trace element concentrations of sanukitoids suggest derivation from an enriched mantle source, the Nd-isotope compositions require derivation from a depleted mantle source. The solution proposed was to postulate the subsequent enrichment of a previously depleted mantle source region. Such a solution is unconstrained and difficult to test.

The relevance of this observed paradox is highlighted by the recent experimental studies of Rapp and Shimizu (1996, 1997) and Rapp *et al.* (1999). These authors propose that sanukitoids are the product of mixing between slab melts and the mantle wedge. In an experimental study of the infiltration of a slab-melt into peridotite, Rapp *et al.* (1999) have modeled the process of slab melt–mantle wedge interaction. Their results show that an assimilation reaction between slab melts and the mantle wedge causes the resultant melt to

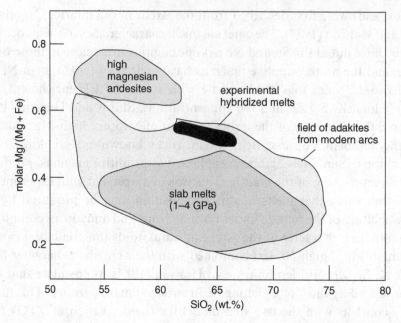

Fig. 6.17. Plot of Mg (molar Mg/(Mg+ Fe) versus wt.% SiO_2 showing the
fields of slab melts, adakites, high magnesian andesites and experimental
compositions produced by the mixing of slab melts with mantle peridotite
(experimental hybridized melts). Melt compositions from the southern Andes
studied by Stern and Kilian (1996) also plot in the hybridized melt field (after
Rapp *et al.*, 1999).

increase in Fe and Mg and decrease in Si and Al relative to the composition of
the initial slab melt. Thus, on a plot of SiO_2 versus Mg# the hybridized melts lie
on a trend of decreasing SiO_2 and increasing Mg#, between the composition of
the initial slab melt (high SiO_2, low Mg#) and sanukitoid magmas (low SiO_2,
high Mg#) (see Fig. 6.17). Whereas the concentration of most trace elements
also increases in the hybridized melts, the trace element ratio Sr/Y does not
change between slab melts and those that contain a mantle component. This
indicates that the process of assimilation does not lead to further Sr/Y fractionation, beyond that in the initial slab melting. These results also show that a
diagram showing SiO_2 content and Mg# (perhaps with the addition of Ni and
Cr – elements not determined in this experimental study) is an adequate
monitor of the degree of slab melt–mantle interaction.

Thus, in this refined version of the standard slab-melting model, slab melts
are seen as end-members in a spectrum of compositions, which through their
hybridization with mantle peridotite, may become progressively enriched in
MgO, Ni, and Cr. This raises the possibility that Archean sanukitoids are the

product of mixing between a slab melt and the overlying mantle. It is likely, therefore, that there is a spectrum of natural compositions between TTGs, which are dominated by the slab-melt contribution, and sanukitoids, dominated by a mantle-wedge contribution. Thus, the similarities and differences between sanukitoids and TTGs may be explained by a model in which "hot" slab melts (TTGs or adakites) interact with the overlying mantle wedge en-route to the surface. In this way, they become enriched in Mg, Ni, and Cr, elements common in the mantle, but maintain some of their slab-melt trace element signatures such as high La/Yb and high Sr/Y ratios. At present, there is no systematic understanding of the timing of sanukitoid formation. In the Superior Province of Canada, Henry *et al.* (1998) show that sanukitoids post-date the emplacement of the main TTGs. In contrast, in the Northern Marginal Zone of the Limpopo Belt, Zimbabwe, diorite inclusions within the TTGs have sanukitoid chemistry, but clearly formed early in the magmatic history of the region (unpublished data) and pre-date the emplacement of the TTGs.

6.7.3 *TTGs containing an older crustal component*

Further evidence for the variability of TTG compositions comes from a recent study by Berger and Rollinson (1997) on TTGs from the Northern Marginal Zone of the Limpopo Belt in Zimbabwe. This trace element and isotopic study showed that TTG-like melts, now preserved as lower crustal granulites, had an unusual lower crustal composition. They are characterized by low K/Rb ratios, high U and Th concentrations, high Th/U ratios, light REE-enriched REE patterns and have high model μ-values. These features may be explained if the TTGs were produced by mixing between a melt derived from a basaltic source produced at around 2700 Ma (with μ, low Th, U, Pb, unfractioned REE, and a low Pb/Nd ratio) and an older crustal component – characterized by a high lead isotope μ-value and an old Nd-model age (older than 3250 Ma), high Th/U, high Th, U, Pb, high Pb/Nd, and fractionated REE. Old crust of this type has not as yet been identified in this part of the Limpopo Belt, although old (3500 Ma) crust is known from the nearby Zimbabwe craton.

In detail, the geochemistry of the Northern Marginal Zone TTGs is a paradox for, on the one hand, the rocks are evidently mixtures, and yet, on the other, they show great chemical homogeneity. This means that they have experienced very thorough mixing (Fig. 6.18). Their present granulite facies mineralogy suggests that they crystallized at the middle to base of the continental crust, indicating that they may have formed in some form of mixing zone near the base of the crust. This relationship is reminiscent of a modern Andean-type continental margin in which new crust is being added to

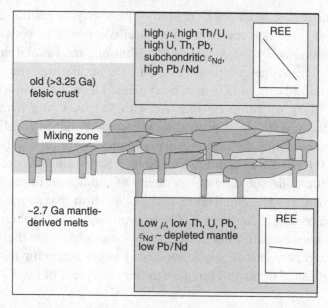

Fig. 6.18. Crust–mantle interaction in the Northern Marginal Zone of the Limpopo Belt is thought to require the mixing of a low μ, low Th, U, and Pb, and low Pb/Nd *c.* 2.7 Ga mantle-derived melt with a high μ, high Th/U, high Th, U, and Pb, and high Pb/Nd >3.25 Ga felsic crust in a mid- to lower-crustal mixing zone at *c.* 2.7 Ga (after Berger & Rollinson, 1997).

pre-existing older continental crust. Thus, evidence from the Limpopo Belt suggests another dimension to TTG mixing (i.e., mixing with old continental crust), further expanding the possible spectrum of TTG-related compositions that may be present in early continental crust. Here, the crucial feature that identifies the older crustal component is its high U/Pb and the resulting enhanced $^{207}Pb/^{206}Pb$. The old Nd model ages are the product of mixing and an artifact of the DM model that was used (Goldstein *et al.*, 1984).

A study by Henry *et al.* (1998) of the late-Archean crustal evolution of the western Superior Province came to a similar conclusion. Neodymium and Pb isotopes in TTG magmas from the western Superior Province indicate that they were derived by mixing between melts derived from a depleted mantle source and from an enriched source, thought to be 3.0–3.2 Ga old felsic crust. The TTGs have age-corrected ε_{Nd} values of between +3.0 and −2.0, and $^{207}Pb/^{204}Pb$ ratios between 14.5 and 14.75. These measurements are consistent with mixing between a continental-crust component ($\varepsilon_{Nd} \sim -2.0$, $^{147}Sm/^{144}Nd = 0.08$, $\mu = 9.0–9.1$, with a minimum extraction age of 3.2 Ga) and a range of end-members derived from a depleted mantle source. In this study, Henry *et al.* (1998) propose, on the basis of the observed Nd/Pb ratios, that the crustal signature represents 1–10% mixing with older sediment, introduced by means of sediment subduction, rather than through the contamination of a TTG melt by pre-existing older continental crust.

Further evidence of an older crustal contribution to Archean TTGs comes from the presence of inherited old zircons. Whereas the evidence for zircon inheritance in some Archean TTG gneiss suites, particularly very ancient ones, is contested (see Whitehouse *et al.*, 1999), the presence of old zircon cores in magmatic zircons is well established from a number of Archean terrains (see, for example, Kinny & Friend, 1997).

A conclusion of a more generic nature serves as a warning for those engaged in trace element studies of Archean TTG suites. Given that some Archean TTGs show evidence of an older crustal component, it follows that geochemical inversion modeling of TTG magma compositions may yield erroneous results. At best, trace element inversion modeling leads to non-unique solutions, but in the case of TTGs containing an older crustal component the tendency will be to infer a trace element-enriched source. This, of course, is not the case and is simply an artefact of the mixing process that took place during TTG genesis.

6.7.4 *Slab fluids and Rb-depletion in granulites*

In a recent study of the problem of element depletion in granulites – the typical rocks of the lower continental crust – Rollinson and Tarney (1998) have shown that element depletion in granulites may be best understood in terms of the process of crust formation. The problem has a long history, but in brief, element depletion in granulites has traditionally been understood in terms of metamorphic processes in the lower crust. However, in many regions the lower crustal rocks in question are TTGs of Archean age. Rollinson and Tarney (1998) found no evidence that uniquely linked the process of element depletion to the processes of granulite facies metamorphism. Instead, they argue that depletion of elements such as Rb, U, and Th in the lower continental crust is linked, more probably, to the process of crust formation. They propose that the observed depletion is a feature of the TTG magmas themselves and pre-dates the metamorphism.

Rubidium depletion is best explained through the loss of Rb from a subducting basaltic slab via a chloride-rich fluid, prior to partial melting. Uranium and Th are thought to be held in a phase such as zoisite in the unmelted portion of the slab during melting. If this is the case, the process of mineral dehydration and fluid release probably took place during Archean subduction, in a manner analogous to modern subducting systems. This finding is contrary to the standard model for Archean crust generation, which emphasizes the process of dehydration melting over the process of dehydration. The comparative rarity of Rb-depleted TTGs suggests that dehydration of the slab was not a common occurrence, and was not typical of Archean subduction. It also indicates that the process of slab melting may have yielded a range of primary melt compositions.

6.8 A multicomponent mixing model for Archean crust generation

In the previous section it was shown that the slab-melting model for Archean crust generation in the early Earth, a model that was popular in the 1980s, is in need of some revision. Here, an extended slab-melting model is proposed in which slab melting is but one end-member of a multicomponent mixing process involving contributions from older continental crust and the mantle. A number of inferences follow from this hypothesis.

(1) Some Archean TTGs are derived directly from slab melting; variable K/Rb ratios, and U and Th concentrations in some TTGs may indicate that there is a range of slab melt compositions.
(2) Some Archean TTGs are mixtures of a slab melt and older continental crust. This may be the product of either sediment subduction or the assimilation of older crust by the slab melt.
(3) Some Archean TTGs are hybrid melts between a slab melt and peridotite from the mantle wedge. Where the peridotite contribution is large (>30% by volume), these melts take the character of sanukitoids.

A range of possible mixing scenarios, governed by the geothermal gradient in the subduction zone, is illustrated by Martin (1999). Where there is a high thermal gradient there is shallow slab melting and minimal mantle wedge interaction. With a lower geothermal gradient, slab melting is deeper and interaction between the slab melt and the mantle wedge is more probable. With a low geothermal gradient, slab dehydration triggers mantle melting and there is no slab melt component. Each of these possibilities could also be combined with crustal mixing.

The mixing model presented here for Archean crust generation has a number of close similarities to the processes described from sites of modern magma generation associated with subduction in either island arcs or active continental margins. Three examples are presented below, from areas of recent subduction-related volcanism, which are illustrative of the processes proposed for the Archean.

First is the recent study of Shinjo (1999) of andesites, high magnesian andesites, and basalts from the Okinawa Trough–Ryuku Arc system, Japan.

Shinjo (1999) showed from the trace element geochemistry that these magmas have a number of different origins, and are probably derived from several different sources, despite being closely associated in space and time. The majority of the lavas in the area are typical calc-alkaline rocks produced by melting in the mantle wedge during the Miocene. However, some high Mg andesites and basalts from the central part of the Okinawa Trough–Ryuku Arc system have geochemical characteristics that are similar to adakites suggesting an origin by slab melting. However, in detail, melt compositions are not identical to those of adakitic melts (moderate MgO, high Ni, Cr, Co, high LILE, low HFSE) leading Shinjo (1999) to propose a model in which they evolved by the mixing of a slab melt with the overlying mantle wedge. The model is consistent with geological arguments for the tectonic evolution of the region.

A second example comes from the work of Hildreth and Moorbath (1988) who described the geochemistry of fifteen andesite–dacite stratovolcanoes from the volcanic front of the southern Volcanic Zone of the Andes in Chile. They documented a systematic change in geochemistry from south to north over a distance of 500 km in a region where the geometry of subduction is known to be constant. The geochemical changes coincide with the thickening of the continental crust from 30–35 km in the south to 50–60 km in the north, and (because the slab depth is constant) the complementary thinning of the mantle wedge. Increasing concentrations in SiO_2, LIL, LREE, and radiogenic Sr, and a decrease in radiogenic Nd along the volcanic front, are interpreted as indicating an increased crustal contribution to the andesitic magmas. Mixing between subcrustal magmas and crustal magmas is thought to take place principally at the crust–mantle transition in zones of melting, assimilation, and storage. The mixing model proposed here has a number of similarities with that proposed for Archean crust generation in the North Marginal Zone of the Limpopo Belt (Berger & Rollinson, 1997), with the difference that in the Archean the subcrustal magmatic contribution was a slab melt, whereas in the Andean example it is a melt of the mantle wedge.

Finally, the great complexity of this process is illustrated by the study by Stern and Kilian (1996) of Holocene adakitic andesites from the southern part of the Andean chain, in Chile, in the Andean Austral Volcanic Zone. This is a region of slow subduction of young ocean crust (12–24 Ma) beneath relatively thin (<35 km) continental crust. While all six volcanic centers studied have adakite-like (slab-melt) characteristics, they show variable Sr/Y, La/Yb, and LIL/LREE ratios, and a range of O, Sr, Nd, and Pb isotopic ratios. The variable chemistry is interpreted as a function of differing contributions from four different sources. These include contributions from the subducted slab, the mantle wedge, subducted sediment, and overlying continental crust. Andesites from Cook Island, the most southerly of the volcanic centers, have the highest slab melt component and are thought to represent a >90% slab melt and <10% mantle melt mixture. Magmas in other volcanic centers are more complex and represent melts of the slab and subducted sediment, which subsequently interacted with the mantle wedge and the overlying continental crust.

Discriminating between the different contributions of slab melt, old continental crust, and the mantle wedge to the process of Archean TTG/sanukitoid genesis and crust formation is an important task for future research in early crustal genesis. At the present, the indications are that old crustal contributions to TTG magmas are best recognized through the use of Pb isotopes and U and Th trace element concentrations. A mantle contribution, on the other hand, may be most easily recognized by the combined use of SiO_2, Ni, and Cr concentrations together with the Mg# (Fig. 6.17).

6.9 Acknowledgements

This chapter was first presented as an inaugural lecture at Cheltenham and Gloucester College of Higher Education in March 1999. Cheltenham and Gloucester College is thanked for their support of my research into Archean crustal evolution over the past two decades. By an interesting coincidence, part of this manuscript was written in the city of Trondheim, Norway, the site of the type trondhjemite locality. Kent Condie, Jan Kramers, and John Tarney are thanked for their helpful reviews.

References

Abouchami, W., Boher, M., Michard, A. and Albarède, F. (1990). A major 2.1 Ga event of mafic magmatism in west Africa: an early stage of crustal accretion. *Journal of Geophysical Research*, **95**(17), 605–17, 629.

Arculus, R. J. (1994). Aspects of magma genesis in arcs. *Lithos*, **33**, 189–208.

Armstrong, R. L. (1981). Radiogenic isotopes: the case for crustal recycling on a near-steady-state no-continental-growth Earth. *Philosophical Transactions of the Royal Society, London*, **A301**, 443–72.

Armstrong, R. L. (1991). The persistent myth of crustal growth. *Australian Journal of Earth Science, Geological Society of Australia*, **38**, 613–40.

Arndt, N. T. and Goldstein, S. L. (1987). Use and abuse of crust formation ages. *Geology*, **15**, 893–5.

Arndt, N. T., Albarède, F. and Nisbet, E. G. (1997). Mafic and ultramafic magmatism. In *Greenstone Belts*, ed. M. DeWit and L. D. Ashwal. Oxford: Oxford University Press, pp. 233–54.

Barker, F. and Arth, J. G. (1976). Generation of trondhjemitic–tonalitic liquids and Archean bimodal trondhjemite–basalt suites. *Geology*, **4**, 596–600.

Barth, M. G., McDonough, W. F. and Rudnick, R. L. (2000). Tracking the budget of Nb and Ta in the continental crust. *Chemical Geology*, **165**, 197–213.

Bennett, V. C., Nutman, A. P. and McCulloch, M. T. (1993). Nd isotopic evidence for transient, highly depleted mantle reservoirs in the early history of the Earth. *Earth and Planetary Science Letters*, **119**, 299–317.

Berger, M. and Rollinson, H. R. (1997). Isotopic and geochemical evidence for crust–mantle interaction during late Archean crustal growth. *Geochimica et Cosmochimica Acta*, **61**, 4809–29.

Bickle, M. J. (1978). Heat loss from the Earth: a constraint on Archean plate tectonics from the relation between geothermal gradients and the rate of plate production. *Earth and Planetary Science Letters*, **40**, 301–15.

Bowring, S. A. and Housh, T. B. (1995). The earth's early evolution. *Science*, **269**, 1535–40.

Bowring, S. A. and Williams, I. S. (1999). Priscoan (4.00–4.03) orthogneisses from northwest Canada. *Contributions to Mineralogy and Petrology*, **134**, 3–16.

Campbell, I. H. and Taylor, S. R. (1983). No water, no granites – no oceans, no continents. *Geophysical Research Letters*, **10**, 1061–4.

Chavagnac, V. (1999). Behaviour of the Sm–Nd isotopic system during metamorphism. *Memoires de Geosciences, Rennes*, **89**, 405.

Christensen, N. I. and Mooney, W. D. (1995). Seismic velocity structure and composition of the continental crust: a global view. *Journal of Geophysical Research*, **100**, 9761–88.

Collerson, K. D. and Kamber, B. (1999). Evolution of the continents and the atmosphere inferred from Th–U–Nb systematics of the depleted mantle. *Science*, **283**, 1519–22.

Collerson, K. D., Campbell, L. M., Weaver, B. L. and Palacz, Z. A. (1991). Evidence for extreme fractionation in early Archean ultramafic rocks from northern Labrador. *Nature*, **349**, 209–14.

Condie, K. C. (1981). *Archean Greenstone Belts*. Amsterdam: Elsevier.

Condie, K. C. (1990). Growth and accretion of continental crust: inferences based on Laurentia. *Chemical Geology*, **83**, 183–94.

Condie, K. C. (1997). *Plate Tectonics and Crustal Evolution*, 4th edn. Oxford: Butterworth-Heinemann.

Condie, K. C. (1998). Catastrophic events in the mantle and episodic growth of continents. *Geological Society of America Annual Meeting, Abstracts with Programs*, **A-345**.

Corfu, F. (1987). Inverse age stratification in the Archean crust of the Superior Province: evidence for infra- and subcrustal accretion from high resolution U–Pb zircon and monazite ages. *Precambrian Research*, **36**, 259–75.

Defant, M. J. and Drummond, M. S. (1990). Derivation of some modern arc magmas by melting of young subducted lithosphere. *Nature*, **347**, 662–5.

Defant, M. J. and Kepezhinskas, P. (2001). Evidence suggests slab melting in arc magmas. *Eos*, **82**, 68–9.

Defant, M. J., Jackson, T. E., Drummond, M. S., de Boer, J. Z., Bellon, H., Feigenson, M. D., Maury, R. C. and Stewart, R. H. (1992). The geochemistry of young volcanism throughout western Panama and southeastern Costa Rica: an overview. *Journal of the Geological Society London*, **149**, 569–79.

DeWit, M. J. and Ashwal, L. D. (1997). *Greenstone Belts*. Oxford: Oxford University Press.

Drummond, M. S. and Defant, M. J. (1990). A model for trondhjemite–tonalite–dacite genesis and crustal growth via slab melting: Archean to modern comparisons. *Journal of Geophysical Research*, **95**, 21 503–21.

Ducea, M. and Saleeby, J. (1998). A case for delamination of the deep batholithic crust beneath the Sierra Nevada, California. *International Geological Review*, **40**, 78–93.

Durrheim, R. J. and Mooney, W. D. (1991). Archean and Proterozoic crustal evolution: evidence from crustal seismology. *Geology*, **19**, 606–9.

Durrheim, R. J. and Mooney, W. D. (1994). Evolution of the Precambrian lithosphere: seismological and geochemical constraints. *Journal of Geophysical Research*, **99**, 15 359–74.

Ellam, R. M. and Hawkesworth, C. J. (1988). Is average continental crust generated at subduction zones? *Geology*, **16**, 314–17.

England, P. and Houseman, G. (1989). Extension during continental convergence with applications to the Tibetan Plateau. *Journal of Geophysical Research*, **94**, 15 561.

Evans, O. C. and Hanson, G. N. (1997). Late- to post-kinematic Archean granitoids of the S.W. Superior Province: derivation through direct mantle melting. In *Greenstone Belts*, ed. M. DeWit and L. D. Ashwal.Oxford: Oxford University Press, pp. 280–95.

Galer, S. J. G. and Goldstein, S. L. (1988). Early mantle depletion and thermal consequences. *Chemical Geology*, **70**, 143.

Galer, S. J. G. and Goldstein, S. L. (1991). Early mantle differentiation and thermal consequences. *Geochimica et Cosmochimica Acta*, **55**, 227–39.

Gao, S., Zhang, B.- R., Jin, Z- M., Kern, H., Luo, T- C. and Zhao, Z.- D. (1998). How mafic is the lower continental crust? *Earth and Planetary Science Letters*, **161**, 101–17.

Gibbs, A. K., Montgomery, A. A., O'Day, P. A. and Erslev, E. A. (1986). The Archean–Proterozoic transition: evidence from the geochemistry of metasedimentary rocks of Guyana and Montana. *Geochimica et Cosmochimica Acta*, **50**, 2125–41.

Gill, J. B., (1981). *Orogenic Andesites and Plate Tectonics*. New York: Springer-Verlag.

Goldstein, S. L., O'Nions, R. K. and Hamilton, P. J. (1984). A Sm–Nd study of atmospheric dusts and particulates from major river systems. *Earth and Planetary Science Letters*, **70**, 221–36.

Goodwin, A. M., (1991). *Precambrian Geology*. London: Academic Press.

Green, M. G., Sylvester, P. J. and Buick, R. (2000). Growth and recycling of early Archean continental crust: geochemical evidence from Coonterunah and Warrawoona Groups, Pilbara craton, Australia. *Tectonophysics*, **322**, 69–88.

Halliday. A. N. (2000). Terrestrial accretion rates and the origin of the Moon. *Earth and Planetary Science Letters*, **176**, 17–30.

Hawkesworth, C. J., Kempton, P. D., Rogers, N. W., Ellam, R. M. and van Calsteren, P. W. (1990). Continental mantle lithosphere and shallow mantle enrichment processes in the Earth's mantle. *Earth and Planetary Science Letters*, **96**, 256–61.

Henry, P., Stevenson, R. K. & Gariepy, C. (1998). Late Archean mantle composition and crustal growth in the western Superior Province of Canada: neodymium and lead isotopic evidence from the Wawa, Quetico and Wabigood subprovinces. *Geochimica et Cosmochimica Acta*, **62**, 143–57.

Hildreth, W. and Moorbath, S. (1988). Crustal contributions to arc magmatism in the Andes of Central Chile. *Contributions to Mineralogy and Petrology*, **98**, 455–89.

Hofmann, A. W., Jochum, K. P., Seufert, M. and White, W. M. (1986). Nb and Pb in oceanic basalts: new constraints on mantle evolution. *Earth and Planetary Science Letters*, **79**, 33–45.

Holbrook, W. S. (1998). Magmatism at volcanic rifted margins: a potential contributor to continental growth. *Geological Society of America Annual Meeting, Abstract with Programs*, **A-244**.

Holbrook, W. S., Lizarralde, D., McGeary, S., Bangs, N. and Diebold, J. (1999). Structure and composition of the Aleutian Island arc and implications for continental crustal growth. *Geology*, **27**, 31–4.

Hurley, P. M. and Rand, J. R. (1969). Pre-drift continental nucleii. *Science*, **164**, 1229–42.

Hynes, A. (2001).Freeboard revisited: continental growth, crustal thickness change and Earth's thermal efficiency. *Earth and Planetary Science Letters*, **185**, 161–2.

Ireland, T. R., Rudnick, R. L. and Spetius, Z. (1994). Trace elements in diamond inclusions from eclogites reveal link to Archean granites. *Earth and Planetary Science Letters*, **121**, 199–213.

Jacob, D. E. and Foley, S. F. (1999). Evidence for Archean ocean crust with low, high field strength element signature from diamondiferous eclogite xenoliths. *Lithos*, **48**, 317–36.

Jaupart, C. and Mareschal, J. C. (1999). The thermal structure and thickness of continental roots. *Lithos*, **48**, 93–114.

Kamber, B. and Moorbath, S. (1998). Initial Pb of the Amitsoq gneiss revisited: implication for the timing of early Archean crustal evolution in west Greenland. *Chemical Geology*, **150**, 19–41.

Kasting, K. F. and Holm, N. G. (1992). What determines the volume of the oceans? *Earth and Planetary Science Letters*, **109**, 507–15.

Kay, R. W. (1978). Aleutian magnesian andesites: melts from subducted Pacific Ocean crust. *Journal of Volcanology Geothermal Research*, **4**, 117–32.

Kinny, P. and Friend, C. R. L. (1997). U–Pb isotopic evidence for the accretion of different crustal blocks to form the Lewisian Complex of northwest Scotland. *Contributions to Mineralogy and Petrology*, **129**, 326–40.

Klein, M., Stosch, H.-G., Seck, H. A. and Shimizu, N. (2000).Experimental partitioning of high field strength and rare earth elements between clinopyroxene and garnet in andesite to tonalitic systems. *Geochimica et Cosmochimica Acta*, **64**, 99–115.

Kramers, J. D. (1987). Link between Archean continent formation and anomalous sub-continental mantle. *Nature*, **325**, 47–50.

Kramers, J. D. (1988). An open-system fractional crystallisation model for very early continental crust formation. *Precambrian Research*, **38**, 281–95.

Kramers, J. D. (1998). Reconciling siderophile element data in the Earth and Moon. W isotopes and the upper lunar age limit in a simple model of homogeneous accretion.*Chemical Geology*, **145**, 461–78.

Kramers, J. D. and Tolstikhin, I. N. (1997). Two terrestrial lead isotope paradoxes, forward transport modelling, core formation and the history of the continental crust. *Chemical Geology*, **139**, 75–110.

Martin, H. (1986). Effects of steeper Archean geothermal gradient on geochemistry of subduction-zone magmas. *Geology*, **14**, 753–6.

Martin, H. (1994). The Archean grey gneisses and the genesis of the continental crust. In *Archean Crustal Evolution, Developments in Precambrian Geology*, ed. K. C. Condie. Oxford: Elsevier, pp. 205–59.

Martin, H. (1999). Adakitic magmas: modern analogues of Archean granitoids. *Lithos*, **46**, 411–29.

McCulloch, M. (1993). The role of subducted slabs in an evolving earth. *Archean Crustal Evolution, Developments in Precambrian Geology*, **115**, 89–100.

McCulloch, M. and Bennett, V. C. (1994). Progressive growth of the Earth's continental crust and depleted mantle: geochemical constraints. *Geochimica et Cosmochimica Acta*, **58**, 4717–38.

McDonough, W. F., (1991). Partial melting of subducted oceanic crust and isolation of its residual eclogitic lithology. *Philosophical Transactions of the Royal Society, London*, **A335**, 407–18.

McKenzie, D. P. (1989). Some remarks on the removal of small melt fractions in the mantle. *Earth and Planetary Science Letters*, **95**, 53–72.

McLennan, S. M., Taylor, S. R., McCulloch, M. T. and Maynard, J. B. (1990). Geochemical and Nd–Sr isotopic composition of deep sea turbidites: crustal evolution and plate tectonic associations. *Geochimica et Cosmochimica Acta*, **54**, 2015–50.

Menzies, M. A. (1989). Cratonic, circumcratonic and oceanic mantle domains beneath the western United States. *Journal of Geophysical Research*, **94**, 7899–915.

Miller, R. G., O'Nions, R. K., Hamilton, P. J. and Welin, E. (1986). Crustal residence ages of clastic sediments, orogeny and continental evolution. *Chemical Geology*, **57**, 87–99.

Moorbath, S., Whitehouse, M. J. and Kamber, B. S. (1997). Extreme Nd-isotope heterogeneity in the early Archean – fact or fiction? Case histories from northern Canada and west Greenland. *Chemical Geology*, **135**, 213–31.

Morgan, P. (1985). Crustal radiogenic heat production and the selective survival of ancient continental crust. *Journal of Geophysical Research*, **90**, C561–70.

Moyen, J-F., Martin, H. and Jayananda, M. (1997). Origine du granite fini-Archeen de Closepet (Inde du sud): apports de la modelisation géochimique du compartement des elements en trace. *Comptes Rendues Academie des Sciences, Earth and Planetary Sciences*, **325**, 659–64.

Nagler, T. F. and Kramers, J. D. (1998). Nd isotopic evolution of the upper mantle during the Precambrian: models, data and the uncertainty of both. *Precambrian Research*, **91**, 233–52.

Nagler, T. F., Kramers, J. D., Kamber, B. S., Frei, R. and Prendergast, M. D. A. (1997). Growth of sub-continental lithospheric mantle beneath Zimbabwe started at or before 3.8 Ga: Re–Os study on chromites. *Geology*, **25**, 983–6.

Nelson, B. K. and DePaulo, D. J. (1985). Rapid production of continental crust 1.7 to 1.9 b.y. ago: Nd isotopic evidence from the basement of the North American mid-continent. *Bulletin of the Geological Society of America*, **96**, 746–54.

Nutman, A. P. (2001). On the scarcity of > 3900 Ma detrital zircons in > 3500 Ma metasediments. *Precambrian Research*, **105**, 93–114.

Nyblade, A. A. (1999). Heat flow and the structure of Precambrian lithosphere. *Lithos*, **48**, 81–91.

Nyblade, A. A. and Pollack, H. N. (1993). A global analysis of heat flow from Precambrian terrains: implications for the thermal structure of Archean and Proterozoic lithosphere. *Journal of Geophysical Research*, **98**, 12 207–18.

O'Hara, M. J. (1968). The bearing of phase equilibria studies in synthetic and natural systems on the origin and evolution of basic and ultrabasic rocks. *Earth Science Reviews*, **4**, 69–133.

O'Nions, R. K. and McKenzie, D. P. (1988). Melting and continent generation. *Earth and Planetary Science Letters*, **90**, 449–56.

Patchett, P. J. and Arndt, N. T. (1986). Nd isotopes and tectonics of 1.9–1.7 Ga crustal genesis. *Earth and Planetary Science Letters*, **78**, 329–38.

Peacock, S. M., Rushmer, T. and Thompson, A. B. (1994). Partial melting of subducting oceanic crust. *Earth and Planetary Science Letters*, **121**, 227–44.

Pearson, D. G. (1999). The age of continental roots. *Lithos*, **48**, 171–94.

Rapp, R. P. (1997). Heterogeneous source regions for Archean granitoids: experimental and geochemical evidence. In *Greenstone Belts*, ed. M. DeWit and L. D. Ashwal. Oxford: Oxford University Press, pp. 256–66.

Rapp, R. P. and Shimizu, N. (1995). Partitioning of REE, Ti, Sr, Y, Cr and Zr between tonalitic–trondhjemitic–granitic melts and eclogite residue at 1–11 GPa: ion microprobe analyses at natural abundance levels. *Eos, Spring Meeting Abstracts*, S296.

Rapp, R. P. and Shimizu, N. (1996). Arc magmatism in hot subduction zones: interactions between slab derived melts and the mantle wedge, and the petrogenesis of adakites and high-magnesian andesites (HMA). *Journal Conference Abstracts*, **1**, 497.

Rapp, R. P. and Shimizu, N. (1997). Trace-element characteristics of pristine and mantle-hybridized slab melts: implications for the petrogenesis of adakite and high-magnesian andesites. *Seventh Annual V. M. Goldschmidt Conf. (Abstracts)*.

Rapp, R. P. and Shimizu, N. (1998). The nature of subduction-derived metasomatism in the upper mantle: dehydration melting of hydrous basalt from 3–12 GPa. *(Abs)*. *Mineralogical Magazine*, **62A**, 1237–8.

Rapp, R. P. and Watson, E. B. (1995). Dehydration melting of metabasalt at 8–32 kbar: implications for continental growth and crust–mantle recycling. *Journal of Petrology*, **36**, 891–931.

Rapp, R. P., Watson, E. B. and Miller, C. F. (1991). Partial melting of amphibolite/ eclogite and the origin of Archean trondhjemites and tonalites. *Precambrian Research*, **51**, 1–25.

Rapp, R. P., Shimizu, N., Norman, M. C. and Applegate, G. S. (1999). Reaction between slab-derived melts and peridotite in the mantle wedge: experimental constraints at 3.8 GPa. *Chemical Geology*, **160**, 335–56.

Reymer, A. and Schubert, G. (1984). Phanerozoic additions to the continental crust and crustal growth. *Tectonics*, **3**, 63–77.

Richardson, S. H., Erlank, A. J. and Hart, S. R. (1985). Kimberlite-borne garnet peridotite xenoliths from old enriched subcontinental lithosphere. *Earth and Planetary Science Letters*, **75**, 116–28.

Rollinson, H. R. (1987). Early basic magmatism in the evolution of Archean high-grade terrains: an example from the Lewisian complex of N. W. Scotland. *Mineralogical Magazine*, **51**, 345–55.

Rollinson, H. R. (1993). *Using Geochemical Data: Evaluation, Presentation, Interpretation*. Harlow: Longman.

Rollinson, H. R. (1995). The birth of the continents – new clues from inclusions in diamonds. *Geology Today*, 240–2.

Rollinson, H. R. (1996). Tonalite–trondhjemite–granodiorite magmatism and the genesis of the Lewisian crust during the Archean. In *Precambrian Crustal Evolution in the North Atlantic Region*, ed. T. S. Brewer. Geological Society Special Publications 112, pp. 25–42.

Rollinson, H. R. (1997). Eclogite xenoliths in West African kimberlites are residues from Archean granitoids. *Nature*, **389**, 173–6.

Rollinson, H. R. (1999). Petrology and geochemistry of metamorphosed komatiites and basalts from the Sula Mountains greenstone belt, Sierra Leone. *Contributions to Mineralogy and Petrology*, **134**, 86–101.

Rollinson, H. R. and Blenkinsop, T. G. (1995). The magmatic, metamorphic and tectonic evolution of the Northern Marginal Zone of the Limpopo Belt in Zimbabwe. *Journal of the Geological Society London*, **151**, 65–75.

Rollinson, H. R. and Fowler, M. B. (1987). The magmatic evolution of the Scourian Complex at Gruinard Bay. In *The Evolution of the Lewisian and Comparable Precambrian High-Grade Terrains*, ed. R. G. Park and J. Tarney. Geological Society Special Publication 27, pp. 57–71.

Rollinson, H. R. and Lowry, D. (1992). Early basic magmatism in the evolution of the Northern marginal zone of the Archean Limpopo Belt. *Precambrian Research*, **55**, 33–45.

Rollinson, H. R. and Tarney, J. (1998). The myth of element depletion during lower crustal metamorphism. *Geological Society of America 1998 Annual Meeting, Abstract with Programmes*, A-394.

Rollinson, H. R. and Windley, B. F. (1980). Geochemistry and origin of an Archean granulite grade tonalite–trondhjemite–granite suite from Scourie, N.W. Scotland. *Contributions to Mineralogy and Petrology*, **72**, 265–81.

Rudnick, R. L. (1995). Making continental crust. *Nature*, **378**, 571–8.

Rudnick, R. L. and Fountain, D. M. (1995). Nature and composition of the continental crust: a lower crustal perspective. *Reviews in Geophysics*, **33**, 267–309.

Rudnick, R. L. and Taylor, S. R. (1987). The composition and petrogenesis of the lower crust: a xenolith study. *Journal of Geophysical Research*, **92**, 13 981–14 005.

Rudnick, R. L., McDonough, W. and O'Connell, R. J. (1998). Thermal structure, thickness and composition of continental lithosphere. *Chemical Geology*, **145**, 395–411.

Rudnick, R. L., Barth, M., Horn, I. and McDonough, W. (2000). Rutile-bearing refractory eclogites: the missing link between continents and depleted mantle. *Science*, **287**, 278–81.

Rushmer, T. (1991). Partial melting of two amphibolites: contrasting experimental results under fluid absent conditions. *Contributions to Mineralogy and Petrology*, **107**, 41–59.

Saunders, A. D., Tarney, J., Kerr, A. C. and Kent, R. W. (1996). The formation and fate of large oceanic igneous provinces. *Lithos*, **37**, 81–95.

Schiano, P., Clocchiatti, R., Shimizu, N., Maury, R. C., Jochum, K. P. and Hofmann, A. W. (1995). Hydrous, silica-rich melts in the sub-arc mantle and their relationships with erupted arc lavas. *Nature*, **377**, 595–600.

Sen, C. and Dunn, T. (1994). Dehydration melting of a basaltic composition amphibolite at 1.5 and 2.0 Gpa: implications for the origin of adakites. *Contributions to Mineralogy and Petrology*, **117**, 394–409.

Shinjo, R. (1999). Geochemistry of high Mg andesites and the tectonic evolution of the Okinawa Trough–Ryukyu arc system. *Chemical Geology*, **157**, 69–88.

Shirey, S. B. and Hanson, G. N. (1984). Mantle-derived Archean monzodiorites and trachyandesites. *Nature*, **310**, 222–4.

Smithies, R. H. (2000). The Archean tonalite–trondhjemite–granodiorite (TTG) series is not an analogue of Cenozoic adakite. *Earth and Planetary Science Letters*, **182**, 115–25.

Staudigel, H., Davies, G. R., Hart, S. R., Marchant, K. M. and Smith, B. M. (1995). Large scale isotopic Sr, Nd and O isotopic anatomy of altered oceanic crust: DSDP/ODP sites 417/418, *Earth and Planetary Science Letters*, **130**, 169–85.

Stein, M. and Goldstein, S. L. (1996). From plume head to continental lithosphere in the Arabian–Nubian shield. *Nature*, **382**, 773–8.

Stein, M. and Hofmann, A. W. (1994). Mantle plumes and episodic crustal growth. *Nature*, **372**, 63–8.

Stern, C. R. and Kilian, R. (1996). Role of the subducted slab, mantle wedge and continental crust in the generation of adakites from the Andean Austral Volcanic Zone. *Contributions to Mineralogy and Petrology*, **123**, 263–81.

Stern, R. A. and Hanson, G. N. (1991). Archean high-Mg granodiorite: a derivative of light rare earth element-enriched monzodiorite of mantle origin. *Journal of Petrology*, **32**, 201–38.

Stevenson, R. K. and Patchett, P. J. (1990). Implications for the evolution of continental crust from Hf isotope systematics of Archean detrital zircons. *Geochimica et Cosmochimica Acta*, **54**, 1683–97.

Stevenson, R. K., Henry, P. and Gariepy, C. (1999). Assimilation-fractional crystallisation origin of Archean sanukitoid suites: Western Superior Province, Canada. *Precambrian Research*, **96**, 83–99.

Sun, S.-S. (1989). Growth of lithospheric mantle. *Nature*, **340**, 509–10.

Sun, S.-S. and McDonough, W. F. (1989). Chemical and isotopic systematics of ocean basalts: implications for mantle composition and processes. In *Magmatism in Ocean Basins*, ed. A. D. Saunders and M. J. Norry. Geological Society of London Special Publication 42, pp. 313–45.

Sylvester, P. J., Campbell, I. H. and Bowyer, D. A. (1997). Niobium/Uranium evidence for early formation of the continental crust. *Science*, **275**, 521–3.

Tarney, J. and Jones, C. (1994). Trace element geochemistry of orogenic igneous rocks and crustal growth models. *Journal of the Geological Society London*, **151**, 855–68.

Taylor, S. R., (1977). Island arc models and the composition of the continental crust. In *Maurice Ewing Series*, Vol. 1, ed. M. Talwani and W. C. Pitmann. American Geophysical Union Monograph, pp. 325–35.

Taylor, S. R. and McLennan, S. M. (1981). The composition and evolution of the continental crust: rare earth element evidence from sedimentary rocks. *Philosophical Transactions of the Royal Society, London*, **A301**, 381–99.

Taylor, S. R. and McLennan, S. M. (1985). *The Continental Crust: Its Composition and Evolution*. Oxford: Blackwell Science.

Taylor, S. R. and McLennan, S. M. (1995). The geochemical evolution of the continental crust. *Reviews of Geophysics*, **33**, 241–65.

Veizer, J. (1983). Geological evolution of the Archean–Early Proterozoic Earth. In *Earliest Biosphere – Its Origin and Evolution*, ed. J. W. Schopf. New Jersey: Princeton University Press, pp. 240–59.

von Huene, R. and Scholl, D. W. (1991). Observations at convergent margins concerning sediment subduction and the growth of the continental crust. *Reviews of Geophysics*, **29**, 279–316.

Voshage, H., Hofmann, A. W., Mazzuchelli, M., Rivalenti, G., Sinigoi, S., Raczek, I. and Demarchi, G. (1990). Isotopic evidence from the Ivrea Zone for a hybrid lower crust formed by magmatic underplating. *Nature*, **347**, 731–4.

Weaver, B. L. and Tarney, J. (1982). Andesitic magmatism and continental growth. In *Andesites*, ed. R. S. Thorpe. Oxford: Oxford University Press, pp. 639–61.

Weaver, B. L. and Tarney, J. (1984). Empirical approach to estimating the composition of the continental crust. *Nature*, **310**, 575–7.

Wedepohl, K. H. (1995). The composition of the continental crust. *Geochimica et Cosmochimica Acta*, **59**, 1217–32.

Wever, T. (1992). Archean and Proterozoic crustal evolution: evidence from crustal seismology. Comment. *Geology*, **20**, 664–5.

White, R. V., Tarney, J., Kerr, A. C., Saunders, A. D., Kempton, P. D., Pringle, M. S. and Klaver, G. T. (1999). Modification of an oceanic plateau, Aruba, Dutch Caribbean: implications for the generation of continental crust. *Lithos*, **46**, 43–68.

White, W. M. (1989). Geochemical evidence for crust-to-mantle recycling in subduction zones. In *Crust/Mantle Recycling at Convergence Zones*, ed. S. R. Hart and L. Gulen. NATO ASI Series 258, pp. 43–58.

Whitehouse, M. J., Kamber, B. S. and Moorbath, S. (1999). Age significance of U–Th–Pb zircon data from early Archean rocks of west Greenland – a reassessment based upon combined ion-microprobe and imaging studies. *Chemical Geology*, **160**, 201–24.

Wilde, S. A., Valley, J. W., Peck, W. H. and Graham, C. M. (2001). Evidence from detrital zircons for the existence of continental crust and oceans on the Earth 4.4 Gyr ago. *Nature*, **409**, 175–8.

Winther, K. T. (1996). An experimentally based model for the origin of tonalitic and trondhjemitic melts. *Chemical Geology*, **127**, 43–59.

Winther, K. T. and Newton, R. C. (1991). Experimental melting of hydrous low-K tholeiite: evidence on the origin of Archean cratons. *Bulletin of the Geological Society of Denmark*, **39**, 213–28.

Wolf, M. B. and Wyllie, P. J. (1994). Dehydration melting of amphibolite at 1 kbar: effects of temperature and time. *Contributions to Mineralogy and Petrology*, **115**, 369–83.

Woodhead, J. D., Greenwood, P., Harmon, R. S. and Stoffers, P. (1993). Oxygen isotope evidence for recycled crust in the source of EM-type ocean island basalts. *Nature*, **362**, 809–13.

Wyborn, L. A. I., Wyborn, D., Warren, R. G. and Drummond, B. J. (1992). Proterozoic granite types in Australia: implications for lower crust composition, structure and evolution. *Transactions of the Royal Society of Edinburgh*, **83**, 201–9.

Wyllie, P. J. and Wolf, M. B. (1993). Amphibolite dehydration-melting: sorting out the solidus. *Geological Society of London Special Publication*, **76**, 405–16.

Wyllie, P. J., Wolf, M. B. and van der Laan, S. R. (1997). Conditions for formation of tonalites and trondhjemites: magmatic sources and products. In *Greenstone Belts*, ed. M. DeWit and L. D. Ashwal. Oxford: Oxford University Press, pp. 256–66.

Yasuda, A., Fujii, T. and Kurita, K. (1994). Melting phase relations of an anhydrous mid-ocean ridge basalt from 3–20 GPa: implications for the behavior of subducted oceanic crust in the mantle. *Journal of Geophysical Research*, **99**, 9401–14.

Zandt, G. and Ammon, C. J. (1995). Continental crust composition constrained by measurements of crustal Poisson's ratio. *Nature*, **374**, 152–4.

7

Structural and metamorphic process in the lower crust: evidence from a deep-crustal isobarically cooled terrane, Canada

MICHAEL L. WILLIAMS AND SIMON HANMER

7.1 Introduction

Knowledge of the composition and structural character of the lower crust is essential for understanding the growth, evolution, and present state of the continental crust. The lower crust is an important geochemical reservoir. Its composition and history are fundamental variables in models for Earth evolution, and play a key role in the structural and tectonic behavior of the continental crust as a whole. Under some circumstances, cool, anhydrous deep crust may be strong and resist deformation, but in young, active orogenic belts, deep-seated flow may accommodate isostatic compensation of the topographic expression of tectonic features (Rutter & Brodie, 1992; Clark & Royden, 2000).

Data on the composition of the deep crust come from a variety of sources including geophysical experiments, geochemical and petrological studies, and inferences drawn from studies of high-pressure (>1.0 GPa) lower crustal terranes. Although all of these types of data provide constraints on the present state of the continental crust, exhumed high-pressure lower crustal terranes allow direct observation of spatial and temporal geological relationships that may be used to understand structural and petrological processes as well as the tectonic evolution of the deep crust through time. Field studies of exposed high-pressure lower crustal terranes provide a detailed picture of large and small-scale structures, deformation mechanisms, and compositional make-up with which to better evaluate the significance of the geophysical data (see also Fountain & Salisbury, 1981).

The application of inferences from high-pressure lower crustal terranes to the understanding of the deep crust requires that we distinguish high-pressure regions that developed within an over-thickened crust from those that evolved near the base of a crust of normal thickness (i.e., terranes that sat close to the

Evolution and Differentiation of the Continental Crust, ed. Michael Brown and Tracy Rushmer. Published by Cambridge University Press. © Cambridge University Press 2005.

continental Moho (Harley, 1989; Percival *et al.*, 1992; Mezger, 1992)). The former may provide insight into the character and processes of the deep crust during transient periods of tectonism, whereas the latter provide insight into the deeper parts of isostatically stable "normal thickness" crust and help to characterize the long-term nature of the lower crust. Similarly, with respect to deep-crustal processes, we must distinguish the overprinting effects of the exhumation process from those features that characterize isostatically stable deep crust. To the degree that we are able to recognize the latter, it may be possible to draw inferences about processes and characteristics of the deep continental crust today, or in the geologic past. In this chapter we focus on high-pressure lower crustal terranes that have cooled isobarically, which may be particularly informative with respect to deep-crustal processes. They are related to the "wide oblique transitions" of Percival *et al.* (1992) or to the "regional granulite terranes" of Mezger (1992), and include terranes that were apparently stable in the deep crust for some period of time before being exhumed by events unrelated to the preserved tectonic and metamorphic fabrics.

One purpose of this chapter is to evaluate the evidence for long-term residence of high-pressure terranes in the deep crust, and the implications for crustal evolution. We focus discussion here principally on the East Athabasca mylonite triangle and its immediate wall rocks in northern Saskatchewan, Canada. Although questions remain regarding the details of the geologic history of parts of this example, as a whole it serves as an illustration of the criteria for, and uncertainties relating to, identifying isobarically cooled terranes and the type of information they provide about the deep crust. The East Athabasca mylonite triangle may be considered as an end-member type of isobarically cooled terrane in that it preserves evidence for high metamorphic temperatures, but essentially constant pressures ($\sim$800–1000$\,^\circ$C, 1.0 GPa) and thus provides a map view of the deep crust. Also, it represents the deeper (high-P) end of the spectrum of such terranes yet recognized (see Harley, 1989), and thus may offer a rare opportunity to directly investigate the lower continental crust.

A second goal of this chapter is to evaluate the expression and implications of heterogeneity in the deep crust. A variety of data types have been interpreted to suggest that the deep crust is heterogeneous in composition and/or in structural character (see below). However, because these conclusions have been based on geophysical data, or on interpreted depth sections, in general it has not been possible to characterize the scale, character, or origin of this heterogeneity. The East Athabasca mylonite triangle provides a map view of at least one variety of deep crust and thus allows a first-order evaluation of its fundamental character. Furthermore, comparison of this terrane with other exposures of high-P rocks from the Canadian Shield allows an evaluation of deep-crustal heterogeneity on a larger scale and also provides insight into the ultimate significance of any one such terrane for understanding the general character of the deep crust.

7.2 Background

Data sources on the nature and composition of the deep crust may be divided into three types: (1) remote geophysical data; (2) xenoliths or magmas that sampled the deep crust, and (3) data from direct observation of high-pressure terranes. There has been significant discussion and debate about the implications of these data (Fountain & Salisbury, 1981; Percival *et al.*, 1992; Percival & West, 1994; Rudnick & Fountain, 1995), and a full summary is beyond the scope of this chapter. Instead, several general observations will be summarized here in order to serve as a basis for comparing and interpreting isobarically cooled deep-crustal terranes such as the East Athabasca mylonite triangle.

7.2.1 Remote data for the deep crust

Seismic refraction data suggest that the deep crust is compositionally variable, ranging from felsic to mafic or ultramafic, and may have a bimodal velocity distribution (Holbrook *et al.*, 1992). However, the bulk lower crustal composition is interpreted to be mafic (Holbrook *et al.*, 1992; Rudnick & Fountain, 1995). Seismic reflection studies reveal that parts of the deep crust are highly reflective, although the origin of the reflectivity has been the subject of much discussion (Mooney & Meissner, 1992). Many workers consider the reflectivity to indicate subhorizontal lithologic layering, perhaps involving underplated or intraplated mafic magma (Warner, 1990), but others invoke metamorphic or tectonic layering (Mooney & Meissner, 1992). Electromagnetic properties of the deep crust may also be highly variable and interpretations have invoked laterally continuous saline fluid, grain boundary carbon film, conducting mineral or melt (Jones, 1992). The thermal state of the continental crust is particularly important for tectonic interpretations because rheological models indicate that typical deep-crustal rocks may vary from relatively strong to relatively weak, depending on their temperature (Rutter & Brodie, 1992; Chapter 3), and transient thermal perturbations may be associated with significant changes in material properties. This may help to explain how nominally strong deep crust may locally flow at relatively fast rates, perhaps driven by surface topography (e.g., Clark & Royden, 2000).

7.2.2 *Tangible samples of the deep crust*

Samples of the deep crust occur as accidental inclusions (xenoliths) in volcanic rocks, but their interpretation remains equivocal. Xenoliths are subject to sampling bias, and questions have been raised regarding systematic differences between studies of xenoliths and direct observation of high-pressure lower crustal terranes (Rudnick, 1992; Rudnick & Fountain, 1995). Many xenolith studies report pressures greater than those typical of high-pressure lower crustal terranes (average of ~ 0.8 GPa), and most xenoliths are mafic in composition, whereas high-pressure lower crustal terranes tend to be more felsic or mixed (Rudnick, 1992; Rudnick & Fountain, 1995). Igneous rocks from presumed deep-crustal sources also suggest that the deep crust is heterogeneous in composition (Farmer, 1992). Because melting would involve a relatively large volume, such rocks may provide a more representative sample of the deep crust than xenoliths, but inferences based on inversion of geochemical data require interpretations of compositional changes during melting and crystallization.

7.2.3 *High-pressure terranes*

Three broad groups of high-pressure ($P > 1.0$ GPa) lower crustal terranes may be distinguished: (1) those that evolved at high pressures within a tectonically thickened crust that underwent thermal relaxation; (2) ultra-high pressure (UHP) terranes that probably reflect underthrusting during subduction and continental collision; and (3) those that experienced high pressures near the base of a crust of more normal thickness and cooled isobarically.

The first two types commonly yield clockwise *P–T* paths associated with near isothermal decompression (England & Thompson, 1984; Harley, 1989; Percival *et al.*, 1992), although some subduction-related terranes yield retrograde paths similar to prograde paths (e.g., Chopin, 2003). Examples include the granulite facies and eclogitic rocks of the Western Gneiss Terrane, Norway (Griffin *et al.*, 1985; Terry *et al.*, 2000), the Western and Eastern Alps (Selverstone, 1985; Henry *et al.*, 1993), and parts of the Grenville Orogen, eastern Canada (see Indares, 1995; Indares *et al.*, 2000). Each of these terranes was exhumed shortly after the orogenic event that led to crustal thickening. These regions provide information about rock properties at deep-crustal conditions, and reveal much about orogenic processes and histories. However, such terranes offer less insight into the present state of isostatically stable continental crust.

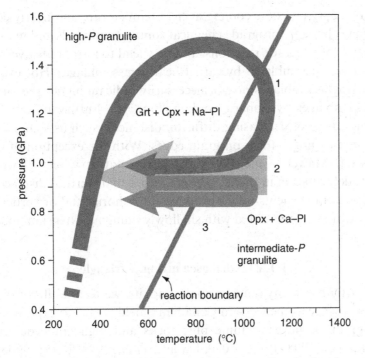

Fig. 7.1. Possible *P–T* paths leading to isobaric cooling in the deep crust. (1) Looping path that might result from crustal thickening followed by exhumation (thermal relaxation) and then isostatic equilibrium. (2) *P–T* path for hot igneous rocks emplaced into the deep crust. (3) *P–T* path for isostatically stable rocks that undergo heating near hot igneous rocks or underplated magma. For further discussion about *P–T* paths and tectonics see Harley (1989) and Ellis (1987).

Isobarically cooled terranes offer a more direct view of the lower crust, potentially in a stable, or steady-state environment (Harley, 1989; Percival, 1989; Rudnick & Fountain, 1995). These are regions that resided in the deep crust for an extended period, prior to exhumation by a tectonic event quite distinct from those that initially buried them (Mezger, 1992; Percival *et al.*, 1992). Several characteristics, in addition to the mineral equilibria indicative of high-pressure conditions, might be expected in isobarically cooled terranes, but the key piece of evidence is an isobaric cooling path. Such a path could be fundamentally clockwise, involving even higher pressures along an earlier part of the path (Harley, 1989; Ellis, 1987), but the essential feature is that, after some initial decompression, the terrane was isostatically stable and cooled in the lower crust (Fig. 7.1).

Isobarically cooled terranes remain near peak metamorphic temperatures for extended periods, so that we might expect some or all of the following features to occur. The slow cooling at high temperature and nearly constant pressure may eliminate prograde chemical zoning in minerals and microstructural evidence of prograde reactions, and may lead to retrograde overprinting of garnet-absent assemblages by garnet-bearing assemblages. However, these garnet-bearing assemblages do not necessarily indicate higher pressure, e.g., Opx–Pl assemblages by isobaric cooling (Fig. 7.1). Furthermore, late-stage textures indicative of decreasing diffusion distances, such as symplectites or narrow coronae, might form on grain edges. With the exception of diapiric structures (cf. Mawer *et al.*, 1997), sub-horizontal flow vectors would be favored by deformation in the deep crust, rather than vertical displacements. Although such movements might produce either horizontal or vertical foliations, they would be associated with shallowly plunging extension lineations.

7.3 East Athabasca mylonite triangle

The East Athabasca mylonite triangle and its western wall rocks in the Churchill Province of the western Canadian Shield (Fig. 7.2) expose a diverse array of granulite facies metamorphic rocks and high grade deformation fabrics (Hanmer, 1994, 1997a, 2000; Hanmer *et al.*, 1994, 1995a; Snoeyenbos *et al.*, 1995; Williams *et al.*, 1995, 2000a, 2000b; Kopf, 1999). The East Athabasca mylonite triangle, Tantato domain of Gilboy (1980), is a northeast tapering zone of penetratively developed granulite facies mylonites, 80 km across at the east–west baseline. It is a part of the Striding–Athabasca mylonite zone (Hanmer *et al.*, 1995b), itself a segment of the Snowbird tectonic zone (Fig. 7.2), the geophysically located boundary between the Rae and Hearne domains that constitute the Western Churchill Province (Goodacre *et al.*, 1987; Hoffman, 1988).

The East Athabasca mylonite triangle is divided into three structural blocks, each with distinct lithologies, structural character, and geologic history (Fig. 7.3) that provide examples of three varieties of deep crust and deep-crustal processes. The northwestern and southeastern blocks, collectively referred to as the lower deck, have subvertical, northeast-striking mylonitic foliation, and a moderately southwest-plunging extension lineation, but have different lithologic associations and kinematic histories. The third block, referred to as the upper deck, has a northeast- to east-striking and south-dipping mylonitic foliation and southwest-plunging extension lineation. It structurally overlies the lower deck and contains early, very high-pressure metamorphic assemblages that are not recorded elsewhere in the East Athabasca mylonite triangle area. Granulite facies wall rocks to the west of the triangle are part of the Rae domain (Fig. 7.2). They have been investigated

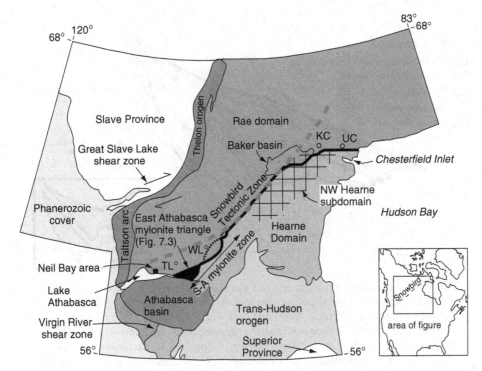

Fig. 7.2. Generalized tectonic map of the western Canadian Shield (after Hanmer *et al.*, 1994). Black line shows the trace of the geophysically defined Snowbird tectonic zone (Hoffman, 1988). Dotted lines are geophysically defined lozenges (Hanmer *et al.*, 1994). Dashed line is trace of possible exhumation-related fault zone (Mahan *et al.*, 2001, 2003). See text for discussion. S–A = Striding–Athabasca; WL = Wholdaia Lake; TL = Thicke Lake; KC = Kramanituar Complex; UC = Uvauk Complex.

on a reconnaissance scale (Lewry & Sibbald, 1977; Gilboy, 1980; Slimmon, 1989), except in several isolated areas where they have been mapped in more detail (Kopf, 1999; Ashton *et al.*, 2000; Card *et al.*, 2000; Williams *et al.*, 2000a). The East Athabasca mylonite triangle is bounded to the east by a broad shear zone, the Legs Lake shear zone that represents the eastern boundary of the large high-*P* terrane and a fundamental structure in the exhumation history of the region.

7.3.1 The northwestern block – the Mary granite and associated rocks

The northwestern block of the East Athabasca mylonite triangle is dominated by *c.* 2.62–2.60 Ga plutonic rocks, most of which were pervasively mylonitized during dextral strike-slip shearing (Hanmer *et al.*, 1994; Hanmer, 1997a). The

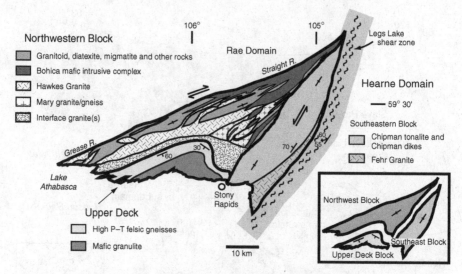

Fig. 7.3. Generalized geologic map of the East Athabasca mylonite triangle (EAmt) and inset map showing the location of the three main structural/ tectonic blocks. (From Williams *et al.* (2000b), Microstructural tectonometamorphic processes and the development of gneissic layering: a mechanism for metamorphic segregation. *Journal of Metamorphic Geology*, **18**, 41–57. © 2000 Blackwell Publishing. Reproduced by permission of Blackwell Publishing.)

block has three main lithologic components: (1) the Bohica mafic complex (Grt–Cpx and Opx–Pl metanorite, metagabbro and meta-anorthosite); (2) the Reeve Lake diatexites (Grt–Sil–Opx mylonitized migmatite), and (3) the Mary granite, all of which were formed and/or magmatically emplaced at *c.* 2.62–2.60 Ga (Hanmer, 1997a; Williams *et al.*, 2000b). The Mary granite, the largest coherent granitoid in the East Athabasca mylonite triangle, ranges from granite to granodiorite in composition. It is characterized by textural variations and strain gradients (Hanmer, 1997a; Williams *et al.*, 2000b). Exposures range from medium- to coarse-grained, equigranular, igneous rocks with no visible foliation or extension lineation, to strongly compositionally banded, granulite facies gneisses, mylonites, and ultramylonites (Fig. 7.4). Low-strain varieties locally cross-cut higher strain equivalents, suggesting that deformation was synchronous with magmatic emplacement.

The metamorphic and microstructural aspects of the Mary granite have been studied in some detail (Hanmer, 2000; Williams *et al.*, 2000b). Two end-member mineral assemblages correspond to low- and high-strain samples, respectively. Low-strain samples contain Ca-rich plagioclase, orthopyroxene, K-feldspar, quartz, and ilmenite. Higher-strain samples contain garnet,

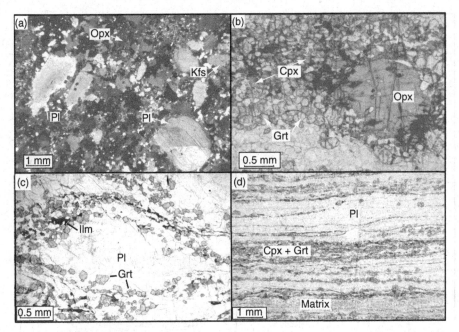

Fig. 7.4. Photomicrographs of rocks from the Mary granite, northwestern block. (From Williams *et al.* (2000b), Microstructural tectonometamorphic processes and the development of gneissic layering: a mechanism for metamorphic segregation, *Journal of Metamorphic Geology*, **18**, 41–57. © 2000 Blackwell Publishing. Modified by permission of Blackwell Publishing.) (a) Photomicrograph of low-strain Mary gneiss (Sample S823b). (b) Photo of relict Opx (igneous?) with mantle of Grt, Cpx, Ilm, and Na-rich Pl. (c) Partially recrystallized plagioclase porphyroclast with garnet in recrystallized mantle. Garnet typically develops in and around recrystallized plagioclase mantles, between Ca-rich (igneous cores) and Na-richer rims. With continued deformation, these domains become Grt–Pl-rich gneissic layers. (d) Composite view of highly strained Mary granite (gneiss) showing Grt–Cpx–Opx domains (former Opx) interlayered with dynamically recrystallized Pl–Grt layers (former Pl phenocrysts). Abbreviations after Kretz (1983).

clinopyroxene, Na-rich plagioclase, quartz, ilmenite, ± hornblende and ortho-pyroxene (Fig. 7.4). Textures and phase compositions indicate a metamorphic reaction involving the production of garnet, clinopyroxene, and a more sodic plagioclase from the original (igneous?) orthopyroxene (or ilmenite) and more calcic plagioclase (Fig. 7.5; Williams *et al.*, 2000b), as follows.

$$\text{Opx} + \text{Pl}_1(\text{Ca-rich}) = \text{Grt} + \text{Cpx} + \text{Pl}_2(\text{Na-rich}). \tag{7.1}$$

Peak metamorphic conditions are estimated to be 700–800 °C at 0.8–1.0 GPa, and the emplacement conditions of the parent Mary granite magma

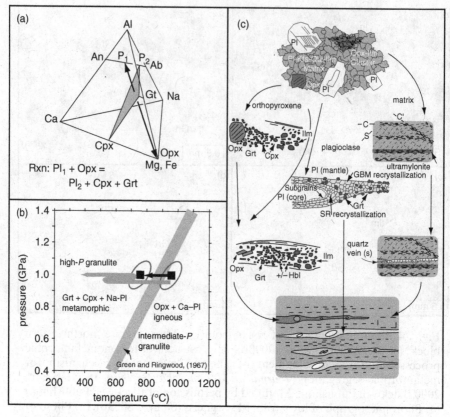

Fig. 7.5. (a) A–C–F–Na phase diagram showing the generalized metamorphic reaction in Hbl-absent Mary granite (P1 = early, igneous Pl; P2 = late, metamorphic Pl). The phase diagram is not strictly thermodynamically valid because all compositional variation cannot be shown (Spear, 1993). (b) Simplified phase diagram showing inferred metamorphic reaction and calculated igneous and metamorphic conditions for Mary granite. Shaded reaction curve is from Green and Ringwood (1967). *P–T* path may involve simple cooling (solid path) from igneous conditions or cooling with subsequent reheating (grey path) perhaps due to later phases of the Mary granite; (c) Schematic diagram showing the development of gneissic fabric in Mary granite (gneiss). The gneissic fabric reflects the combined evolution of three fabric domains: (1) feldspar megacrysts; (2) orthopyroxene megacrysts, and (3) matrix domains. (From Williams *et al.* (2000b), Microstructural tectonometamorphic processes and the development of gneissic layering: a mechanism for metamorphic segregation, *Journal of Metamorphic Geology*, **18**, 41–57. © 2000 Blackwell Publishing. Modified by permission of Blackwell Publishing.)

are interpreted to be approximately 900 °C and 0.8–1.0 GPa (Williams *et al.*, 2000b). The observation that all textural variations from highly strained to essentially unstrained, and from weakly metamorphosed to completely recrystallized, yield similar metamorphic pressures, suggests that metamorphism was isobaric, at least during the interval of Mary granite emplacement. Highly strained or hornblende-rich samples have equilibrium textures, but low-strain samples contain abundant reaction coronae and symplectite textures suggesting that diffusion was more limited during the waning stages of deformation (Williams *et al.*, 2000b).

Equation (7.1) is similar to that investigated experimentally by Green and Ringwood (1967), defining the boundary between the medium- and high-pressure granulite fields. This reaction may be crossed on a path involving either increasing pressure or decreasing temperature. Evidence for diminishing diffusion during waning deformation, the nearly constant calculated pressures at all preserved stages of the reaction, and the lack of evidence for a clockwise *P–T* path (i.e., increasing pressure and/or temperature after intrusion), all indicate metamorphism during isobaric cooling. The Mary granite magma is interpreted to have crystallized as Opx-bearing plutonic rocks in the "intermediate pressure" granulite field, and then to have cooled into the "high-pressure" granulite field during dextral shearing and metamorphism (Fig. 7.5b, c). The progressive evolution from igneous textures to strongly layered gneissic fabrics documents an intimate interaction between metamorphism and deformation associated with flow in the deep crust (Williams *et al.*, 2000b).

No textural or mineralogical evidence has been found for decompression in the Mary granite. Electron microprobe monazite geochronology suggests multiple stages of monazite growth in the northwestern block after the main *c.* 2.6 Ga event, particularly at *c.* 2.45 Ga and 1.8 Ga (Williams & Jercinovic, 2002). Localized reactivation (dextral shearing) occurred near the western margin of the East Athabasca mylonite triangle at *c.* 1.8 Ga (Fig. 7.6), in a high-strain zone locally called the Grease River shear zone (Lafrance & Sibbald, 1997). This localized strain may also be associated with the emplacement of late granitoids, also identified at the western margin of the northwestern block (Hanmer, 1997a).

7.3.2 *The southeastern block–Chipman tonalite and Chipman dike swarm*

The southeastern block of the East Athabasca mylonite triangle is dominated by the poorly dated, apparently Mesoarchean Chipman tonalite, intruded on its eastern flank by the *c.* 2.58 Ga Fehr granite (Hanmer *et al.*, 1994; Hanmer, 1997a; Williams & Davis, unpublished data) (Figs. 7.3, 7.7). The tonalite is

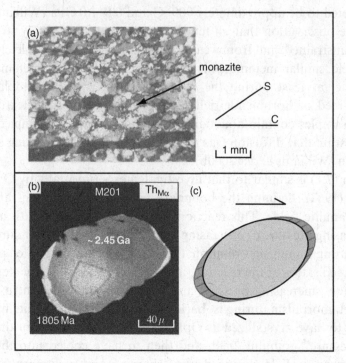

Fig. 7.6. (a) Photomicrograph of late-stage C–S mylonite in Reeve diatexite along eastern edge of northwestern block, East Athabasca mylonite triangle. (b) $Th_{M\alpha}$ compositional map and interpretation (c) of one monazite grain, indicated in Fig. 7.6a. Late dextral shearing is associated with overgrowths on older, *c.* 2.45 Ga, monazite, constraining the late shearing event to approximately 1805 Ma. The 2.45 Ga core domain represents a time of extensive post-kinematic monazite growth in the East Athabasca mylonite triangle. (From Williams & Jercinovic (2002), Microprobe monazite geochronology: putting absolute time into a microstructure analysis, *Journal of Structural Geology*, **24**, 1013–28. © 2002 Elsevier. Modified by permission of Elsevier.) At the time of going to press the original colour version was available for download from www.cambridge.org/9780521066068

divided into structurally and metamorphically defined units that range from hornblende tonalite with a coarse igneous texture and weak deformational fabric, to intensely deformed and annealed hornblende (± garnet) "straight gneiss" (Fig. 7.7a), to garnet–clinopyroxene ribbon mylonite along the northwest margin of the block (Hanmer, 1997a). The Chipman tonalite is intruded by a swarm of mafic dikes, the Chipman dikes, ranging from 1 m to tens of meters thick (Fig. 7.7a) (Hanmer *et al.*, 1994; Williams *et al.*, 1995; Hanmer, 1997a). The main part of the swarm coincides with a positive gravity anomaly, which may indicate the presence of a large mass of mafic material at depth (Hanmer *et al.*, 1994).

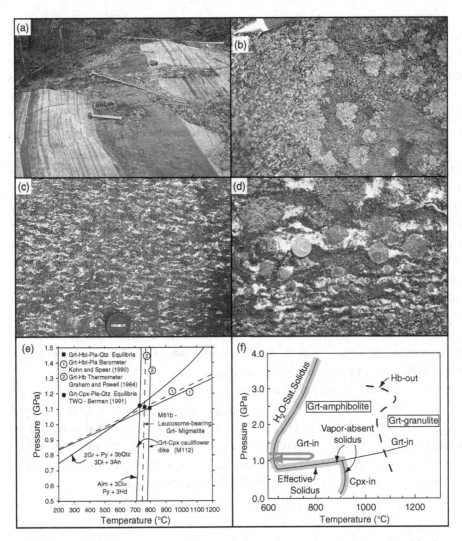

Fig. 7.7. Photographs and interpretations from the southeastern block (Chipman tonalite and Chipman dike swarm). (a) Field photo of Chipman dike in tonalite straight gneiss. Igneous assemblages include Hbl, Pl, Ilm, ± Cpx. (b) Cauliflower dike with metamorphic Grt + Cpx + Ilm clusters. (c) Fine-grained Grt migmatite Chipman dike with small Grt + leucosome segregations. (d) Coarser Grt-migmatite dike. Note that some leucosome has merged into more continuous segregations. (e) Selected equilibria from a Grt–Hbl–Pl migmatite (M61b) and a Grt–Cpx–Hbl–Pl cauliflower texture dike (M112). Grt–Hbl–Pl–Qtz and Grt–Cpx–Pl–Qtz are shown for Sample M112 illustrating consistency of Hbl-bearing and Hbl-absent equilibria. (f) Location of vapor-absent solidus for amphibolite (after Rapp *et al.*, 1991). Note that solidus temperature decreases significantly above 0.8 GPa where Grt is a product phase. Interpreted *P–T* path for the Chipman migmatite dikes is shown for reference (after Williams *et al.*, 1995).

The Chipman dikes were synkinematically emplaced during a regional shearing and shortening event (Williams *et al.*, 1995; Hanmer, 1997a). The oldest dikes were isoclinally folded several times and display multiple foliations. Younger dikes have sharp contacts that truncate older dikes, as well as the fabric in the strongly deformed host tonalite. The Chipman dikes may be separated into two main textural and mineralogical varieties: (1) non-migmatitic dikes in which fine-grained garnet and clinopyroxene occur in spheroidal ("cauliflower") concentrations in a medium- to coarse-grained, hornblende + plagioclase matrix (Fig. 7.7b) (Williams *et al.*, 1995), and (2) migmatitic dikes in which garnet crystals, up to 5 cm diameter, are associated with distinctive quartz–plagioclase (tonalite) leucosome (Fig. 7.7c, d). The migmatitic dikes are particularly interesting with respect to petrologic and structural processes in the deep crust. In dikes with relatively small garnet porphyroblasts (<2 cm diameter), virtually all leucosomes occur adjacent to garnet, with volume proportional to the size of the garnet crystal. Dikes with relatively large garnet–leucosome clusters also contain pods, veins, or gashes filled with leucosome (Williams *et al.*, 1995), interpreted to have coalesced from around a number of garnet crystals. Most of the tonalite leucosomes are aligned with the dominant foliation, but some display classic C–S fabric elements, or asymmetric tails in pressure shadows on garnet.

All of the dikes investigated indicate high-pressure granulite facies conditions, with calculated temperatures on the order of 750–850 °C and pressures of approximately 1.0 GPa (Fig. 7.7e), regardless of mineral assemblage (Williams *et al.*, 1995). Experimental studies of phase relations suggest that at pressures greater than ~0.8 GPa, garnet is an important product phase in the vapor-absent melting of amphibolite by a reaction such as

$$Hbl + Pl = Grt + Cpx + tonalite/trondhjemite\ melt \qquad (7.2)$$

(Rushmer, 1991; Rapp *et al.*, 1991; Beard & Lofgren, 1991; Sawyer, 1991; Wolf & Wyllie, 1993, 1994). Comparisons with experimental data of Wolf and Wyllie (1993, 1994) suggest melting temperatures of 800–1000 °C for the Chipman dikes and pressures greater than 0.8 GPa, consistent with our calculations (Fig. 7.7f). All textural and compositional data indicate that the Chipman dikes provide an example of a process capable of producing tonalitic magma in the lower crust by partial melting of amphibolite at high pressure (Williams *et al.*, 1995; Percival, 1983; Hartel & Pattison, 1996).

The Chipman dikes may be representative of a model for the deep parts of arcs or other regions characterized by syntectonic under-plating or inter-plating of mafic magma. The dikes were synkinematically emplaced into the

Chipman tonalite during sinistral shearing at granulite facies conditions (Hanmer *et al.*, 1994; Williams *et al.*, 1995; Hanmer, 1997a). New dikes were injected while older dikes underwent progressive deformation, metamorphism, and/or partial melting. An important component of heat for metamorphism and partial melting of dikes is interpreted to have come from later dikes and subjacent mafic magma related to the dike swarm. The relatively constant ~1.0 GPa pressures derived from all dikes indicate a period of isobaric deformation and metamorphism, at least during the period of dike emplacement.

Field relations and initial geochronologic data suggested that the dikes were emplaced during the Archean (Hanmer *et al.*, 1994). Although it is likely that multiple ages of dikes are present, new geochronologic data suggest that the predominant generation of dikes, and production of tonalite melt in the dikes, occurred at *c.* 1.9 Ga (Baldwin *et al.*, 2000; Flowers *et al.*, 2002). Furthermore, similar dates from titanite suggest that the dikes cooled rapidly, either because the regional grade of the Chipman tonalite was significantly cooler than local, dike-related temperatures, or because the entire southeastern block was rapidly cooled after metamorphism (Baldwin *et al.*, 2000).

7.3.3 Upper deck

The upper deck is characterized by a pervasive granulite facies mylonitic foliation, with a moderate southward dip, and a strike that curves progressively from northeast to northwest (Hanmer *et al.*, 1994; Hanmer, 1997a, 2000; Fig. 7.3). Taking the mylonite foliation to be parallel to the base of the upper deck, the latter lies structurally above and discordant to the northwestern and southeastern blocks of the lower deck. The upper deck is divided into southern and northern parts that differ somewhat in lithological proportions, but are most obviously distinguished on the basis of their metamorphic histories. It is separated from the lower deck by a domain of generally late syn-tectonic, mylonitic to isotropic, *c.* 2.6 Ga hornblende–biotite "interface granites" (Fig. 7.3) that appear to have been deformed under lower-temperature conditions compared to the adjacent upper deck (Hanmer *et al.*, 1994; Hanmer, 1997a).

High P–T rocks of the Honsval Lake area

(Fig. 7.8a) The predominant rock type of the structurally lower, northern part of the upper deck is a garnetiferous felsic gneiss, with anhydrous assemblages containing garnet, feldspar (alkali or ternary), quartz, kyanite, and rutile, plus accessory graphite, zircon, and chlorapatite. Garnet is Mg-rich, up to Py_{47}.

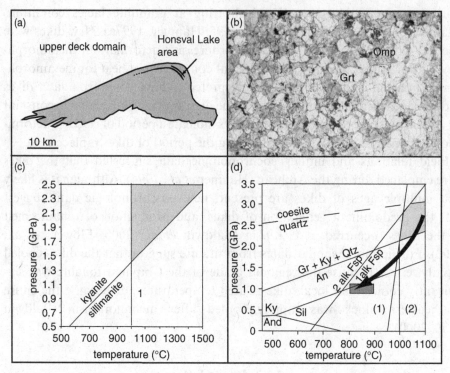

Fig. 7.8. Geologic data from the Honsval Lake area of the upper deck block. (a) Location of two eclogite (Grt–Omp) layers from the northern, structurally lower, part of the upper deck block. The layers may continue to the southwest but are discontinuous. (b) Photomicrograph of eclogite from Honsval Lake area. Some omphacite (Omp) grains are partially replaced by pargasitic amphibole. (c) Approximate Ky–Sil boundary and 1000 °C isotherm (based on feldspar data) constrain pressure to above 1.5 GPa. (d) Proposed early *P–T* path for Honsval Lake gneisses. Temperature constraints – (1) Grt–Cpx thermometry and (2) feldspar solvus data; pressure constraint – An-out reaction (Snoeyenbos *et al.*, 1995; see also Baldwin *et al.*, 2003a, 2004).

Layers of mafic granulite, typically garnet orthopyroxenite and garnet clinopyroxenite, occur within the gneiss, but at least two relatively continuous eclogitic layers with garnet and sodian clinopyroxenite (Fig. 7.8b) have been mapped in the Honsval Lake area. The layers are 3–10 m thick and have been mapped for more than 10 km along strike (Fig. 7.8a) (Snoeyenbos *et al.*, 1995; Baldwin *et al.*, 2000, 2003a, 2004). They have matrix corundum, anorthite–corundum symplectites after kyanite, and kyanite inclusions in garnet. Adjacent to the mafic layers are spectacular sapphirine-bearing rocks with delicate symplectic corundum–anorthite, sapphirine–anorthite and spinel–anorthite microstructures (Baldwin *et al.*, 2003b).

A characteristic feature of the felsic gneisses is the common association of kyanite and ternary feldspar. Because the feldspar has minimum homogenization temperatures of at least 1000 °C, the presence of kyanite requires a pressure of 1.5 GPa or greater (Fig. 7.8c, d) (Snoeyenbos *et al.*, 1995). Similar minimum *P* and *T* estimates are derived from the Honsval Lake eclogite, where Grt–Cpx exchange temperatures of 980 °C are recorded, and the presence of corundum in that assemblage requires ~1.6 GPa (Snoeyenbos *et al.*, 1995). Pressures ranging from 1.6 to 1.0 GPa are recorded in mafic granulite layers (Baldwin *et al.*, 2003a); the range is taken to be the result of varying degrees of re-equilibration during subsequent granulite facies metamorphism. Very high *P–T* conditions occurred in the upper deck at some time before the juxtaposition of the upper deck with the lower deck blocks (Hanmer *et al.*, 1994; Hanmer, 1997a; Baldwin *et al.*, 2000). Initial juxtaposition may have occurred as early as 2.6 Ga (Hanmer, 1997a), but the final juxtaposition of the upper deck with the other domains apparently occurred at 1.9 Ga during metamorphism and mafic dike emplacement (Baldwin *et al.*, 2003a, 2004).

Deep-crustal granulite facies *P–T* conditions (~0.8–1.0 GPa) were recorded at higher structural levels in the upper deck (see below; Kopf, 1999) and within the upper-deck–lower-deck boundary region (Williams *et al.*, 2000a; Williams, unpublished data), but the eclogitic and mafic granulite rocks in the Honsval Lake area seem to preserve an incomplete reworking at granulite facies conditions, presumably due to their anhydrous and refractory nature. High-*P* assemblages in felsic gneisses are well preserved as inclusions in garnet and locally in the matrix; in mafic rocks, high-*P* assemblages are best preserved in relatively low-strain domains.

Upper deck mafic granulites

The structurally higher, southern, part of the upper deck consists of an approximately 5-km thick exposure of layered mafic granulites interleaved with subordinate felsic gneiss (Fig. 7.9a). The latter is similar to that in the underlying southern part, but lacks the very high-pressure, kyanite-bearing assemblages. Mafic granulite, with gabbroic texture locally preserved in low strain windows, exhibits rare internal cross-cutting relationships and contains magmatic zircon, which indicate that it was emplaced during mylonitization at *c.* 2.6 Ga (Hanmer *et al.*, 1994; Hanmer, 1997a). Locally, the mafic granulites preserve spectacular, widespread coronitic reaction textures (Fig. 7.9b; Kopf, 1999). Peak metamorphism and deformation occurred in the high-pressure granulite field (Grt + Cpx + Qtz; ~1.2 GPa at 750–800 °C). A transition to lower pressure granulite (Opx + Pl; 0.8–0.9 GPa at 700–750 °C) involved a phase of

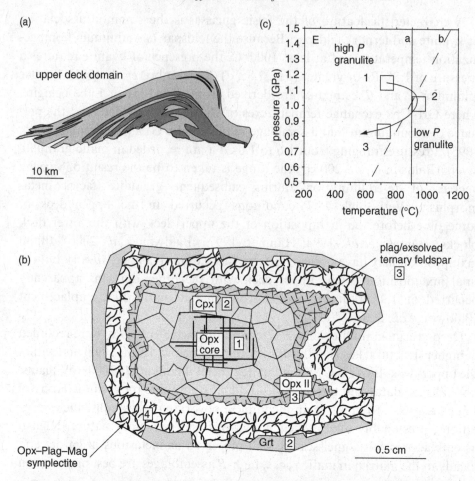

Fig. 7.9. Data from the Axis Lake mafic granulite subdomain of the upper deck block (Fig. 7.3). (a) Generalized location map showing mafic granulite sheets. Note abundance of mafic sheets in the southern, structurally higher part of the block. (b) Schematic sketch of corona textural relationships. Numbers correspond to interpreted sequence of corona minerals (see text for discussion). (c) Interpreted *P–T* history for Axis Lake mafic granulites (see Fig. 7.3). a – feldspar thermometry, b – experimentally determined location of reaction Grt + Cpx = Opx + Pl (Green & Ringwood, 1967). Boxes are calculated conditions for corona stages (after Kopf, 1999). Abbreviations after Kretz (1983).

decompression. This was probably associated with the thrust emplacement of the eclogitic base of the upper deck over the lower deck, and the subsequent development of granulite facies assemblages, followed by isobaric, post-kinematic cooling and Opx–Pl–Mag symplectite development (Fig. 7.9c).

7.3.4 Rae domain – tectonism during exhumation of high-P rocks

Rocks of the East Athabasca mylonite triangle preserve assemblages and fabrics that were developed within the deep crust (~1.0 GPa), but the structural and metamorphic record of eventual exhumation are not well developed. However, rocks to the west of the mylonite triangle preserve evidence of penetrative deformation and metamorphism after 1.9 Ga that records exhumation from the deep crust. Although detailed petrologic and structural work in this part of the Rae domain is limited, results from several small areas are summarized briefly in order to provide context for the evolution of the region in general.

The Neil Bay area (Fig. 7.10a) lies in the Rae domain, ~40 km west of the East Athabasca mylonite triangle. It is dominated by quartzofeldspathic gneisses and granitoid rocks, with subordinate amphibolite and schist. The most abundant rock type is Grt–Opx–Crd–Sil gneiss with granitic leucosomes, interpreted to represent partial melt (Kopf, 1999; Williams *et al.*, 2000a). Most rock types display evidence for an early northwest-oriented tectonic fabric (itself with elements of earlier fabrics) that has been reoriented by open to tight folds with a variably developed, northeast-striking, axial plane foliation, sub-parallel to that in the East Athabasca mylonite triangle. Leucocratic segregations are present along (and continuous between) the two foliations, suggesting that the older northwest-striking fabric was reactivated during the development of the younger northeast-striking fabric.

Based on phase relationships and calculated equilibria, Kopf (1999) interpreted a clockwise synkinematic *P–T* path with peak conditions of ~800–900 °C at ~0.8 GPa, followed by decompression and cooling (Fig. 7.10b). Garnet, generally wrapped by the earlier foliation, grew during high *P–T* biotite dehydration melting (Kopf, 1999). Cordierite, biotite, and sillimanite locally replace garnet and are generally aligned in the northeast-striking, second-generation foliation. They are interpreted to have grown during exhumation (Kopf, 1999; Fig. 7.10b). Monazite that occurs in the matrix and as inclusions in porphyroblasts shows complex compositional and age zonation. Electron microprobe dating of monazite (Williams *et al.*, 1999; Williams & Jercinovic, 2002) yields several distinct clusters of dates (*c.* 1.9, *c.* 1.84, and *c.* 1.80 Ga) that may be correlated with monazite dates from other localities in this region (Williams *et al.*, 2000a; Krikorian *et al.*, 2002) and even more broadly across the Western Churchill Province (e.g., MacLachlan *et al.*, 2000). The oldest dates (*c.* 1.9 Ga), along with local *c.* 2.0 Ga relict cores, come from monazite inclusions in high *P–T* peak metamorphic garnet (Fig. 7.10c). Monazite inclusions in cordierite yield *c.* 1.84–1.85 Ga dates, although

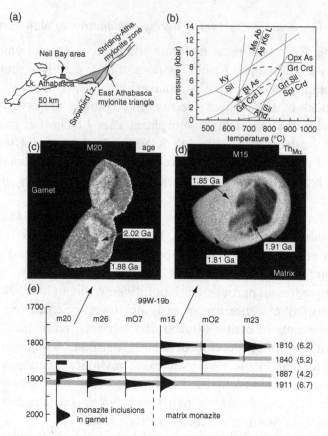

Fig. 7.10. Data from the Neil Bay area, approximately 40 km west of the East Athabasca mylonite triangle. (a) Location of Neil Bay area. (b) Interpreted *P–T* path, based on thermobarometry and phase equilibria (Kopf, 1999). (c) Microprobe age map of monazite inclusion in garnet from sample 99W–19b. Note: monazite contains inclusions of older monazite. (d) $Th_{M\alpha}$ compositional image monazite from matrix of sample 99W19b. (e) Monazite dates from six monazite grains in sample 99W–19b. Gray bars show weighted means from all monazite. (From Williams & Jercinovic (2002), Microprobe monazite geochronology: putting absolute time into a microstructure analysis. *Journal of Structural Geology*, **24**, 1013–28. © 2002 Elsevier. Modified by permission of Elsevier.) Note: monazite inclusions in garnet include only the two older-age populations.

At the time of going to press the original colour version was available for download from www.cambridge.org/9780521066068

they may have grown during heavy alteration of the host. Matrix monazite, associated with biotite and sillimanite, has *c.* 1.81 Ga rims (Fig. 7.10d), with cores of one or more of the older age populations. Taken together, the data suggest that high-temperature and high-pressure metamorphism and penetrative deformation occurred in the interval 1.9–1.84 Ga, probably close to

1.9 Ga, and was followed by decompression and retrograde growth of cordier-ite, sillimanite, and biotite.

Similar results have been obtained at two other localities in the Rae domain, the Thicke Lake and Wholdaia Lake areas, west and north of the East Athabasca mylonite triangle, respectively (Fig. 7.2). Peak metamorphic pres-sures were estimated at *c.* 0.9–1.0 GPa, and in both areas monazite inclusions in peak metamorphic garnet yield *c.* 1.9 Ga dates (Krikorian *et al.*, 2002; Williams, unpublished data). Monazite associated with late-stage amphibolite facies assemblages and with discrete deformation zones yields *c.* 1.8 Ga dates, and suggests that at least some exhumation had occurred by that time (Krikorian *et al.*, 2002; Williams *et al.*, 2000a). No evidence has been found to suggest that these parts of the Rae Province rocks were tectonized under deep-crustal conditions at 2.6 Ga, but Ashton *et al.* (2002) have interpreted medium- to high-grade tectonism in parts of the domain, and other such evidence may have been destroyed during Paleoproterozoic deep-crustal (*c.* 1.9 Ga) and exhumation-related deformation.

7.3.5 Eastern boundary of deep crustal rocks – the Legs Lake shear zone

Metamorphic grade in the southern Hearne domain is lower than in the Rae domain at this latitude (Lewry & Sibbald, 1977, 1980; Hanmer, 1997a; Harper *et al.*, 2001; Aspler *et al.*, 2002), and thus the eastern limit of the East Athabasca mylonite triangle roughly corresponds to the edge of the large deep-crustal terrane. Recent work has documented a 6–8-km wide, west-dipping, amphibolite facies shear zone referred to as the Legs Lake shear zone (Mahan *et al.*, 2001, 2003) that appears to have accommodated at least 20 km of uplift of the high-pressure terrane. The shear zone records at least two major events: (1) early oblique thrusting that placed the high-grade rocks over lower grade rocks to the east, and (2) a more discrete lower grade normal fault zone (roughly corresponding to the Black Lake fault zone of Gilboy, 1980) that, along with similar faults, may record the later part of the exhumation history.

Preliminary geochronology (Mahan & Williams, 2002; Mahan *et al.*, 2003) indicates that the early phase of east-vergent thrusting occurred during the interval from *c.* 1.90 to 1.855 Ga. This appears to be part of a regional-scale exhumation event recorded from Neil Bay to the east that we suggest was driven by plate convergence during the early stages of the Trans-Hudson orogeny. The erosional and/or tectonic denudation processes that were directly responsible for stripping away material at the surface would have operated essentially coeval with uplift in the shear zone. Final exhumation to the surface may have occurred as much as 100 Myr later (*c.* 1.78 Ga) during extensional faulting at more shallow levels. It seems clear that exhumation of the deep crustal rocks was a multistage process, involving both thrust and normal sense shearing and a possible extended residence at mid-crustal levels before final exhumation to the surface (Mahan *et al.*, 2003).

The recognition of thrust-related uplift (and associated exhumation) is an important component in recognizing deep-crustal isobarically cooled terranes, because it provides an answer to a persistent question typically asked about such regions: if the terrane originally resided in the deep crust for a long period of time, what underlies the uplifted high-*P* terrane today? At least part of the answer is that the high-*P* terrane was thrust over the adjacent Hearne domain. The Legs Lake shear zone is considered to be one of a set of structures that regionally accommodated the uplift and exhumation of deep crustal rocks, including a west-vergent fault zone further north, localized on the west side of the high-*P* rocks (Relf & Hanmer, 2000; Berman *et al.*, 2002; Jones *et al.*, 2002) and it may directly correlate with the Virgin River shear zone to the south (Fig. 7.2) (Lewry & Sibbald, 1980; Mahan *et al.*, 2003).

7.3.6 Summary

In summary, the East Athabasca mylonite triangle and a large region in the adjacent Rae domain were resident in the lower crust at 1.9 Ga, and much of the East Athabasca mylonite triangle was apparently resident in the deep crust from at least 2.6 Ga to 1.9 Ga. Based on titanite dates within the East Athabasca mylonite triangle, and monazite dates within the Rae domain and within the Legs Lake shear zone east of the East Athabasca mylonite triangle, exhumation occurred between 1.9 and 1.8 Ga. However, whereas rocks in the East Athabasca mylonite triangle experienced only local deformation during this exhumation, rocks to the west in the Rae domain underwent more penetrative deformation and metamorphism during this event. These later events substantially obscure evidence for the earlier history of the Rae Province rocks. The large-scale partitioning of exhumation-related deformation in the lower crustal terrane is an interesting characteristic. It may be that the extended high-*P*–*T* history of the East Athabasca mylonite triangle left this region relatively dry and strong compared to the adjacent Rae Province, but it is not clear at present that parts of the Rae Province did not also experience an older deep-crustal history (see Ashton *et al.*, 2000, 2002). Exhumation was partly accommodated by early thrusting and later normal faulting along the Legs Lake shear zone and its extensions to the north and south (Mahan *et al.*, 2003).

7.4 Discussion

7.4.1 East Athabasca mylonite triangle: an exhumed, isobarically cooled, high-pressure lower crustal terrane

All three structural blocks of the East Athabasca mylonite triangle experienced granulite facies metamorphism at pressures corresponding to crustal depths of ~ 35 km. However, the preserved assemblages appear to have formed along different deformation and $P–T$ paths, and at different times. The northwestern block and the upper deck both preserve evidence for a Neoarchean granulite facies deformation associated with synkinematic injection of mafic to granitic magmas at c. 2.62–2.60 Ga that advect heat (Hanmer, 1997b). Metamorphic zircon and monazite in the upper deck indicate multiple thermal events that post-date 2.6 Ga, with significant growth of zircon at c. 1.9 Ga (Baldwin et al., 2000, 2003a, 2004). Very high $P–T$ metamorphism in the upper deck occurred during the Paleoproterozoic (c. 1.9 Ga) event and granulite facies metamorphism and deformation probably also occurred in the Neoarchean. The absence of very high pressures in the lower deck suggests that the upper deck was structurally emplaced over the latter, probably along a discrete thrust fault or shear zone, during the transition from eclogite to granulite facies.

Deformation textures in the Neoarchean "interface granites," emplaced just below the upper deck/lower deck interface (Fig. 7.3), and along the east flank of the southeastern block demonstrate that the East Athabasca mylonite triangle had cooled to sub-granulite temperatures by c. 2.60 Ga (Hanmer et al., 1994; Hanmer, 1997a). In the southeastern block of the lower deck, map relationships and outcrop-scale intrusive relationships suggested initially that granulite facies shearing and mafic dike emplacement occurred prior to 2.6 Ga. However, new geochronology from leucosome in synkinematic mafic dikes indicates a granulite facies sinistral shearing event at c. 1.9 Ga. A positive gravity anomaly beneath the southeastern block suggests the presence of a mafic magmatic body that may have advected the heat for the Paleoproterozoic metamorphism. The apparent extension of this anomaly beneath the upper deck implies that the mafic body could have been responsible for the c. 1.9 Ga thermal event there. In summary, the blocks of the East Athabasca mylonite zone were metamorphosed and deformed in the deep crust at c. 2.62–2.60 Ga, where they remained until c. 1.9 Ga when they were subjected to an intense Paleoproterozoic thermal event associated with more localized granulite facies deformation.

Exhumation of the lower crust

A large area in the western Churchill Province, including the Striding Athabasca mylonite triangle, the Rae domain wall rocks, and large areas further north in the Province, record a high-T thermal event at c. 1.9 Ga and exhumation during the period c. 1.9–1.78 Ga. Some workers interpret the high-T event to reflect the emplacement of mafic magma into the deep crust, perhaps associated with a deeper mafic underplate (Sanborn-Barrie *et al.*, 2001; Berman *et al.*, 2002; Baldwin *et al.*, 2003a). Zircon and titanite data from the Chipman dikes in the southeastern block suggest that the c. 1.9 Ga thermal event decayed rapidly (Baldwin *et al.*, 2000), and monazite data from the Legs Lake shear zone indicate that the East Athabasca mylonite triangle was thrusted over the Hearne domain by c. 1.86 Ga (Mahan & Williams, 2002). Elsewhere, in the Chesterfield Inlet area (Fig. 7.2), the Kramanituar complex records rapid tectonic exhumation and cooling of deep-crustal (c. 1.2 GPa) granulite facies rocks at 1.9 Ga (Sanborn-Barrie *et al.*, 2001). However, the processes associated with rapid cooling at c. 1.9 Ga remain uncertain. In contrast, between the East Athabasca mylonite triangle and Chesterfield Inlet, the northwestern Hearne subdomain (Fig. 7.2) locally recorded lower-crustal pressures ($\sim$1.0 GPa) at c. 1.9 Ga (Berman *et al.*, 2000) and was exhumed prior to deposition of the unmetamorphosed, post-1.84 Ga Dubawnt Supergroup (Baker Lake basin (Fig. 7.2); Rainbird *et al.*, 2002). In brief, exhumation of the East Athabasca mylonite triangle appears to be associated with polyphase, broader-scale processes that uplifted lower-crustal rocks from Lake Athabasca to Chesterfield Inlet.

The potential period of exhumation broadly overlaps with the two principal tectonic events in the western Canadian Shield (i.e., the final stages of the 1.97–1.90 Ga Thelon orogeny (e.g., Bowring & Grotzinger, 1992; Hanmer *et al.*, 1992) and much of 1.85–1.80 Ga Trans-Hudson orogeny (e.g., Lucas *et al.*, 1999; Fig. 7.2)). Far-field stresses associated with the latter induced shortening within the southwestern part of the Churchill Province (Ross *et al.*, 2000), and may have been instrumental in driving the contemporaneous exhumation discussed here. A northwest–southeast oriented, co-located magnetotelluric and teleseismic transect across the northern Snowbird tectonic zone at Baker Lake has imaged what appears to be the Rae–Hearne interface dipping south-east beneath the Hearne domain and rising to the surface near the northeast-trending Snowbird tectonic zone (Fig. 7.2; Jones *et al.*, 2000, 2002). If so, this east-dipping exhumation structure in the north may be linked to the west-dipping Legs Lake exhumation structure to the south. By analogy with the Kapuskasing structure (Percival & West, 1994), Paleoproterozoic shortening of the western Churchill Province may have induced activation (or reactivation) of the Rae–Hearne interface as contractional, dip-slip faults that tilted and obliquely exposed the lower crust (Relf & Hanmer, 2000).

Uncertainties in the Isobaric Cooling Model

The model of isobaric cooling and long-term residence in the deep crust of the East Athabasca mylonite triangle, or any other high-*P* terrane, requires additional testing and verification. In other areas where models involving isobaric cooling and counter-clockwise *P–T* paths have been proposed, alternative models with multiple clockwise *P–T* paths may also be viable (Collins & Vernon, 1991; Hand *et al.*, 1992; Selverstone & Chamberlain, 1990). It is particularly difficult to verify that exhumation has not occurred between successive high-*P* tectonic events, or prior to the events that led to final exhumation. However, whereas in modern orogenic belts, clockwise isothermal decompression *P–T* paths might be expected (England & Thompson, 1984), in ancient metamorphic belts (especially polyphase belts) looping *P–T* paths, and isobaric cooling models might be the expected type of *P–T* path in rocks preserved at the present surface. At the end of a given regional metamorphic event, after the crust has regained isostatic equilibrium, metamorphic rocks that reside in the upper crust would be characterized by a full clockwise *P–T* path terminating at low, near surface, pressures, but the lower crust would be characterized by metamorphic rocks that, after initial decompression, cooled isobarically at depth. A later tectonic disturbance that leads to a renewed crustal thickening would be expected to result in exhumation of isobarically cooled levels of the crust. For the remainder of this chapter, we consider some of the implications of the East Athabasca area as a sample of the deep crust of the Canadian Shield.

7.4.2 Nature of the deep crust: implications of the East Athabasca mylonite triangle

The East Athabasca mylonite triangle represents a region of lower crust dominated by intrusive igneous rocks, ranging from mafic to granitic in composition, intruded as synkinematic and synmetamorphic plutons, sheets, or dike swarms. The mafic components are compositionally consistent with derivation by partial melting of upper mantle sources with minor contamination by older continental crust (Hanmer *et al.*, 1994; Williams *et al.*, 1995; Hanmer, 1997a). The felsic and granitic components represent crustal melts generated either in-situ (diatexites), or as magmas that were emplaced into the lower crust just above their sources. These rocks were subsequently modified by deformation and metamorphism during progressive shearing (Fig. 7.1, path 3).

The Mary granite of the northwestern block and the mafic dikes of the southeastern block both in the lower deck provide a detailed picture of the production and evolution of orthogneiss in the deep crust. In the southeastern block, ongoing injection of mafic magma led to partial melting of earlier dikes and their host tonalite, yielding new tonalitic melt and, locally, more felsic melts.

The upper deck had a complex early history involving very high-P metamorphism, followed by decompression (exhumation) to the level of the deep crust, where it experienced injection of voluminous mafic melts, granulite facies conditions, and subsequent isobaric cooling (similar to Path 1, Fig. 7.1). The mechanism and tectonic setting of the early high-P metamorphism are not well known, but all evidence indicates that juxtaposition of the high-P terrane with the other lower crustal blocks occurred within the deep crust. The locally coarse inter-layering of mafic and felsic layers in the upper deck may offer one model for anomalous deep-crustal seismic reflectivity.

In all parts of the East Athabasca mylonite triangle, during the Neoarchean and the Paleoproterozoic, heat advected by igneous intrusions appears to have weakened the lower crust and localized further deformation, thereby facilitating additional intrusion in a feedback relation between deformation, magmatic emplacement, and metamorphism, that led to very pronounced, but regionally heterogeneous, mylonitic fabrics (Hanmer, 1997b; Williams *et al.*, 2000b). A similar relationship between magmatically advected heat and localization of shear zones has been demonstrated along strike in the Chesterfield Inlet area in the Kramanituar complex at *c.* 1.9 Ga (Sanborn-Barrie *et al.*, 2001) and in the Uvauk complex, potentially at *c.* 2.5 Ga and again at *c.* 1.94–1.9 Ga (Mills *et al.*, 2000; Mills, 2001; Fig. 7.2). In both of these cases, pressures of ~1.2 GPa have been documented.

Most importantly, the three blocks of the East Athabasca mylonite triangle were co-resident in the lower crust (with some localized adjustment) from *c.* 2.6 Ga to at least 1.9 Ga. The region, therefore, represents an accessible sample of the Laurentian deep crust during this period, and is a potential analog for one variety of deep crust in relatively stable regions. Perhaps the single most overwhelming feature of the East Athabasca mylonite triangle is its heterogeneity on a broad range of scales. The largest scale is that of the principal blocks themselves, but on a smaller scale, each of the individual blocks preserves a broad range of bulk compositions, deformation fabrics, and metamorphic assemblages, ranging from primary igneous textures and minerals to dramatically recrystallized granulite facies mylonites. If the components of the

East Athabasca mylonite triangle are typical of at least some parts of the deep crust in general, they emphasize the futility of trying to define a single deep-crustal composition or structural configuration, even for a relatively small region.

One particularly notable aspect of the heterogeneity of the East Athabasca area is the variability of igneous rock types and processes. The region is characterized by large sheet-like intrusions, voluminous plutons, discrete dikes and veins, and broad areas of migmatites, all involving a range from mafic to felsic compositions. Even within this small area, it would not be possible to generalize about the manner with which igneous rocks are generated, contaminated, or transported from the deep crust. Insights may be gained from direct observation about specific igneous processes by focusing on individual areas at certain times, but as with other levels of the crust, heterogeneity in space and time appears to be the rule. Had this region been sampled as xenoliths, the East Athabasca mylonite triangle would have produced highly variable assemblages and textures from primary igneous to granulite facies mylonites that might have been mistakenly interpreted as a depth or time sequence. Similarly, it would be dangerous to generalize, from igneous rocks sampled at a higher level, about the composition of this part of the lower crust.

7.4.3 Speculations: how representative is the East Athabasca mylonite triangle?

Perhaps the overriding question in interpreting the East Athabasca mylonite triangle as a sample of the deep crust concerns the degree to which it is representative of other parts of the ancient or present-day lower continental crust. Elements similar to those in the East Athabasca mylonite triangle have also been described from other high-P terranes, including the Kapuskasing uplift, Canada (Percival & Card, 1983; Percival & McGrath, 1986), the Ivrea zone of the southern Alps (Burke & Fountain, 1990; Barboza *et al.*, 1999), the Pikwitonei domain, Manitoba, Canada, (Mezger, *et al.*, 1990) and others (see Fountain, 1989; Percival *et al.*, 1992; Harley, 1989). One of the best-documented examples is the Napier Complex of Enderby Land, Antarctica (Sheraton & Black, 1983; Harley, 1985; Harley & Black, 1987; Sandiford, 1985; Harley & Black, 1997).

In the Napier Complex, a broad variety of rock types is exposed, including tonalitic to granitic orthogneisses interlayered with a variety of metasedimentary rocks and cut by several generations of mafic dikes (Sheraton & Black, 1983). The region underwent two or perhaps three granulite facies tectonothermal events (possibly 2.98, 2.82, 2.48–2.45 Ga (Harley & Black, 1997)) with pressures as high as 1.0 GPa and temperatures as high as from 800 to >1000 °C (Harley, 1985; Sandiford & Powell, 1986; Harley & Motoyoshi, 2000), and distinct *P–T–t* paths may be recorded in different blocks (Harley & Black, 1997). The terrane is interpreted to have reached deep-crustal pressures by 2.84 Ga and to have remained in the deep or middle crust until at least 2.48 Ga and perhaps until *c.* 1.0 Ga (Harley, 1985; Mezger, 1992; Harley & Black, 1987). Thus, it appears to be another sample of the Archean–Proterozoic deep crust that records a similar feedback relation between deformation, high-temperature metamorphism and plutonism to that of the East Athabasca mylonite triangle.

However, as noted previously, the presently exposed northwestern part of the Hearne domain (a tract measuring 500 × 130 km) was also resident in the lower crust at *c.* 1.9 Ga when it experienced metamorphic conditions of *c.* 1.0 GPa at about 650–700 °C (Berman *et al.*, 2000; ter Meer *et al.*, 2000; ter Meer, 2001; Stern & Berman, 2000). These data have been recovered from garnet-bearing pelites and amphibolites intercalated with hornblende–biotite metagranitoid rocks and orthogneisses for which there is no evidence of Paleoproterozoic regional deformation. There is some debate regarding the time at which these rocks first arrived in the lower crust (Aspler *et al.*, 2000, 2001; Berman *et al.*, 2000). However, polyphase, thick-skinned thrust tectonics have been identified in the northwestern Hearne subdomain and dated at *c.* 2.66–2.5 Ga (MacLachlan *et al.*, 2000; Tella *et al.*, 2001), and it is likely that this event resulted in crustal thickening and subsequent equilibration. According to this scenario, the rocks of the northwestern Hearne subdomain would have arrived in the lower crust at the end of the Neoarchean, and resided there until a thermal event related to crust–mantle interaction (magmatic intraplating?) at *c.* 1.9 Ga raised temperatures sufficiently to provoke the metamorphic event during the Paleoproterozoic. The salient point is that during the interval *c.* 2.5–1.9 Ga, the lower crust of the northwestern Hearne subdomain would have resided at temperatures below the ∼650 °C recorded during the ∼1.9 Ga metamorphic event. Even after high-pressure Paleoproterozoic metamorphism, the field aspect of these rocks is that of a relatively homogeneous amphibolite facies region, visually indistinguishable from those of many mid-crustal terranes (Chapter 8).

We speculate that the East Athabasca mylonite triangle may represent one end-member in the spectrum of lower crustal compositions and histories, one that might be called "hot" deep crust. It involved periodic emplacement of a broad range of mafic and felsic magmas, and a close inter-relationship between deformation and granulite facies metamorphism. However, it is clearly part of a regional-scale, localized shear zone system (at least for the later parts of its history), and as such, it could be argued that it is representative of only a small proportion of the Proterozoic deep crust. Alternatively, we would counter that the vertical orientation of the shear zone permitted mantle-derived and associated crustal melts to penetrate further into the lower crust, thereby increasing the chances that they and their associated granulites would eventually be exposed at Earth's surface.

We suggest that the East Athabasca mylonite triangle is an accessible analog of "hot," intra-plated, lowermost continental crust, much as it might appear not far above the Moho. Because of their relatively active tectonic character, such "hot" deep-crustal terranes may have been preferentially sampled by igneous processes and by xenoliths, and may have been preferentially recognized as high-P terranes because of their distinctive high-P–T metamorphic assemblages and strong deformation fabrics. However, if rocks of the northwestern Hearne domain are also representative, other parts of the lower crust appear to have experienced a cooler, more passive residence in the deep crust. Samples of these regions, both as xenoliths and as high-P terranes, may have gone unrecognized because of their non-distinctive mineralogy. Clearly, one key to gaining a more complete understanding of the character and evolution of the deep crust is recognizing more varieties of deep-crustal isobarically cooled terranes and integrating them to provide a more complete picture of the deep crust at particular times in the geologic past.

7.5 Summary and conclusions

Isobarically cooled terranes like the East-Athabasca mylonite triangle are an important end-member type of metamorphic terrane. They provide an opportunity to map geologic relationships and to interpret structural and metamorphic processes that characterize a particular level in the crust. The East Athabasca mylonite triangle is important because it provides an extensive window into the lowermost crust in a tectonically active and reactivated setting. Perhaps the most striking feature of this terrane is the extreme heterogeneity of almost all geologic characteristics, from composition to strain fabrics to tectonic history. A second observation, that would be difficult to realize without this sort of exposure, is the critical role that advected heat from the deeper crust or mantle plays in localizing deformation and metamorphism, and the close interrelationship between deformation, metamorphism, and igneous processes.

Clearly, the generalizations that may be made from any one terrane about crustal behavior are limited. Each region may preserve only a restricted set of tectonic environments. Interpretations and generalizations depend critically on recognizing and characterizing the specific tectonic environment and history of a particular isobarically cooled terrane, and then integrating observations from other such terranes and with information from remote data sets. For investigating large-scale processes such as the complete cycle of generation, transfer, and emplacement of igneous rocks, or the relationship between deep and mid-crustal deformation zones, it will be necessary to link isobarically cooled terranes both vertically and horizontally in order to build a realistic field-based model for the crust as a whole.

7.6 Acknowledgments

MLW is indebted to the Geological Survey of Canada for logistical support during field research in northern Saskatchewan. Field research during 1994 and 1996 and laboratory work at the University of Massachusetts were supported by NSF grants EAR-9106001 and EAR-9406443. The authors thank John Percival, Simon Harley, and Subhas Tella for careful and constructive reviews. SH is especially grateful to Dominique Chardon and staff at the Center Européen de Recherche et d'Enseignement (CEREGE), Université d'Aix-Marseille III, for their hospitality while this chapter was in preparation. Preliminary reviews of this chapter by Sheila Seaman and Kevin Mahan are much appreciated. This is Geological Survey of Canada contribution number 2001088.

References

Ashton, K. E., Hartlaub, R. P., Heaman, L. M. and Card, C. D. (2000). The northeastern Rae Province in Saskatchewan. *Geological Association of Canada Annual Meeting Abstracts*, 696.

Ashton, K. E., Hartlaub, R. P., Heaman, L. M. and Card, C. D. (2002). Neoarchean history of the Rae Province in northern Saskatchewan: insights into Archean tectonism. *Geological Association of Canada Annual Meeting Abstracts*, **27**, 4.

Aspler, L. B., Chiarenzelli, J. R., Cousens, B. L., Davis, W. J., McNicoll, V. J. and Rainbird, R. H. (2000). Intracratonic basin processes from breakup of Kenorland to assembly of Laurentia: new geochronology and models for Hurwitz Basin, Western Churchill Province. *Geological Association of Canada Annual Meeting Abstracts*, 795.

Aspler, L. B., Chiarenzelli, J. R., Cousens, B. L., McNicoll, V. J. and Davis, W. J. (2001). Paleoproterozoic intracratonic basin processes, from breakup of Kenorland to assembly of Laurentia: Hurwitz Basin, Nunavut, Canada. *Sedimentary Geology*, **141–2**, 287–318.

Aspler, L. B., Chiarenzelli, J. R. and McNicoll, V. J. (2002). Paleoproterozoic basement-cover infolding and thick-skinned thrusting in Hearne domain,

Nunavut, Canada: intracratonic response to Trans-Hudson orogen. *Precambrian Research*, **116**, 331–54.

Baldwin, J. A., Bowring, S. A. and Williams, M. L. (2000). U–Pb geochronological constraints on the nature and timing of high-grade metamorphism in the Striding–Athabasca mylonite zone, northern Saskatchewan, Canada. *Geological Association of Canada Annual Meeting Abstracts*, 892.

Baldwin, J. A., Bowring, S. A. and Williams, M. L. (2003a). Petrologic and geochronologic constraints on high-pressure, high-temperature metamorphism in the Snowbird tectonic zone, Canada. *Journal of Metamorphic Geology*, **21**, 81–98.

Baldwin, J. A., Goncalves, P., Williams, M. L. and Bowring, S. A. (2003b). Modelling the decompression P–T–t path in sapphirine granulites associated with eclogites, East Athabasca Mylonite Triangle, northern Saskatchewan, Canada. *Geological Society of America Abstracts with Programs*, **35**, 89–5.

Baldwin, J. A., Bowring, S. A., Williams, M. L. and Williams, I. S. (2004). Eclogites of the Snowbird tectonic zone: petrological and U–Pb geochronological evidence for Paleoproterozoic high-pressure metamorphism in the western Canadian Shield. *Contributions to Mineralogy and Petrology*, doi: 10.1007/s00410–004–0572–4.

Barboza, S. A., Bergantz, G. W. and Brown, M. (1999). Regional granulite facies metamorphism in the Ivrea Zone; is the Mafic Complex the smoking gun or a red herring? *Geology*, **27**, 447–50.

Beard, J. S. and Lofgren, G. E. (1991). Dehydration melting and water-saturated melting of basaltic and andesitic greenstones and amphibolites at 1, 3, and 6.9kb. *Journal of Petrology*, **32**, 365–401.

Berman, R. G. (1991). Thermobarometry using multi-equilibrium calculations: a new technique, with petrological applications. *Canadian Mineralogist*, **29**, 833–55.

Berman, R., Ryan, J. J., Tella, S., Sanborn-Barrie, M., Stern, R., Aspler, L., Hanmer, S. and Davis, W. (2000). The case of multiple metamorphic events in the Western Churchill Province: evidence from linked thermobarometric and in-situ SHRIMP data, and jury deliberations. *Geological Association of Canada Annual Meeting Abstracts*, 836.

Berman, R. G., Pehrsson, S. J., Davis, W. J., Snyder, D. and Tella, S. (2002). A new model for *c.* 1.9 Ga tectonometamorphism in the western Churchill Province: linked upper crustal thickening and lower crustal exhumation. *Geological Association of Canada – Mineralogical Association of Canada, Program with Abstracts*, **27**, 9.

Bowring, S. A. and Grotzinger, J. P. (1992). Implications of new chronostratigraphy for tectonic evolution of Wopmay orogen, northwest Canadian Shield. *American Journal of Science*, **292**, 1–20.

Burke, K. and Fountain, D. M. (1990). Seismic properties of rocks from extended continental crust – New laboratory measurements from the Ivrea zone. *Tectonophysics*, **182**, 119–46.

Card, C. D., Bethune, K. M., Ashton, K. E. and Heaman, L. M. (2000). The Oldman–Bulyea Shear Zone: the Nevins Lake Block–Train Lake Domain Boundary, eastern Rae Province, northern Saskatchewan. *Geological Association of Canada Annual Meeting Abstracts*, 596.

Chopin, C. (2003). Ultrahigh-pressure metamorphism: tracing continental crust into the mantle. *Earth and Planetary Science Letters*, **212**, 1–14.

Clark, M. K. and Royden, L. H. (2000). Topographic ooze: building the eastern margin of Tibet by lower crustal flow. *Geology*, **28**, 703–6.

Collins, W. J. and Vernon, R. H. (1991). Orogeny associated with anticlockwise *P–T*–t paths: evidence from low-*P*, high-*T* metamorphic terranes in the Arunta inlier, central Australia. *Geology*, **19**, 835–58.

Ellis, D. J. (1987). Origin and evolution of granulites in normal thickness crust. *Geology*, **15**, 167–70.

England, P. C. and Thompson, A. B. (1984). Pressure–temperature–time paths of regional metamorphism, Part I: Heat transfer during the evolution of regions of thickened continental crust. *Journal of Petrology*, **4**, 1–30.

Farmer, G. L. (1992). Magmas as tracers of lower crustal composition: an isotopic approach. In *Continental Lower Crust*, ed. D. M. Fountain, R. Arculus and R. W. Kay. Amsterdam: Elsevier, pp. 363–90.

Flowers, R. M., Baldwin, J. A., Bowring, S. A. and Williams, M. L. (2002). Age and significance of the Proterozoic Chipman dike swarm, Snowbird Tectonic Zone, northern Saskatchewan. *Abstracts with Programs: Geological Association of Canada Annual Meeting*, **27**, 35.

Fountain, D. M. (1989). Growth and modification of the lower continental crust in extended terranes: the role of extension and magmatic underplating. In *Properties and Processes of Earth's Lower Crust*, ed. R. F. Mereu, S. Mueller and D. M. Fountain. American Geophysical Union Geophysical Monograph 51, pp. 287–99.

Fountain, D. M. and Salisbury, M. H. (1981). Exposed cross-sections through continental crust: implications for crustal structure, petrology, and evolution. *Earth and Planetary Science Letters*, **5**, 263–77.

Gilboy, C. F. (1980). Bedrock Compilation Geology: Stony Rapids Area (74?), Government of Saskatchewan, Industry and Resources, Geopublications, scale 1:250 000 scale Preliminary Geological Map.

Goodacre, A. K., Grieve, R. A. F., Halpenny, J. F. and Sharpton, V. L. (1987). Horizontal gradient of the Bouguer gravity anomaly map of Canada. In *Canadian Geophysical Atlas*, Map 5, scale 1:10 000 000, Geological Survey of Canada.

Green, D. H. and Ringwood, A. E. (1967). An experimental investigation of the gabbro to eclogite transformation and its petrological applications. *Geochimica et Cosmochimica Acta*, **31**, 767–833.

Griffin, W. L., Austreim, H., Brastad, K., Brynhi, I., Krill, A. G., Krogh, E. J., Mork, M. B. E., Qvale, H. and Torudbakken, B. (1985). High-pressure metamorphism in the Scandinavian Caledonides. In *The Caledonide Orogen – Scandinavia and Related Areas*, ed. D. G. Gee and B. A. Sturt. Chichester: Wiley, pp. 783–801.

Hand, M., Dirks, P. H. G. M., Powell, R. and Buick, I. S. (1992). How well established is isobaric cooling in Proterozoic orogenic belts? An example from the Arunta inlier, central Australia. *Geology*, **20**, 649–52.

Hanmer, S. (1994). Geology, East Athabasca mylonite triangle, Stony Rapids area, northern Saskatchewan, parts of NTS 740-O and 74-P. In *Geological Survey of Canada*, Map 1859A. Geological Survey of Canada, scale 1:100 000.

Hanmer, S. (1997a). Geology of the Striding–Athabasca mylonite zone, northern Saskatchewan and southeastern District of Mackenzie, Northwest Territories. *Geological Survey of Canada Bulletin*, **501**, 1–92.

Hanmer, S. (1997b). Shear zone reactivation at granulite facies: the importance of plutons in the localisation of viscous flow. *Journal of the Geological Society London*, **154**, 111–16.

Hanmer, S. (2000). Matrix mosaics, brittle deformation, and elongate porphyroclasts: granulite facies microstructures in the Striding–Athabasca mylonite zone, western Canada. *Journal of Structural Geology*, **22**, 947–67.

Hanmer, S., Bowring, S., van Breemen, O. and Parrish, R. (1992). Great Slave Lake shear zone, NW Canada: mylonitic record of Early Proterozoic continental convergence, collision and indentation. *Journal of Structural Geology*, **14**, 757–73.

Hanmer, S., Parrish, R., Williams, M. and Kopf, C. (1994). Striding–Athabasca mylonite zone: complex Archean deep crustal deformation in the East Athabasca Mylonite Triangle, N. Saskatchewan. *Canadian Journal of Earth Sciences*, **31**, 1287–1300.

Hanmer, S., Kopf, C. and Williams, M. (1995a). Modest movements, spectacular fabrics in a intracontinental deep-crustal strike–slip fault: Striding–Athabasca mylonite zone, NW Canadian Shield. *Journal of Structural Geology*, **17**(4), 493–507.

Hanmer, S., Williams, M. and Kopf, C. (1995b). Striding–Athabasca mylonite zone: implications for the Archean and Early Proterozoic tectonics of the western Canadian Shield. *Canadian Journal of Earth Sciences*, **32**, 178–96.

Harley, S. L. (1985). Garnet–orthopyroxene bearing granulites from Enderby Land, Antarctica: metamorphic pressure–temperature–time evolution of the Archean Napier Complex. *Journal of Petrology*, **26**, 819–56.

Harley, S. L. (1989). The origins of granulites: a metamorphic perspective. *Geological Magazine*, **126**, 215–331.

Harley, S. L. and Black, L. P. (1987). The Archean geological evolution of Enderby Land, Antarctica. In *Evolution of the Lewisian and Comparable Precambrian High-Grade Terranes*, ed. R. G. Park and J. Tarney. Geological Society Special Publication, pp. 285–96.

Harley, S. L. and Black, L. P. (1997). A revised Archean chronology for the Napier Complex, Enderby Land, from SHRIMP ion-microprobe studies. *Antarctic Science*, **9**, 74–91.

Harley, S. L. and Motoyoshi, Y. (2000). Al zoning in orthopyroxene in a sapphirine quartzite: evidence for > 1120 C UHT metamorphism in the Napier Complex, Antarctica, and implications for the entropy of sapphirine. *Contributions to Mineralogy and Petrology*, **138**, 293–307.

Harper, C. T., Kraus, J., Demmans, C. J., Huebert, C., Coulson, I. M. and Rainville, S. (2001). Phelps Lake project: geology and mineral potential of the Bonokoski Lake area (NTS 64M-11, -12, -13, and -14). *Saskatchewan Geological Survey*, Saskatchewan Energy Mines, 2001–4.2.

Hartel, T. H. D. and Pattison, D. R. M. (1996). Genesis of the Kapuskasing (Ontario) migmatitic mafic granulites by dehydration melting of amphibolite: the importance of quartz to reaction progress. *Journal of Metamorphic Geology*, **14**, 591–611.

Henry, C., Michard, A. and Chopin, C. (1993). Geometry and structural evolution of ultra-high-pressure and high-pressure rocks from the Dora-Maira Massif, Western Alps, Italy. *Journal of Structural Geology*, **15**, 965–81.

Hoffman, P. F. (1988). United Plates of America, the birth of a craton: early Proterozoic assembly and growth of Laurentia. *Annual Review of Earth and Planetary Sciences*, **16**, 543–603.

Holbrook, W. S., Mooney, W. D. and Christensen, N. I. (1992). The seismic velocity structure of the deep continental crust. In *Continental Lower Crust*, ed. D. M. Fountain, R. Arculus and R. W. Kay. Amsterdam: Elsevier, pp. 1–43.

Indares, A. (1995). Metamorphic interpretation of high-pressure-temperature metapelites with preserved growth zoning in garnet, eastern Grenville Province, Canadian Shield. *Journal of Metamorphic Geology*, **13**, 475–86.

Indares, A., Dunning, G. and Cox, R. (2000). Tectono-thermal evolution of deep crust in a Mesoproterozoic continental collision setting; the Manicouagan example. *Canadian Journal of Earth Sciences*, **37**, 325–40.

Jones, A. G. (1992). Electrical conductivity of the continental lower crust. In *Continental Lower Crust*, ed. D. M. Fountain, R. Arculus and R. W. Kay. Amsterdam: Elsevier, pp. 81–143.

Jones, A. G., Snyder, D., Asudeh, I., White, D., Eaton, D. and Clarke, G. (2000). Lithospheric architecture at the Rae–Hearne boundary revealed through magnetotelluric and seismic experiments. *Geological Association of Canada Annual Meeting Abstracts*, 502.

Jones, A. G., Snyder, D., Hanmer, S., Asudeh, I. and White, D. (2002). Magnetotelluric and teleseismic study across the Snowbird Tectonic Zone, Canadian Shield: a Neoarchean mantle suture? *Geophysical Research Letters*, **29**(17), 1829, doi: 10.1029/2002 GL015359, 2002.

Kopf, C. F. (1999). Deformation, metamorphism, and magmatism in the East Athabasca Mylonite Triangle, northern Saskatchewan: implications for the Archean and Early Proterozoic crustal structure of the Canadian Shield. Unpublished Ph.D. thesis, University of Massachusetts, Amherst, MA.

Kretz, R. (1983). Symbols for rock-forming minerals. *American Mineralogist*, **68**, 277–9.

Krikorian, L., Williams, M. L. and Kopf, C. F. (2002). Paleoproterozoic high grade metamorphism in the Wholdaia Lake segment of the Snowbird Tectonic Zone, Northwest Territories. *Geological Association of Canada Annual Meeting Abstracts*, **27**, 64.

Lafrance, B. and Sibbald, T. I. I. (1997). The Grease River shear zone: Proterozoic overprinting of the Archean Tantato Domain. In *Summary of Investigations 1997, Saskatchewan Geological Survey*, Saskatchewan Energy Mines, pp. 132–35.

Lewry, J. F. and Sibbald, T. I. I. (1977). Variation in lithology and tectonometamorphic relationships in the Precambrian basement of northern Saskatchewan. *Canadian Journal of Earth Sciences*, **14**, 1453–67.

Lewry, J. F. and Sibbald, T. I. I. (1980). Thermotectonic evolution of the Churchill Province in northern Saskatchewan. *Tectonophysics*, **68**, 5–82.

Lucas, S. B., Syme, E. C. and Ashton, K. E. (1999). Introduction to Special Issue 2 on the NATMAPShield Margin Project: The Flin Flon Belt, Trans-Hudson Orogen, Manitoba and Saskatchewan. *Canadian Journal of Earth Sciences*, **36**, 1763–5.

MacLachlan, K., Hanmer, S., Davis, W. J., Berman, R. J., Relf, C. and Aspler, L. B. (2000). Complex, protracted, Proterozoic reworking of the Western Churchill Province: the craton that wouldn't grow up? *Geological Association of Canada Annual Meeting Abstracts*, 747.

Mahan, K. H. and Williams, M. L. (2002). Constraints on the timing and metamorphic conditions in the Legs Lake shear zone and uplift of high-pressure rocks in the East Athabasca mylonite triangle, northern Saskatchewan. *Abstracts with Programs: Geological Association of Canada Annual Meeting*, **27**, 72.

Mahan, K. H., Williams, M. L. and Baldwin, J. A. (2003). Contractional uplift of deep crustal rocks along the Legs Lake shear zone, western Churchill Province, Canadian Shield. *Canadian Journal of Earth Sciences*, **40**, 1085–110.

Mahan, K. H., Williams, M. L., Baldwin, J. A. and Bowring, S. A. (2001). Juxtaposition of deep crustal and middle crustal rocks across the Legs Lake shear zone in northern Saskatchewan. *Saskatchewan Geological Survey*, Saskatchewan Energy Mines, 2001–4.2.

Mawer, C. K., Clemens, J. D., Petford, N., Warren, R. G. and Ellis, D. J. (1997). Mantle underplating, granite tectonics, and metamorphic *P–T*-t paths; discussion and reply. *Geology*, **25**, 763–5.

Mezger, K. (1992). Temporal evolution of regional granulite terranes: implications for the formation of lowermost continental crust. In *Continental Lower Crust*, ed. D. M. Fountain, R. Arculus and R. W. Kay. Amsterdam: Elsevier, pp. 447–78.

Mezger, K., Bohlen, S. R. and Hanson, G. N. (1990). Metamorphic history of the Archean Pikwitonei granulite domain and the Cross Lake subprovince, Superior Province, Manitoba, Canada. *Journal of Petrology*, **31**, 483–517.

Mills, A. (2001). Tectonometamorphic investigation of the Uvauk Complex, Nunavut, Canada. *Tectonic Evolution of the Uvauk Complex, Churchill Province, Nunavut Territory, Canada*. Unpublished M.Sc. thesis. Ottawa: Carleton University.

Mills, A., Berman, R., Hanmer, S. and Davis, B. (2000). New insights into the tectonometamorphic history of the Uvauk complex, Nunavut. *Geological Association of Canada Annual Meeting Abstracts*, 733.

Mooney, W. D. and Meissner, R. (1992). Multi-genetic origin of crustal reflectivity: a review of seismic reflection profiling of the continental lower crust and Moho. In *Continental Lower Crust*, ed. D. M. Fountain, R. Arculus and R. W. Kay. Amsterdam: Elsevier, pp. 45–79.

Percival, J. A. (1983). High-grade metamorphism in the Chapleau–Foleyet area, Ontario. *American Mineralogist*, **68**, 667–86.

Percival, J. A. (1989). Granulite terranes and the lower crust of the Superior Province. In *Properties and Processes of Earth's Lower Crust*, ed. R. F. Mereu, S. Mueller and D. M. Fountain. American Geophysical Union Geophysical Monograph 51, pp. 301–10.

Percival, J. A. and Card, K. D. (1983). Archean crust as revealed in the Kapuskasing Uplift, Superior Province, Canada. *Geology*, **11**, 323–6.

Percival, J. A. and McGrath, P. H. (1986). Deep crustal structure and tectonic history of the northern Kapuskasing Uplift of Ontario: an integrated petrological–geophysical study. *Tectonics*, **5**, 553–72.

Percival, J. A. and West, G. F. (1994). The Kapuskasing Uplift: a geological and geophysical synthesis. *Canadian Journal of Earth Sciences*, **31**, 1256–86.

Percival, J. A., Fountain, D. M. and Salisbury, M. H. (1992). Exposed crustal cross sections as windows on the lower crust. In *Continental Lower Crust*, ed. D. M. Fountain, R. Arculus and R. W. Kay. Amsterdam: Elsevier, pp. 317–62.

Rainbird, R. H., Davis, W. J., Stern, R. A., Hadlari, T. and Donaldson, J. A. (2002). Integrated geochronology of the late Paleoproterozoic Baker Lake Group (Dubawnt Supergroup), Baker Lake Basin. *Geological Association of Canada – Mineralogical Association of Canada, Program with Abstracts*, **27**, 95.

Rapp, R. P., Watson, E. B. and Miller, C. F. (1991). Partial melting of amphibolite/eclogite and the origin of Archean trondhjemites and tonalites. *Precambrian Research*, **51**, 1–25.

Relf, C. and Hanmer, S. (2000). A speculative and critical summary of the current state of knowledge of the Western Churchill Province: a NATMAP perspective. *Geological Association of Canada Annual Meeting Abstracts*, 857.

Ross, G. M., Eaton, D. W., Boerner, D. E. and Miles, W. (2000). Tectonic entrapment and its role in the evolution of continental lithosphere: an example from the Precambrian of western Canada. *Tectonics*, **19**, 116–34.

Rudnick, R. L. (1992). Xenoliths – Samples of the lower continental crust. In *Continental Lower Crust*, ed. D. M. Fountain, R. Arculus and R. W. Kay. Amsterdam: Elsevier, pp. 269–316.

Rudnick, R. L. and Fountain, D. M. (1995). Nature and composition of the continental crust: a lower crustal perspective. *Reviews of Geophysics*, **33**, 267–309.

Rushmer, T. (1991). Partial melting of two amphibolites: contrasting experimental results under fluid-absent conditions. *Contributions to Mineralogy and Petrology*, **107**, 41–59.

Rutter, E. H. and Brodie, K. H. (1992). Rheology of the lower crust. In *Continental Lower Crust*, ed. D. M. Fountain, R. Arculus and R. W. Kay. Amsterdam: Elsevier, pp. 201–67.

Sanborn-Barrie, M., Carr, S. D. and Theriault, R. (2001). Geochronological constraints on metamorphism, magmatism and exhumation of deep-crustal rocks of the Kramanituar Complex, with implications for the Paleoproterozoic evolution of the Archean western Churchill Province, Canada. *Contributions to Mineralogy and Petrology*, **141**, 592–612.

Sandiford, M. (1985). The metamorphic evolution of granulites at Fyfe Hills, implications for Archean crustal thickness in Enderby Land, Antarctica. *Journal of Metamorphic Geology*, **3**, 155–78.

Sandiford, M. and Powell, R. (1986). Pyroxene exsolution in granulites at Fyfe Hills, Enderby Land, Antarctica: evidence for 1000 °C metamorphic temperatures in Archean continental crust. *American Mineralogist*, **71**, 946–54.

Sawyer, E. W. (1991). Disequilibrium melting and the rate of melt–residuum separation during migmatization of mafic rocks from the Grenville Front, Quebec. *Journal of Petrology*, **32**, 701–38.

Selverstone, J. (1985). Petrologic constraints on imbrication, metamorphism, and uplift in the SW Tauern Window, Eastern Alps. *Tectonics*, **4**, 687–704.

Selverstone, J. and Chamberlain, C.-P. (1990). Apparent isobaric cooling paths from granulites; two counterexamples from British Columbia and New Hampshire. *Geology*, **18**, 307–10.

Sheraton, J. W. and Black, L. P. (1983). Geochemistry of Precambrian gneisses: relevance for the evolution of the East Antarctic shield. *Lithos*, **16**, 273–96.

Slimmon, W. L. (1989). Bedrock compilation geology: Fond du Lac (NTS 74-O0; *Geological Map*, scale 1:250 000. Saskatchewan Geological Survey, Saskatchewan Energy and Mines.

Snoeyenbos, D. R., Williams, M. L. and Hanmer, S. (1995). Archean high-pressure metamorphism in the western Canadian shield. *European Journal of Mineralogy*, **7**, 1251–72.

Spear, F. S. (1993). *Metamorphic Phase Equilibria and Pressure–Temperature-Time Paths*. Mineralogical Society of America Monograph 1.

Stern, R. A. and Berman, R. G. (2000). Monazite U–Pb and Th–Pb geochronology by ion microprobe, with application to in situ dating of an Archean metasedimentary rock. *Chemical Geology*, **172**, 113–30.

Tella, S., Hanmer, S., Sandeman, H. A., Ryan, J. J., Mills, A., Davis, W. J., Berman, R. G., Wilkinson, L. and Kerswill, J. A. (2001). Geology, Macquoid Lake–Gibson Lake–Akunak Bay area, Nunavut, Canada. In *Geological Survey of Canada*, "A" Series Map, 2008A, scale 1:100 000.

ter Meer, M. (2001). Tectonometamorphic history of the Nowyak Complex, Nunavut, Canada. Unpublished M.Sc. thesis. Ottawa: Carleton University.

ter Meer, M., Berman, R. G., Relf, C. and Davis, W. D. (2000). Tectonometamorphic history of the Nowyak complex, Nunavut, Canada. *Geological Association of Canada Annual Meeting Abstracts*, 949.

Terry, M. P., Robinson, P., Hamilton, M. A. and Jercinovic, M. J. (2000). Monazite geochronology of UHP and HP metamorphism, deformation, and exhumation, Nordoyane, Western Gneiss Region, Norway. *American Mineralogist*, **85**, 1651–64.

Warner, M. (1990). Basalts, water, or shear zones in the lower continental crust? *Tectonophysics*, **173**, 163–74.

Williams, M. L., Hanmer, S., Kopf, C. and Darrach, M. (1995). Syntectonic generation and segregation of tonalitic melts from amphibolite dikes in the lower crust, Striding–Athabasca mylonite zone, northern Saskatchewan. *Journal of Geophysical Research*, **100**(B8), 15 717–34.

Williams, M. L. and Jercinovic, M. J. (2002). Microprobe monazite geochronology: putting absolute time into microstructural analysis. *Journal of Structural Geology*, **24**, 1013–28.

Williams, M. L., Jercinovic, M. J., Kopf, C. F., Baldwin, J. and Bowring, S. A. (2000a). Microprobe monazite geochronology: an essential tool for regional thermo-tectonic studies of the Western Churchill Province. *Geological Association of Canada Annual Meeting Abstracts*, 911.

Williams, M. L., Jercinovic, M. J. and Terry, M. (1999). High resolution "age" mapping, chemical analysis, and chemical dating of monazite using the electron microprobe: a new tool for tectonic analysis. *Geology*, **27**, 1023–6.

Williams, M. L., Melis, E. A., Kopf, C. and Hanmer, S. (2000b). Microstructural tectonometamorphic processes and the development of gneissic layering: a mechanism for metamorphic segregation. *Journal of Metamorphic Geology*, **18**, 41–57.

Wolf, M. B. and Wyllie, P.-J. (1993). Garnet growth during amphibolite anatexis: implications of a garnetiferous restite. *Journal of Geology*, **101**, 357–73.

Wolf, M. B. and Wyllie, P.-J. (1994). Dehydration-melting of amphibolite at 10 kbar: the effects of temperature and time. *Contributions to Mineralogy and Petrology*, **115**, 369–83.

8

Nature and evolution of the middle crust: heterogeneity of structure and process due to pluton-enhanced tectonism

KARL E. KARLSTROM AND MICHAEL L. WILLIAMS

8.1 Introduction

The middle continental crust plays an important role in controlling the behavior of continental lithosphere during and after crust formation. Through much of the life of a continent, this layer forms a mechanically strong zone in the lithosphere, generally stronger than upper or lower crust, and perhaps also stronger than the upper mantle (Jackson, 2002). However, this strength is not inherent to the materials, as middle crustal rocks are typically a mixture of relatively weak metasedimentary and plutonic rocks that contain appreciable quartz and mica. Instead, the strength of the middle crust comes from the fact that pressures are high enough to inhibit brittle failure (manifested by the absence of earthquakes), and average temperatures are below those needed to allow high ductile strain rates. However, the middle crust is not consistently strong through time because of heterogeneity of temperature and composition, and because of changes in properties throughout an orogenic cycle. In particular, heterogeneous advection of heat via magmatism causes both lateral and vertical gradients in temperature and changes in rheology, and magmatism is a primary contributor to the heterogeneity that characterizes the middle crust of many orogens.

Geophysical studies in various continents have revealed the average characteristics of present-day middle crust. Average velocities range from 6.21 ± 0.27 km s^{-1} at 10 km depths to 6.64 ± 0.29 km s^{-1} at 25 km depths (Christensen & Mooney, 1995), corresponding to densities of 2750–2950 kg m^3. The large standard deviation in velocity of about 0.3 km s^{-1} suggests marked heterogeneity, with velocity variation of about 10% and corresponding differences in density. Within individual seismic sections across orogenic belts, reflectivity is typically highly variable vertically and laterally (e.g., Nelson & Project INDEPTH, 1996), emphasizing heterogeneity of fabric.

Evolution and Differentiation of the Continental Crust, ed. Michael Brown and Tracy Rushmer. Published by Cambridge University Press. © Cambridge University Press 2005.

Studies of the focal depth distribution of earthquakes (Chen & Molnar, 1983; Maggi *et al.*, 2000) show that seismicity in most continental areas is concentrated in the upper 8–15 km, with some continents (Africa and Asia) showing seismicity throughout the crust. The base of the main seismogenic region and the transition from brittle to ductile deformation generally takes place across a zone 8–15 km deep (Sibson, 1986) towards the top of what we call middle crust. This brittle–ductile transition zone is thought to be the strongest part of the crust and hence the region that may sustain maximum differential stresses. However, its stress state is not well known. Estimates of relatively low strength (tens of MPa) are based on geodynamic arguments such as the magnitude of stresses that may be generated by ridge push and slab pull (Jarrard, 1986), arguments about weak faults (Lachenbruch & Sass, 1980) or stress drops on earthquakes (Jackson, 2002). Laboratory experiments, commonly conducted on monomineralic materials, suggest higher strength (yield stresses of perhaps hundreds of MPa), but these should be considered maximum values for crustal rocks (Kohlstedt *et al.*, 1995).

Our understanding of the integrated strength of the crustal part of the lithosphere has similar uncertainty. Geophysical data provide a snapshot of today's stable middle crust and are informative to the extent that they give a rough estimate of average composition, structural pattern (horizontally layered to unlayered or vertically layered), and general rheological properties and stress state of the middle crust. However, to understand the processes that operate to form and modify the middle crust, it is instructive to study exposed rocks that have been exhumed from middle crustal depths.

Middle crustal metamorphic rocks (i.e., those that record peak metamorphic pressures of 3–7 kbar) are exposed in every orogen of every age in the world and, combined with the exposures of lower crustal rocks (Bohlen, 1987; Ellis, 1987; Chapter 7), form the crystalline "basement" in all shield and platform areas. These rocks are so ubiquitous, so varied, and so well studied from various petrologic and structural perspectives, that it seems impossible to synthesize their characteristics. Nevertheless, this chapter is motivated by the ideas that: (1) this heterogeneity itself may be one of the most important characteristics of the middle crust, and (2) magmatism plays an important role in controlling and accentuating the heterogeneity.

The first goal in this chapter is to synthesize our empirical observations on the structural and metamorphic character, and evolution of the well-exposed Proterozoic middle crustal rocks of the southwestern US. This synthesis builds on extensive previous work (Karlstrom & Bowring, 1988, 1993; Bowring & Karlstrom, 1990; Williams & Karlstrom, 1997; Karlstrom & Williams, 1998). This region provides an important example, first, of structures that developed in "juvenile" crust that was thickened and stabilized from assembly of mantle-derived materials from 1.78 to 1.65 Ga, and, second, of middle crust that was reactivated in an intracontinental event from 1.45 to 1.35 Ga. Both show the profound importance of pluton-enhanced tectonism on deformational and metamorphic processes, suggesting that middle crustal character is defined less by differences in initial tectonic setting and more by the cumulative products of progressive deformation and magmatism.

As a second goal, we try to generalize our observations about the middle crustal rocks of the southwestern US for understanding middle crustal structure and processes in other contractional orogens. The premise is that the southwestern US has seen, perhaps, a typical cratonic evolution for "first-cycle" continental crust that has grown and stabilized through accretion of arcs and oceanic terranes. In particular, we emphasize: (1) the nature of wide (100 km scale), distributed middle crustal duplexes and tectonic mixing during amalgamation of diverse terranes and crustal provinces; (2) the role of pluton-ism and magma transfer in producing the heterogeneity of material properties and structural fabric, and accentuating the observed segmented geometry and lateral metamorphic field gradients, and (3) a heterogeneity of rheology through time and space during the orogenic cycle that may be a fundamental characteristic of the middle crust.

8.2 Proterozoic of the southwestern US: geologic evolution of juvenile crust

Paleoproterozoic rocks form the crystalline rock basement for a 1200-km wide orogenic belt that extends southwards from the Archean–Proterozoic suture at the Cheyenne Belt in southern Wyoming into northern Mexico (Fig. 8.1). The main outcrop belts are the Rocky Mountains and the Arizona Transition Zone, with adjacent areas covered by 1–3 km of Phanerozoic sedimentary rock. Based on surface and subsurface studies, the Proterozoic orogenic belts extend across the Colorado Plateau (between the outcrop belts), extend across Laurentia and are truncated to the northeast (Labrador) and southwest (California) at late Precambrian rifted margins, and are likely once to have extended further, to Baltica and Australia, respectively (inset to Fig. 8.1; Karlstrom *et al.*, 1999, 2001).

Neodymium and Pb isotopic studies in the southwestern US suggest that this series of belts records accretion of dominantly juvenile continental material from volcanic arcs and oceanic terranes, and the assembly and stabilization of diverse terranes onto North America from 1.8 to 1.6 Ga (Bennett & DePaolo, 1987; Wooden & DeWitt, 1991). There are areas within the orogen with older components, (e.g., the more radiogenic Pb and Nd

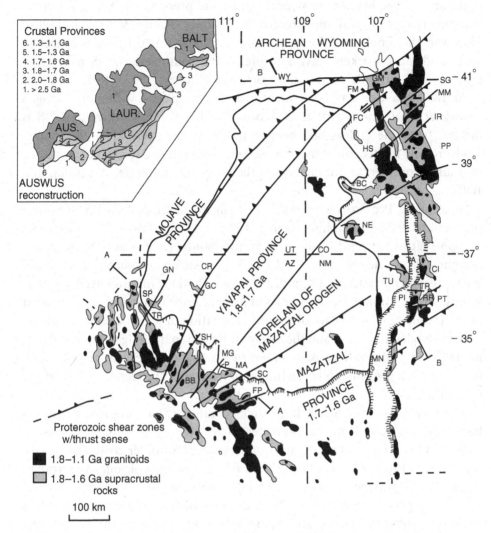

Fig. 8.1. Proterozoic rocks of the southwestern US showing major orogenic provinces and major shear zones. Inset shows the AUSWUS reconstruction of Rodinia with south-younging Proterozoic orogenic belts added to Archean Australia–Laurentia–Baltica. Letters along the cross-section lines in Arizona (A–A′) and Colorado– New Mexico (B–B′) are areas of detailed maps and cross-sections, and metamorphic studies used to constrain the composite cross-section shown in Fig. 8.2. BB – Big Bug block; BC – Black Canyon of the Gunnison; CB – Cheyenne belt; CR – Crystal shear zone; CI – Cimarron Mountains; FC – Fish Creek–Soda Creek shear zone; FM – Farwell Mountain shear zone; FP – Four Peaks; GC – Grand Canyon; GM – Green Mountain block; GN – Gneiss Canyon shear zone; HS – Homestake shear zone; IR – Idaho Springs–Ralston shear zone; MA – Mazatzal thrust belt; MG – Moore Gulch fault; MM – Moose Mountain shear zone; MN – Manzano thrust belt; NE – Needle Mountains; P – Payson ophiolite; PI – Picuris Mountains; PP – Pikes Peak batholith; PT – Pecos thrust; RR – Rincon Range; SC – Slate creek shear zone; SH – Shylock shear zone; SG – Skin Gulch shear zone; SP – Spencer Canyon; TA – Taos range; TR – Travertine Canyon; TU – Tusas Mountains.

signatures of the Mojave Province suggest the presence of Archean crustal material reworked into the Proterozoic orogen (Rämö & Calzia, 1998)). However, the amount of Archean material incorporated into the orogen is small and most workers have interpreted the southwestern US as a type-example of a juvenile accretionary orogen (Windley, 1992, 1993). Crust of 1900–1800 Ma age has been identified in two places within the orogen (Hawkins *et al.*, 1996; Hill & Bickford, 2001), but the southwestern US is distinct from the collisional events typified by the 1900–1800 Ma Trans-Hudson and related orogens, that served to amalgamate Archean cratons into large continental nuclei such as the cores of Laurentia, Australia, and Baltica.

In contrast, Proterozoic orogenic belts in the southwestern US are interpreted to record crustal additions to a 10 000-km long, long-lived, Cordilleran margin that was progressively built on the "southern" (present coordinates) margin of a newly (at *c.* 1.8 Ga) assembled Archean and early Paleoproterozoic nucleus (Karlstrom *et al.*, 2001). Our preferred tectonic setting for accretion of large volumes of juvenile crustal material to a continent is an "Indonesian" model involving differentiation of juvenile arc crust above oceanic subduction zones and the assembly (collisional amalgamation) of arcs, arc fragments, and various oceanic terranes along a convergent margin, similar to what is happening today as Indonesian terranes are accreted to Australia (Hamilton, 1981).

The southwestern US has been divided into several orogenic provinces based on age and isotopic character. The Mojave Province consists of 1.78–1.68 Ga rocks that contain isotopic evidence for older crustal materials; the Yavapai Province consists of 1.78–1.68 Ga juvenile volcanic arc terranes; the Mazatzal province contains *c.* 1.65 Ga continental margin arc rocks and related supracrustal successions. Provinces may be further divided into blocks defined by subvertical shear zones (Fig. 8.1). Like "suspect terranes" (Coney, 1989), the blocks have different tectonic histories or rock assemblages (Karlstrom & Bowring, 1988). Numerous studies have assessed which shear zones (if any) might represent sutures between once-separate plates (e.g., Shaw & Karlstrom, 1999), and seismic studies have helped in trying to link surface shear zones to deep crust and mantle discontinuities (Dueker *et al.*, 2001; Morozova *et al.*, 2002; Karlstrom & CD-ROM Working Group, 2002; Tyson *et al.*, 2002), but defining sutures still remains elusive. Because of the complex deformational history along most shear zones, we prefer to retain the descriptive term "tectonic block" rather than attempt to define paleogeographic terranes (Condie, 1992). This chapter suggests that the actual suture zones at middle crustal levels probably occur at a longer length scale than the

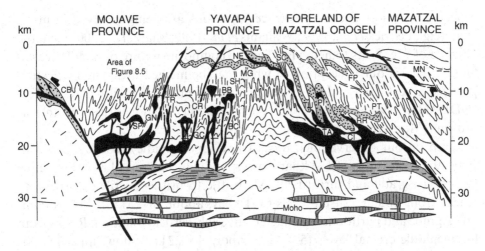

Fig. 8.2. Composite northwest–southeast cross-section of Proterozoic synorogenic crust of the southwestern US inferred from detailed studies of specific isobaric sections. Heavy lines are major tectonic boundaries; light lines are foliation form lines, black is granitoid, gray with horizontal lines is intermediate rocks (tonalites), gray with vertical lines is basaltic underplate; gray with dots is quartzite; cross-hatch is Archean crust. Letters are explained in caption to Fig. 8.1. Box shows the area of Fig. 8.5.

blocks, and consist of zones of tectonic mixing within middle crustal duplexes. Blocks and block-bounding shear zones have complex deformational histories that involve movements during arc accretion, accommodations during various shortening events, and differential uplift (Bowring & Karlstrom, 1990). These polyphase histories tend to obscure the original geometry of the early accretionary structures.

Large subregions of the southwestern US display isobaric, generally syntectonic, middle crustal metamorphism. For example, central Arizona is characterized by 3-kbar pluton-enhanced regional metamorphism (Williams, 1991), New Mexico is characterized by extensive 3.5–4-kbar aluminum silicate triple-point metamorphism (Grambling, 1986; Grambling *et al.*, 1989), and the Grand Canyon is characterized by a path where rocks decompressed from 6 to 3 kbar during the Yavapai orogeny (Ilg *et al.*, 1996). Because various 10–20 km depth regimes are represented, data on deformational and metamorphic styles ·from studies of individual areas may be used to construct a composite cross-section that offers a conceptual framework for understanding middle crustal tectonism (Fig. 8.2). This figure tries to depict the cumulative geologic fabric produced by 1.70, 1.65, and 1.4 Ga tectonism. The cross-section is complicated by the fact that different areas were variably tectonized during the different orogenic episodes. These time–space variations are being resolved in many areas (e.g., Pedrick *et al.*, 1998; Williams *et al.*, 1999; Shaw *et al.*, 2001), but the purpose of Fig. 8.2 is to help visualize the composite nature of middle crustal fabrics. This is our conception of what may be imaged by seismic reflection profiles across the middle crust of young orogens (Nelson & Project INDEPTH, 1996) and the upper crust (exhumed middle crust) of ancient orogens (Morozova *et al.*, 2002).

Based on our observations, rocks from upper-middle crustal levels (with syntectonic pressures of 1–3 kbar, corresponding to depths of 3–10 km) show greenschist grade and thrust-belt-style brittle to brittle–ductile deformation. Rocks from middle crustal levels (3–4 kbar, about 10–15 km) show amphibolite grade and partitioned but penetrative sub-vertical foliation. Rocks from lower-middle crustal levels (5–7 kbar, about 15–22 km) show upper amphibolite grade and a complex network of shallowly and steeply dipping foliation, with both active together during tectonism. There are important subhorizontal movement horizons that may allow partial decoupling of crustal layers (e.g., the Mazatzal and Manzano thrust belts (MA and MN of Fig. 8.2)), and pluton-related detachment horizons (RR of Fig. 8.2; Read *et al.*, 1999). However, our overall interpretation is that there are few thoroughgoing detachments between crustal layers that persist through time or over large regions of the middle crust of the southwestern US and, instead, the crust was heterogeneously shortened and thickened with temporal and spatial transitions in intensity of tectonism due to thermal softening rather than persistent rheologic differences (and decoupling) between blocks or crustal layers.

8.3 Looping *P–T–t–d* paths

By combining metamorphic and structural studies, we suggest that the tectonic history of exposed rocks of the southwestern US may be generalized by an idealized *P–T–t–d* loop (*d* = deformation; Fig. 8.3) that may be different from one tectonic block to another. The addition of deformational history and style (*d*) to *P–T–t* loops helps complete the picture of an evolving thermal–mechanical profile in deforming crust, and emphasizes that strong changes in rheology must have occurred in the history of a single rock to explain the observed heterogeneity of strain and *P–T* conditions. The prograde part of the path for Proterozoic rocks of the southwestern US (and for rocks in many orogens) is poorly constrained, but we use the model for the accretion of arcs to infer that early metamorphism accompanied thrusting and shortening as supracrustal rocks were tectonically buried to depths of 10–20 km.

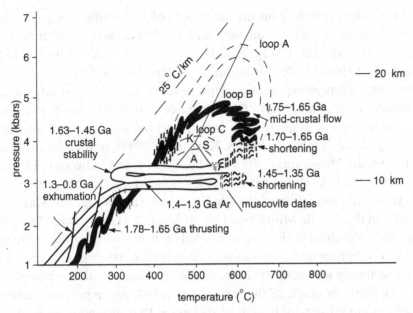

Fig. 8.3. Generalized styles of *P–T–t–d* loops for Proterozoic rocks of the southwestern US. All loops show prograde crustal thickening by upright folding and thrusting. loop (a) Some areas were buried to depths ~20 km (eastern Grand Canyon, northern Colorado); loop (b) other areas reached depths of 15 km (New Mexico, Mojave Province); or loop (c) some blocks were buried to depths of 10 km (central Arizona, Travertine Canyon). Decompression took place via collisional exhumation and all blocks stabilized at about 10 km by 1.6 Ga. After long-term (1.6–1.4 Ga) residence of rocks in the middle crust, renewed tectonism (1.45–1.35 Ga) was accompanied by continued shortening; these thermal spikes involved modest compression or decompression in different areas. Erosional exhumation to the surface took place differentially 1.4–0.8 Ga. K – Kyanite; S – Sillimanite; A – Andalusite.

Figure 8.3 neglects the probability of different prograde *P–T* slopes, the different times of prograde metamorphism (e.g., 1.78–1.73 in the Yavapai Province, and 1.67–1.65 in the Mazatzal Province), and the possibility of low-*T*, high-*P* metamorphism in accretionary prisms, and shows a general prograde path that follows an elevated geothermal gradient. The deformational style during the early prograde part of the path is portrayed as brittle thrusts (e.g., Mazatzal block (MA) of Fig. 8.2) changing to ductile folding and vertical foliation development at depths of about 8 km (2.5 kbar; e.g., in ductile thrust belts). This generalized path progressed to different peak pressures creating different-size loops (Fig. 8.3; Williams & Karlstrom, 1997) for different blocks depending on the magnitude of shortening and geometry of thrusts. The generally warm and weak nature of the crust resulted in pervasive crustal thickening and a very broad region of thickened, but not highly elevated, crust (Bowring & Karlstrom, 1990; Royden, 1996).

Based on overprinting relationships as documented by several studies (Karlstrom, 1989; Ilg *et al.*, 1996), we infer that structures evolved from premetamorphic low-angle thrusts (now preserved only in the lowest grade blocks; e.g., Mazatzal block, Doe & Karlstrom, 1991), to ductile thrust belts (e.g., the Needle Mountains (NE), Four Peaks (FP), and Manzano Mountains (MN) of Fig. 8.2). At somewhat deeper levels (10 km, 3 kbar) in areas such as the Texas Gulch formation of the Big Bug block (BB in Fig. 8.2; Karlstrom, 1989) and in the Needle Mountains (NE of Fig. 8.2; Harris *et al.*, 1987), early thrusts and recumbent folds are overprinted by penetrative northeast-trending subvertical foliation and shear zones that record northwest–southeast horizontal shortening and general shear; this subvertical foliation represents the dominant fabric in much of the southwestern US. An important feature of this deformational history (discussed below) is that decompression of some blocks took place by exhumation involving thrusting and reverse faulting plus erosion during arc collision (e.g., loop A of Fig. 8.3 from eastern Grand Canyon (GC of Fig. 8.2) and loop B of Fig. 8.3 from Spencer Canyon (SP) of western Grand Canyon) at the same time other blocks were being buried by these thrusts (e.g., loop C of Fig. 8.3, from Travertine Canyon (TR of Fig. 8.2); Karlstrom *et al.*, 2002). Deformation at elevated temperatures involved middle crustal flow on both steep and shallow fabrics, as depicted in loop B of Fig. 8.3, with continuing interaction of shortening and fabric reactivation during decompression.

Metamorphic studies indicate that rocks now exposed over a huge region became stabilized at depths near 10 km and resided at this depth between 1.63 and 1.45 Ga, following the Paleoproterozoic orogenies. This period of lithosphere stability is shown in Fig. 8.3 as a path of isobaric cooling without accompanying deformation, which gives the path the first part of its looping character. The extent to which rocks cooled quickly to a normal geothermal gradient remains controversial (as shown in Fig. 8.3), or cooled slowly at elevated temperatures (Thompson *et al.*, 1996; Chamberlain & Bowring, 2000).

Renewed metamorphism and deformation accompanied 1.45–1.35 Ga magmatism (Nyman *et al.*, 1994; Pedrick *et al.*, 1998; Williams *et al.*, 1999). Rocks in many areas at this time were still at depths of about 10–15 km (e.g., in northern New Mexico), as shown by the average pressures of 3–4 kbar of metamorphism in aureoles of 1.4 Ga granites (Nyman & Karlstrom, 1997; MN, RR, PI of Fig. 8.2; Thompson *et al.*, 1996; Read *et al.*, 1999; Williams *et al.*, 1999). This is shown in Fig. 8.3 as isobaric heating, with significant deformation taking place when rocks were thermally softened and were able to record the regional intracratonic stresses (Karlstrom & Humphreys, 1998). This 1.4 Ga deformation locally involved intense transposition and reactivation of older fabrics and significant changes in crustal depth due to thrusting, similar to the thrust exhumation described earlier, but 200 Myr later. For example, in the eastern Grand Canyon area (CR of Fig. 8.2), Ar–Ar studies (Karlstrom *et al.*, 1997) and studies of syntectonic plutons (Nyman & Karlstrom, 1997) document both east- and west-side up-thrust movement on older boundaries. Thus, the 1.4 Ga thermal spike in Fig. 8.3 locally had either positive, neutral, or negative slope depending on the structural position of the thermally softened rocks. Similar intracratonic reactivations at 1.4 Ga, but on subvertical faults, are documented by monazite geochronology and structural studies of the Homestake shear zone of Colorado (HS of Fig. 8.1; Shaw *et al.*, 2001).

Exhumation of middle crustal rocks to near surface took place in a later orogenic event than the ones in which the observed fabrics were created. This is the general case for most rocks exposed in ancient orogens (Ellis, 1987). The exhumation path gives the second part of the overall looping character, as erosional exhumation is likely to have taken place at lower temperature than the prograde path. This is shown in Fig. 8.3 as a decompression/cooling curve labeled 1.3–0.8 Ga. Exhumation took place by erosion of an elevated 1.4 Ga plateau (Karlstrom & Humphreys, 1998) and differential uplift along extensional faults in major pulses at 1.25, 1.1, and 0.8 Ga (Timmons *et al.*, 2001) such that middle crustal rocks across the region had been exhumed too close to the surface by Cambrian time to create the "Great Unconformity" (Powell, 1876). Phanerozoic deformation in the Rockies reactivated earlier basement-penetrating faults (Marshak *et al.*, 2000), generally with little or no penetrative deformation of the basement rocks.

8.4 Plutonism

Granitoid plutons make up ~50% of exposed crust in the southwestern US (Fig. 8.1) and in many orogens. The volume and style of granitoid plutonism changed through time, but, based on the large volume of granitoid, the 10–25-km deep middle crust was apparently a zone of both flux of melt and of final emplacement of granitic magmas (cf. Collins & Sawyer, 1996) and, hence, a sensitive record of dynamically changing crustal columns. The generalizations made by Ilg *et al.* (1996) for the Grand Canyon seem to extend across much of the southwestern US. Earliest granitoids are granodioritic to tonalitic in composition and have calc-alkaline compositions suggestive of derivation as arc plutonic complexes. Ages vary with the particular province (1.78 Ga in the Green Mountain block and parts of the Mojave Province, 1.75–1.72 Ga in the Yavapai Province, to 1.65 Ga in the Mazatzal Province). Where geochronologic data are good, these rocks may be shown to have intruded just slightly older volcanic rocks (Hawkins *et al.*, 1996) and they lack contact aureoles, both features that suggest shallow emplacement in an arc environment.

A compositionally distinct group of plutons was emplaced during peak metamorphism and crustal shortening. These plutons are S-type and A-type granitoids (Anderson & Bender, 1989), and they are interpreted to record lower crustal melting during crustal thickening. These granitoids form a huge magmatic influx in the Mojave and Yavapai provinces at 1.71–1.68 Ga, with continued plutonism until 1.65 Ga, the time of the main pulse of magmatism in the Mazatzal Province. A second major period of A-type magmatism took place at 1.45–1.35 Ga, and a more spatially restricted A-type event took place at 1.1 Ga (Smith *et al.*, 1999). Based on geochemistry, these A-type magmas are interpreted to be the result of mafic underplating of the crust and resulting lower crustal partial melting (Frost & Frost, 1997). Thus, compositionally similar A-type plutonism and related mafic underplating are interpreted to have occurred at several times (1.68 Ga in the Mojave and Yavapai provinces, 1.4 Ga across the region, and 1.1 Ga in the Pikes Peak area) and the cumulative effect of this magmatism is shown schematically in Fig. 8.2.

Syntectonic features have been described in granitoids of each age group. Early arc plutons are generally exposed as strongly foliated sheets subparallel to the earliest layering (S_1). Some appear to have been only weakly deformed (because of their mafic composition and hence relative strength), but even these granitoids contain magmatic fabrics that are parallel to external (S_1) fabrics. The limited size and sheet-like geometry of most of these plutons is unlike the expected preservation of large arc batholiths, and implies syn- and post-emplacement thrust dismemberment of arc plutons during subsequent crustal assembly.

The 1.70–1.68 Ga plutons are variably deformed and are interpreted to be syntectonic relative to peak metamorphism and crustal shortening, based on field evidence such as parallelism of magmatic and solid state fabrics, variably deformed plutonic phases, and syntectonic porphyroblast-matrix relationships in aureoles (e.g., Karlstrom & Williams, 1995, for the area BB of Fig. 8.2). In crust from 20 km depths now exposed in the Grand Canyon, these granitoids form dike swarms (Ilg *et al.*, 1996); in contrast, in crust from 10 km depth, granitoids form large plutons and batholiths (Karlstrom & Williams, 1995). Plutons at 1.4 Ga followed the main period of crust formation by

200 Myr, are not as pervasively foliated as earlier groups, and have been called "anorogenic." However, they also were emplaced synchronously with deformation as recorded preferentially by pluton margins (Nyman *et al.*, 1994; Duebendorfer & Christensen, 1995), contact aureoles (Thompson *et al.*, 1996), dike configurations (Kirby *et al.*, 1995) and interactions between shear zones and adjacent granites (Nyman & Karlstrom, 1997; Selverstone *et al.*, 2000).

Although lithospheric extension has been suggested as the tectonic setting for this A-type magmatism (Emslie, 1978; Hoffman, 1989; Frost & Frost, 1997), structural studies in pluton aureoles (Nyman *et al.*, 1994) suggest a predominance of thrust-sense or transpressional shear, and 1.4 Ga foliations and folds are interpreted by us to be related to a regional intracratonic strain field involving northwest shortening and northeast extension during far-field transpressive deformation. This is compatible with the near continuity of deformation and magmatism, at regional scale, from 1.8 to 1.0 Ga along southern Laurentia, giving rise to models for a long-lived (800 Myr) convergent margin system. Thus, the 1.4 Ga underplating, pluton ascent and emplacement, and thermal softening, and the resulting middle crustal deformation, were all an intracratonic tectonic response to far-field transpression that lasted from 1.5 to 1.3 Ga along a plate margin that has subsequently been overprinted by the Grenville event. Heterogeneous convergence and transpression along this margin culminated with assembly of the Rodinia supercontinent during the Grenville orogeny (Karlstrom *et al.*, 2001). Except for the Pikes Peak pluton, the Grenville event itself had little effect on the geometries of middle crustal rocks in the cross-section of Fig. 8.3, presumably because intracratonic magmatism was not widespread and rocks were at too shallow a level (Smith *et al.*, 1999; Pikes Peak batholith – PP of Fig. 8.1) for appreciable thermal softening.

8.5 Discussion of middle crustal characteristics and processes

8.5.1 Strength profiles

Proterozoic rocks of the southwestern US were pervasively deformed during assembly of juvenile lithosphere from 1.8 to 1.6 Ga (Karlstrom & Bowring, 1988); hence, we infer that the upper mantle layer was not strong enough to prevent whole-lithosphere deformation (Kuznir, 1991), or was decoupled from crustal deformation. Our premise is that the distinctive set of features summarized above qualitatively allows us to characterize middle crustal strength. Traditional lithosphere strength envelopes commonly show pronounced rheologic layering (Molnar, 1992). Upper crustal rocks become stronger line-arly with depth due to pressure-dependent increase in resistance to fracturing and frictional sliding. Lower crustal rocks get exponentially weaker with depth due to thermally activated plastic flow by dislocation creep (Ranalli, 1997). The upper mantle has been considered to be the strongest layer in the litho-sphere because of greater plastic strength of olivine (Molnar, 1992), but this layer may be thermally weakened and/or decoupled from the overlying crust during orogenic events. Alternatively, as suggested by Jackson (2002) based on reassessment of depth of earthquakes, the mantle may not represent a strong zone in the lithosphere and the "jelly sandwich" model for the lithosphere may be incorrect. Regardless, the strongest part of the crust is the middle crustal zone of transition between the brittle upper crust and a ductile lower crust.

The rheology of the brittle–ductile transition is poorly understood because of heterogeneity of rock type, thermal regime, and strain rate, as well as a still incomplete understanding of flow laws for polyphase aggregates (Kohlstedt *et al.*, 1995). Understanding the changes in rheology of the middle crust through time may be more important than deriving a static view of its contribution to crustal strength. Thus, empirical studies of strength profiles, especially of the middle crust, are essential for relating experimental studies to natural defor-mation (e.g., Davidson *et al.*, 1994). How may strength profiles explain the predominance of ductile deformation fabrics, a predominance of vertical foliation, and the extreme heterogeneity of deformational features that we observe in the 8–20 km middle crust?

Figure 8.4 shows a model for the strength of the continental lithosphere during the times of high heat flux and whole lithosphere deformation repre-sented by the 1.75–1.65 Ga tectonism, and also (more locally) during 1.4 Ga tectonism. This strength profile, although based on thermal and strength modeling studies (Ranalli, 1997), and seismic studies (Smith & Bruhn, 1984), is modified to show a family of curves, especially to depict fluid and magma pressures above hydrostatic values, transient high thermal regimes, variable strain rate regimes, and quartz–mica rock compositions, all of which weaken the crust (Fig. 8.4, heavy line). The 10–20-km zone may well be a time-averaged and heterogeneous brittle–ductile transition in quartz-rich crust, but we observe that this zone contains predominantly ductile fabrics suggest-ing that, over time, creep dominates over brittle failure at these depths. Strength curves based on experiments using single rock types, steady state thermal and strain rate regimes, and hydrostatic fluid pressures, significantly overestimate strength of lithosphere in active orogens such that strength envelopes – even complex ones such as the heavy line in Fig. 8.4 – should be viewed as maximum strength estimates.

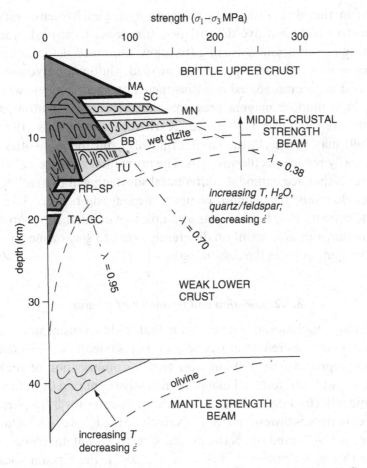

Fig. 8.4. Hypothetical lithospheric strength profile during 1.70–1.65 Ga (and 1.45–1.35 Ga) tectonism in the southwestern US; darker gray pattern = "average" synorogenic strength profile; lighter gray = strength profile at times and places of lower T, higher strain rate. Brittle strength curves (labeled with λ) were derived from Byerlee's law, with variations in pore fluid pressure (and/or magma pressure) P_f where $\lambda = P_f/P_{\text{lithostatic}}$. Ductile strength curves show experimentally derived plastic flow laws for wet quartzite and wet dunite for a geotherm with surface heat flow of 60 mW m^{-1} and a strain rate of 10^{-15} s^{-1} (Kohlstedt *et al.*, 1995); letters correspond to areas shown in Figs. 8.1 and 8.2.

Geometries of structures (vertical foliation, partitioned strain) need to be reconciled with the notion that the 8–15 km may correspond to a strength beam or stress riser. It may be that this zone penetratively shortens at differential stress levels where the upper and lower crust are simultaneously failing brittlely and ductilely. We hypothesize that the middle crustal deforming beam is strain-hardened by upright folding and vertical foliation development somewhat in the same fashion as dislocation tangles harden crystals, and analogous to the penetrative deformation that leads to critical taper in an accretionary prism. Strain softening (to allow continued deformation) takes place during elevated thermal regimes around plutons and zones of melt transfer, and is accommodated by adjustments on shear zone networks. For example, high fluid or magma pressures that accompany pluton-enhanced deformation (e.g., at BB in Fig. 8.2; Karlstrom & Williams, 1995; Read *et al.*, 1999) may reduce brittle strength locally, and thermal softening will simultaneously reduce ductile strength. The horizontal zones serve as boundary structures that accommodate differential shortening above and below the shortening domains, but without complete decoupling between layers. This shear zone network may be viewed as a complex set of interacting slip systems, where simultaneous movement on differently oriented shear zones is required to maintain continuity in the deforming crust.

8.5.2 *Location and geometry of sutures*

An important conclusion of our model is that middle crustal suture zones – rather than being discrete thrust zones, as they sometimes are in the upper crust – are approximately 100-km wide zones of imbrication of rocks from both hanging wall and footwall plates. As recently supported by both geologic and seismic reflection data, there are several candidates for lithospheric-scale boundaries in the southwestern US. These include the Farwell Mountain shear zone (Fig. 8.1 – boundary between the Green Mountain block and the Colorado (Yavapai) province; Tyson *et al.*, 2002), the Crystal shear zone (part of a widely distributed Mojave–Yavapai province boundary; Ilg *et al.*, 1996), the Slate Creek and Moore Gulch shear zones, which bound the Payson ophiolite (Dann, 1991), and the Jemez lineament, where seismic studies suggest the possibility of a south-dipping paleosubduction zone (Karlstrom & CD-ROM Working Group, 2002). The seismic data from the Cheyenne belt (Morozova *et al.*, 2002), Farwell Mountain zone (Tyson *et al.*, 2002), and Jemez lineament (Magnani *et al.*, 2002) show complex crustal geometries that seem compatible with the crustal model proposed here.

The geologic transect in the Grand Canyon provides one of the best exposed and studied of these zones in terms of the middle crust (Fig. 8.5). Here, one might select the Crystal shear zone as a discrete boundary due to changes in Pb isotopic composition, metamorphic grade, and suggestive ophiolite components such as peridotite, pillow basalt, chert, and basaltic dikes (Ilg *et al.*, 1996). However, there are also important tectonic slices to the east that involve ultramafic fragments and chert, and important shear zones to the west that also

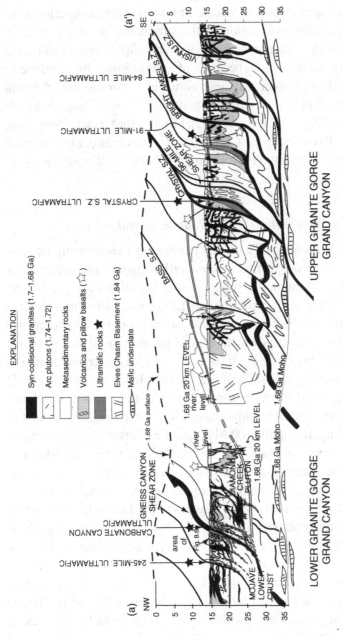

Fig. 8.5. Composite NW–SE cross-section through the Upper and Lower Granite Gorges of the Grand Canyon showing hypothetical geometry of a middle crustal duplex at *c*. 1.68 Ga. The Crystal shear zone may be the most important of a family of middle crustal shear zones that mark the distributed suture zone between the Yavapai and Mojave provinces. Ultramafic rocks within supracrustal schists at miles 84 and 91 are interpreted to be cryptic thrust zones of the distributed duplex. Gneiss Canyon shear zone of the Lower Granite Gorge marks a major zone of collisional exhumation during shortening and cratonization of the crust.

record isotopic, metamorphic, and compositional changes (Karlstrom *et al.*, 2002). Thus, we view the whole zone as the imbricated middle crustal duplex zone that records, over a width of >100 km, disappearance of one or more ocean basins and suturing of disparate blocks. This view seems compatible with the scale of arc systems. The orientation of the subduction system (probably northwest or north–south with northerly or westerly dip) is not constrained firmly by the existing surface data because of overprinting by northwest–southeast shortening (D_2). Numerous shear zones in Colorado and Arizona also contain fragments of ophiolitic material and/or tectonic melange (Cavosie *et al.*, 1999; Shaw *et al.*, 2001; Tyson *et al.*, 2002), but, like the Crystal shear zone, it seems more likely that these zones are parts of folded and dismembered accretionary complexes rather than each shear zone being itself a suture.

8.5.3 *Interplay of subvertical and subhorizontal fabrics*

One interesting conundrum of middle crustal tectonics is that suturing (at least to the extent it is related to subduction systems rather than transcurrent juxta-position) may be initially dominantly subhorizontal yet fabrics are pervasively vertical. At least two mechanisms probably operate. First, early thrust sutures may get steepened and rotated into subvertical orientation due to progressive transposition by shortening, to produce the type of middle crustal duplex shown in Fig. 8.5. Second, early thrust-related sutures may get folded, transposed, and broken up by later shortening and transcurrent faulting. In the southwestern US, we have not yet distinguished the relative importance of these two mechanisms, even in areas relatively unaffected by subsequent 1.4 Ga reactivations, because of intense crustal shortening during the 100 Myr episode of crust formation. Since we have not identified by mapping a folded and faulted, but regionally extensive, early suture zone, nor have large magnitude transcurrent systems been identified (cf. Bergh & Karlstrom, 1992), we prefer the middle crustal duplex model.

One limitation, in terms of looking for sutures related to S_1 fabrics, is that both subvertical (S_2) and subhorizontal (S_1) fabrics moved simultaneously as shown by syntectonic relationships to 1.70–1.68 Ga pegmatites and leuco-somes. For example, in the synform shown in Fig. 8.6, we infer that there is a complex temporal and spatial interaction of crust that is shortening and crust that is shearing horizontally. Also, steep and shallow zones may change through time: flat fabrics may get shortened and transposed as they move into zones of greater strength (cooler zones), and steep fabrics may get rotated into shallow orientation along melt- and fluid-related detachments, especially at depths >15km (e.g., Read *et al.*, 1999). We note that this complex interplay

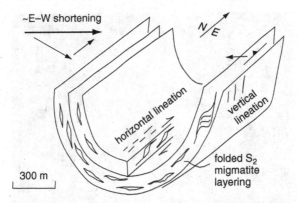

Fig. 8.6. Synform in mixed granitic and supracrustal gneisses in the Lower Granite Gorge (see Fig. 8.5 for location). Melt pods, related to the enclosing 1.7 Ga granitic orthogneiss, show east-side-up shear related to a subvertical lineation on the fold limbs; and top-northeast shear related to a subhorizontal lineation in the fold hinge region. We assume melt pods are all of similar age (1.7 Ga) and that this structure records movement along both S_1 (the pre-existing gneissic layering) and the S_2 (limb regions developing into shear zones) during D_2.

of steep and shallow fabrics, analogous to lithons and cleavage domains at thin section scale, is also a dominant geometry in lower crustal rocks (Chapter 7).

8.5.4 Pluton effects and processes – rates and duration

It is difficult to underestimate the importance of plutonism and magma flux in establishing the geometry, rheology, and thermal structure of the middle crust. Fifty percent of the crust is granite, metamorphic field gradients show temperature variations of 200 °C at the same crustal level that may be related to plutons and dike swarms in the eastern Grand Canyon (shown in Fig. 8.5; Ilg *et al.*, 1996), and granites soften the crust and facilitate deformation, both during shortening (Karlstrom & Williams, 1995) and shearing (Read *et al.*, 1999). Rather than try to resolve the chicken and egg issues surrounding granite emplacement, we view middle crustal deformation, metamorphism, and plutonism as linked systems both during crustal assembly at 1.75–1.65 Ga and during reactivation at 1.45–1.35 Ga. The fact that very different tectonic regimes produced similar tectonic features suggests that the linked processes are important, not the particular plate tectonic setting. Nevertheless, several observations about pluton-enhanced tectonism might reach the status of generalizations.

Brittle deformation takes place around plutons at all (10–25 km) depths in the form of diking, indicating that elevated magma pressure acted like elevated

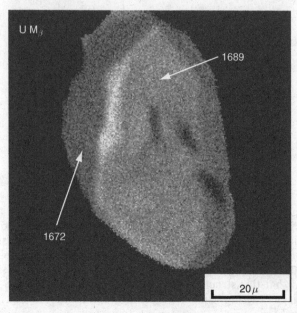

Fig. 8.7. Monazite dates from the Lower Granite Gorge of the Grand Canyon help explain the apparent 20 Ma spread of conventional monazite dates by highlighting discreet pulses of monazite growth within a broader orogenic event. The *c.* 1690 monazite core is interpreted to represent the time of peak metamorphism and anatexis; *c.* 1672 may represent a thermal spike associated with a more distal pluton. Dates were obtained by electron microprobe at the University of Massachusetts (Williams *et al.*, 1999; see also Williams & Jercinovic, 2002). At the time of going to press the original colour version was available for download from www.cambridge.org/9780521066068

pore pressure in lowering the effective normal stress and thus allowing tensile fracturing. Interestingly, in the Grand Canyon, pluton-enhanced tectonism at 1.70–1.68 and 20 km depths took place around dike swarms that we interpret to be zones of melt migration (cf. Clemens & Mawer, 1992). These dikes presumably fed large peraluminous batholiths, such as those that are present at 10 km levels in central Arizona (BB of Fig. 8.2) and in the Lower Granite Gorge. If dike networks were zones of appreciable melt flux, this provides an explanation for the ~200 °C lateral metamorphic field gradients around the dike complexes, as the thermal effect of the dikes would be proportional to the total melt flux rather than the observed granite volume (Ilg *et al.*, 1996). This concept is also compatible with durations of orogenic pulses of 10–20 Myr seen in the range of monazite dates (Fig. 8.7; Hawkins *et al.*, 1996). Ponding of magma in batholiths at depths of 10 km is compatible with observations by other workers (Collins & Sawyer, 1996) and is presumably due to loss of buoyancy of typical granitoid magmas relative to typical host rock bulk density.

Similarly, the 1.4 Ga intracratonic tectonic event seems to be best explained in terms of advection of heat from the mantle by basaltic underplating, lower crustal melting, then ascent and emplacement of granitoids to the middle crust. Thus, the 1.4 Ga plutons, at the scale of the lithosphere, were probably a response to mantle magmatism. However, at the scale of the middle crust, the plutons also acted to focus deformation and metamorphism. The length scale of thermal and deformational aureoles is of the order of 10 km in much of Arizona and Colorado and is directly relatable to heating and thermal softening during pluton emplacement. However, in northern New Mexico, there is a 100-km wide zone with relatively few 1.4 Ga plutons where geochronological studies indicate that rocks reached temperatures >600 °C (Pedrick *et al.*, 1998; Read *et al.*, 1999) and where structural studies indicate that middle crustal fabric was reactivated and transposed at *c.* 1.4 Ga (Williams *et al.*, 1999). These rocks appear to record a steep vertical thermal gradient between an upper-middle crust (Arizona and Colorado) that behaved in a semi-brittle manner, and a flowing middle crust (northern New Mexico) that behaved in a pervasively ductile manner. As plutons of 1.4 Ga are rare in the hot, flowing middle crustal zone in New Mexico, we infer that the zone was below a regional melt-enriched layer somewhat similar to the zone inferred from seismic data beneath Tibet (Nelson & Project INDEPTH, 1996).

The duration of the *c.* 1.4 Ga episode (1.45–1.35 Ga) probably reflects the time scale of the mantle magmatism. At middle crustal levels, 10–20 Ma thermal spikes (at 1.42, 1.4, and 1.38 Ga) were superimposed on this longer event and these are likely to correspond to times of major melt flux and granitoid emplacement. Thus, pluton-enhanced tectonism appears to have characteristic temporal and spatial scales and characteristic styles of metamorphism and deformation even though the tectonic settings were different during the two main orogenic episodes (1.75–1.65 and 1.45–1.35 Ga). The reason for this is that the progressive magmatic and metamorphic processes, and related progressive deformation, tend to obscure original style differences.

8.5.5 *Looping P–T paths*

The concept of looping *P–T* paths, described above, reflects an overall tectonic regime where rocks decompressed during orogenic activity via collisional exhumation of hanging walls, and/or syn-orogenic rebound, but reach stability (in this case at 10-km depths) in the crust at the end of the orogenic episode. This is followed by nearly isobaric cooling and long-term residence in the post-orogenic stable crust. This is compatible with the model (Bowring & Karlstrom, 1990) for development of normal thickness (30–40 km), isostatically stable, crust via thickening of thinner (25 km) arc crust, but without dramatic overthickening. It is also compatible with a model for the development of an orogenic plateau at 1.4 Ga, where pressure did not change markedly (and rocks in different regions record either modest compression or decompression) due to inflation by mafic magma and modest crustal thickening at 1.4 Ga.

The *P–T* loops shown in Fig. 8.3 thus show temperature spikes due to plutonism. These would have had profound effects on crustal rheology. Combined with the idea of magma ponding at 10–15 km, and large-volume magmatism, this suggests that 1.4 Ga tectonism was related to a middle crustal melt-pluton detachment (or flow) layer that separates a brittle–ductile upper crust from a pervasively flowing lower crust. These effects were superimposed on similar styles that formed at 1.70–1.68 Ga, making areas where both events were well developed and difficult to unravel (Pedrick *et al.*, 1998; Williams *et al.*, 1999). Middle crustal rocks in many continental sections may record similar complexity of superimposed tectonic events separated by long-term residence in the deep crust.

8.5.6 *Vertical layering and decoupling in the crust?*

One of the predictions of simplified strength–depth curves is that there may be important decoupling zones above and below a middle crustal strength beam that allows delamination of crust during tectonism (Oldow *et al.*, 1990; Karlstrom & Williams, 1998; McKenzie *et al.*, 2000). Such decoupling may be defined as the absence of mechanical or kinematic communication between layers. However, our view is that there are unlikely to be regional subdivisions between upper, middle, and lower crust that persist through time. Instead, middle crustal mechanical and chemical processes grade upwards and downwards (and change through time) depending mainly on average composition (quartz versus feldspar as the dominant strength-controlling mineral), strain rates, and temperature gradients. In addition, brittle deformation is important at nearly any depth in the middle crust at various times in its evolution (not just the region above the seismogenic zone) due to pluton-enhanced embrittlement (Davidson *et al.*, 1994) and high strain rates (Axen *et al.*, 1998).

The transition to lower crust is commonly thought to be more compositionally controlled (Rudnick & Fountain, 1995) and may involve a change from quartz-rich (imbricated metasedimentary rocks) to plagioclase-rich (basalt underplate and/or melt residue), but in our view is also a transitional boundary in time and space because of duplex imbrication of contrasting lithologies. From our model for the middle crust, we infer that there is generally a significant degree of coupling between lower, middle, and upper crustal zones, and this seems to be supported by the general intact character of similar-age crust and mantle columns based on isotopic studies (Livacarri & Perry, 1993). Thermal spikes may regulate the intermittent coupling of upper, middle, and lower crust as the orogen evolves, but all of these zones appear to record similar contractional strains characterized by an interplay between subvertical and subhorizontal fabrics.

8.6 Summary

Middle crustal rocks of Proterozoic age in the southwestern US are segmented into 10-km scale blocks by shear zones that record multiple movements. The different blocks record different styles, timing, and conditions of tectonism. This segmented structure may have initiated in part as distributed duplex systems during suturing, but has been accentuated by reactivation of weak zones during shortening and erosional exhumation that resulted in preservation of nearly isobaric sections. As a result, these blocks provide a record of middle crustal processes at various (10–25 km) depths and provide an opportunity to construct a synthetic cross-section and to generalize about middle crustal processes.

The middle crust may commonly be the strongest layer in the syn-orogenic crust, but its rheology changes markedly in time and space and traditional strength–depth curves probably overestimate the strength and underestimate the complexity of the brittle–ductile transition. Plutons are an important control on middle crustal rheology because of thermal softening and embrittlement caused by elevated magma pressure. Thus, the middle crust may be both a stress concentrator with significant strength and also capable of flow over time largely because of the changing thermal state induced by magma flux and pluton emplacement. Although much of the middle crust of the southwestern US has subvertical foliation due to crustal shortening, one form of structural heterogeneity that is present at all scales is interplay between domains of sub-vertical fabric and zones of reactivated older shallowly dipping foliation. We view this as a product of reactivation of older fabrics during younger events. Middle crustal rocks are thus a sensitive but complex recorder of continental deformational processes both at the 100 Myr time scale of orogenic events (1.75–1.65 Ga and 1.45–1.35 Ga) and the 10 Myr time scale of orogenic pulses that are interpreted to reflect pluton-enhanced tectonism.

Metamorphic assemblages record clockwise-looping *P–T* paths that resulted from thrust-related thickening (including synorogenic thrust-related exhumation of some blocks), then isobaric cooling following an orogenic cycle. Metamorphic grade is heterogeneous at the 10-km scale, with large lateral temperature gradients (450–700 °C) preserved at relatively shallow depths (10–20 km). These are interpreted to reflect the influence of large volumes of synorogenic granitoid melt that has fluxed-through and solidified in the middle crust. Both length scales (10 km) and time scales (10 Myr) of melt-enhanced tectonism are compatible with size and duration of major granitoid emplacement events.

We conclude that heterogeneity of fabric and rheology in time and space may be the most important characteristic of the middle crust, and magmatism, deformation, and metamorphism are linked systems, with magmatism playing an important role in accentuating and controlling the heterogeneity of the middle crust.

References

Anderson, J. L. and Bender, E. E. (1989). Nature and origin of Proterozoic A-type magmatism in the southwestern United States. *Lithos*, **23**, 19–52.

Axen, G. J., Selverstone, J., Byrne, T. and Fletcher, J. M.(1998). If the strong crust leads, will the weak crust follow? *GSA Today*, **8**, 1–8.

Bennett, V. C. and DePaolo, D. J. (1987). Proterozoic crustal history of the western United States as determined by Neodymium isotopic mapping. *Geological Society of America Bulletin*, **99**, 674–85.

Bergh, S. G. and Karlstrom, K. E. (1992). The Proterozoic Chaparral fault zone of central Arizona; deformation partitioning and escape block tectonics during arc accretion. *Geological Society of America Bulletin*, **104**, 329–45.

Bohlen, S. R. (1987). Pressure–temperature–time paths and a tectonic model for the evolution of granulites. *Journal of Geology*, **95**, 617–32.

Bowring, S. A. and Karlstrom, K. E. (1990). Growth, stabilization, and reactivation of Proterozoic lithosphere in the southwestern United States. *Geology*, **18**, 1203–6.

Cavosie, A. J., Lucky, E., Rogers, S. and Selverstone, J. (1999). Early Proterozoic ophiolite fragments in northern Colorado Front Range? *Geological Society of America Abstracts with Programs*, **31**, A177.

Chamberlain, K. R. and Bowring, S. A. (2000). Apatite–feldspar U–Pb thermo-chronometer: a reliable, mid-range (~450 °C), diffusion-controlled system. *Chemical Geology*, **172**, 173–200.

Chen, W.-P. and Molnar, P. (1983). Focal depths of intracratonic and intraplate earthquakes and their implications for the thermal and mechanical properties of the lithosphere. *Journal of Geophysical Research*, **95**, 12 527–52.

Christensen, N. I. and Mooney, W. D. (1995). Seismic velocity structure and composition of the continental crust: a global view. *Journal of Geophysical Research*, **100**, 9761–88.

Clemens, J. D. and Mawer, C. K. (1992). Granitic magma transport by fracture propagation. *Tectonophysics*, **204**, 339–60.

Collins, W. J. and Sawyer, E. W. (1996). Pervasive granitoid magma transfer through the lower-middle crust during non-coaxial compressional deformation. *Journal of Metamorphic Geology*, **14**, 565–79.

Condie, K. C. (1992). Proterozoic terranes and continental accretion in southwestern North America. In *Proterozoic Crustal Evolution*, ed. K. C. Condie. Amsterdam: Elsevier; *Developments in Precambrian Geology*, **10**, 447–80.

Coney, P. J. (1989). Structural aspects of suspect terranes and accretionary tectonics in western North America. *Journal of Structural Geology*, **11**, 107–25.

Dann, J. C. (1991). Early Proterozoic ophiolite, central Arizona. *Geology*, **19**, 590–3.

Davidson, C., Schmid, S. M. and Hollister, L. S. (1994). Role of melt during deformation in the deep crust. *Terra Nova*, **6**, 133–42.

Doe, M. and Karlstrom, K. E. (1991). Structural geology of an early Proterozoic foreland thrust belt, Mazatzal Mountains, Arizona. In *Proterozoic Geology and Ore Deposits of Arizona*, ed. K. E. Karlstrom. Arizona Geological Society Digest 19, pp. 181–92.

Duebendorfer, E. M. and Christensen, C. (1995). Synkinematic (?) intrusion of the "anorogenic" 1425 Ma Beer Bottle Pass pluton, southern Nevada. *Tectonics*, **14**, 168–84.

Dueker, K., Yuan, H. and Zurek, B. (2001). Thick-structured lithosphere of the Rocky Mountain region. *GSA Today*, **11**, 4–9.

Ellis, D. J. (1987). Origin and evolution of granulites in normal and thickened crust. *Geology*, **15**, 167–70.

Emslie, R. F. (1978). Anorthosite massifs, rapakivi granites, and late Proterozoic rifting of North America. *Precambrian Research*, **7**, 61–98.

Frost, C. D. and Frost, B. R. (1997). Reduced rapakivi-type granites: the tholeiite connection. *Geology*, **25**, 647–50.

Grambling, J. A. (1986). Crustal thickening during Proterozoic metamorphism and deformation in New Mexico. *Geology*, **14**, 149–52.

Grambling, J. A., Williams, M. L., Smith, R. and Mawer, C. K. (1989). *The Role of Crustal Extension in the Metamorphism of Proterozoic Rocks of New Mexico*. Geological Society of America Special Paper 235, pp. 87–110.

Hamilton, W. B. (1981). Crustal evolution by arc magmatism. *Philosophical Transactions of the Royal Society*, London, Series A, 279–91.

Harris, C. W., Gibson, R. G., Simpson, C. and Eriksson, K. A. (1987). Proterozoic cuspate basement-cover structure, Needle Mountains, Colorado. *Geology*, **15**, 950–3.

Hawkins, D. P., Bowring, S. A., Ilg, B. R., Karlstrom, K. E. and Williams, M. L. (1996). U–Pb geochronologic constraints on the Paleoproterozoic crustal evolution of the Upper Granite Gorge, Grand Canyon. *Geological Society of America Bulletin*, **108**, 1167–81.

Hill, B. M. and Bickford, M. E. (2001). Paleoproterozoic rocks of central Colorado: accreted arcs or extended older crust? *Geology*, **29**, 1015–18.

Hoffman, P. F. (1989). Speculations on Laurentia's first gigayear (2.0–1.0 Ga). *Geology*, **17**, 135–8.

Ilg, B. R., Karlstrom, K. E., Hawkins, D. P. and Williams, M. L. (1996). Tectonic evolution of the Paleoproterozoic rocks in the Grand Canyon: insights into middle crustal processes. *Geological Society of America Bulletin*, **108**, 1149–66.

Jackson, J. (2002). Strength of the continental lithosphere: time to abandon the jelly sandwich? *GSA Today*, **12**, 4–10.

Jarrard, R. D. (1986). Relations among subduction parameters. *Reviews of Geophysics*, **24**, 217–84.

Karlstrom, K. E. (1989). *Early Recumbent Folding during Proterozoic Orogeny in Central Arizona.* Geological Society of America Special Paper 235, pp. 155–71.

Karlstrom, K. E. and Bowring, S. A. (1988). Early Proterozoic assembly of tectonostratigraphic terranes in southwestern North America. *Journal of Geology*, **96**, 561–76.

Karlstrom, K. E. and Bowring, S. A. (1993). Proterozoic orogenic history in Arizona. In *Precambrian: Conterminous U. S. Geology of North America, C2*, ed. J. C. Reed, Jr., M. E. Bickford, R. S. Houston, P. K. Link, D. W. Rankin, P. K. Sims and W. R. Van Schmus. Boulder, Col.: Geological Society of America, pp. 188–211.

Karlstrom, K. E. and Humphreys, G. (1998). Influence of Proterozoic accretionary boundaries in the tectonic evolution of western North America: interaction of cratonic grain and mantle modification events. *Rocky Mountain Geology*, **33**, 161–79.

Karlstrom, K. E. and Williams, M. L. (1995). The case for simultaneous deformation, metamorphism, and plutonism: an example from Proterozoic rocks in central Arizona. *Journal of Structural Geology*, **17**, 59–81.

Karlstrom, K. E. and Williams, M. L. (1998). Heterogeneity of the middle crust: implications for strength of continental lithosphere. *Geology*, **26**, 815–18.

Karlstrom, K. E., Åhäll, K. I., Harlan, S. S., Williams, M. L., McLelland, J. and Geissman, J. W. (2001). Long-lived (1.8–0.8 Ga) Cordilleran-type orogen in southern Laurentia, its extensions to Australia and Baltica, and implications for refining Rodinia. *Precambrian Research*, **111**, 5–30.

Karlstrom, K. E. and CD-ROM Working Group (2002). Structure and evolution of the lithosphere beneath the Rocky Mountains: initial results from the CD-ROM experiment. *GSA Today*, **12**, 4–10.

Karlstrom, K. E., Harlan, S. S., Williams, M. L., McLelland, J., Geissman, J. W. and Åhäll. K.-I. (1999). Refining Rodinia: geologic evidence for the Australian–Western U. S. connection in the Proterozoic. *GSA Today*, **9**, 1–7.

Karlstrom, K. E., Heizler, M. T. and Williams, M. L. (1997). $^{40}Ar/^{39}Ar$ muscovite thermochronology within the Upper Granite Gorge of the Grand Canyon. *Eos*, **78**, F784.

Karlstrom, K. E., Ilg, B. R., Williams, M. L., Hawkins, D. P., Bowring, S. A. and Seaman, S. J. (2002). Paleoproterozoic rocks of the Granite Gorges. In *Grand Canyon Geology*, ed. S. S. Beus and M. Morales, 2nd edn. Oxford: Oxford University Press.

Kirby, E., Karlstrom, K. E. and Andronicos, C. (1995). Tectonic setting of the Sandia pluton: an orogenic 1.4 Ga granite in New Mexico. *Tectonics*, **14**, 185–201.

Kohlstedt, D. L., Evans, B. and Maxwell, S. J. (1995). Strength of the lithosphere: constraints imposed by laboratory experiments. *Journal of Geophysical Research*, **100**, 17 587–602.

Kuznir, N. J. (1991). The distribution of stress in the lithosphere: thermorheological and geodynamic constraints. *Philosophical Transactions of the Royal Society, London*, **A337**, 95–107.

Lachenbruch, A. H. and Sass, J. H. (1980). Heat flow and energetics of the San Andreas fault zone. *Journal of Geophysical Research*, **85**, 6185–222.

Livacarri, R. F. and Perry, F. V. (1993). Isotopic evidence for preservation of Cordilleran lithospheric mantle during the Sevier–Laramide orogeny, western United States. *Geology*, **21**, 719–22.

Maggi, A., Jackson, J. A., McKenzie, D. and Priestley, K. (2000). Earthquake focal depths, effective elastic thickness, and the strength of the continental lithosphere. *Geology*, **28**, 495–8.

Magnani, M. B., Levander, A., Miller, K. C. and Eshete, T. (2002). Seismic structure of the Jemez lineament, New Mexico: evidence for heterogeneous accretion and extension in Proterozoic time and modern volcanism. *Eos*, **82**, F865.

Marshak, S., Karlstrom, K. E. and Timmons, J. M. (2000). Inversion of Proterozoic extensional faults: an explanation for the pattern of Laramide and Ancestral Rockies intracratonic deformation, United States. *Geology*, **28**, 735–8.

McKenzie, D., Nimmo, F. and Jackson, J. A. (2000). Characteristics and consequences of flow in the lower crust. *Journal of Geophysical Research*, **105**, 11 029–46.

Molnar, P. (1992). Brace–Goetze strength profiles, the partitioning of strike slip and thrust faulting at zones of oblique convergence and the San Andreas stress–heat flow paradox. In *Fault Mechanics and Transport Properties of Rocks*, ed. B. Evans and T. F. Wong. London: Academic Press, pp. 435–59.

Morozova, E., Wan, X., Chamberlain, K. R., Smithson, S. B., Morozov, I. B. and Boyd, N. K. (2002). Geometry of Proterozoic sutures in the central Rocky Mountains from seismic reflection data: Cheyenne belt and Farwell Mountain structures. *Geophysical Research Letters*, **29**, doi:10.1029/2001GL013819.

Nelson, K. D. and Project INDEPTH (1996). Partially molten middle crust beneath southern Tibet: synthesis of Project INDEPTH results. *Science*, **274**, 1684–7.

Nyman, M. W. and Karlstrom, K. E. (1997). Pluton emplacement processes and tectonic setting of the 1.42 Ga Signal batholith, SW USA: important role of crustal anisotropy during regional shortening. *Precambrian Research*, **82**, 237–63.

Nyman, M. W., Karlstrom, K. E., Kirby, E. and Graubard, C. M. (1994). Mesoproterozoic contractional orogeny in western North America. *Geology*, **22**, 901–4.

Oldow, G. S., Bally, A. W. and Ave Lallemant, H. G. (1990). Transpression, orogenic float, and lithosphere balance. *Geology*, **18**, 991–4.

Pedrick, J. M., Karlstrom, K. E. and Bowring, S. A. (1998). Reconciliation of conflicting models for Proterozoic rocks of northern New Mexico. *Journal of Metamorphic Geology*, **16**, 687–707.

Powell, J. W. (1876). *Exploration of the Colorado River of the West*. Washington, DC: Smithsonian Institution.

Rämö, O. T. and Calzia, J. P. (1998). Nd isotopic composition of cratonic rocks in the southern Death Valley region: evidence for substantial Archean source component in Mojavia. *Geology*, **26**, 891–4.

Ranalli, G. (1997). Rheology of the lithosphere in space and time. In *Orogeny Through Time*, ed. J.-P. Burg and M. Ford. Geological Society (London) Special Publication 121, pp. 19–37.

Read, A., Karlstrom, K. E., Grambling, J. A., Bowring, S. A., Heizler, M. and Daniel, C. (1999). A mid-crustal cross section from the Rincon Range, northern New Mexico: evidence for 1.68 Ga pluton-influenced tectonism and 1.4 Ga regional metamorphism. *Rocky Mountain Geology*, **34**, 67–91.

Royden, L. H. (1996). Coupling and decoupling of crust and mantle in convergent orogens: implications for strain partitioning in the crust. *Journal of Geophysical Research*, **101**, 17 679–705.

Rudnick, R. L. and Fountain, D. M. (1995). Nature and composition of the continental crust: a lower crustal perspective. *Reviews of Geophysics*, **33**, 1–44.

Selverstone, J., Hodgins, M., Aleinikoff, J. N. and Fanning, C. M. (2000). Middle Proterozoic reactivation of an Early Proterozoic transcurrent boundary in the northern Colorado Front Range: implications for ca. 1.7 and 1.4 Ga tectonism. *Rocky Mountain Geology*, **35**, 139–62.

Shaw, C. A. and Karlstrom, K. E. (1999). The Yavapai–Mazatzal crustal boundary in the southern Rocky Mountains. *Rocky Mountain Geology*, **34**, 37–52.

Shaw, C. A., Karlstrom, K. E., Williams, M. L., Jercinovik, M. J. and McCoy, A. M. (2001). Electron microprobe monazite dating of ca. 171–163 Ga and ca. 1.45–1.38 Ga deformation in the Homestake shear zone, Colorado: origin and early evolution of a persistent intracontinental tectonic zone. *Geology*, **29**, 739–42.

Sibson, R. H. (1986). Earthquakes and rock deformation in crustal fault zones. *Annual Reviews of Earth and Planetary Sciences*, **14**, 149–75.

Smith, R. B. and Bruhn, R. L. (1984). Intraplate extensional tectonics of the eastern Basin Range: inferences on structural style from seismic reflection data, regional tectonics, and thermo-mechanical models of brittle–ductile transition. *Journal of Geophysical Research*, **89**, 5733–62.

Smith, D. R., Noblett, J., Wobus, R. A., Unruh, D. and Chamberlain, K. R. (1999). A review of the Pikes Peak batholith, Front Range, central Colorado: a "type example" of A-type granitic magmatism. *Rocky Mountain Geology*, **34**, 289–312.

Thompson, A. G., Grambling, J. A., Karlstrom, K. E. and Dallmeyer, R. D. (1996). Proterozoic metamorphism and $^{40}Ar/^{39}Ar$ thermal history of the 1.4 Ga Priest pluton, Manzano Mountains, New Mexico. *Journal of Geology*, **104**, 583–98.

Timmons, M. J., Karlstrom, K. E., Dehler, C. M., Geissman, J. W. and Heizler, M. T. (2001). Proterozoic multistage (~1.1 and ~0.8 Ga) extension in the Grand Canyon Supergroup and establishment of northwest and north–south tectonic grains in the southwestern United States. *Geological Society of America Bulletin*, **113**, 163–80.

Tyson, A. R., Morozova, E. A., Karlstrom, K. E., Chamberlain, K. R., Smithson, S. B., Dueker, K. G. and Foster, C. T. (2002). Proterozoic Farwell Mountain–Lester Mountain suture zone, northern Colorado: subduction flip and progressive assembly of arcs. *Geology*, **30**, 943–6.

Williams, M. L. (1991). Overview of Proterozoic metamorphism in Arizona. In *Proterozoic Geology and Ore Deposits of Arizona*, ed. K. E. Karlstrom. Arizona Geological Society Digest 19, pp. 11–26.

Williams, M. L. and Jercinovic, M. J. (2002). Microprobe monazite geochronology: putting absolute time into microstructural analysis. *Journal of Structural Geology*, **24**, 1013–28.

Williams, M. L. and Karlstrom, K. E. (1997). Looping *P–T* paths and high-*T*, low-*P* middle crust metamorphism: Proterozoic evolution of the southwestern United States. *Geology*, **105**, 205–23.

Williams, M. L., Karlstrom, K. E., Lanzirotti, A., Read, A. S., Bishop, J. L., Lombardie, C. E., Pedrick, J. N. and Wingstead, M. B. (1999). New Mexico middle crustal cross sections: 1.65 Ga macroscopic geometry, 1.4 Ga thermal structure and continued problems in understanding crustal evolution. *Rocky Mountain Geology*, **34**, 53–66.

Windley, B. F. (1992). Proterozoic collisional and accretionary orogens. In *Proterozoic Crustal Evolution*, ed. K. C. Condie. Amsterdam: Elsevier, pp. 419–46.

Windley, B. F. (1993). Proterozoic anorogenic magmatism and its orogenic connection. *Journal of the Geological Society of London*, **150**, 39–50.

Wooden J. L. and DeWitt, E. D. (1991). Pb isotopic evidence for the boundary between the Early Proterozoic Mojave and central Arizona crustal provinces in western Arizona. In *Proterozoic Geology and Ore Deposits of Arizona*, ed. K. E. Karlstrom. Arizona Geological Society Digest, 19, 27–50.

9

Melting of the continental crust: fluid regimes, melting reactions, and source-rock fertility

JOHN D. CLEMENS

9.1 Preamble

This chapter presents a synthesis of partial melting in Earth's continental crust. However, rather than simply attempting to synthesize all the data from previous experimental and theoretical studies, the chapter takes a definite scientific position that reflects present understanding of crustal melting. Unless otherwise stated, the term "granitic" is used in the broad sense to denote magma or melt compositions ranging from syenogranite through to tonalite. The word "fertility" is used to denote the relative capacity of a given protolith material to produce granitic partial melt, under the specified physicochemical conditions (mainly P, T, and fO_2).

The high temperatures that are required to partially melt crustal rocks and form granitic magmas equate with the conditions of upper amphibolite- to granulite-facies metamorphism. This is one important reason for the inferred intimate connection between the production of granulite-facies rocks, the production and withdrawal of partial melts, and the differentiation of the continental crust (see, for example, Fyfe, 1973; Clemens, 1990; Thompson, 1990). Partial melting may occur in response to various intracrustal processes that occur during tectonic thickening and orogenic collapse of the crust (e.g., Patiño-Douce et al., 1990; Harris & Massey, 1994). However, the thermal requirements of granulite-facies processes generally demand that additional, extra-crustal heat sources be available to drive the reactions.

Thermal modeling has demonstrated that thickened crust (with a normal thermal profile) does not reach the temperatures necessary to partially melt (on time scales of up to 100 Myr) unless large amounts of aqueous fluid are also introduced to depths of 20–40 km (e.g., England & Thompson, 1984). From the analysis of England and Thompson (1984) it appears that the only exceptions to this general rule would occur where the crust has unusually low

Evolution and Differentiation of the Continental Crust, ed. Michael Brown and Tracy Rushmer. Published by Cambridge University Press. © Cambridge University Press 2005.

thermal conductivity ($<2\ Wm^{-1}\ K^{-1}$) combined with very high surface heat flow ($>65\ mW\ m^{-2}$). Thus, except in some migmatite terranes, production of voluminous, mobile, granitic magma commonly involves advection of mantle heat to the continental crust. The most likely vectors of this heat would be underplated or intraplated mafic magmas. Such under-accretion probably represents the major means by which Earth's continents have grown in volume since the Archean (e.g., Rudnick, 1990). The only alternative is to postulate that the crust was unusually enriched in heat-producing elements (e.g., Sandiford *et al.*, 1998; McLaren *et al.*, 1999; Chapter 11). This does seem to be the case in a few specific terranes, but is unlikely to be a general feature.

In any case, the withdrawal of large volumes of granitic magmas from the deep crust, and their emplacement at higher crustal levels, has two major consequences. The first consequence is that the deep crust will be left in a mafitized, partially dehydrated, residual condition (e.g., Brown & Fyfe, 1970; Fyfe, 1973; Clemens, 1990; Thompson, 1990). To become exposed at the surface, such dense rocks would need to undergo a second major tectonic episode, perhaps temporally unrelated to the original partial-melting event, (e.g., the residual metapelites ("stronalites") of the Ivrea Zone). This necessity may contribute to the relative scarcity of such rocks, in comparison with the abundance of their mobile magmatic counterparts. Delamination and foundering into the mantle may also be the fate of some residual lower crust (e.g., Bohlen, 1991). The second consequence is that the upper crust will become enriched in felsic minerals and heat-producing elements. This is probably the major mechanism for post-Archean and ongoing, large-scale crustal differentiation (Vielzeuf *et al.*, 1990).

9.2 Crustal fluid regimes and partial melting reactions

Three potentially important fluid regimes might exist during high-T metamorphism and partial melting. The first is fluid-dominated, where the fluid/rock ratio is relatively high. Here, the composition of the fluid phase controls both the activities of volatile components and the nature of the reactions. The second is rock-dominated, in which the fluid/rock ratio is relatively low. Here, the fluid composition initially controls volatile activities and triggers reactions (e.g., dehydration, decarbonation, or partial melting). However, the fluid then changes its composition in response to dilution with components from devolatilization or to dissolution of fluid components into a melt phase. The third regime is fluid-absent. Here, the rock system may contain tiny traces of fluid in isolated, unconnected pockets, but most grain boundaries are effectively "dry." This means that the only significant reactions that may take place are those that do not require the presence of a free fluid phase.

The relationships between the various kinds of melting reactions, in a theoretical simple system, are illustrated in Fig. 9.1. The P–T plot in Fig. 9.1a shows the various reactions in the two-component system A–H_2O, that may contain an anhydrous mineral A, a crystalline hydrate mineral H, melts of various compositions, and a fluid phase of nearly pure H_2O. All reactions are labeled according to the phase that does not participate. They are: H = A + Fl [subsolidus dehydration (M)]; H + Fl = M [the wet solidus at high P (A)]; A + Fl = M [wet solidus at low P (H)]; and, H = A + M [fluid-absent melting reaction (Fl)]. Note that these reactions all meet at an invariant point, where all phases coexist. The fluid-absent reaction occurs at high T (above the pressure of the invariant point) and has nothing to do with the dehydration reaction, which is unstable at pressures and temperatures higher than the invariant point.

Melting reactions in rocks mostly involve many more phases than in the simple system of Fig. 9.1, and Eggler (1973) gives the definitive theoretical treatment of all main variations on this theme. Eggler (1973) also dealt with the complications (new reactions and singular points) introduced by the possibility that hydrous mineral assemblages could break down, at low P, to produce melts that contain less H_2O than the original hydrous mineral assemblage. At least for biotite reactions, the pressures at which these complications occur are very low – relevant only to contact metamorphism (e.g., Vielzeuf & Clemens, 1992). Here we are discussing deep-crustal, regional metamorphic processes. Natural rock systems behave in a very similar manner to the simple theoretical system above, and it is unnecessary to introduce the various complications in order to understand the fundamental principles involved.

In nature, H would stand for biotite or hornblende, for example, and A would be an assemblage of minerals such as pyroxene and garnet. Except in contact metamorphism (at very low P), micas and amphiboles do not liberate H_2O when they decompose under high-grade conditions; they produce silicate melts (e.g., Clemens, 1990). The only exception is in the presence of a fluid phase with low H_2O activity, where dehydration reactions are, indeed, possible.

As Fig. 9.1a shows, the fluid-absent melting reaction is fixed at any given P by the T. The T–aH_2O diagram in Fig. 9.1b is for the same reactions as in Fig. 9.1a, but plotted at some fixed pressure (P_1) above the invariant point. As is apparent, the fluid-absent melting reaction is pinned at its T as well as its aH_2O. This means that, for any given pressure, the temperature and aH_2O of such a fluid-absent reaction are fixed. Furthermore, since the aH_2O controls

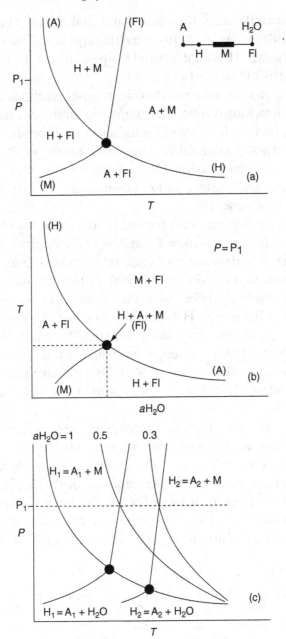

Fig. 9.1. Schematic phase diagrams of a hypothetical system $a\mathrm{H_2O}$, modified from Fig. 1 of Clemens and Petford (1999). In this system, A is an anhydrous mineral and there is also a hydrous mineral H. Diagram (a) is a P–T section showing all the reactions, and an inset showing the chemographic relations. Each of the four divariant fields is labeled to indicate the stable

the H_2O content of the melt, it is clear that T and melt H_2O content cannot vary independently during most crustal melting episodes. Theoretical relationships between the melt H_2O contents and temperatures of fluid-absent melting reactions are further illustrated in Fig. 9.1c.

The preceding analysis assumes that the melting reactions are univariant. Certainly, partial melting reactions in complex natural rocks are likely to have much higher variance (due to solid solution in both product and reactant minerals). This results in reactions that form bands, in P–T or T–aH_2O space. However, as illustrated in Fig. 9.2, even within these bands, T and melt H_2O content are constrained to systematically co-vary, as long as the system remains fluid-absent.

Partial melts (granitic magmas) formed by relatively low-T reactions will be wetter; those formed by high-T reactions will be drier (e.g., Clemens *et al.*, 1997). Fig. 9.1c demonstrates these relationships. Since melt viscosity decreases with increased T and melt H_2O content, there is a natural "buffering" of viscosity, in which the effect of higher T tends to be counterbalanced by the consequent lower melt H_2O content. This buffering will always exist, except in situations of fluid-dominated melting (Clemens & Watkins, 2001). Clemens and Droop (1998) presented an exhaustive analysis of the effects of fluid-dominated, rock-dominated, and fluid-absent conditions on melt production. Readers are referred to this work for details of the different kinds of reactions.

It has often been pointed out that granitic magmas are, in general, H_2O-undersaturated prior to extensive crystallization and differentiation. Given the inferred connection between the granulite-facies and the production of granitic partial melts, we need to decide whether we should deal with both fluid-present and fluid-absent reactions, in order to discuss adequately the subject of partial melting at granulite grade. The next section is an attempt to do this, based on the more detailed arguments presented by Clemens and Watkins (2001).

Fig. 9.1. (cont.)

assemblage. Diagram (b) is a T–aH_2O section, showing the same reactions. The stable assemblages in the divariant fields are labeled, as is the isobarically invariant (three-phase) assemblage at the fluid-absent melting reaction (Fl). Diagram (c) is a P–T section similar to (a), but showing the independent breakdown of two different hydrous minerals (H_1 and H_2). The solidus is shown here at three values of aH_2O ($1.0 = H_2O$-saturated, 0.5, and 0.3). The dehydration reactions intersect the solidus and generate two fluid-absent melting reactions: $H_1 = A_1 + M$ (at lower T) and $H_2 = A_2 + M$ (at higher T). At pressure P_1, the lower-T reaction occurs at higher aH_2O and therefore the melt produced will have a higher H_2O content than the melt formed by the higher-T reaction. See the text for a fuller discussion.

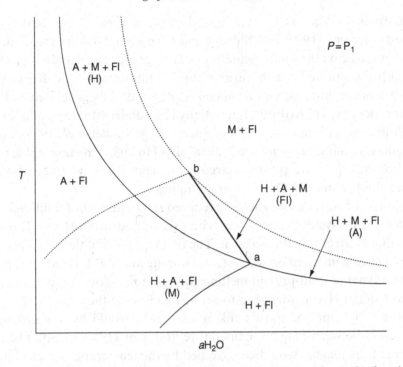

Fig. 9.2. Schematic T–aH_2O phase diagram (similar to Fig. 9.1b) for the hypothetical system A–B–H_2O, where B is a component that goes into solid solution in anhydrous phase A and in hydrous phase H, and is soluble in the melt. No new phases are introduced by the addition of component B. This would be analogous to taking into account Fe–Mg solid solution in micas and pyroxenes, for example. The effect is to make the univariant reactions (lines in Fig. 9.1b) into divariant reactions (fields). The various fields in the diagram are labeled according to which phases are stable, and the fluid-absent reaction is labeled, as in Fig. 9.1. The solid and dashed curves represent the lower- and higher-temperature dehydration and melting reactions – the lower and upper thermal stability limits of the H and A solid solutions, respectively. The fluid-absent melting reaction now plots as the line a–b (instead of the point in Fig. 9.1b). Fluid-absent partial melting of H will begin at point a. As the temperature rises, the residual H will become richer in component B and the melting path will progress toward point b. At this point, all of the hydrate H will have broken down. Note that, within the interval over which H decomposes, T and aH_2O are co-constrained along this single line (a-b) and cannot vary independently.

9.3 The granulite facies: fluid-present or fluid-absent?

As noted above, it is generally agreed that granitic magmas are generated mainly by crustal melting at high temperatures, and that this suggests a genetic connection between granitic magma generation and granulite-facies metamorphism. What is not yet agreed is the nature of the fluid regime involved. Clemens (1990) and Stevens and Clemens (1993) reviewed various lines of evidence on the fluid regime at granulite grade. Their conclusions were that both fluid-present (with various fluid compositions) and fluid-absent behaviour occur, but that mobile magmas are not usually generated by fluid-present processes. The hydrous, but initially H_2O-undersaturated, character of all high-level granitic magmas (e.g., Clemens, 1984; Scaillet *et al.*, 1998), as well as the phase equilibrium evidence for low aH_2O in high-T metamorphic rocks (e.g., Bohlen, 1991) are powerful arguments against the presence of excess aqueous fluid. Thus, aH_2O is $\ll 1$ in granulites.

As stated above, rocks may achieve reduced H_2O activity in a fluid-deficient, rock-dominated system, in a system with excess fluid with $aH_2O < 1$, or in a system that is effectively fluid-absent. The first two possibilities could result in any aH_2O being imposed on the system. Granitic magmas formed by either of these two types of fluid-present melting equilibria therefore could have essentially any initial H_2O content (up to saturation level at the local pressure). In these cases, it would be most unlikely that there would be any systematic co-variation between magma temperature and melt H_2O content. The temperature of the magma would be governed by the temperature reached in the protolith at the moment of magma extraction. The melt H_2O content would be governed by the aH_2O in the fluid or the amount of H_2O initially present in the protolith. Therefore, fluid-present granulite-facies metamorphism implies that granitic magmas could have a range of temperatures (e.g., 800–1000 °C) and H_2O contents that could vary randomly from almost zero to quite high values. There could be hot wet melts, hot dry melts, cool dryish ones, cool wet ones, and any combination in-between. This situation contrasts strongly with the effects of fluid-absent metamorphism, where temperature and melt H_2O content should systematically co-vary such that cooler magmas are wetter and hotter magmas are drier. Furthermore, because of the quasilinear shapes and steep P–T slopes of fluid-absent melting reactions (roughly parallel to the granite solidus at low aH_2O), the buffered T and particularly the melt H_2O content are constrained to co-vary within narrow limits.

Scaillet *et al.* (1998) collated all available experimental data on the temperatures and initial H_2O contents of granitic magmas (volcanic and plutonic), to constrain the viscosities of granitic melts. The data were taken from studies that compared experimentally determined equilibrium-phase relations with petrographic evidence of mineral paragenesis and crystallization sequences. A plot of the data is given in Fig. 9.3, where the distribution of points may be compared to the regions where points would be expected to fall for

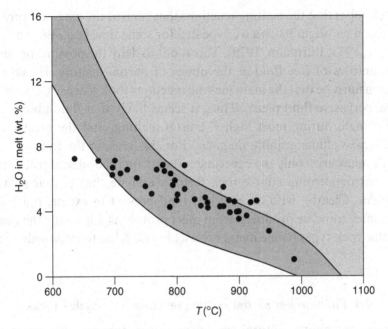

Fig. 9.3. Graph of melt H_2O content versus temperature with the data points of Scaillet *et al.* (1998) plotted. The grey-shaded areas represent the modeled ranges of these parameters in various fluid regimes. Fluid-present melting of intermediate to felsic and pelitic protoliths would result in points scattered across the entire gray region. Fluid-absent melting would limit the range of T and melt H_2O content to the dark grey region. (From Clemens & Watkins (2001), The fluid regime of high-temperature metamorphism during granitoid magma genesis. *Contributions to Mineralogy and Petrology*, **140**, 600–6. © 2001 Springer-Verlag. Reproduced by permission of Springer-Verlag.)

fluid-present and fluid-absent magma genesis. The notable feature here is the strong negative correlation displayed and the narrow band of possible melt H_2O contents and temperatures. The magmas plotted in Fig. 9.3 range in composition from rhyolitic (granitic) to dacitic (tonalitic). They are from a variety of localities spread over five continents and several island arcs. They cover both plutonic and volcanic emplacement styles, including lava flows and pyroclastic deposits. The tectonic regimes range from inter-oceanic island arcs, to Andean margins, continental collision zones and within-plate settings. The ages of the rocks vary from early Paleozoic to late Cenozoic. Therefore, it is reasonable to infer that Fig. 9.3 carries evidence of an important general relationship, common to a wide range of granitic magmas.

From the above discussion, developed more fully in Clemens and Watkins (2001), it is clear that this type of covariation, between T and melt H_2O

content, implies that the melting reactions that created the magmas were fluid-absent. Such an origin has been advocated for some time (e.g., Brown & Fyfe, 1970; Fyfe, 1973; Burnham, 1979). This is not to deny the possible presence of small quantities of free fluid at the outset of partial melting. However, the conclusion must be that the main melting reactions took place in the absence of any free, pervasive fluid phase. Thus, it seems likely that fluid-absent conditions dominate during most high-T crustal melting, and the production of mobile, large-volume granitic magmas. For this reason, the following treatment will summarize only the experimental data on fluid-absent reactions. All conclusions concerning source-rock fertility assume that partial melting is fluid-absent. Clearly, with an aqueous fluid present in excess, much larger melt volumes could be produced from most protoliths, especially the quartzo-feldspathic rock types (some metagreywackes and felsic to intermediate meta-igneous rocks).

9.4 Fluid-absent partial melting reactions in complex rocks

As pointed out above, natural crustal rocks are complex, multicomponent systems, in which melting reactions will usually be multivariant. That is, they will occur over some range of T at any given P, and should be represented as bands instead of lines in P–T space. Nevertheless, the complex reactions involved may be distilled into a relatively small number of major types that depend on the protolith mineralogy and the temperature. Experiments on fluid-absent partial melting have been carried out by Clemens (1981), LeBreton and Thompson (1988), Rutter and Wyllie (1988), Vielzeuf and Holloway (1988), Beard and Lofgren (1991), Patiño-Douce and Johnston (1991), Rapp *et al.* (1991), Rushmer (1991), Wolf and Wyllie (1991, 1994), Skjerlie and Johnston (1992, 1993), Sen and Dunn (1994), Vielzeuf and Montel (1994), Carrington and Harley (1995), Finger and Clemens (1995), Gardien *et al.* (1995, 2000), Patiño-Douce and Beard (1995, 1996), Rapp and Watson (1995), Singh and Johannes (1996a, b), Montel and Vielzeuf (1997), Stevens *et al.* (1997), Patiño-Douce and Harris (1998), Pickering and Johnston (1998), and López and Castro (2001).

The following are simplified examples of the types of fluid-absent melting reactions that have been observed in experiments (at P up to 1.5 GPa) on a range of common crustal rock types. Standard mineral abbreviations are as given by Kretz (1983), with M = H_2O-undersaturated granitic melt.

In metapelitic rocks:
Ms ± Bt + Pl + Qtz = Als ± Grt ± Kfs + M (amphibolite facies); and
Bt + Pl + Als + Qtz = Grt ± Crd ± Kfs + M (upper amphibolite to granulite facies).

In alumina-poor metagreywackes and quartzofeldspathic rocks:
Bt + Pl + Qtz = Hbl + Cpx ± Kfs + M (upper amphibolite to granulite facies).

In alumina-poor metagreywackes and quartzofeldspathic rocks
Bt ± Hbl + Pl + Qtz = Opx + Cpx ± Grt ± Kfs + M (granulite facies).

In peraluminous metagreywackes:
Bt + Pl + Qtz = Opx ± Grt ± Crd ± Kfs + M (granulite facies).

In metabasaltic to meta-andesitic amphibolites (granulite facies):
Hbl + Pl = Opx + Cpx ± Grt + M; or,
Hbl = Opx + Cpx ± Pl ± Grt ± Qtz or Ol + M.

Muscovite reactions occur at relatively low *T* (Storre, 1972; Chatterjee & Froese, 1975; Petö, 1976). The modal abundance of muscovite in amphibolite-facies metapelites normally ranges from 0 to 10 vol.%, but may reach up to about 25 vol.%. Muscovite breakdown at the second sillimanite isograd is thus predicted to produce mainly small melt proportions (<10 vol.%; Clemens & Vielzeuf, 1987) but may produce up to 20 vol.% or more in muscovite-rich protoliths.

The biotite breakdown reactions occur in a similar temperature range, in all rock types except amphibolites, with the mica decreasing markedly in modal proportion at about 850 °C. Hornblende breakdown tends to occur at higher *T* (around 900 °C). K-feldspar is shown as a product in many of the reactions above, and should theoretically be present. Experiments on some simple analog systems have produced K-feldspar as a product phase (e.g., Vielzeuf & Clemens, 1992). However, the experiments cited above show that this phase is commonly absent, due to a variety of chemical effects in the complex, natural rock materials that have been used as starting materials.

9.5 Results of some fluid-absent experiments

This section summarizes some important findings of a variety of experimental studies of fluid-absent partial melting of common crustal rock types. Although there have been other studies, emphasis is placed on experiments on compositions as close as possible to complex natural rocks. The reason for this choice is that phase relations may differ markedly between simple and natural systems. As an illustration, the reader may wish to compare the results of the metapelite experiments mentioned below with those of Carrington and Harley (1995) on model metapelites that lack CaO and Na$_2$O components. There are some major differences in phase relations, and many of these result from the absence of plagioclase components in the simple system. Of course, there are also appreciable differences between various studies using natural rock composi-tions. Appendix 1 contains a tabulation of the major-element chemistry of each experimental rock and mineral mixture for which it has been possible to construct a melt proportion versus T graph.

In all experiments summarized here, finely powdered crystalline or pow-dered glassy starting materials were used. Based on the experience of experi-mental petrologists, such starting materials seem to be the best for obtaining near-equilibrium mineral/melt proportions. The products are therefore likely to have closely approached equilibrium, with reference to mineral species and proportions, and the compositions of the melts and their proportions, even over the relatively short experimental time scales. Some studies have checked for and found high levels of consistency between the results of equivalent experiments carried out with powdered rock and with glassy starting materials (e.g., Montel & Vielzeuf, 1997). However, the results of many experimental studies indicate that certain minerals commonly have disequilibrium composi-tions (e.g., Al contents of pyroxenes and plagioclase compositions); such disequilibrium is also common in natural rocks. Some other studies have used small rock cylinders as starting materials (e.g., Wolf & Wyllie, 1991; Rushmer, 1995). The results of these studies indicate gross chemical disequili-brium, although this sort of experiment may be useful in the study of mechan-ical behavior, textural evolution, and the kinetics of melt segregation.

Many authors do not repeat what appear to be successful experiments, so it is very difficult to make a firm statement on the reproducibility of the reported results. In addition, modal measurement techniques vary considerably between different studies (e.g., visual estimates (rarely), point counting (also rarely), image analysis and mass balance calculations). Thus, figures quoted for melt proportions carry significant but variable errors. Again, it would be very difficult to give precise magnitudes of such errors; most authors attempt no such statistical analysis themselves, although a few quote quite small errors, around 3% relative in one case. However, as a rough approximation, the reader should probably apply a conservative standard error of about 5% (absolute) to all estimates.

In the case of fluid-absent partial melting at given P, T, etc., protolith fertility will depend mainly on: (1) the available H_2O content of the rock (as expressed in the hydrous mineral content); (2) the thermal stability of the hydrous mineral assemblage (the proportion of the hydrous minerals that break down during melting), and (3) the abundance of other phases required for the melting reactions to take place (e.g., quartz and feldspars). Experimental work to date suggests that these fluid-absent melting reactions proceed rapidly, once the required T has been reached. Thus, it seems unlikely that the kinetics of melting reactions would limit fertility of a source rock, although disequilibrium may well affect the trace element and isotopic characteristics of the melts. The reader is referred to Clemens and Vielzeuf (1987) and Thompson (1988) for more detailed discussions of the issue of protolith fertility.

9.5.1 Metapelites

Studies of fluid-absent melting in metapelites by Vielzeuf and Holloway (1988), Patiño-Douce and Johnston (1991), Pickering and Johnston (1998) and Patiño-Douce and Harris (1998) provide interesting data. At 1.0 GPa, significant quantities of melt (>10 vol.%) were present only at temperatures above 800 °C. In the study by Vielzeuf and Holloway (1988), the biotite-out temperature is $\sim$850 °C or so. At higher temperatures, all the H_2O in the starting material is present in the melt phase. Consequently, melt production in this study is characterized by a strong pulse that corresponds to biotite breakdown around 860 °C. In contrast, the study by Patiño-Douce and Johnston (1991) showed biotite persisting to a temperature of $\sim$975 °C. In this study, melt appeared at 800 °C and the proportion of melt increased, in an approximately linear fashion, with increasing temperature. Figure 9.4 shows a comparison of the various melt production curves. There are large variations in overall fertility, in the patterns of melt production as a function of T, and in the upper T limit of biotite stability. The differences appear to depend on the bulk rock TiO_2, Na_2O, and H_2O contents, and $aSiO_2$. For example, the biotite-rich starting material of Patiño-Douce and Johnston (1991) contained an unusually low plagioclase content for a metapelite (reflected in its high $MgO + FeO$ and K_2O, and low $CaO + Na_2O$) and is rather aluminous (with about 24 wt.% Al_2O_3, calculated on an anhydrous basis).

The compositions of melts produced from the two-mica rocks were all strongly peraluminous, similar to some S-type granitic rocks. However, melts of the muscovite schist (Patiño-Douce & Harris, 1998) were leucogranitic, even at 900 °C. A number of these studies also report experiments at pressures above and below 1 GPa. As would be predicted (see later) the melt proportions developed at any given T are greater at lower P. The wide variations in fertility (apparent in Fig. 9.4) mean that it is difficult to generalize about the fertility of metapelites. Differences in rock chemistry lead to wide variations in melting behavior. This will also be seen to be the case for most other classes of rock types. It should be possible to develop rock fertility models that take account of bulk-rock chemical and mineralogical variables, but this will not be attempted here (Vielzeuf & Schmidt, 2001).

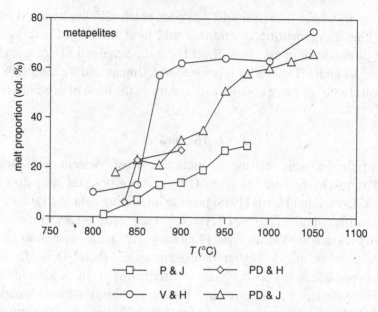

Fig. 9.4. Melt-production curves for natural metapelites at 1 GPa. Vielzeuf and Holloway (1988) – sodic two-mica metapelite (circles); Patiño-Douce and Johnston (1991) – Na-poor two-mica metapelite (triangles); Pickering and Johnston (1998) – two-mica metapelite (squares), and Patiño-Douce and Harris (1998) – muscovite schist (diamonds). Note that the Vielzeuf and Holloway (1988) data at 800 and 850 °C are visual estimates only and that this data set is in weight percentage rather than volume percentage.

The experiments of LeBreton and Thompson (1988) were on an "average pelite" mixture of natural Bt, Pl, Ky, and Qtz. The data do not permit construction of a melt production curve but, at 860 °C, only 6 vol.% of a peraluminous granitic melt was produced. The main interval of biotite breakdown was evidently not reached in this study.

Stevens (1995) and Stevens *et al.* (1997) used synthetic Qtz–Pl–Bt–Sil metapelites to systematically investigate the effects of variations in Mg# ($=100Mg/$ $(Mg+Fe)$) and the presence of Ti on partial melting behavior at 0.5 and 1 GPa. Their bulk compositions contained around 26 wt.% Al_2O_3, far in excess of the average metapelite (about 19 wt.%; Taylor & McLennan, 1985). This high alumina content ensured sillimanite saturation throughout the melting interval investigated. As long as we view the results as modeling Al_2SiO_5-saturated rocks, the excess of sillimanite should not affect the outcome. The aAl_2O_3 will be the same in any sillimanite–quartz-saturated composition, regardless of the actual quantities of quartz and sillimanite present.

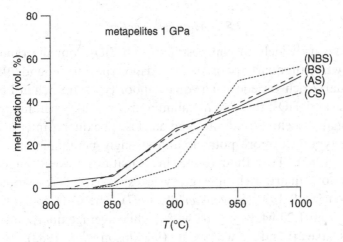

Fig. 9.5. Experimentally determined melt production curves of synthetic metapelites, modified from Fig. 3b of Stevens (1995). AS, BS, and CS are Ti-free compositions with Mg#s of 50, 60, and 80, respectively, and NBS is the analog of BS with Ti-bearing biotite. See also Stevens *et al.* (1997).

Figure 9.5 shows melt production versus T for the experiments of Stevens *et al.*, at a pressure of 1.0 GPa. The melting pulse (between 850 and 900 °C) for the Ti-free rocks is relatively subtle compared with the huge spike reported by Vielzeuf and Holloway (1988). However, the rock with Ti-bearing biotite (NBS) is much closer to a natural pelitic rock in composition, and it displays a pronounced melting pulse, shifted to higher T compared to the Ti-free rocks. This illustrates the enhanced thermal stability of Ti-bearing micas (cf. Patiño-Douce, 1993) and shows that strong melting pulses (over limited T ranges) are to be expected during the partial melting of metapelites, at least at 1 GPa.

Given that none of the experimental metapelites is compositionally a good model for the average post-Archean shale (see Table 2.9 of Taylor & McLennan, 1985), and the current lack of a model that accounts for compositional variations, we must have some reservations about interpreting the results. However, it appears that fluid-absent partial melts from metapelitic sources are dominantly peraluminous, potassic granites (*sensu stricto*). Total FeO + MgO contents increase with T, reaching maximum values of around 4 wt.%. Pelitic protoliths with lower Mg# tend to produce more mafic partial melts. This is most probably because the solubility of Fe in granitic melts is higher than that of Mg. Breakdown of more Fe-rich biotites will result in the formation of melts with somewhat higher Fe contents and slightly smaller amounts of magnesian mafic minerals in the solid products.

9.5.2 Metagreywackes

Greywackes vary widely in composition, with SiO_2 contents ranging from around 58 wt.% (in quartz-poor volcaniclastic types) to around 79 wt.% (in the most quartz-rich varieties). The quartz-poor types are rich in FeO, MgO, CaO, Na_2O, and TiO_2; they are metaluminous or weakly peraluminous, and compositionally resemble andesites and dacites. The quartz-intermediate and quartz-rich types are more potassic and strongly peraluminous (Taylor & McLennan, 1985). Two fluid-absent investigations have been conducted using Ca-poor, quartz-rich metagreywackes (Vielzeuf & Montel, 1994; Montel & Vielzeuf, 1997; Stevens et al., 1997). The CaO contents of these rocks were from 1.20 wt.% to 1.67 wt.%, while average quartz-intermediate greywackes have around 2.5 wt.% CaO (Dickinson et al., 1982).

Vielzeuf and Montel (1994) found that at 1.0 GPa, fluid-absent melting began at ~900 °C. This suggests that fluid-absent melting in metagreywackes is unlikely to occur in collisional settings, where temperatures do not normally reach this level, unless some tectonic process such as slab detachment or lithosphere delamination allows asthenospheric heat to reach the lower crust. Montel and Vielzeuf (1997) studied the same composition, at a variety of pressures, and found that the melting interval broadens considerably in *T* at higher *P*. In agreement with theory (see later), the melt proportion produced at any *T* is lower at higher *P*. A striking tendency noted by Montel and Vielzeuf (1997) is that curves showing the melt proportion as a function of *T* flatten progressively with increasing *P*. Thus, the melting becomes more smoothly progressive, and there is less of a melt-production pulse. The cause of this change in behavior is linked to changes in the subsolidus mineralogy as a function of *P*. For example, phengitic muscovite appears in the sub-solidus assemblage of this rock at higher pressures. This causes melting to begin at lower *T*, flattening the melt production curve and broadening the melting interval. The difference in fertility between pelites and greywackes increases at higher *P*, with pelites being more fertile above 0.8 GPa. All melts from the metagreywacke were peraluminous, potassic leucogranites, with total FeO = MgO ≤ 3 wt.%.

Patiño-Douce and Beard (1995) studied a synthetic mixture of natural biotite, plagioclase, and quartz. Although this composition was described by the authors as a relatively magnesian biotite–plagioclase–quartz (BPQ) tonalite (Mg# = 52), it more closely resembles a relatively magnesian metagreywacke in which the mineralogical proportions have been formulated to maximize its fertility. At a pressure of 3 GPa, the solidus was at about 850 °C, but at 1.5 GPa melting did not begin until 930 °C. The biotite was

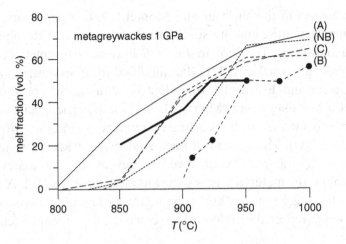

Fig. 9.6. Experimentally determined melt production curves of synthetic metagreywackes, modified from Fig. 3d of Stevens (1995). A, B, and C are Ti-free compositions with different Mg#s of 50, 60, and 80, respectively, and NB is the analog of B with Ti-bearing biotite. See also Stevens *et al.* (1997). The additional curves are the melt production trends for the greywacke (Mg# = 23) of Patiño-Douce and Beard (1996) (thick line) and Patiño-Douce and Beard (1995) (dashed line with points).

consumed within 50 °C of the solidus at both pressures, with the production of 50 to 60 vol.% of strongly peraluminous ($Al_2O_3/(CaO + Na_2O + K_2O) > 1.1$) granitic melt. The melt production curve for this material (at 1.5 GPa) is presented in Fig. 9.6.

Stevens (1995) and Stevens *et al.* (1997) conducted fluid-absent partial melting experiments, at 0.5 and 1.0 GPa, on a range of synthetic magnesian metagreywackes, in the system $K_2O–Na_2O–CaO–FeO–MgO–Al_2O_3–SiO_2–H_2O$, at temperatures appropriate to granulite-facies metamorphism. These authors also included data on the effects of additional TiO_2 and of fO_2. The metagreywackes were made from the same starting materials as the metapelite study by the same authors, and described above, but lacked sillimanite. In both the metapelites and metagreywackes, the onset of anatexis occurred under similar conditions. Thus, under relatively reducing (low fO_2) fluid-absent conditions, all similar natural metasedimentary granulites should contain evidence of partial melting, provided that metamorphic temperatures reached at least 830 °C. Increasing Mg# from 50 to 80 raises the temperature of the fluid-absent solidus by only about 30 °C. With increasing Mg#, the maximum temperature at which biotite and melt coexist increases, and the divariant melting interval expands. The addition of a TiO_2 component substantially extends the thermal stability of biotite, but has no discernible effect on the solidus. As in the Vielzeuf and Montel (1994) experiments, the melt compositions vary little, and are similar to those of natural strongly peraluminous leucogranites. The plot in Fig. 9.6 illustrates the melt production curves for these greywackes at 1.0 GPa (modified from Stevens, 1995).

Patiño-Douce and Beard (1996) studied the fluid-absent partial melting behavior of a less magnesian (Mg# = 23) and more calcic biotite metagreywacke. Melts formed from this starting material were granitic, with up to 3 wt.% (FeO + MgO). The pattern of melt production, with increasing T at 1 GPa (Fig. 9.6), is similar to that observed by Stevens (1995) and Stevens *et al.* (1997), although the material seems somewhat less fertile overall. A probable explanation for this lower fertility is the relatively low quartz content of the starting material (reflected in its lower SiO_2 content; see appendix, Section 9.13).

9.5.3 *Intermediate to felsic quartzofeldspathic rocks*

Rutter and Wyllie (1988), Skjerlie and Johnston (1992), Singh and Johannes (1996a, b) and Gardien *et al.* (2000) studied natural metatonalite compositions. Rutter and Wyllie (1988) provide no chemical analysis of their starting material. The unusual experimental design of Singh and Johannes (1996a, b) (a large plagioclase crystal surrounded by a quartz–biotite mixture) does not permit calculation of the effective starting composition, although it must have been broadly tonalitic. Gardien *et al.* (1995) studied BPQ gneisses (one with additional muscovite present). These authors intended the two muscovite-free starting materials (BPQI and BPQII) to represent ("graywackes or compositionally intermediate metavolcanic and plutonic rocks;" Gardien *et al.*, 1995, p. 15 581); BPQI has 30 vol.% and BPQII 26.8 vol.% biotite. However, with SiO_2 contents of 65 wt.% to 70 wt.%, their K_2O/Na_2O values of 0.95 and 1.26, respectively, are far higher than those of greywackes (Taylor & McLennan, 1985, Table 8). These rocks are evidently meta-igneous – granodiorite and tonalite. The third composition (BPQM) studied by Gardien *et al.* (1995) has 13 vol.% Bt and 15.3 vol.% Ms. It was intended to model a two-mica metapelite. However, it contains about 70 wt.% SiO_2, the Al_2O_3 content is rather lower than a typical pelite, and the Na_2O content is much higher (leading to a K_2O/Na_2O of only 1.2 – very much lower than a typical pelite value of around 3.0). Although its chemistry is broadly granitic, the mineralogy of this particular starting material does not seem to correspond well with any natural metaigneous or metasedimentary rock type. It is included here for comparison. Figure 9.7 illustrates the differing melt production curves of the tonalitic rocks. Gardien *et al.* (2000) did not provide sufficient melt percentage data to plot. However, the melt proportions that

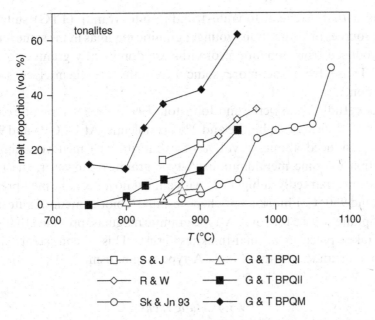

Fig. 9.7. Compilation of experimentally determined melt production curves (at ~1 GPa) for tonalitic rocks. Data are from Rutter and Wyllie (1988) – open diamonds, Skjerlie and Johnston (1992, 1993) – open circles, Singh and Johannes (1996a, b) – open squares, and the three rocks studied by Gardien *et al.* (1995) – other symbols.

they record for temperatures up to 875 °C remain low – around 5 vol.%. The differences in the curves in Fig. 9.7 are caused by variations in the ratios of biotite to hornblende, and perhaps variations in quartz contents and plagio-clase compositions in the parent tonalites. Note that the Rutter and Wyllie (1988) and Skjerlie and Johnston (1992, 1993) data (open diamonds and open circles, respectively, in Fig. 9.7) show strongly stepped melt production curves. Such steps represent narrow temperature intervals over which most of the hydrous mineral breaks down. Skjerlie and Johnston (1993, p. 799) write: "A major increase in the melt fraction takes place from 950 to 975°C and is related to a major pulse of biotite breakdown."

Metatonalites are less fertile than metagreywackes, but may still yield tens of percent of melt at attainable temperatures (up to 1000 °C). Recent work (Watkins, 1999) concentrated on the fluid-absent partial melting behavior of more sodic hornblende and biotite tonalites with 10 to 12 vol.% mafic hydrates and SiO_2 contents of 69 to 70 wt.% – typical of many felsic grey gneisses. Results show that such rocks are particularly infertile as potential magma sources, even with experimental temperatures as high as 900 °C. This casts some doubt on the idea that Archean tonalite–trondhjemite–granite (TTG) suite rocks form the sources of voluminous younger granitic magmas in the same terranes. Melts produced from tonalitic protoliths are dominantly granitic to grano-dioritic. Those from the more sodic protoliths are themselves sodic in composition.

The rock studied by Skjerlie and Johnston (1992, 1993) is a tonalite containing 20 vol.% fluorine-rich biotite and 2% hornblende. At 1.0 GPa and 975 °C, this rock produced about 20 vol.% of a fluorine-rich melt very similar in composition to some metaluminous A-type granites. However, the fertility decreased very markedly at higher P, with almost no detectable melt present at 1.4 GPa and 950 °C. Fluorine-enriched (residual?) biotite metatonalite may be an appropriate source for some A-type granitic magmas, provided that partial melting takes place in normal-thickness crust. This is consistent with the extensional tectonic settings of most A-type magmatism.

9.5.4 Mafic rocks

Work on mafic (greenschist-, amphibolite- and eclogite-facies) protoliths has been carried out by Beard and Lofgren (1991), Rapp *et al.* (1991), Rushmer (1991), Sen and Dunn (1994), Wolf and Wyllie (1994), Patiño-Douce and Beard (1995), Rapp and Watson (1995) and López and Castro (2001). The common residual assemblage involves two pyroxenes, and garnet appears at pressures between about 0.5 and 1.0 GPa, depending on the whole-rock Mg# and the melting T.

Beard and Lofgren (1991) studied the fluid-undersaturated partial melting behavior of a number of greenschist-facies metabasaltic to meta-andesitic rocks, at pressures of 0.1, 0.3, and 0.69 GPa, at temperatures from 800 to 1000 °C. Strictly the experiments could not be described as fluid-absent, since dehydration of greenschist-facies minerals would have occurred during the experiments, prior to partial melting. However, there would certainly have been no fluid present after the onset of melting, and the authors report measurable melt appearing only at about 850 °C. Figure 9.8 shows the melt production curve for their basaltic sample 478, at a pressure of 0.69 GPa. Most melt production occurs at $T < 950$ °C. The melts produced were granodioritic to trondhjemitic in composition, with total FeO + MgO contents in the range of 1.5 to 11.5 wt.%.

Working at a variety of pressures (between 0.8 and 3.2 GPa), Rapp *et al.* (1991) investigated the fluid-absent partial fusion of a variety of natural amphibolites (three that were like mid-ocean ridge basalt (MORB) and one alkali basalt), at temperatures up to 1075 °C. Melt proportions were 10–40

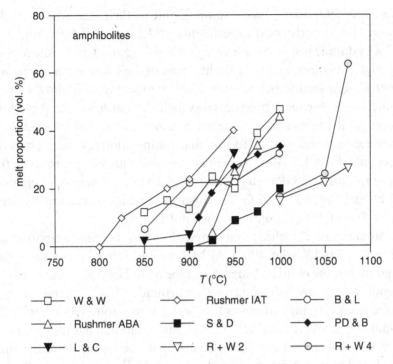

Fig. 9.8. Experimentally determined melt production curves for basaltic amphibolites. Data are from Beard and Lofgren (1991) – open circles, Rushmer (1991) – open diamonds (island arc tholeiite) and triangles (alkali basalt), Wolf and Wyllie (1994) – open squares, Sen and Dunn (1994) – filled squares, Patiño-Douce and Beard (1995) – filled diamonds, Rapp and Watson (1995) rock 2 – open inverted triangles, rock 4 – filled circles, and López and Castro (2001) – filled inverted triangles. Note that the Rapp and Watson (1995) data are at $P = 0.8$ GPa.

vol.%, and all melts were tonalitic to trondhjemitic, with total $FeO + MgO$ contents mostly in the range of 1 to 7 wt.%. Garnet was present in the residues at $P > 0.8$ GPa, and this produced melts markedly depleted in HREE, similar to Archean TTG rocks.

Rushmer (1991) performed her experiments at a pressure of 0.8 GPa. She found that an alkaline metabasaltic amphibolite (ABA) started to produce melt under fluid-absent conditions at about 925 °C, with melt proportion rising rapidly, to over 25 vol.% by 950 °C, and reaching about 45 vol.% at 1000 °C. With a tholeiitic amphibolite (IAT), melting began at T as low as 800 °C and the melt proportion rose steadily to around 35 vol.% at 950 °C. The higher fertility of IAT correlates with its higher content of hydrous minerals. The melts produced were tonalitic to trondhjemitic in composition. Melt production curves for both rocks are shown in Fig. 9.8.

Using a low-SiO_2 alkali basaltic amphibolite as their starting material, Sen and Dunn (1994) performed experiments at 1.5 and 2.0 GPa, and 850 to 1150 °C. Crystalline residues were always eclogitic (garnet and sodic pyroxene) at these high pressures, and only slight traces of melt formed at temperatures up to 900 °C. The section below, on predictive modeling, will show why fluid-absent melting at higher *P* produces less melt than at lower *P*. Melts formed here were granitic (*sensu lato*) at the lower temperatures and evolved through granodioritic or trondhjemitic to tonalitic compositions at the highest temperatures studied. At 1.5 GPa the melts were systematically more sodic (lower K_2O/Na_2O) than at 2 GPa, suggesting than TTG magmas may generally be derived by melting at $P \leq 1.5$ GPa. A melt production curve for the Sen and Dunn (1994) rock (at 1.5 GPa) is shown in Fig. 9.8.

Fluid-absent partial melting of a simple (Hbl–Pl) tholeiitic amphibolite was studied by Wolf and Wyllie (1994). At 1.0 GPa and 750 °C, there was a trace of melt present, but the main melting stage appears to begin at about 850 °C. The hornblende was completely consumed at around 950 °C. The melt compositions were similar to high-alumina tonalites, up to about 900 °C, but became quartz-dioritic above about 950 °C and high-alumina basaltic at 1000 °C. Figure 9.8 shows a melt production curve for this rock, derived from the longest duration experiments (mostly 4 and 8 days duration). Wolf and Wyllie (1994) ascribed the oscillations in the melt production curve to changes in reaction stoichiometry and proximity to the solidus – which is believed to have a flat dP/dT slope at lower *T* before it becomes steeply positive at higher *T* (Wyllie & Wolf, 1993).

The synthetic quartz amphibolite used by Patiño-Douce and Beard (1995) is an unusual composition with an andesitic SiO_2 content, but the rest of the chemical composition does not closely resemble any known andesitic rock. This material began melting at about 850 °C at 0.3 GPa, with melt proportions rising to 20 to 30 vol.% at 975 °C. Figure 9.8 shows the melt production curve for the 1 GPa experiments. Melts are all peraluminous granodiorites.

In a range of olivine tholeiitic starting materials, Rapp and Watson (1995) observed granitic melts at around 5 vol.% melting, trondhjemitic at 5 to 10 vol.%, and granodioritic to tonalitic and quartz-dioritic with melt proportions increasing up to 40 vol.% (at $T > 1000$ °C) – a finding similar to Sen and Dunn (1994). Melts at $T > 1100$ °C were high-Al basalt in composition. The experiments were at pressures between 0.8 and 3.2 GPa, and crystalline residues varied from being pyroxene-plagioclase granulites at 0.8 GPa to garnet granulites and eclogites at higher pressures. An important conclusion from this study is that Archean TTG rocks could have been formed by partial melting of hydrous metabasaltic protoliths at temperatures of 1000 to 1100 °C and pressures above 1.2 GPa. Figure 9.8 shows the melt production curves for two of the Rapp and Watson (R & W) rocks (rocks 2 and 4) at 0.8 GPa. Rocks 1 and 3 are not considered here. Rock 1 is at a low grade of metamorphism and its experimental melting behavior is likely to have been influenced by the release of volatiles from hydrous phases. Rock 3 is migmatitic and has evidently already suffered some melt loss, so that its composition may be unrepresentative of basaltic amphibolites that lack a migmatitic (fluid-present melting?) history. Geochemically, rocks 2 and 4 are E-Mid Ocean Ridge Basalt (E-MORB) tholeiites, with rock 2 somewhat more magnesian (appendix, Section 9.12). As Fig. 9.8 shows, rock 4 is far more fertile than rock 2, at $T > 1050\,°C$. If this is due to the relatively small difference in Mg# (47 in rock 4 as against 50 in rock 2), this parameter must be considered as critically important in determining the fertility of mafic amphibolites. However, there are other significant chemical differences between the two rocks and fertility is likely to be a complex function of a number of these parameters.

López and Castro (2001) took a MORB-derived amphibolite as their starting material. Using an innovative incremental heating experimental technique, they tracked the position of the solidus between 0.4 and 1.4 GPa. They found epidote as a stable phase on the solidus at $P > 1$ GPa. Garnet assumed this position at higher P. The general shape of the solidus was found to be similar to previous determinations, but the temperature step between the high- and low-P regions was found to be less significant. At all pressures investigated, there was ≤ 5 vol.% melt produced at $T \leq 900\,°C$, and a sharp increase to > 30 vol.% between 900 and 950 °C. This sharp rise in melt proportion corresponds to breakdown of Pl and Hbl, and production of pyroxenes, suggesting a reaction such as $Hbl + Pl = Grt + Opx + Cpx + M$, for P up to 1 GPa. All melts were tonalitic in composition. Figure 9.8 shows the melt production curve for this rock at a pressure of 0.6 GPa.

9.5.5 Lithological mixtures

One model for the production of certain types of metaluminous granitic magmas is that they could be derived by partial melting of mixed sources (younger mantle-related rocks and older crustally recycled materials). This would certainly be one way of interpreting the isotopic evidence (but see also the following section of this review). Skjerlie and Patiño-Douce (1995) determined the fluid-absent partial melting behavior of a garnet amphibolite in contact with a metapelite. This work was designed to model the melting behavior of a terrane with interbedded mafic and pelitic rocks. Examples of such rocks are found at the margins of some mafic plutons, where numerous dikes penetrate surrounding pelitic wall rocks. However, large regions with this sort of geology appear to be uncommon. Therefore, it is unlikely that the deep crust would contain large terranes with such lithological characteristics.

One of the gaps in our present knowledge of partial melting is knowledge of the time scales involved, and therefore of the equilibration volumes (characteristic diffusion length scales) in partially molten crustal rocks. If these volumes were generally small in rocks undergoing fluid-absent partial melting, extensive interaction between melts formed in adjacent lithological layers would not be favored. Although there is growing evidence of small equilibration volumes during solid-state metamorphism (Carlson, 2002), at present no definite statement may be made concerning partially molten rock systems. Another potential problem is that chemical interactions would only occur if the partial melts were retained in their source regions; there is evidence that most such melts are segregated rapidly and withdrawn to higher structural levels (e.g., Sawyer, 1991, 1994; Bea, 1996). Nevertheless, the Skjerlie and Patiño-Douce (1995) results are interesting.

The experiments were conducted at 1.0 GPa and 900 to 950 °C. The metapelite was the same plagioclase-poor rock used by Patiño-Douce and Johnston (1991), and produced about 15 vol.% melt at 900 °C. The amphibolite produced about 40 vol.% melt at this T. When these two materials were run as a diffusion couple, there were two significant effects. First, both the pelitic and amphibolitic ends produced more melt than when they were run separately. The pelite produced about 5 vol.% more, and the amphibolite up to 30 vol.% more melt (at the interface). Diffusion of Na from the pelite into the amphibolite seems to have been the major factor that enhanced fertility. The second effect is that the whole couple became mineralogically zoned, with metasomatic effects particularly marked in the amphibolite end. The authors suggested that such effects might be far more common than petrologists have hitherto recognized.

McCarthy and Patiño-Douce (1997) pursued a different model for the formation of granitic magmas with "mixed" isotopic characteristics. They partially melted a biotite metapelite in contact with high-alumina basalt at temperatures between 950 and 1150 °C, at 1.1 GPa. This series of experiments was designed to model diffusive and reactive interactions between underplated basaltic magma and pelitic crust. The tacit assumption was that the time and surface area effects would allow interactions in nature similar to those in the experiments. As in the case of the amphibolite–pelite experiments described above, there is some doubt that this physical situation would occur in nature over any appreciable rock volume. Nevertheless, over the temperature range investigated, between 0 and 37 vol.% of weakly peraluminous tonalitic melt formed at the basalt end of the couple. At the pelitic end, between 38 and 71 vol.% of highly peraluminous granitic melt formed. The melts coexisted with a variety of residual minerals, ranging from Al_2SiO_5 near the pelite, through $Grt + Opx$ near the interface to Cpx within the basaltic zone. CaO and Na_2O rapidly diffused into the pelite zone, and H_2O, K_2O, SiO_2 and Al_2O_3 into the basalt end.

9.6 The question of high-K calc-alkaline granitic magmas

An interesting and vexed question surrounds the origin of the high-K, meta-luminous, calc-alkaline granitic rocks that are volumetrically important components of post-orogenic magmatic suites. Many geologists believe that these rocks may be formed by the partial melting of mafic source rocks, by AFC processes involving mantle-derived magmas and assimilated crust, or by mixing between crustal and mantle melts. Skjerlie and Patiño-Douce (1995) produced high-K melts in their amphibolite–pelite diffusion-couple experiments. However, Roberts and Clemens (1993, 1995, 1997) showed that it is most likely that these magmas were crustally derived, and that AFC models are usually rather poorly constrained, yielding ambiguous results. Melting experiments demonstrate that metabasaltic protoliths produce melts of normal-K calc-alkaline and trondhjemitic character, and only those melts derived from metatonalitic and high-K andesitic protoliths produce high-K, calc-alkaline granite magmas (Roberts & Clemens, 1993). Thus, these authors suggested that a viable mechanism for producing high-K, calc-alkaline granite magmas would be partial melting of already K-rich meta-andesitic rocks previously accreted to the crust. Large volumes of appropriate andesitic source rocks exist in many magmatic arcs.

More recent experiments (Clemens & Petford, 1999) have confirmed that single-stage partial melting of typical mature arc crust may be all that is required to impart the observed chemical and isotopic signatures to high-K calc-alkaline rocks. The work of Villaseca *et al.* (1999) is also interesting in this regard. These authors studied the geochemical and isotopic characteristics of deep crustal residual xenoliths in mafic dikes from central Spain. They demonstrated that the deep continental crust of the region is compositionally unlike any of the rocks exposed at the surface. Upper crustal granites here (usually modeled as mixtures of basaltic magma and melt from exposed migmatites or gneisses) are more simply modeled as partial melts of the deep crust. Thus, it is inadvisable to assume that the specific rocks cropping out in a region are representative of the regional crust as a whole, or that the protoliths of exposed granitic magmas are necessarily represented anywhere at the present Earth's surface. Without additional physical evidence from the granitic rocks, it also seems inappropriate to apply magma mixing or contamination models to the problem of the genesis of metaluminous granitic suites that have mildly evolved crustal isotope ratios.

9.7 Predictive modeling of protolith fertility

Clemens and Vielzeuf (1987) used the relationship between the H_2O content of a magma source rock and the solubility of H_2O in a fluid-absent partial melt to erect a model that could be used to predict the amount of melt that would be formed from a given source rock at any particular P and T. Essentially, the H_2O solubility model of Burnham (1981) was combined with an assumption of the nature of the rock solidus as a function of P, T, and aH_2O to calculate these results. The simple relationship used is: melt wt% $= 100 \times$ wt% H_2O in source rock/wt% H_2O in melt at P, T. This equation assumes that the hydrous mineral is completely exhausted in the melting reaction (cf. Patiño-Douce & Johnston, 1991), and that H_2O availability is the limiting factor in rock fertility. There are examples of fluid-absent melting limited by K_2O or Na_2O contents of the protoliths (e.g., Pickering & Johnston, 1998) or the plagioclase content (Patiño-Douce & Johnston, 1991). Thus, the above equation may only be used to calculate the maximum melt proportion, unless it is feasible to estimate the proportion of the hydrous phase that breaks down at a given T. Johannes and Holtz (1990) produced a version of this model that used experimentally constrained (instead of calculated) $P–T$ loci for the solidi of various feldspar–quartz–H_2O systems at varying aH_2O. The Johannes and Holtz (1990) and Clemens and Vielzeuf (1987) versions predict essentially identical melt proportions for a given source rock and melting conditions. The general conclusion from this predicted modeling is that metagreywackes may represent the optimum rock compositions for production of large quantities of granitic melt. Pelites are also highly fertile (sometimes more so than greywackes), whereas other rock types are generally far less fertile, and usually require higher temperatures to produce significant melt volumes.

9.8 The fates of partial melts

Partial melts may form in the presence or absence of free fluid phases. If fluids are present, the system may be either rock-dominated (fluid-undersaturated) or fluid-dominated (fluid-saturated), leading to melts with a wide range of H_2O contents. However, the evidence presented earlier suggests that those melts that escape to form granitic plutons, higher in the crust, are formed mainly by fluid-absent reactions.

Once formed, a partial melt may or may not chemically segregate from its complementary residue. The terrane may then evolve along either a clockwise or an anticlockwise *P–T* path, involving (ideally) either quasi-isothermal decompression or quasi-isobaric cooling. With this variety of possible processes, it seems likely that a variety of products could result. Clemens and Droop (1998) attempted to track the physical and mineralogical consequences of the different combinations of the above variables. Their analysis showed that there are indeed many different possible outcomes, although space does not allow inclusion of the details here. One important conclusion of the Clemens and Droop (1998) work is that it may be feasible to constrain some of the initial variables (e.g., fluid regime, extent of melt segregation and *P–T* path) simply by examining the mineralogy and texture of an exhumed migmatite or residual granulite. This emphasizes the importance of sound petrographic observation. The interested reader is referred to Holness and Clemens (1999) and Clemens and Holness (2000) for a recent example of the advances in understanding petrogenesis that may be gained primarily through petrography.

9.9 A word of caution

The granodiorite melt chemistry obtained in the experiments of Patiño-Douce and Beard (1995) is significant because it means that granites with at least one geochemical characteristic similar to S-types (Chappell & White, 1974; Chappell, 1984) may be derived through partial melting of non-sedimentary metaluminous protoliths. Indeed, there are examples of cordierite-bearing "S-type" granites that have uncomfortably low $^{87}Sr/^{86}Sr$ initial ratios (e.g., Flood & Shaw, 1977). Also, there is mounting evidence that isotopic disequilibrium may lead to shifts in Sr and Nd isotope ratios during partial melting (e.g., Ayers & Harris, 1997; Knesel & Davidson, 1999). Such shifts commonly appear to give the magmas more isotopically evolved signatures than their source rocks. This possibility should engender some caution in the interpretation of source characteristics from one or two isolated geochemical parameters of a particular granite. Peraluminous granitic rocks may not always have evolved metasedimentary protoliths. Conversely, mantle-like initial isotope ratios need not indicate direct mantle input into a magma system. Another point of note is that the Patiño-Douce and Beard (1995) synthetic quartz amphibolite, and the model metatonalite/greywacke studied by the same authors, have essentially the same solidus *T*, which increases with *P*. This means that a wide variety of crustal rocks (metapelites, metagreywackes, metatonalites, and metabasalts) may begin melting (in the absence of a fluid phase) by biotite and hornblende breakdown at about 850 °C. Fertility decreases with P, as predicted by Clemens and Vielzeuf (1987).

9.10 Conclusion

Common crustal rock types may be partially melted at T as low as 650 °C, if sufficient H_2O-rich fluid is present. However, most crustally derived granitic magmas are formed by fluid-absent melting reactions, and this means that major melt production only occurs at temperatures above about 800 °C. For a wide range of rock types, significant partial melting occurs at temperatures of about 850 °C and, for more magnesian, hornblende-bearing metabasalts or meta-andesites, at temperatures of about 900 °C. These thermal requirements suggest extra-crustal heat sources. That is, magmatic underplating is of major importance to the occurrence of much of the world's granitic magmatism, probably even in the Archean. A local alternative would be to postulate that the source crust was unusually enriched in heat-producing elements. Amounts of fluid-absent melt will vary as a function of P, T and mineral (especially hydrous mineral) content in the protolith (as well as other chemical factors). Common rock types (metapelites, metagreywackes, meta-andesites and some amphibolites) may yield 10 to 50 vol.% of H_2O-undersaturated melt at attainable crustal temperatures. Up to about 1000 °C, the compositions of the melts vary from leucogranitic to trondhjemitic and granodioritic to tonalitic, mainly depending on the chemistry and mineralogy of the source rocks and the T of melting.

9.11 Acknowledgements

This chapter was ably reviewed by D. R. Baker, D. Johnston, and an anonymous individual. Together with the astute editorial comments of Michael Brown, these reviews served materially to improve the breadth, readability, and intelligibility.

References

Ayres, M. and Harris, N. (1997). REE fractionation and Nd-isotope disequilibrium during crustal anatexis: constraints from Himalayan leucogranites. *Chemical Geology*, **139**, 249–69.

Bea, F. (1996). Controls on the trace element composition of crustal melts. *Transactions of the Royal Society of Edinburgh: Earth Sciences*, **87**, 33–41.

Beard, J. S. and Lofgren, G. E. (1991). Dehydration melting and water-saturated melting of basaltic and andesitic greenstones and amphibolites at 1, 3, and 6.9 kb. *Journal of Petrology*, **32**, 365–401.

Bohlen, S. R. (1991). On the formation of granulites. *Journal of Metamorphic Geology*, **9**, 223–9.

Brown, G. C. and Fyfe, W. S. (1970). The production of granitic melts during ultrametamorphism. *Contributions to Mineralogy and Petrology*, **28**, 310–18.

Burnham, C. W. (1979). Magmas and hydrothermal fluids. In *Geochemistry of Hydrothermal Ore Deposits*, ed. H. L. Barnes. 2nd edn. New York: John Wiley & Sons.

Burnham, C. W. (1981). The nature of multicomponent aluminosilicate melts. *Physics and Chemistry of the Earth*, **13**, 197–229.

Carlson, W. D. (2002). Scales of disequilibrium and rates of equilibration during metamorphism. *American Mineralogist*, **87**, 185–204.

Carrington, D. P. and Harley, S. L. (1995). Partial melting and phase relations in high-grade metapelites – an experimental petrogenetic grid in the KFMASH system. *Contributions to Mineralogy and Petrology*, **120**, 270–91.

Chappell, B. W. (1984). Source rocks of I- and S-type granites in the Lachlan Fold Belt, southeastern Australia. *Philosophical Transactions of the Royal Society*, **A310**, 693–707.

Chappell, B. W. and White, A. J. R. (1974). Two contrasting granite types. *Pacific Geology*, **8**, 173–4.

Chatterjee, N. D. and Froese, E. (1975). A thermodynamic study of the pseudobinary join muscovite–paragonite in the system $KAlSi_3O_8$–$NaAlSi_3O_8$–Al_2O_3–SiO_2–H_2O. *American Mineralogist*, **60**, 985–93.

Clemens, J. D. (1981). *Origin and Evolution of some Peraluminous Acid Magmas*. Unpublished Ph.D. thesis. Melbourne: Monash University.

Clemens, J. D. (1984). Water contents of intermediate to silicic magmas. *Lithos*, **17**, 273–87.

Clemens, J. D. (1990). The granulite–granite connexion. In *Granulites and Crustal Differentiation*, ed. D. Vielzeuf and P. Vidal. Dordrecht: Kluwer Academic Publishers, pp. 25–36.

Clemens, J. D. and Droop, G. T. R. (1998). Fluids, *P–T* paths and the fates of anatectic melts in the Earth's crust. *Lithos*, **44**, 21–36.

Clemens, J. D. and Holness, M. B. (2000). Textural evolution and partial melting of arkose in a contact aureole: a case study and implications. *Electronic Geosciences*, **5**(4).

Clemens, J. D. and Petford, N. (1999). Granitic melt viscosity and silicic magma dynamics in contrasting tectonic settings. *Journal of the Geological Society London*, **156**, 1057–60.

Clemens, J. D. and Vielzeuf, D. (1987). Constraints on melting and magma production in the crust. *Earth and Planetary Science Letters*, **86**, 287–306.

Clemens, J. D. and Watkins, J. M. (2001). The fluid regime of high-temperature metamorphism during granitoid magma genesis. *Contributions to Mineralogy and Petrology*, **140**, 600–6.

Clemens, J. D., Droop, G. T. R. and Stevens, G. (1997). High-grade metamorphism, dehydration and crustal melting: a reinvestigation based on new experiments in the silica-saturated portion of the system $KAlO_2$–MgO–SiO_2–H_2O–CO_2 at $P \leq 1.5$ GPa. *Contributions to Mineralogy and Petrology*, **129**, 308–25.

Dickinson, W. R., Ingersoll, R. V., Cowan, D. S., Helmold, K. P. and Suczek, C. A. (1982). Provenance of Franciscan graywackes in coastal California. *Geological Society of America Bulletin*, **93**(2), 95–107.

Eggler, D. H. (1973). Principles of melting of hydrous phases in silicate melt. *Carnegie Institute of Washington Yearbook*, **72**, 491–5.

England, P. and Thompson, A. B. (1984). Pressure–temperature–time paths of regional metamorphism. Part I: heat transfer during the evolution of regions of thickened continental crust. *Journal of Petrology*, **25**, 894–928.

Finger, F. and Clemens, J. D. (1995). Migmatization and "secondary" granitic magmas: effects of emplacement of "primary" granitoids in Southern Bohemia, Austria. *Contributions to Mineralogy and Petrology*, **120**, 311–26.

Flood, R. H. and Shaw, S. E. (1977). Two "S-type" granite suites with low initial $^{87}Sr/^{86}Sr$ ratios from the New England Batholith, Australia. *Contributions to Mineralogy and Petrology*, **61**, 163–73.

Fyfe, W. S. (1973). The granulite facies, partial melting and the Archean crust. *Philosophical Transactions of the Royal Society, London*, **A273**, 457–61.

Gardien, V., Thompson, A. B., Grujic, D. and Ulmer, P. (1995). Experimental melting of biotite + plagioclase + quartz ± muscovite assemblages and implications for crustal melting. *Journal of Geophysical Research*, **B100**, 15 581–91.

Gardien, V., Thompson, A. B. and Ulmer, P. (2000). Melting of biotite + plagioclase + quartz gneisses: the role of H_2O in the stability of amphibole. *Journal of Petrology*, **41**, 651–66.

Harris, N. and Massey, J. (1994). Decompression and anatexis of Himalayan metapelites. *Tectonics*, **3**, 1537–46.

Holness, M. B. and Clemens, J. D. (1999). Partial melting of the Appin "Quartzite" driven by fracture-controlled H_2O infiltration in the aureole of the Ballachulish Igneous Complex, Scottish Highlands. *Contributions to Mineralogy and Petrology*, **136**, 154–68.

Johannes, W. and Holtz, F. (1990). Formation and composition of H_2O-undersaturated granitic melts. In *High-temperature Metamorphism and Crustal Anatexis*, ed. J. R. Ashworth and M. Brown. London: Unwin Hyman, pp. 87–104.

Knesel, K. M. and Davidson, J. P. (1999). Sr isotope systematics during melt generation by intrusion of basalt into continental crust. *Contributions to Mineralogy and Petrology*, **136**, 285–95.

Kretz, R. (1983). Symbols for rock-forming minerals. *American Mineralogist*, **68**, 277–9.

LeBreton, N. and Thompson, A. B. (1988). Fluid-absent (dehydration) melting of biotite in metapelites in the early stages of crustal anatexis. *Contributions to Mineralogy and Petrology*, **99**, 226–37.

López, S. and Castro, A. (2001). Determination of the fluid-absent solidus and supersolidus phase relationships of MORB-derived amphibolites in the range 4–14 kbar. *American Mineralogist*, **86**, 1396–403.

McCarthy, T. C. and Patiño-Douce, A. E. (1997). Experimental evidence for high-temperature felsic melts formed during basaltic intrusion of the deep crust. *Geology*, **25**, 463–6.

McLaren, S., Sandiford, M. and Hand, M. (1999). High radiogenic heat-producing granites and metamorphism – an example from the western Mount Isa inlier, Australia. *Geology*, **27**, 679–82.

Montel, J.-M. and Vielzeuf, D. (1997). Partial melting of metagreywackes, Part II. Compositions of minerals and melts. *Contributions to Mineralogy and Petrology*, **128**, 176–96.

Patiño-Douce, A. E. (1993). Titanium substitution in biotite: an empirical model with applications to thermometry, O_2 and H_2O barometries, and consequences for biotite stability. *Chemical Geology*, **108**, 133–62.

Patiño-Douce, A. E. and Beard, J. S. (1995). Dehydration-melting of biotite gneiss and quartz amphibolite from 3 to 15 kbar. *Journal of Petrology*, **36**, 707–38.

Patiño-Douce, A. E. and Beard, J. S. (1996). Effects of P, $f(O_2)$ and Mg/Fe ratio on dehydration melting of model metagraywackes. *Journal of Petrology*, **37**, 999–1024.

Patiño-Douce, A. E. and Harris, N. (1998). Experimental constraints on Himalayan anatexis. *Journal of Petrology*, **39**, 689–710.

Patiño-Douce, A. E. and Johnston, A. D. (1991). Phase equilibria and melt productivity in the pelitic system: implications for the origin of peraluminous granitoids and aluminous granulites. *Contributions to Mineralogy and Petrology*, **107**, 202–18.

Patiño-Douce, A. E., Humphreys, E. D. and Johnston, A. D. (1990). Anatexis and metamorphism in tectonically thickened continental crust exemplified by the Sevier hinterland, western North America. *Earth and Planetary Science Letters*, **97**, 290–315.

Petö, P. (1976). An experimental investigation of melting relations involving muscovite and paragonite in the silica-saturated portion of the system $K_2O–Na_2O–Al_2O_3–SiO_2–H_2O$. *Progress in Experimental Petrology*, **3**, 41–5.

Pickering, J. M. and Johnston, A. D. (1998). Fluid-absent melting behavior of a two-mica metapelite. *Journal of Petrology*, **39**, 1787–804.

Rapp, R. P. and Watson, E. B. (1995). Dehydration melting of metabasalt at 8–32 kbar: implications for continental growth and crust–mantle recycling. *Journal of Petrology*, **36**, 891–931.

Rapp, R. P., Watson, E. B. and Miller, C. F. (1991). Partial melting of amphibolite/ eclogite and the origin of Archean trondhjemites and tholeiites. *Precambrian Research*, **51**, 1–25.

Roberts, M. P. and Clemens, J. D. (1993). Origin of high-potassium, calc-alkaline, I-type granitoids. *Geology*, **21**, 825–8.

Roberts, M. P. and Clemens, J. D. (1995). Feasibility of AFC models for the petrogenesis of calc-alkaline magma series. *Contributions to Mineralogy and Petrology*, **121**, 139–47.

Roberts, M. P. and Clemens, J. D. (1997). Correction to Roberts and Clemens (1995) "Feasibility of AFC models for the petrogenesis of calc-alkaline magma series". *Contributions to Mineralogy and Petrology*, **128**, 97–9.

Rudnick, R. (1990). Continental crust – growing from below. *Nature*, **347**, 711–12.

Rushmer, T. (1991). Partial melting of 2 amphibolites – contrasting experimental results under fluid-absent conditions. *Contributions to Mineralogy and Petrology*, **107**, 41–59.

Rushmer, T. (1995). An experimental deformation study of partially molten amphibolite – application to low-melt fraction segregation. *Journal of Geophysical Research*, **B100**, 15 681–95.

Rutter, M. J. and Wyllie, P. J. (1988). Melting of vapor-absent tonalite at 10 kbar to simulate dehydration-melting in the deep crust. *Nature*, **331**, 159–60.

Sandiford, M., Hand, M. and McLaren, S. (1998). High geothermal gradient metamorphism during thermal subsidence. *Earth and Planetary Science Letters*, **163**, 149–65.

Sawyer, E. W. (1991). Disequilibrium melting and the rate of melt-residuum separation during migmatization of mafic rocks from the Grenville Front, Quebec. *Journal of Petrology*, **32**, 701–38.

Sawyer, E. W. (1994). Melt segregation in the continental crust. *Geology*, **22**, 1019–22.

Scaillet, B., Holtz, F. and Pichavant, M. (1998). Phase equilibrium constraints on the viscosity of silicic magmas – 1. Volcanic-plutonic association. *Journal of Geophysical Research*, **B103**, 27 257–66.

Sen, C. and Dunn, T. (1994). Dehydration melting of a basaltic composition amphibolite at 1.5 and 2.0 GPa: implications for the origin of adakites. *Contributions to Mineralogy and Petrology*, **117**, 394–409.

Singh, J. and Johannes, W. (1996a). Dehydration melting of tonalites. 1. Beginning of melting. *Contributions to Mineralogy and Petrology*, **125**, 16–25.

Singh, J. and Johannes, W. (1996b). Dehydration melting of tonalites. 2. Compositions of melts and solids. *Contributions to Mineralogy and Petrology*, **125**, 26–44.

Skjerlie, K. P. and Johnston, A. D. (1992). Vapor-absent melting at 10-kbar of a biotite-bearing and amphibole bearing tonalitic gneiss – implications for the generation of A-type granites. *Geology*, **20**, 263–6.

Skjerlie, K. P. and Johnston, A. D. (1993). Fluid-absent melting behavior of an F-rich tonalitic gneiss at mid-crustal pressures: implications for the generation of anorogenic granites. *Journal of Petrology*, **34**, 785–815.

Skjerlie, K. P. and Patiño-Douce, A. E. (1995). Anatexis of interlayered amphibolite and pelite at 10 kbar – effect of diffusion of major components on phase-relations and melt fraction. *Contributions to Mineralogy and Petrology*, **122**, 62–78.

Stevens, G. (1995). Compositional controls on partial melting in high-grade metapelites; a petrological and experimental study. PhD. thesis. Manchester: University of Manchester.

Stevens, G. and Clemens, J. D. (1993). Fluid-absent melting and the roles of fluids in the lithosphere: a slanted summary? *Chemical Geology*, **108**, 1–17.

Stevens, G., Clemens, J. D. and Droop, G. T. R. (1997). Melt production during granulite-facies anatexis: experimental data from "primitive" metasedimentary protoliths. *Contributions to Mineralogy and Petrology*, **128**, 352–70.

Storre, B. (1972). Dry melting of muscovite + quartz in the range $P_s = 7$ kb to $P_s = 20$ kb. *Contributions to Mineralogy and Petrology*, **37**, 87–9.

Taylor, S. R. and McLennan, S. M. (1985). *The Continental Crust: its Composition and Evolution*. Geoscience Texts. Oxford: Blackwell Scientific Publications.

Thompson, A. B. (1988). Dehydration melting of crustal rocks. *Rendiconti della Societa Italiana di Mineralogia e Petrologia*, **43**, 41–60.

Thompson, A. B. (1990). Heat, fluids and melting in the granulite facies. In *Granulites and Crustal Differentiation*, ed. D. Vielzeuf and P. Vidal. Dordrecht: Kluwer Academic Publishers, pp. 37–58.

Vielzeuf, D. and Clemens, J. D. (1992). Fluid-absent melting of phlogopite + quartz: experiments and models. *American Mineralogist*, **77**, 1206–22.

Vielzeuf, D. and Holloway, J. R. (1988). Experimental determination of the fluid-absent melting reactions in the pelitic system. Consequences for crustal differentiation. *Contributions to Mineralogy and Petrology*, **98**, 257–76.

Vielzeuf, D. and Montel, J. -M. (1994). Partial melting of metagreywackes. 1. Fluid-absent experiments and phase relationships. *Contributions to Mineralogy and Petrology*, **117**, 375–93.

Vielzeuf, D. and Schmidt, M. W. (2001). Melting relations in hydrous systems revisited: application to metapelites, metagreywackes and metabasalts. *Contributions to Mineralogy and Petrology*, **141**, 251–67.

Vielzeuf, D., Clemens, J. D., Pin, C. and Moinet, E. (1990). Granites, granulites and crustal differentiation. In *Granulites and Crustal Differentiation*, ed. D. Vielzeuf and P. Vidal. Dordrecht: Kluwer Academic Publishers, pp. 59–86.

Villaseca, C., Downs, H., Pin, C. and Barbero, L. (1999). Nature and composition of the lower continental crust in Central Spain and the granulite–granite linkage: implications from granulitic xenoliths. *Journal of Petrology*, **40**, 1465–96.

Watkins, J. M. (1999). Lewisian granulite-facies metamorphism: regional event or local contact effect? *Journal of Conference Abstracts*, **4**, 700.

Wolf, M. B. and Wyllie, P. J. (1991). Dehydration-melting of solid amphibolite at 10 kbar: textural development, liquid interconnectivity and applications to the segregation of magmas. *Contributions to Mineralogy and Petrology*, **44**, 151–79.

Wolf, M. B. and Wyllie, P. J. (1994). Dehydration-melting of amphibolite at 10 kbar – the effects of temperature and time. *Contributions to Mineralogy and Petrology*, **115**, 369–83.

Wyllie, P. J. and Wolf, M. B. (1993). Amphibole dehydration-melting: sorting out the solidus. In *Magmatic Processes and Plate Tectonics*, ed. H. M. Pritchard, T. Alabaster, N. B. W. Harris and C. R. Neary. Geological Society of London Special Publication **76**, pp. 405–16.

9.12 Appendix

9.12.1 Compositions of experimental starting materials (normalized 100 wt % anhydrous)

	Metapelite V+H 88	Metapelite PD+J 91	Metapelite S et al. 97 AS	Metapelite S et al. 97 BS	Metapelite S et al. 97 CS	Metapelite S et al. 97 NBS	Metapelite P+J 98	Metapelite PD+H 98 MS
SiO_2	65.78	58.62	58.72	59.48	59.97	58.61	78.66	75.86
TiO_2	0.84	1.29	0.00	0.00	0.00	0.64	0.51	0.36
Al_2O_3	18.53	23.75	26.66	26.47	26.43	27.89	11.42	14.40
FeO^*	6.40	8.78	6.63	5.13	3.08	4.96	3.46	2.42
MnO	0.09	0.17	0.01	0.01	0.01	0.01	n.d.	0.13
MgO	2.49	2.78	3.48	4.40	5.90	3.64	1.25	0.67
CaO	1.55	0.41	0.92	0.87	0.93	0.87	0.76	0.95
Na_2O	1.70	0.49	0.96	1.01	0.99	0.87	1.45	2.79
K_2O	2.62	3.71	2.61	2.63	2.69	2.53	2.49	2.42
Mg#	41	36	48	60	77	57	39	33

	Metapelite PD+H 98 MBS	Metagreywacke V+M	Metagreywacke PD+B 95	Metagreywacke PD+B 96	Metagreywacke S et al. 97 A	Metagreywacke S et al. 97 B	Metagreywacke S et al. 97 C	Metagreywacke S et al. 97 NB
SiO_2	68.81	71.44	64.37	63.44	67.24	68.28	70.00	67.07
TiO_2	0.81	0.71	2.54	2.85	0.00	0.00	0.00	0.88
Al_2O_3	16.69	13.23	12.49	12.12	12.78	12.51	12.44	14.50
FeO^*	5.65	4.97	7.92	11.91	8.93	6.85	4.01	6.60
MnO	0.14	0.06	0.10	0.20	0.01	0.01	0.01	0.01
MgO	1.93	2.41	4.77	1.87	4.81	6.09	8.15	5.03
CaO	1.09	1.70	2.13	2.14	1.28	1.21	1.29	1.20
Na_2O	1.40	3.01	2.03	2.04	1.33	1.40	1.36	1.18
K_2O	3.48	2.46	3.65	3.46	3.63	3.66	3.74	4.53
Mg#	38	46	52	22	49	61	78	58

	Metatonalite	Metatonalite	Metatonalite	?	Metabasalt	Metabasalt	Metabasalt	Metabasalt
	Sk+J 92/93	G et al. BPQ I	G et al. BPQ II	G et al. BPQM	B+L 91 478	R 91 ABA	R91 IAT	W+W 94
SiO_2	68.55	65.42	69.70	71.40	53.15	49.96	52.98	49.34
TiO_2	0.52	0.68	0.83	0.39	1.76	1.29	1.02	0.41
Al_2O_3	14.95	16.61	12.68	15.21	15.49	16.68	16.72	14.88
FeO^*	4.69	4.65	6.48	3.72	11.94	9.35	8.92	8.56
MnO	0.06	0.05	0.12	0.09	0.22	0.18	0.11	0.20
MgO	1.74	1.39	2.59	1.42	5.36	7.59	7.70	10.91
CaO	2.94	3.90	1.80	1.82	9.33	11.01	9.12	14.58
Na_2O	4.49	3.63	2.57	2.70	2.58	3.48	3.17	1.02
K_2O	2.06	3.45	3.24	3.24	0.16	0.45	0.27	0.10
Mg#	40	35	42	40	44	59	61	69

	Metabasalt	Metabasalt	Metabasalt	Meta-andesite?	Metabasalt
	S+D 94	R+W 95 2	R+W 95 4	PD+B 95	L+C 2001
SiO_2	47.21	49.68	48.81	61.38	49.56
TiO_2	1.21	2.11	1.22	1.73	1.62
Al_2O_3	15.11	17.41	14.54	11.48	16.14
FeO^*	13.18	10.93	14.12	8.03	11.03
MnO	0.26	0.21	0.19	0.20	0.22
MgO	8.31	6.20	7.03	6.81	7.23
CaO	11.36	9.87	11.27	7.72	10.79
Na_2O	2.53	3.37	2.62	1.93	3.32
K_2O	0.81	0.21	0.19	0.71	0.09
Mg#	53	50	47	60	54

*total Fe as FeO, and analyses F- and P$_2$O$_5$-free.

Source:

Note:

n.d. = not determined.

starting composition not given for Rutter and Wyllie (1988) and indeterminate for Singh and Johannes (1996a, b).

V + H = Vielzeuf and Holloway (1988).

Sk + J 92/93 = Skjerlie and Johnston (1992, 1993).

PD + J 91 = Patiño-Douce and Johnston (1991).

G *et al.* = Gardien *et al.* (1995, 2000).

S *et al.* 97 = Stevens *et al.* (1997).

B + L 91 = Beard and Lofgren (1991).

P + J 98 = Pickering and Johnston (1998).

R 91 = Rushmer (1991)

PD + H 98 = Patiño-Douce and Harris (1998)

W + W 94 = Wolf and Wyllie (1994)

V + M = Vielzeuf and Montel (1994), Montel and Vielzeuf (1997)

S + D 94 = Sen and Dunn (1994)

PD + B 95 = Patiño-Douce and Beard (1995)

R + W 95 = Rapp and Watson (1995)

PD + B 96 = Patiño-Douce and Beard (1996)

L + C 2001 = López and Castro (2001)

10

Melt extraction from the lower continental crust of orogens: the field evidence

MICHAEL BROWN

10.1 Introduction

In this chapter, the focus is on melt extraction from protoliths of supracrustal origin that have been taken into a lower crustal setting and metamorphosed to uppermost amphibolite and granulite facies conditions, and magma transport through plastically deforming crust (cf. Brown, 1994a; Sawyer, 1994). By lower crustal setting, I mean to depths below the transition to viscous anatectic flow (Handy et al., 2001) during orogenic deformation, which is a dynamic transition (cf. Brown & Solar, 1999) that does not correspond to the post-orogenic (static) definition of the lower continental crust based on geophysical data (e.g., Rudnick & Gao, 2003; Chapter 2). In collisional orogenesis, the role of mantle-derived basalt as a heat source for melting of the lower continental crust is generally cryptic compared to its role in arc magmatism (e.g., Jackson et al., 2003; Chapter 5), and radiogenic heat production in crust that is thickened by contractional deformation and/or replacement of lithospheric mantle by asthenospheric mantle may each have an important role in achieving the temperatures necessary for melt generation (cf. Loosveld & Etheridge, 1990; Jamieson et al., 1998, 2002; Molnar et al., 1998; Barboza et al., 1999; O'Reilly et al., 2001).

Melt extraction, magma transport, and pluton emplacement in orogens represent a significant and irreversible mass transfer from lower to middle and upper crust (e.g., Brown et al., 1995a), that has led over time to development of a differentiated continental crust. This mass transfer involves a multitude of physical and chemical processes that operate at several different length and time scales. For example, melting occurs at distributed multiphase grain boundaries in lower crustal rocks (millimeter length scale and cubic millimeters of melt), whereas plutons have horizontal dimensions of several tens of kilometers and represent large volumes of melt (from $\sim 10^3$ to 10^4km^3)

Evolution and Differentiation of the Continental Crust, ed. Michael Brown and Tracy Rushmer. Published by Cambridge University Press. © Cambridge University Press 2005.

aggregated at a shallower crustal level. Thus, melt extraction and magma transport, from segregation to emplacement, is a process with a length scale that spans more than seven orders of magnitude (by volume more than 21 orders of magnitude).

In many cases, the size, shape, and spatial distribution of upper crustal plutons probably reflect aspects of melting in the lower crust (e.g., Cruden & McCaffrey, 2001; Chapter 13), and extraction and emplacement are presumed complementary processes between which there is a feedback relation modulated by the ascent mechanism (e.g., Brown & Solar, 1999). Fertile crustal rocks have the potential to yield 10–50 vol.% melt (Chapter 9) at temperatures of the metamorphic peak, which suggests that source volumes vary from about 10 times to only 2 times the volume of a pluton (Brown, 2001a, b). Thus, a major issue in melt extraction is focusing flow of segregated melt in the source into structures that will allow transport out of the anatectic zone through subsolidus crust to the site of pluton emplacement via the small number of feeders implied by geophysical data (e.g., Vigneresse, 1999; Chapter 13).

Many lower crustal migmatites and granulites have bulk rock geochemistries that demonstrate they are residual with respect to a granite melt component, which is presumed lost from the anatectic zone (e.g., Schnetger, 1994; Solar & Brown, 2001; Guernina & Sawyer, 2003), although some early crystallized (cumulate) products and interstitial liquid may remain to form leucosome (typically <10 vol.%) that commonly is inferred to record the fossil melt flow network (e.g., Sawyer, 1987; Brown, 1994a; Brown & Rushmer, 1997; Sawyer, 2001; Marchildon & Brown, 2003). The leucosome represents the integrated residue of melt that mostly passed through the point of observation in the lower crust, since melt extraction implies a flux of melt throughout the melting zone (Brown, 2004). Local space-making processes due to deformation acting on heterogeneous and/or anisotropic crustal protoliths provide the mechanistic link between porous flow of melt located at grain boundaries and melt migration out of the source to shallower crustal levels via dikes and/or shear zones (Brown, 1994a; Sawyer, 1994). Furthermore, the preservation of preanatectic structures such as compositional layering in residual lower crust implies drainage of melt without loss of structural coherence (Guernina & Sawyer, 2003), and indicates that the melt fraction did not exceed the rheological critical melt percentage or transition to melt-dominated behavior (Sawyer, 1994; Rosenberg & Handy, 2005). This is consistent with the notion of a low melt fraction "window" of extractable melt, in the range 6–14 vol.%, based on an optimal ratio of permeability to melt viscosity (Barboza & Bergantz, 1996; Chapter 14).

The generation and transport of granite magma from the lower crust to the middle and upper crust involves a sequence of events limited by heat flow into the protolith (Rutter & Neumann, 1995; Chapter 11). Although the early stage of melt segregation involving diffusive mass transfer is a slow process, perhaps operating on time scales that approach a million years given the many variables that affect diffusion rates and efficiencies (e.g., Costa *et al.*, 2003; Dohmen & Chakraborty, 2003), advective mass transfer will dominate as the melt fraction approaches the melt percolation threshold (Vigneresse *et al.*, 1996; Rosenberg & Handy, 2005), consistent with times of transport from veins to dikes and from dikes to plutons of tens of years to tens of thousands of years (e.g., Harris *et al.*, 2000). Overall, orogenic magmatism is commonly short-lived and episodic in relation to the time scales of orogenesis (e.g., Solar *et al.*, 1998; Ducea, 2001).

To a first approximation, the volume of granite emplaced in the upper crust may be presumed equal to the volume of melt extracted from the lower crust. Although the fluxes associated with melt extraction, ascent, and emplacement may vary with time, most probably they are balanced at any particular time via a feedback relation (Brown, 2004; Chapter 13). However, in the exposed crust of deeply eroded orogens, where evidence of melting is preserved in residual migmatites and granulites, there is no evidence of magma chambers of any kind related to the inferred melting event. Based on outcrop evidence, many authors have inferred that melt accumulated in a network that fed drainage channels (ascent conduits) by pervasive macroscopic porous flow (e.g., Brown & Solar, 1999; Weinberg, 1999; Sawyer, 2001; Marchildon & Brown, 2003; Olsen *et al.*, 2004). Alternatively, there may be a mechanism, such as mobile melt-filled hydrofractures, by which melt batches aggregate to increase in size and ascend through the crust (e.g., Bons *et al.*, 2001, 2005), or the source may ascend en masse by bulk instability flow (Weinberg & Podlachikov, 1994; Olsen *et al.*, 2004).

Numerical, analog, and physical modeling have led to the suggestion that melt extraction is a self-organized critical (SOC) phenomenon (e.g., Petford & Koenders, 1998; Vigneresse & Burg, 2000; Bons *et al.*, 2001; Bons & van Milligen, 2001), similar to that proposed for basalt melt extraction from the shallow mantle beneath ocean spreading centers based on spatial statistical analysis of dunite-filled conduits (Braun & Kelemen, 2002). Further, the size–frequency relationships of plutons and their spacing, which are similar to those of volcanoes (Shaw & Chouet, 1991; Pelletier, 1999), strongly suggest that magmatic systems are self-organized from the bottom up (Cruden & McCaffrey, 2001; Chapter 13). However, the underlying physical mechanism by which this is accomplished is not adduced, nor is the relationship to the observed patterning of anatectic lower crust addressed satisfactorily. This highlights one of the unsatisfying aspects of SOC theory – that it is only the outcome of assumed process parameters that may be calculated. None the less, the expectation generated by this modeling and statistical analysis motivates a more quantitative approach to mapping of anatectic terrains and in the application of spatial statistical measures to relict melt flow networks in the future than has generally been the case in the past.

To improve our understanding of magmatic systems and the process by which crustal differentiation occurs, we must establish the spatial and temporal characteristics of melt segregation, migration, and extraction, and link these to realistic models of magma ascent. Thus, the aims in this chapter are: (1) to review selectively aspects of melt segregation, migration, and extraction to identify common features among the microstructures and mesoscale patterning of residual migmatites and granulites; (2) to describe in detail one example of a fossil melt flow network, and (3) to develop a general model for melt extraction from, and transport through, the lower crust of collisional orogens.

10.2 Observations and inferences concerning melt extraction

10.2.1 Interpretation of the evidence in anatectic terrains

Any evaluation of whether melt-related structural features preserved in exposed lower continental crust represent what intuition leads us to infer they represent – namely, melt accumulation networks and magma ascent conduits (e.g., Brown, 1994a) – must rely on the weight of circumstantial evidence. Active crustal scale melt flow is inferred if the bulk rock chemistry of anatectic migmatites and granulites reveals residual compositions in comparison with protolith compositions, and the difference is consistent with net melt loss (e.g., Schnetger, 1994; Sawyer *et al.*, 1999; Solar & Brown, 2001; Guernina & Sawyer, 2003). Melt loss is also implied if peritectic minerals survive without significant retrogression during exhumation (Powell & Downes, 1990; Brown, 2002; White & Powell, 2002). On this basis, I consider the topology of deformation band networks filled with leucosome and petrographically continuous granite in dikes to be remnant evidence of active crustal-scale melt flow, rather than the preserved record of a nonproductive system that failed to support melt flow.

Hydrate-breakdown melting reactions are necessary to generate the melt volumes associated with upper crustal granites (Chapter 9). Such reactions involve a volume increase (Brown, 1994b). As a result, natural systems may not operate at a constant superimposed pressure, but may be transiently overpressured. Pressure in any system is equalized by deformation, and herein lies one of the potential driving forces for melt segregation and extraction. However, melt-bearing rocks are weak, perhaps only able to support flow stresses up to 1 MPa (Chapter 11). There are three potential responses to volume increase attendant on melting.

The first response is really an implicit assumption of the schematic prograde *P–T* paths that are drawn in metamorphic geology, which cross hydrate-breakdown melting reactions without deviation. These paths suggest that the crust maintains pressure across the reaction; thus, by inference the environment must be presumed to deform to accommodate the volume increase. Second, the crust may behave as a closed system, in which case, a transient increase in pressure must occur in order to maintain constant volume; however, this increase is limited by the strength of supra-solidus crust, so that, for continued melting, the integrity of the closed system is unlikely to be maintained. Therefore, the third possibility – that the system will be open – is the most likely general response, in which case the volume change required for constant pressure is accommodated by melt loss, perhaps after a small transient increase in pressure.

The implication is clear – melt generated by breakdown of micas in metapelites and metagreywackes and amphibole in amphibolites during the transition to granulite-facies mineral assemblages will escape. This is consistent with the residual composition of these lithologies in granulite-facies terranes (e.g., Guernina & Sawyer, 2003; Schnetger, 1994; Solar & Brown, 2001). In this chapter, net melt loss is assumed to be the common situation based on examples in which melt loss has been demonstrated, but melt loss must be evaluated as part of any specific case study.

10.2.2 Microstructural (grain-scale) features of residual migmatites and granulites

The grain-scale distribution and connectivity of silicic melt is of primary importance in determining the physical properties (e.g., permeability, rheology, and effective diffusivity) of melt-bearing lower continental crust. At low melt fraction under isotropic stress, permeability is imposed by intergranular pore geometry, which is controlled by the wetting properties of the melt; these are a function of relative interfacial energies at solid–melt joins. Silicic melt wets matrix minerals in crustal rocks (Laporte & Watson, 1995; Laporte & Provost, 2000), and melt is predicted to occupy pores where three or more grains join, and to be absent along boundaries where only two grains meet.

However, in natural rocks, microstructural equilibrium generally is not achieved due to continuous grain growth driven by reduction of surface energy, which is an important factor that affects grain-scale melt distribution (Walte *et al.*, 2003).

The transport properties of melt-bearing crustal rocks strongly influence, and are influenced by, deformation, even at the grain scale. For example, observations from deformation experiments on analog materials show evidence of dynamic wetting of grain boundaries, which enhances early interconnection of melt, and inter-granular fracture and grain-boundary sliding (Rosenberg & Handy, 2000, 2001), all features that are expected to occur during syntectonic melting of the crust. In melt-bearing crust, it takes 5–10 vol.% melt to wet approximately 90% of the grain boundaries, and melt is expected to be interconnected at these melt fractions (e.g., Vigneresse *et al.*, 1996; Laporte *et al.*, 1997; Lupulescu & Watson, 1999; Rosenberg & Handy, 2003; Wark *et al.*, 2003).

Observations of microstructures in residual migmatites and granulites provide evidence of the former presence of melt at the grain-scale (e.g., Fig. 10.1; see also Brown *et al.*, 1999; Brown, 2001a, 2001b; Sawyer, 2001). Melting and crystallization from melt may be inferred using the following criteria: igneous microstructure of leucosomes in migmatites and granulites (e.g., Cuney & Barbey, 1982; Vernon & Collins, 1988); mineral pseudomorphs after grain-boundary melt films and pockets (in contact migmatites (e.g., Holness & Clemens, 1999; Marchildon & Brown, 2001, 2002), in regional migmatites (e.g., Brown & Dallmeyer, 1996; Brown, 2002; Marchildon & Brown, 2003) and in residual granulites (e.g., Sawyer, 1999, 2001; Guernina & Sawyer, 2003)); magmatic rims on sub-solidus cores of grains (e.g., rational faces, overgrowths of different compositions (e.g., Marchildon & Brown, 2001, 2002; Sawyer, 2001)); spatial distribution of like and unlike phases (i.e., mineral phase distribution (Ashworth & McLellan, 1985)); and annealed microfractures (Watt *et al.*, 2000; Marchildon & Brown, 2001, 2002).

10.2.3 Mesostructural (outcrop scale) features of residual migmatites and granulites

In addition to an interconnected porosity, diffusive and advective mass transfer requires a driving force for transport and space to accommodate the transferred mass. Based on leucosome distribution, melt is inferred to have been progressively concentrated into a developing network of structures (cf. Brown, 1994a; Brown & Rushmer, 1997), including layer-parallel foliation planes, transverse structures such as interboudin partitions and shear

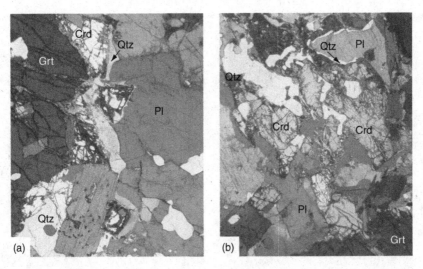

Fig. 10.1. Microstructural features of residual granulite facies migmatite, Southern Brittany, France. (a) Peritectic cordierite (Crd) that occurs adjacent to earlier-formed peritectic garnet (Grt) is separated from plagioclase (Pl) by quartz (Qtz); the plagioclase has subhedral to euhedral terminations that penetrate the quartz, which forms pseudomorphs after pockets of residual melt (width of field of view 5 mm). (b) Peritectic cordierite (Crd) adjacent to earlier-formed peritectic garnet (Grt), associated with subhedral plagioclase (Pl) and quartz (Qtz); note quartz film along plagioclase cordierite grain boundary at top of view of view (width of field of view 5 mm).

surfaces, and axial surfaces and hinges of folds, with the exact form of the network being controlled by the style of deformation. Studies of these natural examples indicate that fabric-parallel and transverse leucosome stromata form by mass transfer down gradients in pressure (van der Molen, 1985a, b; Powell & Downes, 1990; Maaløe, 1992; Brown *et al.*, 1995b).

In layered (anisotropic) protoliths, differences in the rheology of layers lead to differences in the response to applied stress imposed by orogenic deformation. As a result, pressure gradients form between layers, which drives segregation, concentrating melt preferentially in some layers rather than others (Brown *et al.*, 1995b), and additional space is created as inter-layer interfaces dilate and other transverse structures form; mesoscale melt migration follows the changing location of dilatant sites in a feedback relation (Brown & Solar, 1998a; Brown, 2004). Thus, the scale of compositional layering and/or the strength of the anisotropy and the strain field generally control sites of early melt concentration (e.g., Brown *et al.*, 1999; Brown, 2001a, b).

Fig. 10.2. View to the Northnortheast of subhorizontal and subvertical surfaces through concordant sheets of granite in residual stromatic migmatite within a regional-scale zone of apparent flattening strain in the Acadian orogen of west-central Maine, USA (Swift River, Roxbury, Maine; lens cap for scale).

The role of anisotropy in controlling melt transport has been suggested by many studies (e.g., D'Lemos *et al.*, 1992; Brown *et al.*, 1995b, 1999; Collins & Sawyer, 1996; Oliver & Barr, 1997; Brown & Solar, 1998a; Sawyer *et al.*, 1999; Sawyer, 2001; Guernina & Sawyer, 2003; Marchildon & Brown, 2003). The consequences of this control vary with orientation of the fabric elements. In western Maine, USA, in residual migmatites that formed at depths of around 15 km, the form of magma ascent conduits mimics the apparent strain ellipsoid recorded by steeply inclined host rock metamorphic fabrics. Thus, in zones of apparent flattening strain, the ascent conduits are recorded by foliation-parallel sheet-like granites (e.g., Fig. 10.2), whereas in zones of apparent constrictional strain, the ascent conduits are recorded by lineation-parallel moderate to steeply plunging rod-like granites (Brown & Solar, 1999; Solar & Brown, 2001). In contrast, where the fabric elements are shallow, such as might be expected in lower crustal rocks, melt may move predominantly laterally, particularly if the overlying sub-solidus units are impermeable to melt, until a steep structure such as a shear zone is intersected to allow ascent. This may have happened in the St Malo migmatite belt, France

(D'Lemos *et al.*, 1992) and in migmatites in southern Brittany, France (this chapter). However, where magma ascent is controlled strongly by a major steeply inclined crustal-scale shear zone, a linear alignment of plutons in the upper crust may not necessarily reflect a linear source region in the partially molten lower crust (e.g., the Peri–Adriatic Fault System (Rosenberg, 2004)).

10.3 Ascent mechanisms

Diking by brittle fracture is not considered to be a viable mechanism for extracting melt from melt-bearing crustal sources (e.g., Rubin, 1998), although brittle fracture is a possible mechanism for second-stage ascent of fractionated melt from a magma chamber into the surrounding sub-solidus crust (e.g., Weinberg, 1996). Alternative mechanisms for ascent of magma through the crust include mobile melt-filled hydrofractures (Dahm, 2000; Bons *et al.*, 2001, 2005) and various mechanisms to bridge porous grain-boundary flow to channeled flow in ductile crust (e.g., Brown, 1994a; Collins & Sawyer, 1996; Brown & Solar, 1998a; Sawyer *et al.*, 1999; Weinberg, 1999; Leitch & Weinberg, 2002; Olsen *et al.*, 2004).

The slow rates of migration of viscous silicic melt through rock pores and shear-enhanced compaction of residue probably require that melt drained in a dike propagation event must have been previously extracted from pores and reside in an accumulation network, and there must be sufficient hydraulic head to drive ascent. In general, tabular intrusions are favored in nature (Emerman & Marrett, 1990; Rubin, 1995; Weinberg, 1999), and initiating dikes by ductile fracture avoids the problems and assumptions inherent in brittle fracture. By "ductile," I mean fracturing associated with distributed inelastic deformation at the fracture tip (cf. Rutter, 1986), but no particular stress–strain relationship or deformation mechanism is implied (cf. Eichhubl, 2004). A model for melt extraction from lower continental crust based on shear localization in ductile rocks and magma transport through ductile fractures is developed in the last section of this chapter.

The anatectic zone is an over-pressured system with supra-magmastatic fluid pressure gradients and spatial and temporal variations in permeability. Permeability has a power law dependence on melt fraction (Wark & Watson, 1998), which may lead to compaction-generated flow instabilities, manifested as self-propagating high-porosity structures (Connolly & Podladchikov, 1998; Rabinowicz & Vigneresse, 2004). These high-porosity structures may grow by influx of melt, which may lead to fully segregated melt flow (Richardson *et al.*, 1996; Podladchikov & Connolly, 2001). Thus, flow in high-porosity structures (waves) may represent an intermediate mechanism bridging pervasive or per-colative flow to channeled or segregated flow.

Diapirism involves ascent of melt plus residue and the process generally implies partial convective overturn of the lower continental crust; it is not generally considered to be a viable process for emplacement of upper crustal plutons (Emerman & Marrett, 1990; Clemens & Mawer, 1992; Rubin, 1993), although subsequent modeling contradicts this view (Weinberg & Podladchikov, 1994; Burov *et al.*, 2003). Diapirism may occur in the lower crust (Barbey *et al.*, 1999; Teyssier & Whitney, 2002), and it is a mechanism by which (some) plutons are inferred to have been emplaced in the deep crust in continental arcs (e.g., Miller & Patterson, 1999); diapirism is generally expected to have been an important process in the Archean (e.g., Bloem *et al.*, 1997; Collins & van Kranendonk, 1998).

10.4 An example from the Variscides of western France

In western France, the Variscan cycle is interpreted to have comprised Ordovician rifting to form an ocean basin, Silurian subduction and ocean basin closure, and Devonian–Mississippian collision between Gondwana and Armorica that involved close to orthogonal sinistral transpressive deformation (Brun & Burg, 1982; Pin & Peucat, 1986; Burg *et al.*, 1987; Audren, 1990; Jones, 1991; Dias & Ribeiro, 1995; Matte, 2001). Pennsylvanian intra-continental displacement involved highly oblique dextral transtensive deformation and exhumation asso-ciated with the emplacement of leucogranites (e.g., Faure & Pons, 1991; Gapais *et al.*, 1993; Geoffroy, 1993; Burg *et al.*, 1994; Brown & Dallmeyer, 1996).

10.4.1 Geology of the Domaine Sud-Armoricain

The rocks discussed here are part of the Domaine Sud-Armoricain (DSA; Fig. 10.3), a northwest–southeast-trending belt of deformed and metamorphosed supra-crustal rocks now exposed as a core of migmatitic gneiss domes (Brun & Burg, 1982; Peucat, 1983; Gapais *et al.*, 1993; Brown & Dallmeyer, 1996; Johnson & Brown, 2004). The migmatites in the central part of the DSA, around Vannes and St Nazaire (Fig. 10.3), are suitable for the study of melt extraction for several reasons. First, the metamorphic history is well con-strained (Johnson & Brown, 2004), and the migmatites in particular have been the subject of recent research (e.g., Marchildon & Brown, 2003; Brown, 2005). Second, the migmatites are exposed within a large area, implying that the results will be representative of lower crustal processes in general rather than of local interest only. Third, the outcrops around Port Navalo and Petit

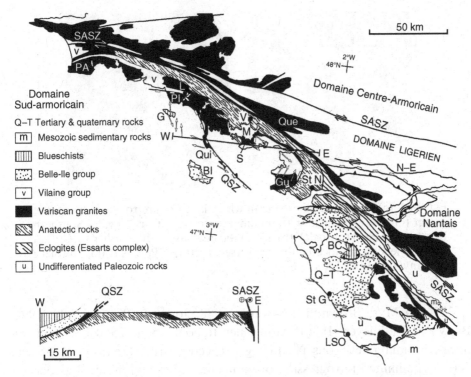

Fig. 10.3. Geological map of the Domaine Sud-Armoricain (DSA), a Northwest–Southeast-trending belt of supracrustal rocks deformed and metamorphosed during the Paleozoic Variscan orogeny. The southern branch of the South Armorican Shear Zone (SASZ) separates the DSA from the three domains to the northwest (unornamented with domain names); together these four domains comprise the Variscan belt of western France. Granites: PA – Pont l'Abbé; PL – Ploemur; QUI – Quiberon; QUE – Questembert; S – Sarzeau; GU – Guérende. Place names: L – Lorient; G – Ile de Groix; BI – Belle Ile; V – Vannes; M – Golfe du Morbihan; St N – St Nazaire; BC – Bois de Céné; St G – St Gildas; LSO – les Sables d'Orlonne; QSZ – Quiberon Shear Zone. The west–east sketch section is modified after Gapais *et al.* (1993).

Mont in Morbihan (Fig. 10.3) comprise wave-cut platforms that are approximately perpendicular to foliation and parallel to lineation, and these platforms are backed by steep cliffs that enable a three-dimensional perspective to be gained of the mesoscopic structures in the migmatites and associated dikes of granite. Fourth, the migmatites do not record evidence of penetrative post-Variscan deformation, so that structures occupied by leucosome and granite are inferred to have been melt-bearing based on preserved igneous microstructures.

The migmatites preserve a mineralogical record of metamorphism and melting along an overall clockwise *P–T* path (peak *P–T* of 9 kbar, 800 °C at

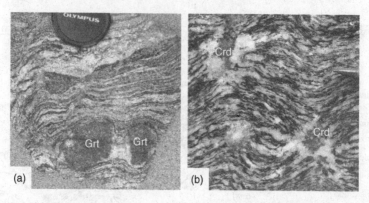

Fig. 10.4. (a) Stromatic migmatite in which largely unretrogressed peritectic garnet (Grt) is associated with minimal leucosome (Port Navalo, Southern Brittany; lens cap for scale). (b) Stromatic migmatite with cordierite (Crd) in minimal leucosome located in transverse structures (Port Navalo, Southern Brittany; tip of pen for scale).

c. 325–320 Myr; Brown, 1983; Jones, 1988; Jones & Brown, 1989, 1990; Brown & Dallmeyer, 1996), with generation of about 25 vol.% melt via mica-consuming reactions producing peritectic garnet (Johnson & Brown, 2004). The limited retrogression of peritectic garnet and minimal associated leucosome (Fig. 10.4a) led Johnson and Brown (2004) to infer that approximately 60 vol.% of the melt produced at the metamorphic peak was lost from the system. The retrograde evolution (Fig. 10.5) began with erosion-controlled conductive cooling, recorded by the partial reaction of garnet with melt to form biotite and sillimanite. Conductive cooling was interrupted by about 3 kbar of decompression that involved a second episode of melt production evidenced by the presence of centimetric peritectic cordierite in minimal leucosome (Fig. 10.4b) located in apparent dilational sites (cf. Bons, 1999). The migmatites are traversed by centimetric to metric dikes of granite. Decompression was followed by rapid cooling below the solidus at *c.* 305–300 Myr (Brown & Dallmeyer, 1996). The metamorphic evolution may have taken several millions to several tens of millions of years (Brown & Dallmeyer, 1996), with much of this evolution being supra-solidus; this is consistent with residence times for melt in anatectic crust of other studies (e.g., Rubatto *et al.*, 2001; Montero *et al.*, 2004).

Early deformation is recorded by isoclinal folds of compositional layering (S_0) and an associated axial planar foliation (S_1), which is typically parallel to compositional layering. The trend of the composite S_0–S_1 defines the structural grain of the area, striking predominantly northwest–southeast, and dipping steeply, except in the hinge zone of later folds. A second generation

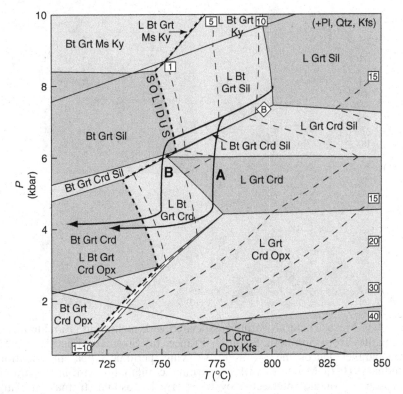

Fig. 10.5. A *P–T* pseudosection for the model system MnO–Na$_2$O–CaO–K$_2$O–FeO–MgO–Al$_2$O$_3$–SiO$_2$–H$_2$O for an average metapelite composition with 10% of bulk MnO (the remainder assumed sequestered in residual garnet) and assuming loss of 15 mol.% melt of granite composition. This diagram is appropriate only for consideration of the retrograde phase equilibria in the migmatites. Two alternate *P–T* paths consistent with petrological and thermobarometric constraints are shown in black. The darkest fields are quinivariant (F = 5). For further details, see Johnson and Brown (2004). (From Johnson & Brown, 2004, Quantitative constraints on metamorphism in the Variscides of Southern Brittany – a complementary pseudosection approach. *Journal of Petrology*, **45**, 1237–59. © 2004 Oxford University Press. Reproduced by permission of Oxford University Press.)

of northwest–southeast-trending regional synformal and antiformal fold structures (Cogné, 1960; Audren, 1987) with sub-horizontal hinge lines and sub-vertical axial surfaces accounts for the variable dip of S$_0$–S$_1$. During the early part of a third deformation, extension related to the initiation of decompression resulted in widespread apparent boudinage (Fig. 10.6a). Open to tight, generally asymmetric folds with steep hinge lines and subvertical axial surfaces produced during the latter part of the third deformation, account for

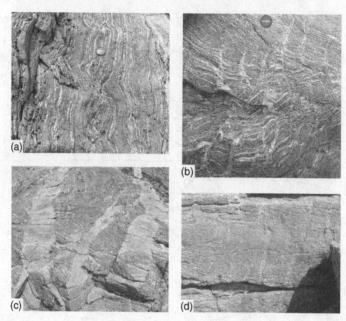

Fig. 10.6. Mesostructural features associated with melting in granulite facies migmatites from Southern Brittany, France. (a) Stromatic migmatite with layer-parallel leucosome stromata showing periodic apparent foliation boudinage (Petit Mont, Arzon). (b) Stromatic migmatite with layer-parallel leucosome stromata intersected by transverse leucosome stromata infilling shear bands (Port Navalo). (c) View approximately northwest of 0.5-m wide dikes of granite in stromatic migmatite (Petit Mont, Arzon). (d) Steeply oriented surface parallel to foliation (note subhorizontal elongation lineation) to show vertical extent of leucosome in transverse structures (Petit Mont, Arzon).

the scatter in strike of the S_0–S_1 foliation. This last phase of deformation is interpreted to have been co-eval with dextral displacement along the north-west–southeast-trending transcurrent South Armorican Shear Zone (SASZ; Jégouzo, 1980). A subhorizontal quartz–feldspar aggregate elongation lineation is ubiquitous throughout the area; it is particularly well defined on foliation-parallel surfaces. All three phases of deformation fold migmatitic layering (cf. Jones, 1988, 1991), from which it is inferred that migmatites of the DSA hosted melt continuously from prior to or during the first deformation until dike emplacement.

The high-grade metamorphic core is bounded to the north by the SASZ, and is separated from overlying units composed of lower-grade rocks to the southwest by shallow-dipping detachments and associated crustally derived granites (Gapais *et al.*, 1993). S–C mylonitic fabrics defined by a flat-lying schistosity, a single set of dominant shear bands and a strong mineral elongation lineation, are common within these leucogranites, particularly in their lower parts (e.g., Bouchez *et al.*, 1981; Audren, 1987; Gapais *et al.*, 1993). These features are interpreted to indicate that granite emplacement was syn-kinematic (cf. Gapais, 1989), and S–C fabrics indicate normal-sense displacement. The magma source for these leucogranites is interpreted to be the associated melt-depleted anatectic rocks of the DSA (Bernard-Griffiths *et al.*, 1985). Leucogranites emplaced along the SASZ yield Rb–Sr whole-rock isochron ages of *c.* 330–320 Myr and $^{40}Ar/^{39}Ar$ ages on mica of *c.* 310–300 Myr, which suggest emplacement and crystallization were coeval with the peak of metamorphism and subsequent retrograde evolution to the south (Brown, 2005). In contrast, leucogranites within the shallow-dipping detachments of the DSA yield Rb–Sr whole-rock isochron ages of *c.* 305–300 Myr, consistent with $^{40}Ar/^{39}Ar$ ages on muscovite and coeval with rapid cooling in the migmatites below the detachments (Brown, 2005).

10.4.2 *Former melt-bearing structures*

Leucosome- and granite-filled structures are inferred to record the melt flow network; these structures include millimeter to centimeter-scale foliation-parallel and transverse leucosome stromata (Fig. 10.6b) and centimeter to millimeter scale generally foliation-discordant dikes of granite (Fig. 10.6c). On the outcrop, leucosome in migmatite shows no evidence of solid-state deformation and in petrographic thin sections, magmatic microstructures (e.g., rational faces of feldspar and quartz or feldspar films inferred to pseudomorph inter-granular melt) are common. The lack of evidence of dislocation creep, demonstrated by the preservation of magmatic microstructures and the limited plastic deformation of quartz, indicates that crystallization of melt was not followed by significant penetrative deformation. These observations are interpreted to indicate that a network of connected melt-bearing structures, now represented by a network of foliation-parallel and transverse leucosomes, existed in these rocks until late in the deformation history.

On sections parallel to compositional layering, foliation, and lineation, transverse leucosome stromata are perpendicular to the lineation and continuous over tens of centimeters (Fig. 10.6d), consistent with apparent extension and the formation of melt-bearing transverse structures (apparent interboudin partitions and shear surfaces) being synchronous with development (or reorientation) of the lineation. The trend of this lineation varies with the orientation of the dominant foliation (S_0–S_1) on which it is observed; it is folded together with the stromatic leucosomes, some of which carry peritectic cordierite, suggesting that late folding was synchronous with

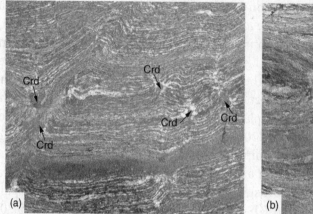

Fig. 10.7. Evidence for melt loss from stromatic migmatite, Southern Brittany, France. (a) Cordierite (Crd) in transverse structures either is associated with minimal leucosome (e.g., center right) or appears to be independent of any residual leucosome (e.g., center left) (Port Navalo). (b) Pucker structure in which layering is "sucked" into apparent interboudin partitions reflecting melt loss from these sites; note thickening of leucosome stromata in neck of structure (Petit Mont, Arzon).

decompression melting and melt loss (cf. Barraud *et al.*, 2004). Centimetric peritectic cordierite dominates leucosome in many transverse structures (Fig. 10.7a), which connotes an important contribution by diffusive mass transfer in the local concentration of melt (e.g., Powell & Downes, 1990). The absence of significant melt-present retrogression of the cordierite, together with the limited volume of leucosome (Fig. 10.7a), which is less than the volume of melt predicted by mass balance, indicates loss of melt from these sites (e.g., Johnson & Brown, 2004). In some cases, cordierite in host migmatite is isolated from leucosome because all melt has migrated away from these sites (Fig. 10.7a); this conclusion is further supported by pucker structures without leucosome (cf. Kriegsman, 2001) and thickening of layering in necks against leucosome-depleted transverse structures (Fig. 10.7b).

Using quantitative data on the one- and two-dimensional distribution of inferred melt-bearing structures, Marchildon and Brown (2003) investigated the transitions from fabric-parallel to network to conduit flow in an attempt to understand the evolution of the architecture (connectivity and permeability) of melt-bearing systems. They obtained information about thickness and spacing distributions of layer-parallel leucosome stromata from 1-D line traverses across layering on flat-surface exposures of stromatic migmatite. The layer-parallel leucosome stromata represent an early-formed, prograde structure, as demonstrated by residual peritectic garnet wrapped by foliation (Fig. 10.4a);

this is interpreted as a flat-lying lower crustal fabric/layering that was reactivated during decompression melting.

Cumulative leucosome thickness versus distance along the traverse, and log-log plots of leucosome cumulative frequency versus thickness or spacing are useful in discriminating between different possible types of thickness or spacing distributions (e.g., Gillespie *et al.*, 1993, 1999; Loriga, 1999). Based on an analysis of synthetic datasets for different vein distribution types, Gillespie *et al.* (1999) argue that fractal distributions will yield a step-wise plot of cumulative thickness versus distance, because of inherent clustering, and will define linear trends (with slope related to the fractal dimension, *D*) in log cumulative frequency versus log thickness or spacing plots. On the other hand, random vein distributions will yield smooth and regular cumulative thickness versus distance plots, and will define concave downward patterns of log cumulative frequency versus log thickness and spacing (Gillespie *et al.*, 1999).

In Fig. 10.8, the plots of cumulative leucosome thickness versus distance along the traverse are essentially linear, whereas the downward concavity of arrays on the plots of log cumulative frequency versus log thickness or spacing indicates a limited range of leucosome thicknesses (generally between 1 mm (the limit of resolution) to 10 mm (upper limit 20–30 mm)), and shows that frequency decreases rapidly with increasing thickness. These features are inconsistent with scale-invariance.

The volume of leucosome (40–60%) in these outcrops far exceeds that expected at the calculated *P–T* conditions (Johnson & Brown, 2004). Leucosomes are interpreted to represent the product of crystallization of melt during an effective melt flux down a temperature gradient (cf. Solar & Brown, 2001). That is, the volume of melt transported through structures infilled by leucosome was considerably larger than the volume of leucosome, but the leucosome is the early crystallized (cumulate) part of the larger melt flux trapped during transport through this crustal level.

The lack of clustering of stromata in the data of Marchildon and Brown (2003) is inconsistent with an origin by fracturing, since fractures tend to be clustered or have regularly spaced distribution around a single large fracture. However, an origin by flux of melt along pre-existing anisotropy is consistent with the scale-dependent nature of the data and is preferred (cf. Brown *et al.*, 1995b).

Qualitative observation of inferred melt-bearing structures in mutually perpendicular two-dimensional flat-surface exposures from the same outcrop reveals anisotropy of the leucosome network related to the sub-horizontal stretching lineation. On lineation parallel surfaces, traces of stromatic leucosomes are continuous, and the leucosomes have smooth, essentially straight,

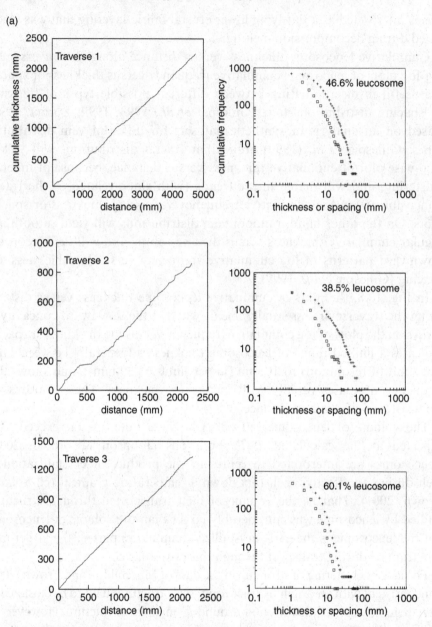

Fig. 10.8. Plots to show results of line' traverses measured across stromatic migmatite exposed as subhorizontal surfaces approximately perpendicular to foliation and parallel to elongation lineation. For six line traverses, two types of diagram are presented: a "staircase" plot, which shows the cumulative leucosome thickness with distance along the traverse, and a plot of log cumulative frequency of leucosome versus log thickness (squares) or spacing (crosses) of leucosomes. In these latter diagrams, the number of leucosomes with thickness (or leucosome spacing) greater than the thickness (or spacing) on the abscissa is plotted for every increment of thickness or spacing from the

(b)

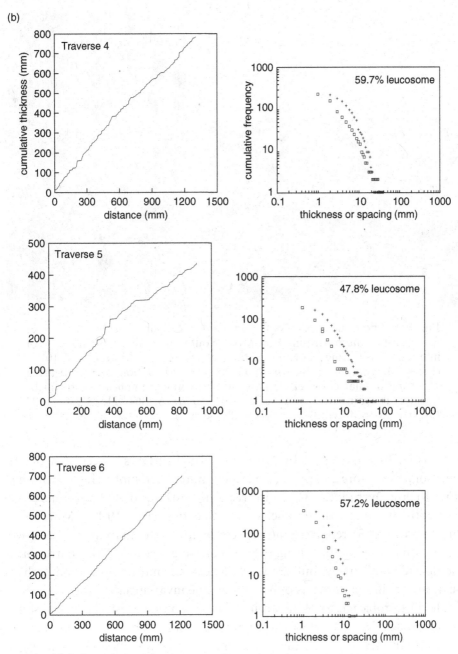

Fig. 10.8. (cont.)

minimum to the maximum values. The calculated percentage of layer-parallel leucosome in each traverse is given. Traverses 1, 2, 4, and 5 are from Petit Mont, and traverses 3 and 6 are from Port Navalo (Southern Brittany). (From Marchildon & Brown, 2003, Spatial distribution of melt-bearing structures in anatectic rocks from Southern Brittany: implications for melt-transfer at grain-to orogen-scale. *Tectonophysics*, **364**, 215–35. © 2003 Elsevier. Reproduced by permission of Elsevier.)

Fig. 10.9. Photo mosaics of subhorizontal (a) and subvertical (b) exposures from coastal outcrop around Petit Mont (Southern Brittany). Binary maps of inferred melt-bearing structures in these exposures are shown in Fig. 10.10. (From Marchildon & Brown, 2003, Spatial distribution of melt-bearing structures in anatectic rocks from southern Brittany: implications for melt-transfer at grain- to orogen-scale. *Tectonophysics*, **364**, 215–35. © 2003 Elsevier. Reproduced by permission of Elsevier.)

edges (Fig. 10.9a), whereas on lineation normal surfaces, traces of stromatic leucosomes are more complex, with less continuity and more irregular outlines (Fig. 10.9b). Overall, the 3-D anisotropy suggests a prolate leucosome structure with the long axis parallel to the lineation (Fig. 10.10). Marchildon and Brown (2003) report results of semiquantitative analysis of these two-dimensional distributions using the box counting method, which corroborate the inferred anisotropy, but show that leucosome morphology (and perhaps distribution) in this case has only limited scale invariance.

The anisotropic nature of the stromatic leucosomes suggests that permeability in the plane of the foliation was larger parallel to lineation than perpendicular to it, which further supports the synchroneity of deformation and leucosome topology. If this interpretation is correct, then it may be inferred that melt flowed in the plane of the foliation parallel to the lineation to transverse structures in response to gradients in melt pressure and/or dilation-compaction pumping, driven by volume change and deformation. However, the growth of peritectic phases in deformation bands, particularly transverse structures

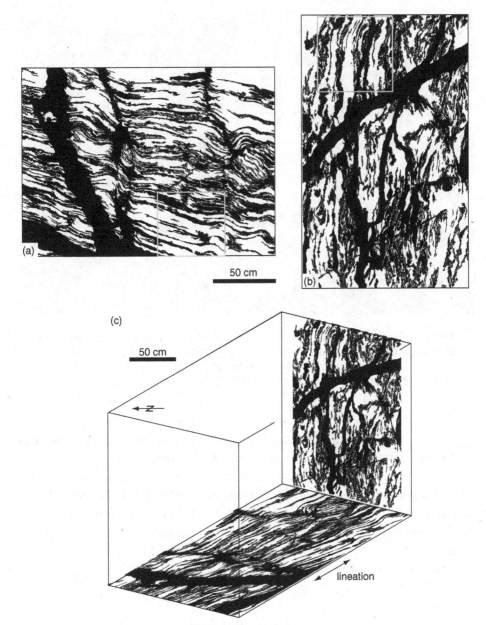

Fig. 10.10. Binary maps for the inferred melt-bearing structures in subhorizontal (a) and subvertical (b) exposures shown in Fig. 10.9. (c) Composite three-dimensional reconstruction of distribution of inferred melt-bearing structures from two-dimensional binary maps. Lineation direction is shown. Gray boxes in (a) and (b) show areas used for the box-counting method analysis by Marchildon and Brown (2003). (From Marchildon & Brown, 2003, Spatial distribution of melt-bearing structures in anatectic rocks from southern Brittany: implications for melt-transfer at grain- to orogen-scale. *Tectonophysics*, **364**, 215–35. © 2003 Elsevier. Reproduced by permission of Elsevier.)

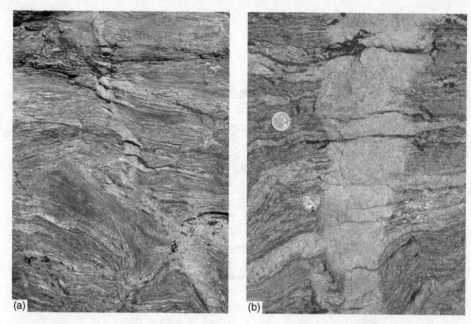

Fig. 10.11. Petrographic continuity (mineralogy, mode, grain size and microstructure) of leucosome and granite (Petit Mont, Arzon, Southern Brittany). (a) and (b) Discordant centimetric dikes of granite that exhibit petrographic continuity with leucosome in stromatic migmatite.

(Fig. 10.7a), connotes an important contribution by diffusive mass transfer in the local concentration of melt (cf. Powell & Downes, 1990; Hand & Dirks, 1992), particularly in the early stages and probably at melt fractions of less than about 7–8 vol.% (cf. Vigneresse *et al.*, 1996; Rosenberg & Handy, 2003). Using the equation for dimensionless melt transport (MT) number of Olsen *et al.* (2004) to evaluate porous flow versus bulk flow for melt fractions appropriate to each melting event in the DSA yields values for MT <1, consistent with macroscopic porous flow in both cases.

10.4.3 Granite dikes

Centimeter- to meter-wide granite dikes are abundant at outcrop; these dikes commonly cross-cut structures in the migmatites (Fig. 10.11a), and they may intersect (Fig. 10.12). The volumetric importance of the dikes varies in space, but they may represent up to 20% of the area of an outcrop (Fig. 10.6c). Individual dikes are planar and typically not deformed, indicating that emplacement post-dated penetrative deformation. The nature of the contacts

Fig. 10.12. Subhorizontal surface through stromatic migmatite to show granite dikes that intersect in a common steeply-inclined "pipe-like" channel structure; although the dikes are discordant at outcrop scale, there is only limited evidence for cross-cutting relationships at the vein-to-grain scale, suggesting that material now in the dikes was melt-bearing at the same time as the material in the host rocks (Petit Mont, Arzon, Southern Brittany).

between granite in the dikes and leucosome in the host migmatites is particularly instructive in distinguishing emplacement mechanisms.

The microstructure of the granites is equigranular to porphyritic, with euhedral to subhedral feldspar and anhedral quartz. Within individual dikes, spatial variation in grain size is limited and finer-grained margins generally are not observed, suggesting approximate thermal equilibrium between melt in the dike and host melt-bearing crust. The modal mineralogy and microstructure of granite in dikes are indistinguishable from the modal mineralogy and microstructure of leucosome in the host migmatites (Fig. 10.11b). This petrographic continuity is interpreted to mean that these structures once hosted a continuous melt-bearing network, and suggests that the material in leucosomes and in the dikes underwent final crystallization at the same time. This inference does not mean that leucosomes (or necessarily granite in dikes) represent liquid compositions; leucosomes in the migmatites

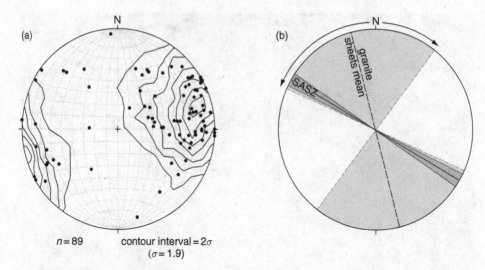

Fig. 10.13. (a) The orientation of granite dikes measured in outcrops around Petit Mont and Port Navalo, Southern Brittany (equal area lower hemisphere projection). (b) Geometric relationship between the strike orientation of granite dikes (light gray ornament) and the SASZ (dark gray ornament).

comprise residual minerals (quartz and feldspar cores), early crystallized (cumulate) material (feldspar rims) and material crystallized from derived liquid.

The dikes demonstrate that melt extraction and ascent through the crust had occurred at and below the crustal level exposed. The dikes show a range of orientations, with a concentration about a trend approximately northnorthwest and a dip steeply to moderately to the west, defining a point maximum (Fig. 10.13a). Audren (1987) interpreted similar results as reflecting the synchroneity of dike formation with dextral strike-slip deformation along the SASZ, the northnorthwest trend being at a low angle to the short axis (i.e., minimum principal finite stretch) of the regional finite strain ellipsoid associated with this deformation.

The geometrical relationship between the overall strike of granite dikes and the orientation of the SASZ is shown in Fig. 10.13b. Dike emplacement is along a preferred direction independent of anisotropy, which suggests the orientation is controlled by stress and this, together with the macroscopic fracture-like discontinuity infilled by the granite, characterizes the process of dike formation as a fracture phenomenon. Petrographic continuity between leucosome in the deformation band networks and granite in the dikes suggests

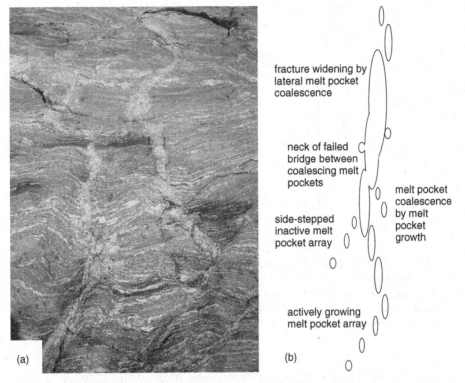

Fig. 10.14. (a) Centimetric discordant dikes of granite with zigzag form at tips (bottom of image; Petit Mont, Arzon, Southern Brittany). (b) Sketch (after Eichhubl & Aydin, 2003) to illustrate the mechanism of formation of ductile opening-mode fractures by pore growth and melt pocket coalescence. Switching between left- and right-stepping direction leads to zigzag propagation paths.

that these ascent conduits formed by opening-mode failure along zones of localized porosity increase, and the zigzag propagation path at the tips of some of the thinner dikes (Fig. 10.14a, b) points to ductile fracture as the likely mode of formation of these conduits (Brown, 2004).

It has been argued that in the transition from grain-scale melt flow to channelized flow in veins the direction of melt flow will be controlled by elements of the metamorphic fabric, particularly the foliation and lineation (Brown & Solar, 1998a; Brown *et al.*, 1999). In zones of highly oblique transtensive deformation, melt flow is expected to be either sub-horizontal along the lineation in the plane of the foliation (Brown & Solar, 1998a) or sub-vertical in discordant structures (cf. Sibson, 1996; Sibson & Scott, 1998). Marchildon and Brown (2003) have argued that flow of melt associated with decompression in the DSA migmatites was along the layering parallel to the

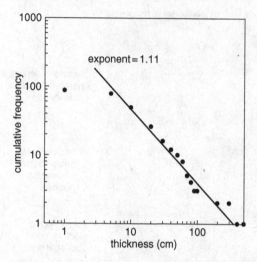

Fig. 10.15. Cumulative frequency plots of dike thickness ($n = 87$) from a traverse along the coastline around Port Navalo Plage and le Petit Mont, Morbihan. The number of dikes wider than a particular value is plotted against that value, and a best-fit line through the data above 10 cm wide defines a power law relationship with an exponent of 1.11

quartz–feldspar mineral aggregate elongation lineation to transverse structures and dikes. Upward transport of melt is inferred to have been principally through dikes (Fig. 10.6c) or backbone structures such as the pipe-like structure formed by intersection of coeval dikes (Fig. 10.12).

The thickness of dikes >10 cm wide shows a power law distribution with an exponent of 1.11 (Fig. 10.15), which suggests the dikes may be scale-invariant (Brown, 2005). The largest dikes measured along the traverse were 3 and 5.5 m wide, within the range of critical dike widths for flowing magma to advect heat faster than conduction through the walls and avoid freezing close to the source; dikes of this size each may transport $> 1000 \, \mathrm{km}^3$ of magma in *c.* 1200 yr, assuming a horizontal length of 1 km (Clemens, 1998). This fugitive melt is inferred to have fed crustally derived plutons at higher crustal levels, such as the Ploemeur, Quiberon, Sarzeau, and Guérande granites.

10.4.4 Early melt loss

An evaluation of melt loss prior to exhumation and core complex formation is problematic due to the flux of melt through any point in the system, but melt loss is implied by the preservation of peritectic garnet in the migmatites (Johnson & Brown, 2004). How this melt was expelled from the stromatic migmatites is speculative, due to overprinting by the cordierite-producing decompression melting event during doming, but the granites along the SASZ, which were emplaced a few million years prior to those along the extensional detachments, may have been sourced from these migmatites. This implies a large component of lateral flow, which is consistent with transport of melt along a shallowly oriented foliation in the lower crust to the SASZ, and with the geometry of plutons that have roots in the SASZ (e.g., Vigneresse & Brun, 1983; Martelet *et al.*, 2004).

10.4.5 *Migmatite–granite relations*

In the DSA, the exposed lower crustal unit is narrow in relation to its length and is elongate parallel to the SASZ (Fig. 10.3), which is a major crustal-scale tectonic discontinuity; leucogranite melt ascent and emplacement was diachronous across the SASZ. To the northwest side, older plutons are spatially related to the SASZ, which could have been a major backbone element for early melt extraction from anatectic lower crust. During emplacement, leucogranite magma expanded into the Domaine Center-Armoricain to form plutons in low-mean-pressure magma traps at the shoulder of the SASZ (Vigneresse & Brun, 1983; Weinberg *et al.*, 2004). Along the SASZ, granites occur as steeply oriented tabular plutons (Jégouzo, 1980). To the southwest side of the SASZ, younger granites occur as thin subhorizontal tabular plutons emplaced along horizons now reactivated as extensional detachments (Gapais *et al.*, 1993).

The system of structures observed in lower crustal migmatites is inferred to have accommodated large-scale melt transport to feed upper Carboniferous two-mica leucogranites. Melt loss from lower crust probably occurred in two stages, with early melt extraction via the SASZ and later melt extraction via dikes associated with decompression and core complex formation. The importance of the two-mica granites in focusing deformation along regional-scale shear zones and detachments has been noted by a number of authors (e.g., Strong & Hanmer, 1981; Lagarde *et al.*, 1990; Gapais *et al.*, 1993; Brown & Dallmeyer, 1996).

The migration with time of the focus of Carboniferous leucogranite magmatism from earlier ascent and emplacement spatially related to the SASZ (Strong & Hanmer, 1981) to later ascent through the migmatites via dikes and emplacement at detachments between units (Gapais *et al.*, 1993) is inferred to reflect the change in geodynamics from sinistral transpression to dextral transtension. This migration implies that early melt flow was along shallowly oriented foliation toward the SASZ. In its present configuration, the SASZ does not extend downward to the Moho, but has been displaced to the north-east by thrusting (Bitri *et al.*, 2003). At the time of earlier Carboniferous leucogranite magmatism, it is possible that the SASZ was rooted in the supra-solidus lower crust and that melt escape from the lower crust occurred via this structural conduit. This is consistent with the emplacement of leuco-granite along the SASZ coeval with the peak of metamorphism and the retro-grade evolution in the migmatites to the south.

In contrast, the synchroneity among the second melting event recorded in the DSA lower crustal migmatites, represented by the cordierite-bearing gran-ite-leucosome network, and the two-mica granites, such as the Ploemeur, Quiberon, Sarzeau, and Guérande plutons emplaced along extensional detachments at higher crustal levels (Brown & Dallmeyer, 1996; Brown, 2005), suggests that there was melt transfer directly from the anatectic zone, represented by the lower crustal migmatites, via dikes to the sites of emplace-ment of those plutons. The inference is that upper Carboniferous exhumation of the DSA was facilitated by magma trapped at horizons that became reacti-vated as extensional detachments (Gapais *et al.*, 1993), and segregation, ascent, and emplacement of crustally derived leucogranite magma was inti-mately linked to regional-scale late orogenic transtensive deformation and core complex formation.

10.4.6 *Core complex formation*

In the DSA, the lower crustal units are exposed in structural culminations, interpreted to be the result of transtension that allowed the middle and lower crustal units to rise up, initiating several dome-like structures (Audren, 1987; Tirel *et al.*, 2003) and imposing boudin-like structures at the crustal scale (Gapais *et al.*, 1995). As the weak melt-bearing crust of the DSA responded to the regional transtension by rising, so the migmatitic layering and fabric were rotated from flat lying in the lower crust to the present steep orientation on the dome margins.

The trigger for the second melting episode is inferred to be the change from sinistral transpression to dextral transtension. Extension due to a change in kinematics of this type may induce incipient sites of dilation, in which case the small decrease in P that occurs at spaced intervals along competent layers in extension is likely to initiate the biotite-breakdown melting reaction involving cordierite as a peritectic product (Fig. 10.5). This is inferred to have been what has happened in the DSA migmatites, since the peritectic cordierite is located in transverse structures that must have been sites of melting to generate the cordierite, and also sites of melt loss to preserve the cordierite. The production of melt occurs over a very small drop in *P* for this particular reaction, effectively reducing further the density throughout the volume of suprasolidus orogenic crust implied by the exposed lower crustal migmatite unit in the DSA, and promoting the initial rise of this unit during transtensive deformation. This initiated a feedback relation between lower crustal doming and melting, and between dome amplification and melt extraction and emplacement in developing extensional detachments.

In such a tectonic scenario, melting occurs in response to approximately isentropic decompression by the lower continental crust (i.e., the decompression is essentially adiabatic (it occurs without significant change in volumetric heat) and is assumed to be reversible (the expansion of the rising lower crust is frictionless); as a consequence, the temperature drop is minimal (but reflects the change in internal energy due to work done by expansion)). In these circumstances, a zone of melting may be developed if the rate of decompression is fast enough to preclude significant conductive heat loss from the melting zone. A dimensionless number, the Peclet number (Pe), may be used to evaluate the importance of advection versus conduction, where $Pe = (D.V.\rho.C_P)/k$, and D is the characteristic length (10^4 m in the DSA (approximately equivalent to 3 kbar decompression; see Johnson & Brown, 2004)), V is the velocity (0.01 m yr^{-1} in the DSA (average exhumation rate derived from 3 kbar decompression in 1 Myr, see Brown & Dallmeyer, 1996)), ρ is the mass density (for supracrustal rocks with low melt fraction $\sim$2700–2800 kg m^{-3}), C_P is the specific heat capacity at constant pressure ($\sim$1000 J kg^{-1}/K), and k is the thermal conductivity ($\sim$2.0 W m^{-1}/K for supra-crustal rocks at lower crustal conditions). For the DSA, Pe is $\sim$4, which is >1 but not $\gg$1, suggesting that advection is marginally dominant, and indicating that an assumption of adiabaticity is fragile. None the less, so-called "decompression melting" appears to have been viable in this case, consistent with observations from the field relations and interpretations of the petrologic relations.

The increase in volume associated with the melting reaction in combination with doming in response to the dextral transtensive deformation is inferred to have ensured interconnection of the deformation band network. The fabric parallel element of this network is inferred to have been reactivated, and the non-coaxial shear component of the deformation facilitated extraction of the melt generated during decompression. By this stage, ductile opening mode fractures were able to propagate to enable melt extraction. This process probably is terminated once sufficient melt has accumulated at the base of the upper crust to allow detachment of upper crustal units from the lower crustal unit, which enabled rapid tectonic exhumation and fast cooling of the lower crustal source, and arrested reaction and froze any residual melt.

Marchildon and Brown (2003) suggested that the melt extracted during the early ductile extension stage of the decompression event accumulated at a shallower crustal level along the lower contact of the Belle-Ile Group to facilitate detachment of the upper crustal units. This level in the Variscan crust is inferred to have been close to the upper Carboniferous brittle–ductile transition zone (e.g., Cagnard *et al.*, 2004); the strength peak in the crust at this level may represent a natural magma trap, unless the melt begins ascent with a magmastatic head sufficient to break through this barrier (Brown & Solar, 1998b). Displacement related to the detachment is recorded by the late-magmatic–early sub-solidus S–C fabrics present in the Quiberon, Sarzeau, and Guérande plutons (Bouchez *et al.*, 1981; Audren, 1987; Gapais *et al.*, 1993; Brown & Dallmeyer, 1996).

10.5 A paradigm for melt extraction from the lower continental crust of orogens

Melting occurs in response to a significant increase in the crustal or lithospheric heat flow, by thickening or by thinning, respectively, which has the consequence of weakening the crust or lithosphere and activating deformation. This simple fact explains the general observation that melting and deformation are inter-related in the differentiation of the continental crust, and that melt extraction, ascent and emplacement commonly occur in actively deforming segments of orogens. Melt segregation and extraction is a two-stage process in which melt migrates from grain boundaries to veins and from veins to dikes or shear zones (cf. Chapter 11). Thus, the sequence of mesoscale leucosome- and granite-filled structures described from lower continental crust records a transition from accumulation to draining, and melt is inferred to have been extracted by upward transport along steeply oriented conduits. This conclusion is common to multiple studies (e.g., Collins & Sawyer, 1996; Oliver & Barr, 1997; Brown & Solar, 1999; Sawyer *et al.*, 1999; Marchildon & Brown, 2003), and although the specific details vary from example to example, the general conclusion is robust.

In isotropic materials with randomly oriented pores of melt, analytical (Sleep, 1988) and numerical (Simakin & Talbot, 2001a) solutions and finite element modeling (Simakin & Talbot, 2001b) indicate that melt-filled veins form parallel to σ_1 and perpendicular to σ_3 by closure of pores perpendicular to σ_1 and expansion of pores perpendicular to σ_3. It is argued that melt segregation into vein networks may lead to sheet-like bodies of melt of sufficient size that buoyant ascent becomes possible. Also bottom to top directionality (vertical fracture alignment) may emerge from non-linear evolution of an isotropic fracture network formed by thermal expansion (Petford & Koenders, 1998). Such models may apply where lower crustal protoliths are isotropic and heating is rapid, which might be the case in the lower parts of arcs during the subduction period of the orogenic cycle, although Rubin (1993, 1995) has argued that melt-filled fractures are unlikely to propagate out of a melt-bearing source, unless the source is already pre-warmed by diking from deeper levels.

Brittle fracture has dominated the view of melt extraction from the anatectic zone, and magma transport through the crust structurally below the brittle–viscous transition zone, but is unlikely realistically to describe these processes because of the ductile behavior of rocks at the prevailing temperatures and pressures of the middle and lower crust. The concentration on brittle fracture most probably is a consequence of the direct application to nature of experimentally derived mechanical properties determined at artificially fast strain rates of about $10^{-5} s^{-1}$, leading to an expectation of melt-enhanced embrittlement in the anatectic environment as a general model (e.g., van der Molen & Paterson, 1979; Renner *et al.*, 2002). Although there are examples in which brittle processes appear to have operated, such as at Milford Sound, New Zealand (Daczko *et al.*, 2001), melt extraction from the source and magma transport through the overlying crust generally are more likely to be the result of processes of shear localization and ductile fracture (Brown, 2004). The physics of ductile fracture is controlled by different length and time scales in comparison with brittle fracture, which explains a number of features of melt-bearing lower-crustal magma systems.

10.5.1 Grain boundaries to veins

Under equilibrium conditions in isotropic crust, melting begins at multiphase grain junctions that include quartz and feldspar, and a hydrate phase. However, Earth's crust is anisotropic and in a state of stress, and variations in bulk composition and grain size influence the sites at which melting begins (Brown & Solar 1998a). Furthermore, the orientation of inequant fabric-forming hydrate phases may determine the shape of low-melt-fraction melt pockets (Sawyer, 2001), the fabric may determine the topology of melt flow (Brown *et al.*, 1999), and the positive volume change associated with hydrate-breakdown melting may contribute to the driving force for deformation and melt escape (Brown, 1994b).

The chemography of melting in the crust has been considered in Chapter 9. Melting may begin at sites of lower or higher P, once the initial thermal overstep is close to that required to overcome the activation energy for the melting reaction. For water-present melting, where the dP/dT of reaction is negative, small increases in P may promote the start of reaction, whereas for the volumetrically more important hydrate-breakdown melting, where the dP/dT of reaction is positive, small decreases in P may promote the start of reaction (Fig. 10.5). Furthermore, for incongruent melting, the solid products of the reaction may have difficulty nucleating, so that once melting is established at a particular site, it is energetically favorable for melting to continue at that site and for the solid products to grow there.

Mesoscale concentration of melt may occur in one of two ways (Powell & Downes, 1990; Hand & Dirks, 1992; Maaløe, 1992; Brown *et al.*, 1995b; Watson, 2001). First, mass transfer may be by diffusional fluid flow – chemical migration through a matrix by a dissolution–diffusion–crystallization process in the matrix through which it moves (e.g., in metasedimentary protoliths with compositional anisotropy, formation of leucosome stromata with melanocratic selvedges in which the grain size is coarser than that for the same mineral in the matrix; and, for hydrate dissolution under supra-solidus conditions, concentration of leucosome around a peritectic phase or in deformation bands that include and connect several grains of a peritectic phase). Second, mass transfer may be by advective fluid flow – physical migration by flow through the matrix without chemical interactions (e.g., preferential movement of melt through stromata with high melt fraction to draining conduits, driven by gradients in melt pressure).

Average rates of diffusive mass transfer yield characteristic distances that may approach 1 m on time scales around a million years, but there are many variables that affect rates and efficiencies of diffusion (e.g., Costa *et al.*, 2003; Dohmen & Chakraborty, 2003). In contrast, advective fluid flow is several orders of magnitude faster (e.g., Harris *et al.*, 2000). Both processes may be inferred to have occurred in any one suprasolidus episode, based on evidence in individual outcrops of anatectic rocks, and commonly these processes act in concert. Coupling between mechanical and physicochemical aspects of rock behavior is significant in the anatectic environment, but the complex interplay among melt fraction, melt segregation, bulk strength contrasts, mechanical properties of different minerals and the melt phase, and time, mean that the rheology is difficult to describe by a single flow law (e.g., Berger & Kalt, 1999).

In partially molten lower continental crust, gradients in melt fraction will affect density and buoyancy forces, and may necessitate separated flow of solids and melt rather than a bulk response. However, what happens to such a mixture of solids and melt, in which the potential range of viscosities may approach 14 orders of magnitude and each end member has a different rheology (e.g., power law with $n = 3$ for the solid and Newtonian with $n = 1$ for the liquid), will depend upon the level of applied stress, the differential stress, the effective stress and the strain rate, the melt fraction, and changes in the proportion of melt to solid with time, and whether localization occurs. At high strain rates, the contrast in effective viscosity is reduced, and the bulk response most likely will be coupled flow (e.g., dominated by the matrix or at high melt fraction behavior as a pseudofluid), whereas at low strain rates the rheology may be dominated by one of the phases, and phase separation (strain rate compatibility) or strain localization (stress compatibility) may occur (Vigneresse & Burg, 2002; Vigneresse, 2004).

Based on examples of the DSA migmatites, it is clear that the key to understanding melt migration is an understanding of the nature of strain localization in melt-bearing crust, and how localization controls melt extraction. Discordant veins of leucosome in residual migmatite and granulite commonly have been interpreted as resulting from brittle fracture due to embrittlement as a result of high melt pore pressure (e.g., van der Molen & Paterson, 1979). However, fracture toughness analysis of crystalline solids indicates that there is a change in failure from brittle to ductile modes with increasing temperature (e.g., Hirth & Tullis, 1994; Hirsch & Roberts, 1997; Stockhert et al., 1999). At middle and lower crustal depths, below the brittle–viscous transition zone, lower strain rates, higher temperatures and confining pressures and the presence of a volatile or melt phase due to dehydration reactions or hydrate-breakdown melting reactions, promote ductile processes.

The beginning of anatexis in the lower crust increases weakening and leads to positive feedback between deformation and melting. The feedback is initiated because sites of early melting are generally stress-induced, so that melting leads to localization of strain at these sites, which in turn enhances spatial variations in mean stress. As the melt fraction increases, weakening leads to formation of a network of deformation bands nucleated on the local concentrations of melt (cf. Grujic & Mancktelow, 1998). The preferential occurrence of peritectic minerals in deformation bands, in spite of a homogeneous distribution of reactant phases in the matrix, supports the inference that, in general, deformation was localized by the concentration of melt (higher porosity) at these sites.

This view contrasts with the study by Hand and Dirks (1992), who also inferred that strain partitioning and partial melting are coupled processes, but who suggested that pre-existing mesoscopic crenulations or high strain zones controlled the sites of initial incongruent melting. In this case, the authors attributed selection of sites of initial melting to an additional strain-induced free energy component that enabled overstepping of the activation energy for

melting in the axial zone of the crenulations. However, growth of the leucosomes along the axial surfaces was ascribed to control by the local stress field,
with gradients in differential stress driving melt migration and elongation
perpendicular to σ_1.

For layers or elongate bands of differing rheology in extension, mean stress
is higher in the weaker material, but in shortening it is lower. Thus, in extension, the resulting difference in mean stress drives melt from layers of greater
melting, which are weaker, down gradients in melt pressure to stronger layers
(cf. Brown *et al.*, 1995b). Also, for an individual melt-bearing elongate shear
band that is stretching and thinning, mean stress is higher in the band and melt
will be expelled (cf. Mancktelow, 2002). Conversely, for an individual melt-
bearing elongate shear band that is shortening and thickening, mean stress is
lower in the band and melt is sucked into it. Finally, fluid pressure enables melt
to exploit planes of weakness in anisotropic protoliths (e.g., by dilating foliation and/or interfaces between layers of contrasting composition), in a fashion
similar to that described by Barraud *et al.* (2004) from analog experiments.

The structural control on sites of leucosome concentration is consistent with
gradients in mean stress as the principal driving force for matrix compaction,
melt segregation and meter-scale melt flow (Chapter 11). Thus, melt may be
pumped in cyclic fashion along shallowly oriented fabrics by a dilation–
compaction mechanism, moving through fabric-parallel segments via connecting
transverse structures. Also, melt may be pumped through shear systems from
thinning to thickening shear bands, and as shear bands rotate into the field of
finite extension, increasingly melt is expelled from the source.

The rate of volumetric strain by melt expulsion cannot exceed the rate at
which the solid matrix deforms by shear-enhanced compaction, so the rate
of melt expulsion is controlled by the rate of matrix compaction or the rate
of volumetric strain is limited by the rate of melt expulsion. If the solid matrix
is able to compact at a higher rate than the rate at which melt may be expelled,
then a higher distortional strain will be the result, requiring melt-bearing shear
bands to develop or more distributed flow to occur (Rutter, 1997). Overall, the
positive feedback relation enables interconnected networks of deformation
bands to evolve to an optimum structure, controlled by the rates of strain
accumulation (and melt production) in relation to rates of matrix deformation
and melt expulsion.

Ideal deformation band networks are composed of three kinds of structure
(Fig. 10.16a): shear bands, characterized by a dominant shear displacement
gradient that is accompanied by porosity reduction or compaction, or porosity
increase or dilation (Antonellini *et al.*, 1994); compaction bands, characterized
by a localized porosity reduction or compaction and lacking macroscopic

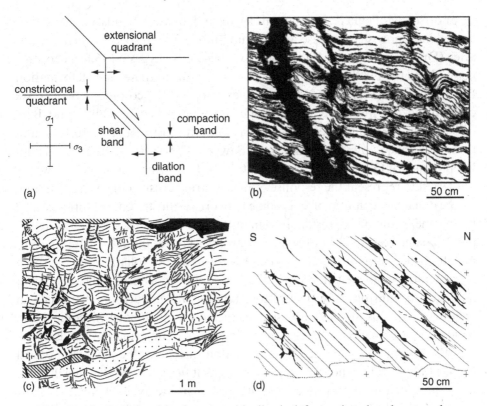

(a)

(b) 50 cm

(c) 1 m

(d) 50 cm

Fig. 10.16. Relationship between idealized deformation band networks and networks of leucosome stromata in residual migmatites. (a) Idealized diagram (after Du Bernard *et al.*, 2002, Dilation bands: a new form of localized failure in granular media. *Geophysical Research Letters*, **29**, 24 2176, doi:1029/2002GL015966. © 2002 American Geophysical Union. Modified by permission of American Geophysical Union) to show the coexistence of shear, compaction, and dilation bands (σ_1 and σ_3 are greatest and least far-field principal stresses, respectively). (b) Binary image of subhorizontal surface perpendicular to foliation and parallel to lineation to show patterning of leucosomes, Petit Mont, Arzon, Southern Brittany. (From Marchildon & Brown, 2003, Spatial distribution of melt-bearing structures in anatectic rocks from southern Brittany: implications for melt-transfer at grain- to orogen-scale. *Tectonophysics*, **364**, 215–35. © 2003 Elsevier. Reproduced by permission of Elsevier.) (c) Detail from outcrop map to show the interrelationships between stromatic leucosomes parallel to fabric and transverse leucosomes in apparent interboudin partitions and shear bands. (From Oliver & Barr, 1997, The geometry and evolution of magma pathways through migmatites of the Halls Creek Orogen, Western Australia. *Mineralogical Magazine*, **61**, 3–14. © 1997 Mineralogical Society. Reproduced by permission of Mineralogical Society.) (d) Outcrop map (vertical face, parallel to the stretching lineation) to show the distribution of leucosomes in a strongly melt-depleted metagreywacke. Thin, continuous leucosomes are oriented parallel to the layering and wider discordant leucosomes occur

shear offset (Mollema & Antonellini, 1996); and, dilation bands, characterized by localized porosity increase and opening-mode with respect to their boundaries (Du Bernard *et al.*, 2002). Brown (2004) has suggested that leucosome networks in the DSA migmatites may be analogous to these ideal deformation band networks (Fig. 10.16b). Further, examples of leucosome networks are common in the literature, and include those described by Oliver and Barr (1997) from the Halls Creek Orogen in Western Australia (Fig. 10.16c), and those described by Guernina and Sawyer (2003) from the Ashuanipi Subprovince of Quebec, Canada (Fig. 10.16b). The formation of such networks is inferred to represent the response of melt-bearing crust to shear deformation.

The dynamic evolution of leucosome networks is illustrated by fabric-parallel leucosome stromata in residual migmatites and granulites. Assuming layer-parallel extension, foliation planes (or stronger layers) initially dilate to accumulate melt, but as transverse structures form, so the foliation (or melt-rich layer) contracts as melt flows into the developing transverse structures down gradients in fluid pressure. I infer that fabric-parallel leucosome stromata in melt-depleted crustal rocks are structures loosely analogous to compaction bands (Fig. 10.16). As an inter-connected network of deformation bands develops, the system will approach the percolation threshold, the permeability will change rapidly with concomitant increase in melt flow rates, and melt will drain from the system if appropriate conduits for ascent are available or conditions are appropriate for these to form. In the residue, a mix of residual and early crystallized (cumulate) phases with only minimal melt is retained as leucosome stromata in the deformation bands because shear and compactive deformation has expelled most of the melt from these sites into ductile opening-mode fractures and out of the local source volume as dikes (cf. Solar & Brown, 2001; Guernina & Sawyer, 2003).

10.5.2 *Crustal scale ascent*

A critical point is reached at some combination of melt fraction and distribution, probably contemporaneously with the developing ductile deformation

Fig 10.16. (cont.)

in shear bands and apparent interboudin partitions to form a netlike array of melt channels in the outcrop. (From Guernina & Sawyer, 2003, Large-scale melt-depletion in granulite terranes: an example from the Archean Ashuanipi Subprovince of Quebec. *Journal of Metamorphic Geology*, **21**, 181–201. © 2003 Blackwell Publishing. Reproduced by permission of Blackwell Publishing.)

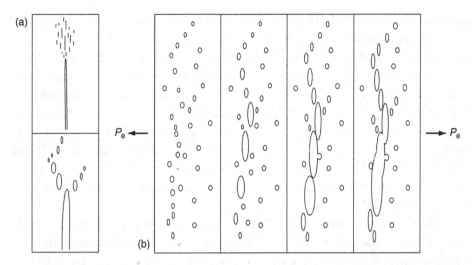

Fig. 10.17. (a) Damage distribution at fracture tips; upper figure shows frontal damage typical of brittle fracture and lower figure shows side-lobe damage typical of ductile fracture (after Eichhubl, 2004). (b) Sequence of formation of ductile fracture by growth of melt pockets and coalescence of melt pockets; fractures are inferred to grow with the long axis perpendicular to the least principal compressive stress, as modified by P_{melt} (P_e = effective pressure, or $\sigma_3 - P_{melt}$; modified after Eichhubl, 2004).

band network achieving the percolation threshold, where conduits of a size that allows melt to flow out of the local source volume are accessed or are formed, and upward buoyancy-driven transport of melt through the crust is possible. For this to occur, it is likely that the melt fraction will be in the range 0.1–0.2; such a conclusion is consistent with estimates of the melt fraction in the crust beneath the Central Andes and the Tibetan Plateau (e.g., Schilling & Partzsch, 2001), and the reduction in viscosity necessary to model channel extrusion of weak crust in the Himalayas (e.g., Beaumont et al., 2004). Exactly how crustal-scale ascent happens varies with tectonic regime, which is exemplified by the DSA case study.

In the brittle fracture model for melt extraction and magma transport, small melt-filled fractures propagate, coalesce, and drain into larger developing crustal-scale features that propagate as brittle cracks to the upper crust, draining the source and transporting a large volume of magma through a small number of individual dikes (e.g., Lister & Kerr, 1991; Rubin, 1995; Petford et al., 2000). This view is based on linear-elastic fracture mechanics in which inelastic deformation associated with breaking bonds by inter-granular and trans-granular fracture in a small process zone ahead of the fracture tip (Fig. 10.17a), with or without chemical weakening (sub-critical brittle fracture or brittle fracture, respectively), is driven by the magnified stresses at the macroscopic fracture tip (e.g., Eichhubl, 2004). However, in the deeper crust, below the brittle-to-viscous transition zone, thermally activated flow processes lead to extensive inelastic deformation and the blunting of crack tips, and process similar to those in creep failure of ceramics at high homologous temperature under low rates of loading (e.g., nucleation, growth, and coalescence of pores by predominantly diffusive deformation mechanisms and side lobe damage (Fig. 10.17a)), seem particularly relevant to failure of anatectic lower crust, leading to ductile fracture (cf. Eichhubl, 2004).

In supra-solidus lower crust, fracture propagation takes place by development and coalescence of melt-filled pores ahead of the fracture tip, with fracture opening involving extensive inelastic deformation and diffusive mass transfer (cf. Regenauer-Lieb, 1999; Eichhubl *et al.*, 2001; Eichhubl & Aydin, 2003; Eichhubl, 2004). This coupling between fracture formation and mass transfer is significant for both opening- and shearing-mode fracturing in a lower crustal environment above the solidus, where a large amount of plastic strain may accumulate before fracture. In this environment, the differential stresses are low, the effective mean stress is low, the cohesive strength is low (relative to the strength of the grains) at melt fractions greater than about 7–8 vol.%, and the rapid exhumation of the lower crust means that the loading rate is fast, which are the conditions necessary for ductile failure (Peter Eichhubl, personal communication, 2003). In isotropic protoliths, orientation of growing pockets of melt perpendicular to the direction of minimum principal compressive stress (Fig. 10.17b, effective stress is tensile due to melt pressure) is likely to be due to diffusive mass transfer (Wilkinson, 1981), and perhaps has some similarity to pore alignment and channelization produced in response to gradients in fluid composition in experiments (Wark & Watson, 2002). However, in anisotropic protoliths, ductile opening-mode fractures propagate from melt-filled transverse structures (e.g., Fig. 10.13a).

Rubin (1993) has shown that dike growth is governed primarily by the ratio of the elastic response to the viscous response of the host rock. As viscosity contrast is reduced with increasing temperature and proportion of melt, the viscous response dominates, although the resulting intrusions are dike-like. As has been shown from the DSA migmatites, there are several features that characterize the nature of the fracture mechanism in ductile crust (Brown, 2004, 2005). These features include the nature of the tips and the geometry of granite-filled fractures close to the propagating tip, which are blunter

 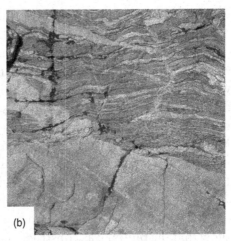

(a) (b)

Fig. 10.18. (a) View from cliff-top of sub-horizontal wave cut platform to show several dikes (tape measure for scale). Note the thinner dike crossing diagonally between the two thicker dikes, and particularly what happens at the dike tip in the center of the field of view, which is enlarged in (b) to show the blunt nature of the dike tip.

than expected for brittle fracture (Fig. 10.18) and which sometimes zigzag (Fig. 10.13a) in a manner similar to ductile fractures rather than forming the en echelon arrays typical of brittle fracture (Eichhubl & Aydin, 2003). Another important feature is the petrographic continuity with stromatic leucosomes in the migmatitic host rock, which suggests ductile fracture rather than the discontinuous boundaries indicative of brittle fracture. In ductile fracture, the tendency for oblique fracture growth due to side lobe damage controlled by the stress distribution at the fracture tip (Hirsch & Roberts, 1997) is suppressed by the far field stresses that force macroscopic fracture propagation along a characteristic zigzag path parallel to the greatest principal compressive stress. This is consistent with the dike orientation data retrieved from the DSA migmatites, which also imply that the effects of compositional layering and fabric anisotropy on macroscopic fracture orientation are weak. However, given the variation in leucosome from layer to layer I infer that spacing of conduits probably was controlled by the strong intrinsic permeability anisotropy in some layers and the weak extrinsic permeability anisotropy between layers, and consequent variable fluid focusing resulting in a larger collection zone for each conduit (Eichhubl & Boles, 2000). Draining of melt from the deformation band network and capture of small conduits by large conduits will be favored by the lower melt pressure in the larger conduits (cf. Sleep, 1988). However, it remains to be determined what changes occur with decreasing depth as melt-bearing ductile fractures propagate towards colder subsolidus rocks.

Whether there is continuous draining of melt from the anatectic zone (Sawyer, 1994) or a single melt-draining event from a collection zone, or whether there is a build-up of melt pressure before periodic activation of a draining conduit in a cyclic fashion as suggested by Handy *et al.* (2001) must be determined on a case-by-case basis (e.g., Solar & Brown, 2001; Marchildon & Brown, 2003; Glazner *et al.*, 2004). However, there is increasing evidence of common ascent conduits for different batches of melt (e.g., Brown & Solar, 1999; Solar & Brown, 2001; Sawyer & Bonnay, 2003). Such a view is supported by detailed isotope geochemistry from some crustally derived plutons where data from representative samples are interpreted to show that the pluton was constructed incrementally from multiple batches of isotopically distinct melt (e.g., the Phillips pluton in Maine, northern Appalachians; Pressley & Brown, 1999). Furthermore, such a conclusion is consistent with the occurrence of clearly discordant, but petrographically continuous, dikes arrested at the same structural level in the crust as ductile opening-mode fractures aborted during formation in the second melt loss event from the DSA migmatites (Figs. 10.6c, 10.11a and 10.13a). Thus, at a particular structural level in the lower crustal source, newly forming conduits and conduits transporting melt through that level may occur together (cf. Braun *et al.*, 1996).

In lower crust that is flowing laterally, melt migration will be sub-horizontal, using foliation and layering, and forming deformation band networks due to shear localization. Although speculative, this may have been the situation for the early melting event in the DSA. Where the lithosphere-scale SASZ intersected this melt source in the lower crust, melt was sucked into the shear zone, which probably was able to thicken as sinistral transpressive deformation changed to dextral transtension. In contrast, in the case of the second melting event in the DSA, melt extraction was associated with upward doming of the lower crustal source, and in this case diking by ductile opening-mode fracture was the preferred mode of melt ascent to the upper crust.

10.5.3 *Relationship to alternative mechanisms of ascent*

The model proposed in this chapter combines elements of pervasive migration with diking by ductile fracture (Eichhubl & Aydin, 2003). There is no a-priori reason why the dikes described from the DSA migmatites cannot be the physical expression of mobile hydrofractures frozen in ascent, or successive porosity waves passing through the anatectic zone.

10.6 Self-organization

Self-organization is a fundamental property of natural systems, including orogens, driven far from equilibrium by a constant energy flux (e.g., Hodges, 2000). The segregation, extraction, and ascent of magma in orogens have been postulated to be self-organized, resulting from non-linear interactions and feedback relations among deformation and melting (Brown & Solar, 1998a, b). Sources of non-linearity include spatial and temporal variation in melting, melt production, and melt distribution, and the combination of linear and non-linear rheologies in the melting environment. Thus, deformation during anatexis induces strain partitioning at all scales, increasing the vorticity within the weaker phase, which promotes melt redistribution and bulk softening (shear thinning), and further concentrates strain in a positive feedback relation that induces increasing non-coaxial deformation and accelerates melt displacement. As a result, the large-scale interaction between melting and deformation, and granite magma ascent and emplacement, have been proposed in several studies (e.g., Brown & Dallmeyer, 1996; Brown & Solar, 1998a, b; this chapter), although such an intimate feedback relation may not always be the case (e.g., Fitzsimons, 1996). However, the studies to date have provided only qualitative descriptions of the interactions and feedback relations, and a quantitative analysis of how these non-linear effects interact in the differentiation of heterogeneous, anisotropic, lower continental crust during orogenesis has not been undertaken.

One feature of self-organized phenomena is that these are driven systems that may achieve a critical state, which may lead to "avalanches" with a fractal (power-law) frequency–size distribution (Turcotte, 1999, 2001). Thus, if melt extraction reflects a critical state in melt-bearing lower crust, we might expect the distribution of melt batch sizes and/or melt flow channels to be fractal. Whereas the system as a whole may converge to a critical state that allows melt escape, fractal (scale–invariant) distribution of melt batches (or leucosome-filled structures) has not yet been demonstrated satisfactorily (Tanner, 1999; Marchildon & Brown, 2003; Bons *et al.*, 2005). This may reflect the residual nature of the drained source in depleted anatectic migmatite and granulite terrains. In a recent example of modeling melt segregation from a homogeneous isotropic mush subject to both compactive and shear deformation, Rabinowicz and Vigneresse (2004) demonstrate that instabilities will generate parallel melt-filled veins in the σ_1–σ_2 plane with size and spacing close to the compaction length, which is expected to be metric or sub-metric. How this modeling relates to observations from field examples of exposed lower continental crust, where the tectonically induced metamorphic fabric clearly exerts the primary control over leucosome orientation and spacing, is not addressed by these authors.

10.7 Conclusions

Melt extraction from the lower continental crust of orogens involves the migration of melt down gradients in pressure over distances of millimeters by porous flow along grain boundaries to form layer or fabric-parallel leucosome stromata. With increasing melt fraction, melt moves distances of tens of centimeters along these stromata to transverse structures (e.g., dilatant and shear bands), which form by shear localization; these structures become interconnected to form an accumulation network. Melt is extracted from these accumulation networks by flow down gradients in pressure to ductile openingmode fractures and shear zones, where ascent is buoyancy-driven. Extraction occurs at some critical combination of melt fraction and distribution, most probably as the developing accumulation network reaches the percolation threshold. The partially molten lower continental crust of orogens is driven far from thermal and mechanical equilibrium in relation to the adjacent subsolidus crust, so that melt extraction is via emergent structures that are probably self-organized, although this has not yet been demonstrated satisfactorily. Thus, the next important step in studies of melt extraction from lower continental crust is to characterize numerically the temporal geometric evolution (self-organized criticality) of residual migmatites and granulites, and typical time scales for draining melt from lower crustal sources.

10.8 Acknowledgements

I acknowledge J. L. Vigneresse, R. F. Weinberg, and E. W. Sawyer for discussion and comments, and I thank B. Reno for help with the figures.

References

Antonellini, M. A., Aydin, A. and Pollard, D. D. (1994). Microstructure of deformation bands in porous sandstones at Arches National Park, Utah. *Journal of Structural Geology*, **16**, 941–59.

Ashworth, J. R. and McLellan, E. L. (1985). Textures. In *Migmatites*, ed. J. R. Ashworth. Glasgow: Blackie & Son, pp. 180–203.

Audren, C. (1987). Evolution structurale de la Bretagne méridionale au Paléozoique. *Société Géologique et Minéralogique de Bretagne Mémoire*, **31**.

Audren, C. (1990). Evolution tectonique et métamorphique de la chaîne varisque en Bretagne méridionale. *Schweizerische Mineralogische und Petrographische Mittelunge*, **70**, 17–34.

Barbey, P., Marignac, C., Montel, J. M., Macaudière, J., Gasquet, D. and Jabbori, J. (1999). Cordierite growth textures and the conditions of genesis and emplacement of crustal granitic magmas: the Velay Granite Complex (Massif Central, France). *Journal of Petrology*, **40**, 1425–41.

Barboza, S. A. and Bergantz, G. W. (1996). Dynamic model of dehydration melting motivated by a natural analogue: applications to the Ivrea–Verbano zone, northern Italy. *Transactions of the Royal Society of Edinburgh*, **87**, 23–31.

Barboza, S. A., Bergantz, G. W. and Brown, M. (1999). Regional granulite facies metamorphism in the Ivrea zone: is the mafic complex the smoking gun or a red herring? *Geology*, **27**, 447–50.

Barraud, J., Gardien, V., Allemand, P. and Grandjean, P. (2004). Analogue models of melt-flow networks in folding migmatites. *Journal of Structural Geology*, **26**, 307–24.

Beaumont, C., Jamieson, R. A., Nguyen, M. H. and Medvedev, S. (2004). Crustal channel flows: I. Numerical models with applications to the tectonics of the Himilayan-Tibetan orogen. *Journal of Geophysical Research*, **109**, B06406, doi:10.1029/2003JB002809.

Berger, A. and Kalt, A. (1999). Structures and melt fractions as indicators of rheology in cordierite-bearing migmatites of the Bayerische Wald (Variscan Belt, Germany). *Journal of Petrology*, **40**, 1699–719.

Bernard-Griffiths, J., Peucat, J. -J., Sheppard, S. and Vidal, P. (1985). Petrogenesis of Hercynian leucogranites from the southern Armorican Massif; contribution of REE and isotopic (Sr, Nd, Pb and O) geochemical data to the study of source rock characteristics and ages. *Earth and Planetary Science Letters*, **74**, 235–50.

Bitri, A., Ballèvre, M., Brun, J. -P., Chantraine, J., Gapais, D., Guennoc, P., Gumiaux, C. and Truffert, C. (2003). Imagerie sismique de la zone de collision hercynienne dans le suds-est du massif Armorican (Projet Armor 2/Géofrance 3D). *Comptes Rendus Geosciences*, **335**, 969–79.

Bloem, E. J. M., Dalstra, H. J., Ridley, J. R. and Groves, D. I. (1997). Granitoid diapirism during protracted tectonism in an Archean granitoid–greenstone belt, Yilgarn Block, Western Australia. *Precambrian Research*, **85**, 147–71.

Bons, P. D. (1999). Apparent extensional structures due to volume loss. *Proceedings of the Estonian Academy of Sciences, Geology*, **48**, 3–14.

Bons, P. D. and van Milligen, B. P. (2001). New experiment to model self-organized critical transport and accumulation of melt and hydrocarbons from their source rocks. *Geology*, **29**, 919–22.

Bons, P. D., Dougherty-Page, J. and Elburg, M. A. (2001). Stepwise accumulation and ascent of magmas. *Journal of Metamorphic Geology*, **19**, 627–33.

Bons, P. D., Arnold, J., Elburg, M. A., Calder, J. and Soesoo, A. (2005). Melt accumulation from partially molten rocks. *Lithos*, **78**, 25–42.

Bouchez, J. -L., Guillet, P. and Chevalier, F. (1981). Structures d'écoulement liées à la mise en place du granite de Guérande (Loire-Atlantique, France). *Bulletin of the Société géologique de France*, **23**, 387–99.

Braun, M. G. and Kelemen, P. B. (2002). Dunite distribution in the Oman ophiolite: implications for melt flux through porous dunite conduits. *Geochemistry, Geophysics, Geosystems*, **3**, Article 8603.

Braun, I., Raith, M. and Kumar, G. R. R. (1996). Dehydration-melting phenomena in leptynitic gneisses and the generation of leucogranites: a case study from the Kerala Khondalite Belt, southern India. *Journal of Petrology*, **37**, 1285–305.

Brown, M. (1983). The petrogenesis of some migmatites from the Presqu'île de Rhuys, southern Brittany, France. In *Migmatites, Melting and Metamorphism*, ed. M. P. Atherton and C. D. Gribble. Nantwich: Shiva Publishing, pp. 174–200.

Brown, M. (1994a). the generation, segregation, ascent and emplacement of granite magma: the migmatite-to-crustally-derived granite connection in thickened orogens. *Earth-Science Reviews*, **36**, 83–130.

Brown, M. (1994b). Melt segregation mechanism controls on the geochemistry of crustal melts. *V. M. Goldschmidt Conference, Mineralogical Magazine*, **58A**, 124–25.

Brown, M. (2001a). Crustal melting and granite magmatism: key issues. *Physics and Chemistry of the Earth (A)*, **26**, 201–12.

Brown, M. (2001b). Orogeny, migmatites and leucogranites: a review. *Proceedings of the Indian Academy of Science*, **110**, 313–36.

Brown, M. (2002). Prograde and retrograde processes in migmatites revisited. *Journal of Metamorphic Geology*, **20**, 25–40.

Brown, M. (2004). The mechanism of melt extraction from lower continental crust of orogens. *Transactions of the Royal Society of Edinburgh: Earth Sciences*, **95**, 35–48.

Brown, M. (2005). Synergistic effects of melting and deformation: an example from the Variscan belt, western France. In *Deformation Mechanisms, Rheology and Tectonics: from Minerals to the Lithosphere*, ed. D. Gapais, J. P. Brun and P. R. Cobbold. Geological Society Special Publication **243**, pp. 205–25.

Brown, M. and Dallmeyer, R. D. (1996). Rapid Variscan exhumation and role of magma in core complex formation: southern Brittany metamorphic belt, France. *Journal of Metamorphic Geology*, **14**, 361–79.

Brown, M. and Rushmer, T. (1997). The role of deformation in the movement of granite melt: views from the laboratory and the field. In *Deformation-enhanced Fluid Transport in the Earth's Crust and Mantle*, ed. M. B. Holness. The Mineralogical Society Series: 8. London: Chapman and Hall, pp. 111–44.

Brown, M. and Solar, G. S. (1998a). Shear zone systems and melts: feedback relations and self-organization in orogenic belts. *Journal of Structural Geology*, **20**, 211–27.

Brown, M. and Solar, G. S. (1998b). Granite ascent and emplacement during contractional deformation in convergent orogens. *Journal of Structural Geology*, **20**, 1365–93.

Brown, M. and Solar, G. S. (1999). The mechanism of ascent and emplacement of granite magma during transpression: a syntectonic granite paradigm. *Tectonophysics*, **312**, 1–33.

Brown, M., Rushmer, T. and Sawyer, E. W. (1995a). Introduction to Special Section: mechanisms and consequences of melt segregation from crustal protoliths. *Journal of Geophysical Research*, **100**, 15 551–63.

Brown, M., Averkin, Y. A., McLellan, E. L. and Sawyer, E. W. (1995b). Melt segregation in migmatites. *Journal of Geophysical Research*, **100**, 15 655–79.

Brown, M. A., Brown, M., Carlson, W. D. and Denison, C. (1999). Topology of syntectonic melt flow networks in the deep crust: inferences from three-dimensional images of leucosome geometry in migmatites. *American Mineralogist*, **84**, 1793–818.

Brun, J.-P. and Burg, J.-P. (1982). Combined thrusting and wrenching in the Ibero-Armorican arc: a corner effect during continental collision. *Earth and Planetary Science Letters*, **61**, 319–32.

Burg, J.-P., Balé, P., Brun, J.-P. and Girardeau, J. (1987). Stretching lineation and transport direction in the Ibero-Armorican arc during the Siluro-Devonian collision. *Geodynamica Acta*, **1**, 71–87.

Burg, J.-P., van den Driessche, J. and Brun, J.-P. (1994). Syn- to post-thickening extension in the Variscan Belt of Western Europe: modes and structural consequences. *Géologie de la France*, **1994**, 33–51.

Burov, E., Jaupart, C. and Guillou-Frottier, L. (2003). Ascent and emplacement of buoyant magma bodies in brittle-ductile upper crust. *Journal of Geophysical Research*, **108**, B4, 2177, doi:10.1029/2002JB001904.

Cagnard, F., Gapais, D., Brun, J.-P., Gumiaux, C. and van den Driessche, J. (2004). Late pervasive crustal-scale extension in the south Armorica Hercynian belt (Vendée, France). *Journal of Structural Geology*, **26**, 435–49.

Clemens, J. D. (1998). Observations on the origins and ascent mechanisms of granitic magmas. *Journal of the Geological Society*, **155**, 843–51.

Clemens, J. D. and Mawer, C. K. (1992). Granitic magma transport by fracture propagation. *Tectonophysics*, **204**, 339–60.

Cogné, J. (1960). Schistes cristallins et granites en Bretagne méridionale: le domaine de l'Anticlinal de Cornouaille. *Mémoire du Service de la Carte Géologique de France*, 1–382.

Collins, W. J. and Sawyer, E. W. (1996). Pervasive magma transfer through the lower-middle crust during non-coaxial compressional deformation: an alternative to dyking. *Journal of Metamorphic Geology*, **14**, 565–79.

Collins, W. J. and van Kranendonk, M. J. (1998). Partial convective overturn of Archean crust in the East Pilbara Craton, Western Australia: driving mechanisms and tectonic implications. *Journal of Structural Geology*, **20**, 1405–24.

Connolly, J. A. D. and Podladchikov, Y. Y. (1998). Compaction-driven fluid flow in viscoelastic rock. *Geodynamica Acta*, **11**, 55–84.

Costa, F., Chakraborty, S. and Dohmen, R. (2003). Diffusion coupling between trace and major elements and a model for calculation of magma residence times using plagioclase. *Geochimica et Cosmochimica Acta*, **67**, 2189–200.

Cruden, R. and McCaffrey, K. J. W. (2001). Growth of plutons by floor subsidence: implications for rates of emplacement, intrusion spacing and melt-extraction mechanisms. *Physics and Chemistry of the Earth, Part A: Solid Earth and Geodesy*, **26**, 303–15.

Cuney, M. and Barbey, P. (1982). Mise en evidence de phenomanes de cristallisation fractionnée dans les migmatites. *Comptes Rendus de l'Acadaìnie des Sciences, Paris*, **295**, 37–42.

Daczko, N. R., Clarke, G. L. and Klepeis, K. A. (2001). Transformation of two-pyroxene hornblende granulite to garnet granulite involving simultaneous melting and fracturing of the lower crust, Fiordland, New Zealand. *Journal of Metamorphic Geology*, **19**, 549–62.

Dahm, T. (2000). On the shape and velocity of fluid-filled fractures in the earth. *Geophysical Journal International*, **142**, 181–92.

Dias, R. and Ribeiro, A. (1995). The Ibero-Armorican arc: a collision effect against an irregular continent? *Tectonophysics*, **246**, 113–28.

D'Lemos, R. S., Brown, M. and Strachan, R. A. (1992). Granite magma generation, ascent and emplacement within a transpressional orogen. *Journal of the Geological Society London*, **149**, 487–90.

Dohmen, R. and Chakraborty, S. (2003). Mechanism and kinetics of element and isotopic exchange mediated by a fluid phase. *American Mineralogist*, **88**, 1251–70.

Du Bernard, X., Eichhubl, P. and Aydin, A. (2002). Dilation bands: a new form of localized failure in granular media. *Geophysical Research Letters*, **29**, 24 2176, doi:1029/2002GL015966.

Ducea, M. (2001). The California arc: thick granitic batholiths, eclogitic residues, lithospheric-scale thrusting, and magmatic flare-ups. *GSA Today*, November, 4–10.

Eichhubl, P. (2004). Growth of ductile opening-mode fractures in geomaterials. In *The Initiation, Propagation, and Arrest of Joints and Other Fractures*, ed. J. W. Cosgrove and T. Engelder. London: Geological Society Special Publication 231, pp. 11–24.

Eichhubl, P. and Aydin, A. (2003). Ductile opening-mode fracture by pore growth and coalescence during combustion alteration of siliceous mudstone. *Journal of Structural Geology*, **25**, 121–34.

Eichhubl, P. and Boles, J. R. (2000). Focused fluid flow along faults in the Monterey Formation, Coastal California. *Geological Society of America Bulletin*, **112**, 1667–79.

Eichhubl, P., Aydin, A. and Lore, J. (2001). Opening-mode fracture in siliceous mudstone at high homologous temperature – effective on surface forces. *Geophysical Research Letters*, **28**, 1299–302.

Emerman, S. H. and Marrett, R. (1990). Why dikes? *Geology*, **8**, 231–3.

Faure, M. and Pons, J. (1991). Crustal thinning recorded by the shape of the Namurian-Westphalian leucogranite in the Variscan Belt of the Northwest Massif Central, France. *Geology*, **7**, 730–3.

Fitzsimons, I. C. W. (1996). Metapelitic migmatites from Brattstrand Bluffs, East Antarctica – metamorphism, melting and exhumation of the mid crust. *Journal of Petrology*, **37**, 395–414.

Gapais, D. (1989). Shear structures within deformed granites – mechanical and thermal indicators. *Geology*, **17**(12), 1144–7.

Gapais, D., Lagarde, J. -L., Le Corre, C., Audren, C., Jégouzo, P., Casas Sainz, A. and van den Driessche, J. (1993). La zone de cisaillement de Quiberon: témoin d'extension de la chaîne varisque en Bretagne méridionale au Carbonifère. *Comptes Rendus de l'Acadamie des Sciences, Paris*, **316**, 1123–9.

Gapais, D., van den Driessche, J., Brun, J. -P. and Richer, C. (1995). Pervasive ductile spreading and boudinage in extending orogens; evidence from the French Variscan Belt. *TERRA Abstracts*, **7**, 121.

Geoffroy, L. (1993). Tectonique tardi-varisque en failles normales ductiles en Vendée Littorale, Massif Armoricain. *Comptes Rendus des Seances de l'Academie des Sciences*, **317**, 1237–43.

Gillespie, P. A., Howard, C. B., Walsh, J. J. and Watterson, J. (1993). Measurement and characterisation of spatial distributions of fractures. *Tectonophysics*, **226**, 113–41.

Gillespie, P. A., Johnston, J. D., Loriga, M. A., McCaffrey, K. J. W., Walsh, J. J. and Watterson, J. (1999). Influence of layering on vein systematics in line samples. In *Fractures, Fluid Flow and Mineralization*, ed. K. J. W. McCaffrey, L. Lonergan and J. J. Wilkinson. London: Geological Society Special Publication 155, pp. 35–56.

Glazner, A. F., Bartley, J. M., Coleman, D. S., Gray, W. and Taylor, R. Z. (2004). Are plutons assembled over millions of years by amalgamation from small magma chambers. *GSA Today*, **14**, 4–11.

Grujic, D. and Mancktelow, N. S. (1998). Melt-bearing shear zones: analogue experiments and comparison with examples from southern Madagascar. *Journal of Structural Geology*, **20**, 673–80.

Guernina, S. and Sawyer, E. W. (2003). Large-scale melt-depletion in granulite terranes: an example from the Archean Ashuanipi Subprovince of Quebec. *Journal of Metamorphic Geology*, **21**, 181–201.

Hand, M. and Dirks, P. H. G. M. (1992). The influence of deformation on the formation of axial-planar leucosomes and the segregation of small melt bodies within the migmatitic Napperby Gneiss, central Australia. *Journal of Structural Geology*, **14**, 591–604.

Handy, M. R., Mulch, A., Rosenau, N. and Rosenberg, C. L. (2001). The role of fault zones and melts as agents of weakening, hardening and differentiation of the continental crust: a synthesis. In *The Nature and Tectonic Significance of Fault Zone Weakening*, ed. R. E. Holdsworth, J. Magloughlin, R. J. Knipe, R. A. Strachan and R. C. Searle. London: Geological Society Special Publication 186, pp. 305–32.

Harris, N., Vance, D. and Ayers, M. (2000). From sediment to granite: time scales of anatexis in the upper crust. *Chemical Geology*, **162**, 155–67.

Hirsch, P. B. and Roberts, S. G. (1997). Modelling plastic zones and the brittle–ductile transition. *Philosophical Transactions of the Royal Society, London*, **355**, 1991–2002.

Hirth, G. and Tullis, J. (1994). The brittle–plastic transition in experimentally deformed quartz aggregates. *Journal of Geophysical Research*, **99** (B4), 11 731–47.

Hodges, K. (2000). Tectonics of the Himalaya and southern Tibet from two perspectives. *Geological Society of America Bulletin*, **112**, 324–50.

Holness, M. B. and Clemens, J. D. (1999). Partial melting of the Appin Quartzite driven by fracture-controlled H_2O infiltration in the aureole of the Ballachulish Igneous Complex, Scottish Highlands. *Contributions to Mineralogy and Petrology*, **136**, 154–68.

Jackson, M. D., Cheadle, M. J. and Atherton, M. P. (2003). Quantitative modeling of granitic melt generation and segregation in continental crust. *Journal of Geophysical Research*, **108**, B7, 2332, doi:10.1029/2001JB001050.

Jamieson, R. A., Beaumont, C., Fullsack, P. and Lee, B. (1998). Barrovian regional metamorphism: where's the heat? In *What Drives Metamorphism and Metamorphic Reactions?*, ed. P. J. Treloar and P. J. O'Brien. London: Geological Society Special Publication 138, pp. 23–45.

Jamieson, R. A., Beaumont, C., Nguyen, M. H. and Lee, B. (2002). Interaction of metamorphism, deformation and exhumation in large convergent orogens. *Journal of Metamorphic Geology*, **20**, 9–24.

Jégouzo, P. (1980). The South Armorican Shear Zone. *Journal of Structural Geology*, **2**, 39–47.

Johnson, T. and Brown, M. (2004). Quantitative constraints on metamorphism in the Variscides of southern Brittany – a complementary pseudosection approach. *Journal of Petrology*, **45**, 1237–59.

Jones, K. A. (1988). The metamorphic petrology of the Southern Brittany Migmatite Belt, France. Ph.D. dissertation, Kingston University.

Jones, K. A. (1991). Paleozoic continental margin tectonics of southern Armorica. *Journal of the Geological Society London*, **148**, 55–64.

Jones, K. A. and Brown, M. (1989). The metamorphic evolution of the southern Brittany Migmatite Belt. In *Evolution of Metamorphic Belts*, ed. J. S. Daly,

R. A. Cliff and B. W. D. Yardley. London: Geological Society Special
  Publication 43, pp. 501–5.
Jones, K. A. and Brown, M. (1990). High-temperature 'clockwise' *P–T* paths and
  melting in the development of regional migmatites: an example from southern
  Brittany, France. *Journal of Metamorphic Geology*, **8**, 551–78.
Kriegsman, L. M. (2001). Quantitative field methods for estimating melt production
  and melt loss. *Physics and Chemistry of the Earth, Part A: Solid Earth and
  Geodesy*, **26**, 247–53.
Lagarde, J. -L., Brun, J. -P. and Gapais, D. (1990). Formation des plutons granitiques
  par injections et expansion latérale dans leur site de mise en place: une alternative
  au diapirisme en domaine épizonal. *Comptes Rendus des Seances de l'Academie
  des Sciences*, **310**, 1109–14.
Laporte, D. and Provost, A. (2000). Equilibrium geometry of a fluid phase in a
  polycrystalline aggregate with anisotropic surface energies: dry grain boundaries.
  Journal of Geophysical Research, **105**, 25 937–53.
Laporte, D. and Watson, E. B. (1995). Experimental and theoretical constraints on
  melt distribution in crustal sources – the effect of crystalline anisotropy on melt
  interconnectivity. *Chemical Geology*, **124**, 161–84.
Laporte, D., Rapaille, C. and Provost, A. (1997). Wetting angles, equilibrium melt
  geometry, and the permeability threshold of partially molten crustal protoliths.
  In *Granite: From Segregation of Melt to Emplacement Fabrics*, ed. J. L. Bouchez,
  D. H. W. Hutton and W. E. Stephens. Dordrecht: Kluwer Academic Publishers,
  pp. 31–54.
Leitch, A. M. and Weinberg, R. F. (2002). Modelling granite migration by mesoscale
  pervasive flow. *Earth and Planetary Science Letters*, **200**, 131–46.
Lister, J. R. and Kerr, R. C. (1991). Fluid-mechanical models of crack propagation
  and their application to magma transport in dykes. *Journal of Geophysical
  Research*, **96**, 10 049–77.
Loosveld, R. J. H. and Etheridge, M. A. (1990). A model for low-pressure facies
  metamorphism during crustal thickening. *Journal of Metamorphic Geology*,
  8, 257–67.
Loriga, M. A. (1999). Scaling systematics of vein size: an example from the
  Guanajuato mining district (Central Mexico). In *Fractures, Fluid Flow and
  Mineralization*, ed. K. J. W. McCaffrey, L. Lonergan and J. J. Wilkinson.
  London: Geological Society Special Publication 155, pp. 57–67.
Lupulescu, A. and Watson, E. B. (1999). Low melt fraction connectivity of granitic
  and tonalitic melts in a mafic crustal rock at 800 °C and 1 GPa. *Contributions to
  Mineralogy and Petrology*, **134**, 202–16.
Maaløe, S. (1992). Melting and diffusion processes in closed-system migmatization.
  Journal of Metamorphic Geology, **10**, 503–16.
Mancktelow, N. S. (2002). Finite-element modelling of shear zone development in
  viscoelastic materials and its implications for localisation of partial melting.
  Journal of Structural Geology, **24**, 1045–53.
Marchildon, N. and Brown, M. (2001). Melt segregation in late syn-tectonic anatectic
  migmatites: an example from the Onawa Contact Aureole, Maine, U.S.A.
  Physics and Chemistry of the Earth (A), **26**, 225–9.
Marchildon, N. and Brown, M. (2002). Grain-scale melt distribution in two contact
  aureole rocks: implication for controls on melt localization and deformation.
  Journal of Metamorphic Geology, **20**, 381–96.

Marchildon, N. and Brown, M. (2003). Spatial distribution of melt-bearing structures in anatectic rocks from southern Brittany: implications for melt-transfer at grain-to orogen-scale. *Tectonophysics*, **364**, 215–35.

Martelet, G., Calcagno, P., Gumiaux, C., Truffert, C., Bitri, A., Gapais, D. and Brun, J.-P. (2004). Integrated 3D geophysical and geological modeling of the Hercynian Suture Zone in the Champtoceaux area (south Brittany, France). *Tectonophysics*, **382**, 117–28.

Matte, P. (2001). The Variscan collage and orogeny (480–290 Ma) and the definition of the Armorica microplate: tectonic approach. *Terra Nova*, **13**, 122–8.

Miller, R. B. and Patterson, S. R. (1999). In defense of magmatic diapirs. *Journal of Structural Geology*, **21**, 1161–73.

Mollema, P. N. and Antonellini, M. A. (1996). Compaction bands: a structural analog for anti-mode I cracks in Aeolian sandstone. *Tectonophysics*, **267**, 209–28.

Molnar, P., Houseman, G. A. and Conrad, C. P. (1998). Rayleigh–Taylor instability and convective thinning of mechanically thickened lithosphere; effects of non-linear viscosity decreasing exponentially with depth and of horizontal shortening of the layer. *Geophysical Journal International*, **133**, 568–84.

Montero, P., Bea, F., Zinger, T. F., Scarrow, J. H., Molina, J. F. and Whitehouse, M. (2004). 55 million years of continuous anatexis in Central Iberia: single-zircon dating of the Peña Negra Complex. *Journal of the Geological Society London*, **161**, 255–63.

Oliver, N. H. S. and Barr, T. D. (1997). The geometry and evolution of magma pathways through migmatites of the Halls Creek Orogen, Western Australia. *Mineralogical Magazine*, **61**, 3–14.

Olsen, S. N., Marsh, B. D. and Baumgartner, L. P. (2004). Modeling mid-crustal migmatite terrains as feeder zones for granite plutons: use of dimensionless melt transport number. *Transactions of the Royal Society of Edinburgh: Earth Sciences*, **95**, 49–58.

O'Reilly, S. Y., Griffin, W. L., Poudjom Djomani, Y. H. and Morgan, P. (2001). *GSA Today*, **11**(4), 4–10.

Pelletier, J. D. (1999). Statistical self-similarity of magmatism and volcanism. *Journal of Geophysical Research, Solid Earth and Planets*, **104**, 15 425–38.

Petford, N. and Koenders, M. A. (1998). Self-organization and fracture connectivity in rapidly heated continental crust. *Journal of Structural Geology*, **20**, 1425–34.

Petford, N., Cruden, A. R., McCaffrey, K. J. W. and Vigneresse, J.-L. (2000). Granite magma formation, transport and emplacement in the Earth's crust. *Nature*, **408**, 669–73.

Peucat, J.-J. (1983). Géochronologie des roches métamorphiques (Rb–Sr et U–Pb). *Mémoire de la Société Géologique et Minéralogique de Bretagne*, **28**, 1–158.

Pin, C. and Peucat, J.-J. (1986). Ages des épisodes de métamorphisme paléozoique dans le Massif central et le Massif Armoricain. *Bulletin de Sociéte Géologique de France*, **8**, 461–9.

Podladchikov, Yu. Yu. and Connolly, J. A. D. (2001). The transition from pervasive to segregated melt flow in ductile rock. *EUG XI, Journal of Conference Abstracts*, **6**(1), 814.

Powell, R. and Downes, J. (1990). Garnet porphyroblast-bearing leucosomes in metapelites: mechanisms, phase diagrams and an example from Broken Hill. In *High Temperature Metamorphism and Crustal Anatexis*, ed. J. R. Ashworth and M. Brown. London: Unwin Hyman, pp. 105–23.

Pressley, R. A. and Brown, M. (1999). The Phillips Pluton, Maine, USA: evidence of heterogeneous crustal sources, and implications for granite ascent and emplacement mechanisms in convergent orogens. *Lithos*, **46**, 335–66.

Rabinowicz, M. and Vigneresse, J.-L. (2004). Melt segregation under compaction and shear channeling: application to granitic magma segregation in a continental crust. *Journal of Geophysical Research*, **109**, B04407, doi:10.1029/2002JB002372.

Regenauer-Lieb, K. (1999). Dilatant plasticity applied to Alpine collision: ductile void growth in the intraplate area beneath the Eifel volcanic field. *Journal of Geodynamics*, **27**, 1–21.

Renner, J., Evans, B. and Hirth, G. (2002). On the rheologically critical melt fraction. *Earth and Planetary Science Letters*, **181**, 585–94.

Richardson, C. N., Lister, J. R. and McKenzie, D. (1996). Melt conduits in a viscous porous matrix. *Journal of Geophysical Research*, **101**, 20 423–32.

Rosenberg, C. L. (2004). Shear zones and magma ascent: a model based on a review of the Tertiary magmatism in the Alps. *Tectonics*, **23**, TC3002, doi: 10.1029/2003TC001526.

Rosenberg, C. L. and Handy, M. R. (2000). Syntectonic melt pathways during simple shearing of a partially molten rock analogue (norcamphor-benzamide). *Journal of Geophysical Research*, **105**, 3135–49.

Rosenberg, C. L. and Handy, M. R. (2001). Mechanisms and orientation of melt segregation paths during pure shearing of a partially molten rock analog (norcamphor-benzamide). *Journal of Structural Geology*, **23**, 1917–32.

Rosenberg, C. L. and Handy, M. R. (2005). Experimental deformation of partially melted granite revisited: implications for the continental crust. *Journal of Metamorphic Geology*, **23**, 19–28.

Rubatto, D., Williams, I. S. and Buick, I. S. (2001). Zircon and monazite response to prograde metamorphism in the Reynolds Range, central Australia. *Contributions to Mineralogy and Petrology*, **140**, 458–68.

Rubin, A. M. (1993). Dikes *vs.* diapirs in visco-elastic rock. *Earth and Planetary Science Letters*, **117**, 653–70.

Rubin, A. M. (1995). Propagation of magma-filled cracks. *Annual Review of Earth and Planetary Sciences*, **23**, 287–336.

Rubin, A. M. (1998). Dike ascent in partially molten rock. *Journal of Geophysical Research*, **103**, 20 901–19.

Rudnick, R. L. and Gao, S. (2003). The composition of the continental crust. In *Treatise on Geochemistry*, ed. H. D. Holland and K. K. Turekian. Oxford: Elsevier, Vol. 3, pp. 1–64.

Rutter, E. H. (1986). On the nomenclature of mode of failure transitions in rocks. *Tectonophysics*, **122**, 381–7.

Rutter, E. H. (1997). The influence of deformation on the extraction of crustal melts: a consideration of the role of melt-assisted granular flow. In *Deformation-enhanced Fluid Transport in the Earth's Crust and Mantle*, ed. M. B. Holness. London: Chapman and Hall, Mineralogical Society Series 8, pp. 82–110.

Rutter, E. H. and Neumann, D. H. K. (1995). Experimental deformation of partially molten Westerly granite, with implications for the extraction of granitic magmas. *Journal of Geophysical Research*, **100**, 15 697–715.

Sawyer, E. W. (1987). The role of partial melting and fractional crystallization determining discordant migmatite leucosome compositions. *Journal of Petrology*, **28**, 445–73.

Sawyer, E. W. (1994). Melt segregation in the continental crust. *Geology*, **22**, 1019–22.

Sawyer, E. W. (1999). Criteria for the recognition of partial melting. *Physics and Chemistry of the Earth (A)*, **24**, 269–79.

Sawyer, E. W. (2001). Melt segregation in the continental crust: distribution and movement of melt in anatectic rocks. *Journal of Metamorphic Geology*, **18**, 291–309.

Sawyer, E. W. and Bonnay, M. (2003). Melt segregation and magma movement in the crust. *Geophysical Research Abstracts*, **5**, 02458.

Sawyer, E. W., Dombrowski, C. and Collins, W. J. (1999). Movement of melt during synchronous regional deformation and granulite-facies anatexis, an example from the Wuluma Hills, central Australia. In *Understanding Granites: Integrating New and Classical Techniques*, ed. A. Castro, C. Fernandez and J. -C. Vigneresse. London: Geological Society, Special Publication 168, pp. 221–37.

Schilling, F. R. and Partzsch, G. N. (2001). Quantifying partial melt fraction in the crust beneath the central Andes and the Tibetan Plateau. *Physics and Chemistry of the Earth*, **A26**, 239–46.

Schnetger, B. (1994). Partial melting during the evolution of the amphibolite- to granulite-facies gneisses of the Ivrea Zone, northern Italy. *Chemical Geology*, **113**, 71–101.

Shaw, H. R. and Chouet, B. A. (1991). Fractal hierarchies of magma transport in Hawaii and critical self-organization of tremor. *Journal of Geophysical Research*, **96**, 10 191–207.

Sibson, R. H. (1996). Structural permeability of fluid-driven fault–fracture meshes. *Journal of Structural Geology*, **18**, 1031–42.

Sibson, R. H. and Scott, J. (1998). Stress oblique/fault controls on the containment and release of over pressured fluids: examples from gold-quartz vein systems in Juneau, Alaska, Victoria, Australia and Otago, New Zealand. *Ore Geology Reviews*, **13**, 293–306.

Simakin, A. and Talbot, C. (2001a). Transfer of melt between microscopic pores and macroscopic veins in migmatites. *Physics and Chemistry of the Earth*, **A 26**, 363–7.

Simakin, A. and Talbot, C. (2001b). Tectonic pumping of pervasive granitic melts. *Tectonophysics*, **332**, 387–402.

Sleep, N. H. (1988). Tapping of melt by veins and dikes. *Journal of Geophysical Research*, **93**, 10 255–72.

Solar, G. S. and Brown, M. (2001). Petrogenesis of migmatites in Maine, USA: possible source of peraluminous leucogranite in plutons. *Journal of Petrology*, **42**, 789–823.

Solar, G. S., Pressley, R. A., Brown, M. and Tucker, R. D. (1998). Granite ascent in convergent orogenic belts: testing a model. *Geology*, **26**, 711–14.

Stockhert, B., Brix, M. R., Kleinschrodt, R., Hurford, A. J. and Wirth, R. (1999). Thermochronometry and microstructures of quartz – a comparison with experimental flow laws and predictions on the temperature of the brittle–plastic transition. *Journal of Structural Geology*, **21**, 351–69.

Strong, D. F. and Hanmer, S. K. (1981). The leucogranites of southern Brittany; origin by faulting, frictional heating, fluid flux and fractional melting. *Canadian Mineralogist*, **19**, 163–76.

Tanner, D. C. (1999). The scale-invariant nature of migmatites from the Oberpfalz, NE Bavaria and its significance for melt transport. *Tectonophysics*, **302**, 297–305.

Teyssier, C. and Whitney, D. L. (2002). Gneiss domes and orogeny. *Geology*, **30**, 1139–42.

Tirel, C., Brun, J. -P. and Burov, E. (2003). Thermo-mechanical modelling of extensional gneiss domes. *Deformation Mechanisms, Rheology and Tectonics.* St Malo, France: Géosciences Rennes, Abstract Volume 160.

Turcotte, D. L. (1999). Self-organized criticality. *Reports on Progress in Physics*, **62**, 1377–429.

Turcotte, D. L. (2001). Self-organized criticality: does it have anything to do with criticality and is it useful? *Nonlinear Processes in Geophysics*, **8**, 193–6.

van der Molen, I. (1985a). Interlayer material transport during layer-normal shortening, I, the model. *Tectonophysics*, **115**, 275–95.

van der Molen, I. (1985b). Interlayer material transport during layer-normal shortening, II, boudinage, pinch-and-swell and migmatite at Søndre Strømfjord Airport, west Greenland. *Tectonophysics*, **115**, 297–313.

van der Molen, I. and Paterson, M. S. (1979). Experimental deformation of partially melted granite. *Contributions to Mineralogy and Petrology*, **70**, 299–318.

Vernon, R. H. and Collins, W. J. (1988). Igneous microstructures in migmatites. *Geology*, **16**, 1126–9.

Vigneresse, J. -L. (1999). Intrusion level of granitic massifs along the Hercynian belt: balancing the eroded crust. *Tectonophysics*, **307**, 277–95.

Vigneresse, J. -L. (2004). Rheology of a two-phase material with applications to partially molten rocks, plastic deformation and saturated soils. In *Flow Processes in Faults and Shear Zones*, ed. G. I. Alsop, R. E. Holdsworth, K. J. W. McCaffrey and M. Hand. London: Geological Society, Special Publication 224, pp. 79–94.

Vigneresse, J. -L. and Brun, J. -P. (1983). Les leucogranites armoricains marqueurs de la deformation regionale; apport de la gravimetrie. *Bulletin de la Societe Geologique de France*, **25**, 357–66.

Vigneresse, J. -L. and Burg, J. P. (2000). Continuous *vs.* discontinuous melt segregation in migmatites: insights from a cellular automaton model. *Terra Nova*, **12**, 188–92.

Vigneresse, J. -L. and Burg, J. -P. (2002). Non-linear feedback loops in the rheology of cooling–crystallizing felsic magma and heating–melting felsic rock. In *Deformation Mechanisms, Rheology and Tectonics: Current Status and Future Perspectives*, ed. S. de Meer, M. Drury, H. de Bresser and G. Pennock. London: Geological Society Special Publication 200, pp. 275–92.

Vigneresse, J. -L., Barbey, P. and Cuney, M. (1996). Rheological transitions during partial melting and crystallization with application to felsic magma segregation and transfer. *Journal of Petrology*, **37**, 1579–600.

Walte, N. P., Bons, P. D., Passchier, C. W. and Koehn, D. (2003). Disequilibrium melt distribution during static recrystallization. *Geology*, **31**, 1009–12.

Wark, D. A. and Watson, E. B. (1998). Grain-scale permeabilities of texturally equili-brated, monomineralic rocks. *Earth and Planetary Science Letters*, **164**, 591–605.

Wark, D. A. and Watson, E. B. (2002). Grain-scale channelization of pores due to gradients in temperature or composition of intergranular fluid or melt. *Journal of Geophysical Research*, **107** (B2), doi:10.1029/2001JB000365.

Wark, D. A., Williams, C. A., Watson, E. B. and Price, J. D. (2003). Reassessment of pore shapes in microstructurally equilibrated rocks, with implications for permeability of the upper mantle. *Journal of Geophysical Research*, **108**, doi:10.1029/2001JB001575.

Watson, E. B. (2001). Aspects of fluid/rock microstructure that might affect rock rheology: an experimentalist's historical perspective. *Geological Society of America, Abstracts with programs*, **33**(6), 50.

Watt, G. R., Oliver, N. H. S. and Griffin, B. J. (2000). Evidence for reaction-induced microfracturing in granulite facies migmatites. *Geology*, **28**, 327–30.

Weinberg, R. F. (1996). The ascent mechanism of felsic magmas: news and views. *Transactions of the Royal Society of Edinburgh: Earth Sciences*, **87**, 95–103.

Weinberg, R. F. (1999). Mesoscale pervasive felsic magma migration: alternatives to dyking. *Lithos*, **46**, 393–410.

Weinberg, R. F. and Podladchikov, Y. Y. (1994). Diapiric ascent of magmas through power law crust and mantle. *Journal of Geophysical Research*, **99**, 9543–60.

Weinberg, R. F., Sial, A. N. and Mariano, G. (2004). Close spatial relationship between plutons and shear zones. *Geology*, **32**, 377–80.

White, R. W. and Powell, R. (2002). Melt loss and the preservation of granulite facies mineral assemblages. *Journal of Metamorphic Geology*, **20**, 621–32.

Wilkinson, D. S. (1981). A model for creep cracking by diffusion-controlled void growth. *Materials Science and Engineering*, **49**, 31–9.

11

The extraction of melt from crustal protoliths and the flow behavior of partially molten crustal rocks: an experimental perspective

ERNEST H. RUTTER AND JULIAN MECKLENBURGH

11.1 Introduction

Other chapters in this book deal with melting reactions and the fertility of crustal protoliths (Chapter 9), patterning of melt-depleted lower crustal rocks (Chapter 10) and the ascent and emplacement of magma to form plutons (Chapter 13). In this chapter we are concerned with physical processes in the source region – where melt-producing reactions lead to a rock permeated with viscous fluid – the ways in which the strength of this rock is affected by a developing melt fraction, the mechanisms by which the fluid collects into bodies that may ascend to higher crustal levels, and the mechanical properties of partially molten crustal rocks in those circumstances where melt is unable to escape.

The principal forces to which the partially molten system responds are: (1) gravitationally induced forces, due primarily to the density difference between the melt phase and the matrix of solid crystals; (2) tectonic forces, that produce either elastic or permanent distortion of the rock mass, and (3) surface tension forces that arise between melt and the crystalline matrix. The most important physical properties of the components of the system are: (1) the viscosity of the melt phase, which varies principally with temperature and content of dissolved water or other volatiles; (2) the rheological properties of the crystalline framework of grains (whereas the melt fraction is sufficiently small that they remain inter-connected), and (3) the permeability of the crystalline framework to the flow of the melt phase through it. The permeability may be due either to the inter-granular network of connected pores, and/or to the development of a network of variously connected cracks or veinlets, or to compositional layering developed prior to melting or as a result of the melting process itself. Thus, the melt distribution may be heterogeneous on a number of length scales. We will assume that porosity may be equated to the melt fraction. This is equivalent to assuming that the solid matrix is fully saturated with melt.

Evolution and Differentiation of the Continental Crust, ed. Michael Brown and Tracy Rushmer. Published by Cambridge University Press. © Cambridge University Press 2005.

The mechanical behavior of a porous solid saturated with fluid at low temperatures is treated by soil mechanics. Although the fluid is a viscous melt at high temperature, we will argue that useful parallels may be drawn between soil mechanics and the behavior of partially molten rocks. However, there are important differences. The solid framework of a weak rock or soil at low temperature shows little or no dependence of its strength on elapsed time or rate of deformation. However, time-dependent effects do arise from the viscous flow of water into or out of the pore spaces, with consequent effects on porosity and mechanical strength. These considerations apply also to partially molten rocks, but in addition partially molten rocks may display a range of deformation mechanisms that are strongly rate-dependent in their own right, because the solid framework of crystalline grains is at high temperature. Thus, the strength of the rock will decrease as deformation rate decreases. Whereas the compaction of a soil takes place almost exclusively by relative motion of the grains with respect to each other, thereby changing the packing density, crystalline rocks at high temperature may compact by a number of mechanisms. They may fracture at the grain scale, or individual crystals may become internally plastically deformed by dislocation motion. Distortion and compaction of the grain framework also may take place by diffusive mass transfer through the melt by dissolution from more highly stressed grain contacts and concomitant crystallization of the same or new phases in the melt-filled spaces.

Much of what we know about the flow behavior of partially molten rocks comes from high pressure/temperature deformation experiments on small samples of natural or synthetic materials. Experimental data may be applied to models of natural processes of deformation and melt extraction, in order to estimate the rates of those processes, and to obtain insights that might be tested against field observations. It is important always to remain aware of the fact that processes in partially molten rocks in nature take place on a range of length scales, from the granular to kilometric. Experimental studies only provide information about processes on the grain scale, or where heterogeneities of stress, mineralogy, or melt distribution may develop over the range of the specimen length. The link between experiment and nature must be made via mathematical modeling or, where appropriate, analog modeling.

For some physical mechanisms of flow of partially molten rocks, it is possible to derive flow and compaction/melt-extraction laws entirely from physical first principles, although one aims to test such laws against the results of experiments. A major aim of this chapter is to summarize the kind of information that has been obtained from such mechanical experiments. We will then discuss the theoretical framework for the interpretation of the results of experiments, and the ways in which these ideas may be applied to natural deformation of partially molten crustal rocks and the extraction of granitoid melts as the first step in pluton formation.

11.2 Deformation mechanisms

Deformation of partially molten rocks in which the matrix of solid grains forms a mechanically interconnected framework is expected to involve distortion of the framework by one or some combination of mechanisms (1)–(3) below (also Fig. 11.1). Whenever deformation involves a succession of physical

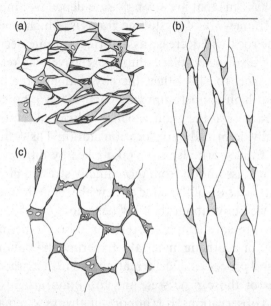

Fig. 11.1. Sketches to illustrate the main features at the grain scale of the deformation mechanisms likely to be important in partially molten rocks. In each case, the compression direction is horizontal, strain path is irrotational, grains are white and the melt phase is shaded gray. (a) Cataclastic-frictional deformation, characterized by grain fragmentation without internal distortion, and frictional sliding between fragments. There is a strong preferred orientation of cracks parallel to the maximum compressive stress direction. (b) Intracrystalline deformation of the solid phases, resulting in a shape-preferred orientation of grains, and probably also a crystallographic-preferred orientation. (c) Granular flow. The fundamental strain accumulation process is slippage of grains with respect to one another. Locking asperities may be removed to allow slippage either by diffusive transfer through the melt phase, leading to formation of overgrowths on grains or neocrystallization in the melt, or by fracture of locking points if grains have become sintered together. If the grains tend originally to tabular habits, a shape-preferred orientation of the grains would be expected. (b) and (c) are likely to be more important than (a) in nature.

processes, the slowest one is expected to control the overall rate. Rosenberg (2001) provides a comparative review of deformation mechanisms in naturally and experimentally deformed partially molten granitic rocks.

(1) Fracturing of the framework of grains, with frictional sliding of the fragments. The crystal structure of the fragments remains undistorted, but the rock mass may change shape, and may be accompanied by volumetric dilatation or compaction. Compaction may be brought about by closer packing of fragments, or concentrating small, spalled fragments into melt-filled spaces between larger grains. Alternatively, frictional sliding of whole grains may occur without fracturing. Because the resistance to deformation arises principally through resistance to fracture and frictional forces, the strength is expected to be dependent strongly on the difference between confining pressure and melt pressure (i.e., effective pressure) but only weakly sensitive to large changes in deformation rate or temperature change. Variations in effective pressure may arise through dilatancy or compaction; hence, rate dependencies on strength may arise (at least on the time scale of experiments) according to the resistance to viscous flow of the melt between grains. All experimental studies of deformation of partially molten granitic rocks at low to moderate confining pressures (<500 MPa) and high strain rates ($\varepsilon > 10^{-6}$ s^{-1}) have induced this type of deformation (e.g., Arzi, 1978; van der Molen & Paterson, 1979; Paquet & François, 1980; Paquet et al., 1981; Rutter & Neumann, 1995).

(2) Intracrystalline plastic deformation of the constituent grains of the framework. Individual crystals distort internally, leading to shear deformation of the rock mass and compaction if the melt phase may be expelled (drained conditions)

$$\dot{\varepsilon} = C\sigma^n \exp(-H/RT), \tag{11.1}$$

in which $\dot{\varepsilon}$ is the strain rate, C is an empirical constant, σ is the flow stress, n is an empirical constant that defines the strain rate sensitivity to stress, $\mathrm{d}\log\dot{\varepsilon}/\mathrm{d}\log\sigma$ (commonly about 4, so that strength is more sensitive to flow rate than in (1) above), T is temperature in K, R is the gas constant, and H is the heat of activation (activation enthalpy) for flow. H effectively defines the temperature sensitivity of the flow rate (at constant stress), according to

$$H = -R\frac{\mathrm{d}\log_e\dot{\varepsilon}}{\mathrm{d}T^{-1}}. \tag{11.2}$$

Intracrystalline plastic processes are commonly characterized by $H > 2 \times 10^5$ kJ mol^{-1}, which means that strain rate is relatively sensitive to temperature change at constant stress. Dell'Angelo et al. (1987), Dell'Angelo and Tullis (1988) and Gleason et al. (1999) have shown that, during deformation of partially molten granitic rocks at 1.5 GPa confining pressure in solid-media apparatus, the quartz grains in the matrix flow by intracrystalline plasticity.

(3) Granular flow with various accommodation mechanisms. Granular flow is comparable geometrically to the process of frictional sliding of unfractured grains

forming a porous aggregate (cf. the flow of cohesionless sand), except that the rate of the process is determined by the kinetics of removal of the asperities or sticking points at grain contacts.

Paterson (2001) considers in detail several processes for the accommodation of the intergranular displacements (accommodation processes). For the case where congruent pressure-melting may occur at highly stressed contacts between grains (e.g., water-ice), the deformation may be rate-limited by the viscous flow of new melt along intergranular contacts to lower-pressure regions. For the case of incongruent pressure-melting or transfer of "dissolved" components of the solid phase in a supposed intergranular melt phase, the rate may be controlled either by the rate of diffusion of the grain component(s) in the melt (diffusion control) or by interface kinetics (reaction control), the latter meaning the rate of detachment of solid phase components at the source (stressed grain contact region) or reattachment at the sink (reprecipitation site).

Hirth and Kohlstedt (1995b) described this process in olivine aggregates pervaded by basaltic melt. However, from transmission electron microscopy (TEM) they reported no wetting of stressed grain interfaces by melt, and concluded that the acceleration of flow (relative to melt free aggregates of olivine of the same grain size) was due to short-circuit diffusive transfer across melt filled pores to the reprecipitation site, with solid state diffusive transfer of material in the stressed interfaces to the edge of the melt filled pores.

These types of process are expected to follow a flow law of the form

$$\dot{\varepsilon} = C\sigma \exp(-H/RT)d^{-m}, \tag{11.3}$$

in which grain size, d, is introduced, raised to a power, m, which might take a value ideally of about 3. The length scale introduced by the grain size is assumed to represent the distance between sources and sinks for diffused matter. The expected linear relationship between strain rate and stress means that it should be favored over other, non-linear processes by relatively low strain rates and/or low stresses. Therefore, it should be least amenable to experimental study, because care would have to be taken to avoid operation of the other processes that are more easily activated at experimentally accessible strain rates (Fig. 11.2). The combination of high temperatures with low strain rates for the deformation of partially molten rocks in nature should favor a diffusion accommodated granular flow process, and small grain sizes should make it even more competitive relative to fracture or intracrystalline plasticity (Dell'Angelo et al., 1987).

The effective operation of diffusion accommodated granular flow depends on whether melt films may exist or be generated in stressed grain boundaries. This issue is discussed further below, but a competing process may be sintering

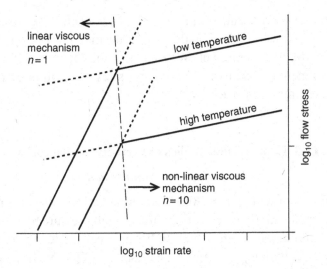

Fig. 11.2. Schematic illustration of how a high numerical value of the stress exponent in a constitutive flow law may make it difficult experimentally to access deformation mechanisms characterized by linear viscosity. Increasing stress at high strain rates preferentially accesses the more nonlinear process. Even increasing temperature may not always help if the boundary between the two processes is not very strain rate-sensitive (as here). The linear viscous process can be accessed only by substantially lowering the strain rate. Provided the linear viscous process is diffusion-controlled, it may alternatively be favored by enhancing kinetic factors such as adding water to increase diffusivity or by reducing grain size.

and neck growth at grain contacts. This process is driven by the surface energy differences between liquid and solid phases that attempt to remove small curvature interfaces between solids and liquid. Transported and redeposited solid to form "necks" may be derived from the walls of pore spaces or by diffusive transfer along melt-bearing interfaces, or both. The process reduces intergranular contact stresses, inhibiting intergranular diffusion creep and causing hardening. Granular flow may still take place, but may become dependent on the failure of the sintered contacts by fracture rather than by purely diffusive processes. Although little is known of high-temperature fracture processes, especially in the presence of a melt phase, we may anticipate that the geometric stress intensification that occurs at crack tips will require a non-linear relation between applied stress and overall strain rate. Thus, a process like this may supervene in high strain rate experiments which otherwise might be expected to show diffusive transfer creep (Fig. 11.2).

In each of the above processes, the relative movement of framework grains requires some viscous flow of the melt phase. At high-melt viscosities, high-applied stresses, and fine grain sizes, it is possible that the overall rate of deformation might be controlled by the flow of the melt phase, as pore (= melt) spaces distort, dilate, or compact. Whereas this may be important in studies at some experimental strain rates and grain sizes, it is unlikely to be so at geological rates and grain sizes.

11.3 Experimental studies of partially molten rocks

11.3.1 Experimental methods

Rather few (on the order of ten to twenty) experimental studies of the mechanical behavior of partially molten crustal rocks under high pressure have been carried out. A similar number of experimental studies has been made of partially molten ultrabasic rocks, containing basaltic melts. Despite the fact that basaltic melts are generally much less viscous than granitoid melts, these studies also have some relevance for understanding the behavior of crustal rocks during partial melting.

Experimentalists have tended to approach the behavior of fully or partially molten rocks from either end of a melt fraction spectrum. Those who are concerned with the viscosity of melt phases and the effect of adding increasing volumes of solid-phase grains must employ methods such as falling sphere or rotating cylinder viscometry. In either case, a container must be used to support the melt, especially if viscosity is low. At the other end of the spectrum are those who are concerned primarily with the effect of a small volume fraction of melt on rheological behavior. Here, axisymmetric compression of cylindrical samples or various direct shear configurations may be used. Where the melt phase is very viscous, this approach may be used even with melt fractions in excess of 40%, but owing to the tendency to approach the problem from either the 100% solid or 100% liquid ends of the spectrum, relatively few data are available for behavior in the middle of the range.

For experiments carried out on "relatively solid" samples, cylindrical test pieces are used, up to 1 cm diameter and 2 cm in length. They may be samples cored from natural rocks or fabricated by hot isostatic pressing of powders. In the latter case, a refractory solid phase is mixed with a proportion of glass of appropriate composition, which melts at the temperature of the experiment. A small percentage of water may be added to the sample, or it may be tested "dry." If hydrous phases (e.g., micas or amphiboles) are present initially in the sample, they may break down at test temperatures to yield a melt into which the water has dissolved. Samples must generally be sealed inside a ductile metal sleeve, to separate the confining pressure medium from the sample, and to prevent loss of melt (and water if added to the experiment).

Two types of testing machine have been used for deformation experiments on partially molten rocks. These methodologies are reviewed by Tullis and Tullis (1986), Paterson (1990) and Green and Borch (1990). In solid medium machines, a ductile solid is used to apply the confining pressure. A disadvantage of this approach is that the confining medium has some intrinsic shear strength, and it may not be possible to measure accurately the low strengths of the partially molten rock sample if melt fractions are large. This disadvantage has been overcome to some degree through the use of confining media (e.g., salts or glasses) that melt around the specimen under the temperature and pressure conditions of the experiment. In these types of apparatus, high confining pressures (>1 GPa) may be applied to the specimen. In gas medium machines, an inert gas such as argon is used to apply the confining pressure. Such machines have high resolution of load measurement, but are normally used only to confining pressures of about 500 MPa. Even with this approach, it is difficult to obtain reliable strength data below about 15 MPa differential stress, as the load supported by the ductile metal jacket around the sample begins to be a significant fraction of the load supported by the sample.

In most studies of partially molten rocks, the specimen has been totally sealed, so neither melt nor vapor phase may be lost from the sample, in which case the sample is said to be *undrained*. This means that deformation is constrained to occur under constant volume conditions, once any compaction of initial voids has taken place, even though melt may move from one place to another within the sample. Local variations in melt pressure may be induced during an experiment if the dominant deformation mechanism permits elastic intergranular dilatation. However, if the shear stresses involved in the flow are much less than the confining pressure, pressure fluctuations in the melt will be only small fractions of the confining pressure. During undrained deformation, therefore, there will be almost zero effective confining pressure (applied confining pressure minus pore fluid (melt) pressure) on the sample, and any strengthening effect of the total confining pressure will be near absent. At smaller melt fractions, when the shear strength of the sample may become commensurate with the confining pressure, any dilatancy that occurs may cause a substantial drop in pore pressure and a corresponding rise in effective pressure, which will cause strengthening (Renner *et al.*, 2000). In most deformation experiments on partially molten granitic systems, an equilibrium melt fraction has not been attained during the experiment. The continuous creation of melt will tend to offset dilatancy hardening effects that might otherwise arise, but this factor in the interpretation of the results of deformation experiments has not yet been subjected to systematic study.

It is possible to configure the sample so that it is effectively *drained*. In the case of rocks deformed with water as the pore fluid, this condition would be attained by connecting the sample void spaces to a pore fluid pressure system via a hollow loading piston. In this case, the solid framework of the sample is free to compact or dilate, according to the active deformation mechanism. Pore fluid flows into or out of the sample accordingly, such that pore pressure is maintained constant at a value less than the confining pressure, and any desired effective pressure condition may be imposed. A fully drained (melt pressure $= 0$) condition may be brought about in a partially molten rock by draining excess melt into a porous but strong solid "sink" at one end of the sample, ensuring that there is sufficient pore space available to accommodate more than the total amount of melt produced (Rutter & Neumann, 1995). Future studies of deformation of partially molten rocks will require drained experiments with independent control of the melt pressure, for only in this way may the kinetics of compaction of the solid framework be studied in a controlled way, with concomitant expulsion of the melt phase (Viskupic *et al.*, 2001; Renner *et al.*, 2003).

11.3.2 *Experimental studies on partially molten granitic and related rocks*

Some experimental studies on partially molten granitic rocks have used natural rocks deformed in gas-confining medium apparatus at moderate confining pressures ($\sim$300 MPa) (e.g., Arzi, 1978; van der Molen & Paterson, 1979; Paquet & François, 1980; Paquet *et al.*, 1981; Auer *et al.*, 1981; Rutter & Neumann, 1995). In all of these studies, deformation of the matrix of solid grains, at least at high strain rates, was by some combination of grain fracturing and sliding between grains or fragments, with some possible but unknown amount of strain by diffusive mass transfer creep. With the exception of a small number of the experiments of Rutter and Neumann (1995), all of these experiments were undrained. Thus, in these experiments, effective confining pressures were not sufficiently high to inhibit fracture, nor were mean pressures on the solid phases high enough to favor intracrystalline plastic flow.

Mechanical data from high-pressure (solid medium, 1500 MPa confining pressure at 900 °C) experiments have been reported for natural granitoid rocks by Dell'Angelo and Tullis (1988) and Dell'Angelo *et al.* (1987). Only in such high-pressure experiments has dislocation creep been described as the mechanism for the deformation of the matrix of solid grains, although (principally microstructural) evidence was presented for a transition to diffusion creep at low stresses, provided the melt fraction was above a certain minimum amount. Unconfined creep experiments have been reported for fine-grained, partially molten synthetic laboradorite polycrystals with up to 12 vol. % melt (Dimanov *et al.*, 1998), and Lejeune and Richet (1995) described axisymmetric creep experiments to define the rheology of viscous melts of aluminous enstatite and lithium disilicate with various crystallite volume fractions from 0 up to 60%.

The amount of melt in experimental studies on granitic aggregates has been controlled either by the addition of different amounts of water (Arzi, 1978; van der Molen & Paterson, 1979; Dell'Angelo *et al.*, 1987; Dell'Angelo & Tullis, 1988) or by temperature, according to the water released from phyllosilicate mineral breakdown reactions (Paquet & François, 1980; Paquet *et al.*, 1981; Rutter & Neumann, 1995). These effects must be borne in mind when evaluating the results of experiments involving a range of melt fractions. In contrast to previous work on granites, the melt fraction in experiments on partially molten olivine rocks could be held constant over a range of temperatures because it was produced by the addition of a measured amount of basalt glass to the olivine aggregate (e.g., Hirth & Kohlstedt, 1995a,b).

It is either difficult or impossible to maintain constant viscosity of the melt phase through a series of experiments with different melt fractions. Either the melt viscosity decreases at constant temperature with increasing water content, or at constant water content the viscosity decreases with increasing temperature (Fig. 11.3). However, the range of melt viscosities encountered as water

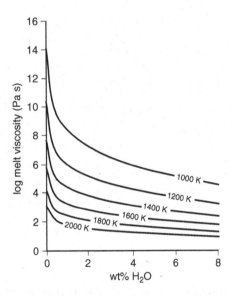

Fig. 11.3. Relationship between temperature, *T*, water content and melt viscosity, η, for haplogranitic melts according to the fit of Hess and Dingwell (1996) of the form $\log \eta = a + b/(T - c)$, where *a*, *b*, and *c* are functions of water content. Viscosity is particularly sensitive to small variations in water content at water contents less than 1 wt.%.

content is varied is typically much greater than the effect of large variations of temperature. Paradoxically, the viscosity of the melt produced in samples with no water added may increase with increasing temperature. This happens when hydrous minerals present (such as biotite) and any free water are consumed first by the fluid-present melting reactions and then by the fluid-absent incongruent melting reactions. Thus, the water content of the melt will rise initially but then decrease with progressive melting after all the hydrous phases are used up. Fluid-absent dehydration melting of a biotite gneiss due to a melting reaction involving biotite is likely to produce about 15 vol.% melt. Thus, the water content of the melt will increase up to this point and the melt will display low viscosity. Further melting with increasing temperature but no further water added causes viscosity to rise, owing to the strong effect of water on viscosity relative to temperature.

11.3.3 Strength of partially molten granitic rocks at laboratory strain rates

The experimental data of Rutter and Neumann (1995) for Westerly granite at 250 MPa total confining pressure and under undrained conditions show that, at constant water content, increasing temperature leads to a dramatic decrease in strength (Fig. 11.4) which is due to some combination of the way in which

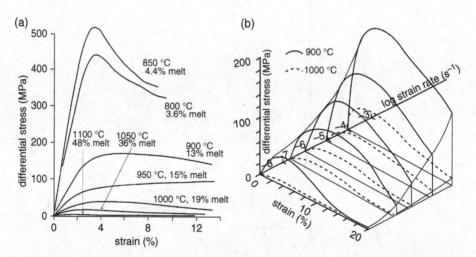

Fig. 11.4. Experimental data on the mechanical properties of partially molten Westerly granite (Rutter and Neumann, 1995) to illustrate the effects of: (a) melt fraction, temperature, and (b) strain rate on strength. Increasing melt fraction through 48% results in a very large (500-fold) decrease in strength and increase in macroscopic ductility. A 10^4-fold decrease in strain rate reduces strength 5-fold.

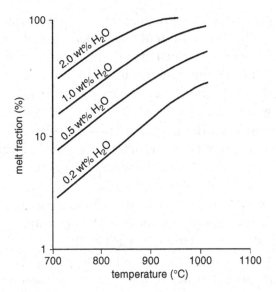

Fig. 11.5. Calculated melt fractions versus temperature for different water contents for a fertile quartzofeldspathic protolith at 500 MPa melt pressure (after Clemens & Vielzeuf, 1987).

temperature reduces the strength of the solid matrix, increases the melt fraction, and decreases the viscosity of the melt. At 850 °C, with 4.4 vol.% melt, the rock supports a differential stress of 500 MPa, but by 1100 °C, with 48 vol.% melt the strength has dropped 500-fold. These melt fractions are only about 60% of the expected equilibrium melt fractions for the temperatures concerned (Clemens & Vielzeuf, 1987; Fig. 11.5), owing to the relatively large grain size of the samples used, the high viscosity of the melt (at a maximum 0.3 wt.% water) and the sluggishness of the melting reactions. Melt fraction, rather than temperature, seems to be the most important control on strength at high strain rates. Up to about 10 vol.% melt, failure involves localization of a cataclastic fault, but at higher melt-fractions the rock flows in a macroscopically wholly ductile manner.

The strength of partially molten granite also decreases with decreasing strain rate (experiments of van der Molen & Paterson (1979) on Delegate aplite, and those of Rutter & Neumann (1995) on Westerly granite). For Westerly granite, the experimental data are described (over the experimentally accessible range) by

$$\dot{\varepsilon} = 10^{-5.3}\sigma^{2.9}\exp(-510 \times 10^3/RT). \tag{11.4}$$

This "flow law" is no more than a fit to experimental data, and will not tolerate much extrapolation outside its range. This is because of variable proportions of cataclasis and granular flow-accommodated deformation across that range, and the melt fraction and melt viscosity varied with temperature. These factors are also included in the apparent activation enthalpy for flow. Nevertheless, extrapolation is likely to lead to conservatively high values for strength at low strain rates. Hence, there is a clear implication that partially molten granite with 10–15 vol.% melt at 800–900 °C, will support a flow stress of less than 1 MPa at a strain rate of $\sim 10^{-11}$ s^{-1}, a rather fast strain rate by geological standards. In contrast, we would expect about 20 MPa to be required for the intracrystalline plastic flow of quartz at the same conditions (e.g., Luan & Paterson, 1992). Based on available experimental data, only at lower melt fractions and at lower temperatures (about 700 °C) would we expect deformation of partially molten granitic rocks to be strong enough to activate intracrystalline plastic flow of the solid phases. Otherwise, migmatitic crustal rocks should be extremely weak, probably even weaker than suggested above if we invoke diffusion-accommodated granular flow.

11.3.4 Diffusion-accommodated granular flow

Given that we might infer diffusion-accommodated granular flow to be potentially the most important flow process in partially molten granitic rocks, how may it be investigated? Of all deformation mechanisms, diffusion-accommodated flow is the only one that potentially allows the form of the constitutive flow law to be predicted from theoretical considerations, provided the geometric aspects of the rate controlling process may be adequately described. It is also possible to devise experimental approaches dedicated to the isolation of this process.

Theoretical studies

Rutter (1997) and Paterson (2001) have attempted to erect simple models for granular flow. Both are based on the earlier model of Paterson (1995) for granular flow of a water-saturated aggregate. Both model variants predict similar results at the order of magnitude level. The only difference between them is that Rutter (1997) predicts deformation rate $\propto$ where V is the molar volume of the diffusing component of the solid phase, whereas Paterson (2001) predicts deformation rate $\propto V^2 c'$, in which c' is the concentration of diffusing component in the melt. These differences are due to the fact that, in the former case, the apparent activation enthalpy for creep contains an extra term to account for the energy of formation of a diffusible point defect in the melt phase by thermal fluctuations.

The approach allows the distortional and compactive (or dilatational) components of strain rate to be described separately, and for both undrained (constant total volume except for elastic distortions that allow pore pressure variations – no gain or loss of pore fluid) and drained behavior (compaction with fluid loss allowed). The results depend on the type of stress state imposed, but since we are initially interested in testing such models against the results of axisymmetric compression tests, it is appropriate to assume principal stresses follow $\sigma_1 > \sigma_2 = \sigma_3$. Thus, the deformation may be described by the rate of axisymmetric shortening $\dot{\varepsilon}_1$ and the volumetric strain rate $\dot{\phi}$, which is a measure of the rate of flow of melt into or out of the pore spaces (neglecting any elastic volume changes in the solid and assuming no progressive melting). Paterson (2001) investigates three possible rate controlling mechanisms: (1) by viscous flow of melt; (2) by diffusive transfer of dissolved components of the solid phases away from locking points between grains, and (3) control by the kinetics of the dissolution or reprecipitation reactions. Paterson argues that diffusive transfer control is likely to be the most important under geological conditions.

For the open (drained) system, Rutter (1997) predicts

$$\dot{\varepsilon}_1 = \frac{C_1 C_2 C_3 V D \phi^r [3(\sigma_1 - \sigma_2) + (\sigma_3 - p)]}{R T d^2}, \tag{11.5}$$

and

$$\dot{\phi} = \frac{C'_1 C'_2 C'_3 V D \phi^r [(\sigma_1 - \sigma_3) + 3(\sigma_3 - p)]}{3 R T d^2}, \tag{11.6}$$

in which p is melt pressure, T is temperature in K, d is grain size, and exponent r is taken to be 2 (based on Archie's law, that relates diffusive transport rate through pore fluids to porosity). Constants C_1, C_2, and C_3 (and their corresponding primed values) describe the geometry of the grain interfaces and their packing. The primed values are given by

$$C'_1 \approx 6; C'_2 \approx \frac{1}{4}\left[\pi \Big/ \left(1 - 0.4\phi^{2/3}\right)\right]^{1/2}; C'_3 \approx \left(1 - 0.4\phi^{2/3}\right)^{-1}. \tag{11.7}$$

We will assume $C_1 C_2 C_3 \approx C'_1 C'_2 C'_3$. The term in square brackets in Eq.(11.6) is the effective mean pressure. Renner *et al.* (2003) showed experimentally that compaction rate is proportional to effective mean pressure in melt-saturated peridotite.

In the formulations of Eqs. (11.5) and (11.6), the effective thickness of an assumed melt layer existing in stressed grain boundaries does not appear explicitly because a lower effective bulk diffusivity is assumed, proportional to ϕ^r. Although depending on the melt film thickness and grain packing geometry, this approximation is likely to lead to some overestimation of deformation rate by Eqs. (11.5) and (11.6) that can be resolved only by experiments. The question of whether stressed intergranular melt films may generally exist is debated widely. In their experimental samples, Hirth and Kohlstedt (1995b) could not detect basaltic melt films between stressed olivine grains (see also Hiraga *et al.*, 2002). On the other hand, melt-assisted sintering is a widely employed technique in the fabrication of several industrial ceramics (e.g., Mortensen, 1997), in which stable, siliceous intergranular melt phases of the order of 1 nm equilibrium thickness have been detected by high resolution TEM (e.g., Clarke, 1987). Drury and Fitzgerald (1996) inferred from TEM observations the existence of nanoscale silicic melt films in polymineralic ultrabasic rocks. The equilibrium thickness of such melt films is expected to depend strongly on the mineralogy of adjacent grains and their orientation relationships when they are mineralogically identical.

The stress components $(\sigma_1 - \sigma_3)$ and $(\sigma_3 - p)$ may be understood primarily as those driving distortion of the aggregate in the absence of effective pressure, and melt flow in the absence of stress difference, respectively. Thus, in the drained condition, $(\sigma_1 - \sigma_3)$ only increases slightly the rate of melt extraction relative to $3(\sigma_3 - p)$ (i.e., differential stress has a small effect on melt flow kinetics unless $(\sigma_1 - \sigma_3) \geq 3(\sigma_3 - p)$). Given the weakness of partially molten granite, this seems unlikely. However, the real role of deformation in enhancing melt flow depends on the distance over which pressure gradients are developed (i.e., the gradients $d(\sigma_1 - \sigma_3)/dz$ and $d(\sigma_3 - p)/dz$, as discussed further below).

In order to evaluate these models, it is necessary to know how the melt fraction ϕ and the diffusion coefficient D vary with temperature and water content. These must be constrained by fits to empirical data, which may be rather approximate, especially for extrapolation outside the range of the experiments. Chekmir and Epel'baum (1991) report the following relation between viscosity (η, in Pascals per second) and diffusivity (D of SiO_2 in granitic melts, in square meters per second)

$$\log_{10} D = -0.58 \log_{10} \eta - 10.633. \tag{11.8}$$

The viscosity of dry, silica-rich melts is very high ($\sim 10^8$ Pa s^{-1} at 1100 °C, Fig. 11.3) compared to that of basalt melt (~ 1 Pa s) at the same temperature. This is because silica-rich melts are highly polymerized, with few cations to break up the polymerization. The structured nature of silica-rich melts is reflected in the small value of entropy change on melting. On the other hand, total saturation of the melt with water may reduce the viscosity by up to 6 orders of magnitude at the lower temperatures in the suprasolidus range, by breaking up silica tetrahedra, and even 1 wt.% water may reduce viscosity by 3 orders of magnitude (Persikov *et al.*, 1990).

Over the temperature and viscosity ranges relevant for viscous flow of melts in the earth, viscosity may be significantly non-Arrhenian; the apparent activation enthalpy for viscous flow may vary significantly with temperature. For haplogranitic melts, Hess and Dingwell (1996) propose an empirical description of the form

$$\log_{10} \eta = a + b/(T - c), \tag{11.9}$$

where *a*, *b*, and *c* are functions of water content (ψ wt.%) and *T* is in K. Through fitting to a wide range of experimental data they obtained $a = -3.545 + 0.833 \log_e (\psi)$, $b = 9601 - 2368 \log_e (\psi)$ and $c = 196 + 32.25 \log_e(\psi)$. The predictions of this model (± 0.46 log units of viscosity) are shown in Fig. 11.3.

From the data of Clemens and Vielzeuf (1987), melt fraction may be related to water content and temperature T_c (in Celsius) for a "fertile" protolith (Fig. 11.5):

$$\phi = 10^{-3.68} \psi^{0.9} \exp(0.00875 T_c). \tag{11.10}$$

Using Eqs. (11.8) and (11.9), the predictions of Eq. (11.5) are shown graphically in Fig. 11.6 for a grain size of 1 mm and for a range of melt fractions – although in nature the actual melt fraction for a given composition would be fixed by water content and temperature (e.g., by an expression like Eq. (11.10)). The relative sensitivities of strain rate to melt percentage, water content (affecting melt viscosity) and temperature may be seen, for a fixed flow stress of 1 MPa. Strain rates for intracrystalline plastic flow are also shown, calculated from the "average quartzite" model of Paterson and Luan (1990). Intracrystalline plasticity is expected only to become competitive with diffusion-accommodated granular flow at relatively low temperatures in the suprasolidus regime, low water contents and low melt percentages, if grain size is increased and/or stress is increased.

Experimental studies focusing on granular flow processes

To test the above theoretical flow laws against experiment it is necessary to use synthetic samples for mineralogical simplicity, to control the grain size of the solid phase and ideally to control the melt fraction independently of temperature and water content. This type of approach has been employed successfully in studies of olivine rocks with small amounts of Mid-Ocean Ridge Basalt (MORB) melt (e.g., Hirth & Kohlstedt, 1995a, b). Experiments should also be carried out either undrained (as has typically been done previously) or melt

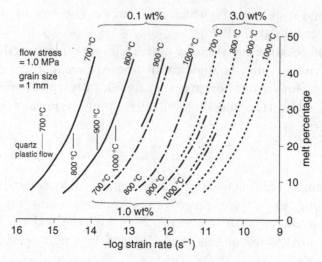

Fig. 11.6. Predictions of the model for the deformation of partially molten granite by diffusion-accommodated granular flow, showing particularly the effects of increasing temperature, water content (shown as 0.1, 1.0, and 3.0 wt.%) and melt percentage on distortional strain rate. Melt pressure assumed equal to confining pressure. All calculations made assuming a grain size of 1 mm and a constant flow stress of 1 MPa. Because the flow is linear-viscous, all curves shift one decade in strain rate to the right as stress is increased 10-fold. Assuming the rheologically critical melt percentage (RCMP) to lie at 50% melt, all curves would take a sharp deflection to the right (to much higher strain rates) above 50% melt. In all cases, over the 0–50% melt-fraction range, strength decreases by more than 100-fold. The short vertical lines at the left show strain rates expected at the same flow stress for intracrystalline plastic flow of quartz (based on the Paterson & Luan (1990) "average" flow law for quartzite). At all temperatures, the partially molten rock flows substantially faster with accommodation by diffusive mass transfer than by intracrystalline plasticity. The difference becomes more pronounced at higher temperatures and at lower stresses.

pressure should be controlled independently of confining pressure (drained). Here, we describe briefly a preliminary attempt at this approach under undrained conditions, using a peralkaline (SiO_2–Al_2O_3–Na_2O) melt phase (Mecklenburgh & Rutter, 2003).

Samples were fabricated with quartz as the only solid phase, mixed with the peralkaline glass formed by fusing alumina, silica, and sodium carbonate in appropriate proportions at atmospheric pressure. The eutectic temperature of the dry albite–quartz system is at 1076 °C at 300 MPa (Wen & Nekvasil, 1994), but deformation experiments must be performed at slightly higher temperatures than this (1100 and 1150 °C were used). This is to prevent albite crystallizing, but at the price of causing silica of the solid phase to dissolve in the melt, increasing melt fraction and its viscosity. The latter effect may be offset partially with addition of a higher sodium fraction, which also helps compensate for sodium volatilization during initial fusion. Electron microprobe analysis of $[SiO_2/(SiO_2 + Al_2O_3)]$ in the melt after deformation showed an average increase from an initial 80, to 87 wt.%, and melt fractions measured after each experiment were always greater than the proportion of glass initially added. Adding water has the effect of lowering the solidus temperature and decreasing the melt viscosity.

Relatively large quartz grain sizes (around 100 μm, compared to 10 μm in olivine + MORB experiments) must also be used with dry siliceous melts, so that viscous flow between grains was not impeded to the extent that it might become deformation-rate controlling in experiments that were aimed at studying diffusion-accommodated granular flow. Unfortunately, large grain size leads to problems of impingement grain fracture during initial pressurization and the early stages of hot pressing when the desired melt fraction is less than 45 vol.% (the porosity of a cold-pressed granular aggregate) and void spaces exist, hence confining pressure has to be incremented in a stepwise fashion during initial fusion and compaction to minimize fracturing. Large grain size unfortunately also mitigates against diffusive mass transfer-controlled flow, although for high melt-fraction samples it is the (smaller) diameter of the contact points between grains that is more important than the mean grain size.

Figures 11.7 and 11.8 show typical experimental results for dry samples. Stress–strain curves at 1100 °C (Fig. 11.7) show characteristics of steady state ductile flow, and how increase in melt fraction and decrease of strain rate both reduce strength. Effects of strain rate variations are also illustrated in Fig. 11.8. The average flow stress decreases with stress in constant strain rate experiments according to a power law, with the stress exponent approximately 3.5 (c.f., the behavior of Westerly granite, Eq. (11.4)). The predicted flow behavior according to the diffusion-accommodated granular flow theory (Eq. (11.5)) for the dry system is also illustrated in Fig. 11.8. It lies outside the range of the constant strain-rate experiments performed to date, and even beyond their extension to lower strain rates using stress-relaxation testing. Addition of water also reduced strength, and stress relaxation tests showed a faster rate of decrease of strength with strain rate when wet, even though the experiments were run at a lower temperature of 980 °C (to maintain a small temperature difference with respect to the reduced eutectic temperature for the wet melt). The apparent stress exponent is still about 2 rather than the predicted unity for diffusion-controlled flow, but the experimental strain rates are probably still too high to evaluate the predictions of the theory. Tests on samples with larger

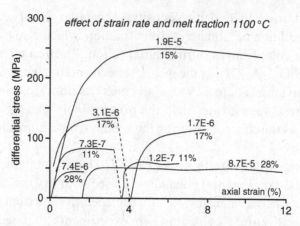

Fig. 11.7. Experimental data (undrained) at 300 MPa confining pressure and 1100 °C illustrating the effects of strain rate and melt fraction on the behavior of "synthetic" granite of 60–100 μm quartz grain size with an albite-quartz melt phase. Against each stress–strain curve is shown strain rate and melt fraction (% ±4%). The weakening effect of strain rate reduction at comparable melt fractions is apparent, and of increasing the melt fraction to 28%. Dashed lines link segments of strain rate stepping tests. (After Mecklenburgh & Rutter, 2003.)

added water fractions and smaller grain sizes will probably be required adequately to evaluate the theory.

Microstructures observed in these experimental samples are not straightforward to interpret (e.g., Fig. 11.9). There is some fracturing of grains, although it is difficult to separate fracturing during high-temperature deformation from that produced earlier, during hot pressing. Open porosity is totally absent, and only a slight preferred orientation of elongate grains was produced in samples deformed at high differential stresses (Mecklenburgh & Rutter, 2003). Grain boundaries are characterized by uneven overgrowths of new quartz and/or solution pitting, but unequivocal evidence of indentation of one grain by another is lacking. Large numbers of small, new quartz grains are produced in the melt. These are rounded or slightly faceted. Most grain interfaces appear to contain melt, although it is impossible from scanning electron microscope images to say whether the main loadbearing contacts contain melt. In these experiments, typically 10% or more strain was accumulated; hence, the microstructures appear to be consistent with most strain being accommodated by sliding of grains over one another. It is tentatively suggested that, given the non-linearity observed in the log stress versus log strain rate behavior, deformation rate in these experiments may be controlled by brittle rupture of sintered grain contacts (Mecklenburgh & Rutter, 2003).

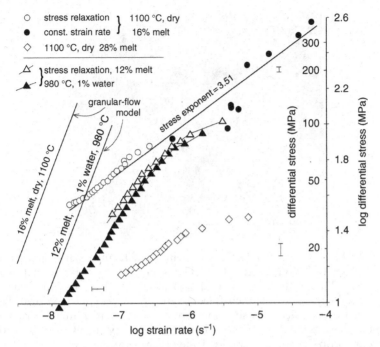

Fig. 11.8. Experimental data (undrained) at 300 MPa confining pressure illustrating the behavior of "synthetic" granite of 60–100 μm quartz grain size with an albite-quartz melt phase. "Dry" data at 1100 °C with 16 and 28% melt (constant strain rate and stress relaxation experiments) show a fairly steady rate of reduction of strength with reducing strain rate corresponding to a stress exponent of about 3.5. The strength at the higher melt fraction is much reduced. For 12% melt but with 1 wt.% water added and at a lower temperature, the more rapid rate of strength reduction corresponds to a stress exponent of about 2. These data are compared with predictions of the theoretical model for diffusion-accommodated granular flow (Rutter, 1997, stress exponent = 1) with corresponding melt fractions and water contents. (After Mecklenburgh & Rutter, 2003.)

11.4 Discussion

11.4.1 Comparability between partially molten rocks and water saturated soils

The mechanics of a water saturated granular aggregate such as a soil or a weakly cemented rock may be expected to be comparable to the mechanics of partially molten rocks. Soil deformation under undrained conditions is effectively a constant volume deformation, but under drained conditions the matrix of solid grains may compact, expelling fluid, or dilate, sucking fluid or dropping the fluid pressure. However, whereas volumetric compaction of soil takes

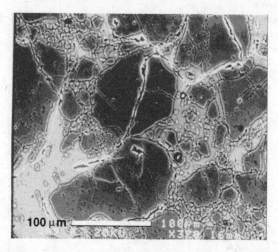

Fig. 11.9. Microstructure of an experimentally deformed synthetic quartz–quartz-albite melt aggregate (dry). The specimen was axisymmetrically shortened 12% in the vertical direction at 1100 °C at a strain rate of 8×10^{-6} s^{-1}. There is little melt-filled fracturing of the quartz grains (dark gray), zero vapor-filled porosity, and growth of new, small quartz grains in the melt phase (light gray, 28% volume). The empty (black) cracks in the quartz are probably quench cracks. Time at temperature = 50 h.

place only by geometric rearrangement of the rigid solid particles, in a partially molten rock at high temperature the individual grains of the rock matrix may deform permanently, with the possibility of total compaction and melt elimination. Compaction may take place driven by gravity, due to the density contrast (about 400 kg m^3) between melt and solids, or it may be driven by tectonic forces.

Soils behave in an elastic-plastic manner; thus, they deform only elastically and recoverably at combinations of shear and normal stress that lie below the yield surface, at which deformation occurs at whatever is the imposed rate. They also obey the law of effective stress, whereby all effective normal stress components, σ'_i (where i designates the ith component) are reduced from the applied normal stress components, σ_i, by an amount proportional to the pore pressure, p:

$$\sigma'_i = \sigma_i - \kappa p. \tag{11.11}$$

The constant of proportionality, κ, is unity when describing permanent failure, or is less than unity when describing elastic deformations. Note that shear stress components are unaffected by changes in pore fluid pressure. This law is expected to apply to partially molten rocks also, although this has not yet been demonstrated experimentally. Thus, the effective stress state may be changed

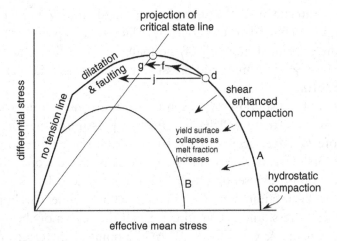

Fig. 11.10. Schematic representation of the yield surface for a porous rock. The yield surface forms a closed loop, the size of which shrinks as porosity increases (e.g., from curve A to curve B). The projection of the critical state line is the projection of the crest of successive yield loops (e.g., small circle at intersection with crest of loop A) at different porosities (Muir Wood, 1990). To the left of this line (Hvorslev region) the deformation is dilatant, and tends to result in localized faulting. To the right it is compactive (Roscoe region) and tends to result in distributed, ductile flow. Yield is possible even on the abscissa if the effective hydrostatic pressure is sufficient to initiate pore collapse and any pore fluid may be expelled. Variations in pore fluid (melt) pressure cause effective pressure changes and, hence, behavioral changes. For example, from the small circle labeled d on curve A, progressive compaction during failure may lead to a displacement of effective stress to the left (path f) until isovolumetric failure is achieved at g, or pore pressure rise (path j) at constant differential stress, perhaps from melt influx from beneath, may displace the stress state to the left until dilatant yield (faulting) is produced.

as a result of changes in pore pressure (that may be related to drainage, for example), with consequent effects on the mechanical behavior.

The yield surface for a soil or porous rock is usually represented on a diagram that plots differential stress ($\sigma_1 - \sigma_3$) against effective mean stress ($\sigma - p$), where $\sigma = (\sigma_1 + \sigma_2 + \sigma_3)/3$ (Fig. 11.10). These are the components of a *stress vector*, and for vectors below the yield surface, no permanent deformation occurs at all. At low effective mean stress, strength increases with effective mean stress (the Hvorslev region), and this region corresponds to Mohr–Coulomb failure behavior, which is accompanied by dilatancy and a tendency for deformation to occur by localized brittle shear failure. This causes fluid pressure to fall, thereby tending to increase effective stress. The stress vector moves to the right, below the yield surface, inhibiting further deformation unless more differential stress is applied or until fluid pressure is restored by drainage into the dilatant region (dilatancy hardening). The Hvorslev region is bounded to the left by a line with slope $= 3$, which corresponds to the presumed inability of the material to sustain effective tensile stresses. In this region of high fluid pressure, extensional, fluid-filled fractures are likely to form.

It is reasonable to suppose that even under hydrostatic conditions (points along the abscissa), there will be an effective pressure at which failure by isotropic pore collapse occurs. Thus, the yield surface must loop around from the Hvorslev region, through a region of negative slope (the Roscoe region), to the point representing hydrostatic compaction. The Roscoe region corresponds to shear failure but involving compaction, under combined hydrostatic and shear stress, and is sometimes called the region of shear-enhanced compaction (Curran & Carroll, 1979). At the point at the crest of the yield surface, where the slope is zero, dilatancy and compaction are balanced, and distortional flow occurs at constant volume (critical state). Granular materials require higher effective stresses to induce hydrostatic compaction if porosity is low; hence, the size of the yield surface increases as porosity decreases (Fig. 11.10). Therefore, as yield occurs in the compactive region, the porosity decreases, causing the yield surface to expand (strain hardening). Thus, a higher combination of stresses (a longer stress vector) must be applied to keep the deformation going. Such local hardening should tend to spread the failure uniformly throughout the material and, hence, to promote macroscopic ductility. Conversely, dilatant deformation (in the Hvorslev region) should tend to result in localization of the flow into a shear band.

Failure changes the properties of the deforming material by changing the porosity. Changes in pore pressure, according to whether the solid framework dilates or compacts, consequently change the effective stress state. Hence, during deformation, the tip of the stress vector is forced to migrate over the evolving yield surface. In the context of the axisymmetric loading configuration under confining pressure σ_3, as is used in experimental studies, the change in pore fluid pressure, δp, in response to a change in confining pressure or differential stress is given by Skempton's (1954) equation

$$\delta p = B\delta\sigma_3 + A\delta(\sigma_1 - \sigma_3). \tag{11.12}$$

For a fully saturated aggregate, $B = 1$, but A may be either positive or negative, according to whether loading by differential stress induces dilatation or compaction, which is, in turn, largely determined by the magnitude of the effective mean pressure (e.g., Smith, 1990). The magnitude and sign of A may evolve during a single deformation episode. Thus, the magnitude of A describes the capacity for local gradients in shear stress to induce corresponding variations in pore pressure, and for tectonic stress gradients to provide a driving force for melt migration.

For rocks and soils at low temperature, the size of the yield surface does not change with time or strain rate, but with strain as porosity evolves. However, at high temperatures, thermally activated creep will cause the effective size of the yield surface to vary with strain rate, becoming very small for slow deformations (i.e., time-dependent isotropic compaction may occur under very small effective pressures, such as <1 MPa). This implies that local variations of shear stress may be only of the same order, or less, and this is a measure of the contribution that tectonic stresses may make to the driving force for melt migration. It is interesting to compare this with the magnitude of gravity induced pressure gradients that drive melt migration. Local melt pressure differences driving flow are likely to be on the order of

$$\delta p = (\rho_s - \rho_f)gz, \tag{11.13}$$

where ρ_s is the density of the melt free solid, and ρ_f is the density of the crystallite-free melt, g is the gravitational acceleration, and z is the height of the melt-containing column of crust, such as the length of a dike. Thus, for typical crustal rock density contrasts, $dp/dz \sim 0.003$ MPa m^{-1}. The longer a vertical dike becomes, the greater will be the driving force that pushes melt into the dike base. If the flow strength of a partially molten rock is around 1 MPa, it is likely that tectonic stress differences driving melt migration will be of the same order. If a gradient of 1 MPa is developed over a distance of the order of from 1 to 10 m, this would represent a substantial (from 100 to 1000 times) enhancement relative to the gravity induced pressure gradient. But if that variation is spread over 1 km or more, then the influence of tectonic stress as a driving force for melt migration may be no more, and probably much less, than the gravity induced gradient.

11.4.2 Viscosity of the melt phase and rheologically critical melt percentage

If, in an experiment, a partially molten rock like Westerly granite could support 1 MPa differential stress at a laboratory strain rate of 10^{-5} s^{-1}, with 48 vol.% of the rock mass molten, this would correspond to a bulk viscosity on the order of 10^{11} Pa s. Thus, although such a rock would be weak by solid-state flow standards, it would remain a strong rock compared to the viscosity of the permeating liquid (at most 10^8 Pa s). Lejeune and Richet (1995) showed how, at low crystallite fractions, effective viscosity of a melt–crystal suspension increased by approximately ×10 in association with an increase in crystallite fraction to 42 vol.% (for an aluminous enstatite melt with a crystallite-free

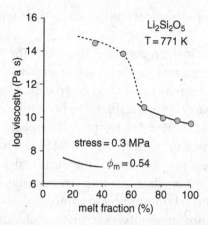

Fig. 11.11. Experimental data of Lejeune and Richet (1995) that illustrate particularly effectively the concept of RCMP, at about 55% melt, for lithium silicate. At high melt fractions, the stiffening effect of increasing crystallite fraction may be described well by the Einstein–Roscoe equation with a critical melt fraction of 54% and an exponent of 2.5. At lower melt fractions (below 54%) there is a dramatic strengthening (RCMP) corresponding to impingement of crystallites and transition to a more nonlinear bulk rheology. At even lower melt fractions, there is still an implied strong sensitivity of viscosity to melt fraction, in the same way as observed for partially molten granitic rocks.

viscosity of 4×10^9 Pa s). In this regime (Fig. 11.11), the behavior is described fairly well by the Einstein–Roscoe equation (Roscoe, 1952):

$$\eta = \eta_0 \left[1 - \frac{1 - \phi}{1 - \phi_m} \right]^{-q}, \tag{11.14}$$

in which η is viscosity, η_0 is viscosity without suspended particles, ϕ the volume fraction of melt, q is an exponent of order 2.5 and ϕ_m is the melt volume fraction for a packed aggregate of grains that may support load, for which the porosity depends on crystallite size and shape distribution. In this range, the viscosity of the suspension also remains linearly viscous (i.e., the strain rate increases linearly with applied shear stress). However, between a crystallite volume fraction $[(1 - \phi) \times 100]$ of 40–60 vol.%, the viscosity was observed to increase by 3 orders of magnitude (Fig. 11.11), deviating sharply from the predictions of the Einstein–Roscoe equation. This jump corresponds to what has been termed the rheologically critical melt percentage (RCMP), the melt fraction beyond which the contiguity of the solid framework of grains breaks down, and grains begin to be carried in the viscous melt. The linearity of stress versus strain rate also breaks down, with strain rate becoming more sensitive to small changes in stress.

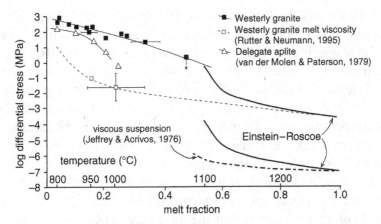

Fig. 11.12. Experimental data of Rutter and Neumann (1995) for partially molten Westerly granite and of van der Molen and Paterson (1979) for Delegate aplite showing the large decrease in strength observed with increasing melt-fraction in both cases. For Westerly granite, melt fraction was increased by increasing the temperature. In the case of Delegate aplite, it was increased by increasing water content at a constant temperature of 800 °C. Although the strength of the aplite decreases more rapidly with melt fraction, note the logarithmic strength scale, and the fact that both rocks weaken by about 500-fold over the observed melt-fraction range. Note that the lowest strength data point for the aplite was determined from reconstituted crushed rock, and that the melt viscosity varies much more over the observed melt fraction range in the case of the aplite than for the granite. Open squares show upper bounds on the viscosity of the Westerly granite melt, extrapolated to its high temperature value, when the rock would be expected to be fully molten. Bold, continuous curves show bulk viscosity calculated for melt–crystal suspensions using the Einstein–Roscoe equation, assuming in each case a critical melt fraction of 50 and 100% melt viscosities of 10^4 and 3×10^7 Pa s for Delegate aplite and Westerly granite melts, respectively. Bulk viscosity variation experimentally determined (Jeffrey & Acrivos, 1976) for a suspension with liquid viscosity of 10^4 Pa s is also shown for comparison. The Westerly granite data therefore are compatible with an RCMP of about 50% melt, or perhaps at a slightly lower melt fraction in the case of Delegate aplite.

The relation between the flow stress and melt fraction for partially molten grantoids at low melt fractions as observed by Rutter and Neumann (1995) and van der Molen and Paterson (1979), compared to the viscosity of the pure melt, is shown in Fig. 11.12. The results of van der Molen and Paterson (1979) for Delegate aplite, involving an acceleration of rate of fall in flow strength with melt fraction, even when plotted on a logarithmic strength scale, were interpreted by them in terms of the existence of an RCMP, at a melt fraction of only 25–35 vol.%. Note that the substantial strength drop (approximately × 500) with increasing melt fraction that is observed in both the above sets of experiments, is smaller than the further implied strength drop (from $\times 10^3$ to 10^4) as solid grain contiguity is lost at a melt fraction of about 45–50 vol.%. The trend at low melt fractions for Westerly granite suggests an RCMP may lie at a melt fraction not less than 48 vol.%, which would be consistent with the results of Lejeune and Richet (1995). In their experiments, van der Molen and Paterson (1979) increased water content in order to increase melt fraction at constant temperature (800 °C), whereas in the Rutter and Neumann (1995) experiments, water content remained constant and melt fraction was increased by increasing temperature. In the former case, melt viscosity was much lower than in the latter, and decreased much more strongly with increasing water content than it would with increasing temperature.

Renner *et al.* (2000) suggest that the difference between these two sets of results may be attributable to the effects of dilatancy hardening (Brace & Martin, 1968; Rutter, 1972). A combination of low permeability, high pore fluid viscosity and high strain rate during rock deformation by a mechanism that may involve dilatancy (such as cataclasis and granular flow), may lead to the pore pressure falling transiently, with concomitant strengthening of the rock (dilatancy-hardening). A lower strain rate, favoring less dilatancy, and allowing more time for the viscous pore fluid to flow in response to local pore volume fluctuation, may allow the pore pressure to remain fully effective everywhere at all times. Thus, the high strength of partially molten granite at low melt fractions (when permeability is low) may be due to a high effective confining pressure, and the rapid decrease in strength with increasing melt fraction may be due to progressively decreasing effective confining pressure, until it becomes zero when melt may flow freely around grains. Delegate aplite, with its lower melt viscosity, would be more likely to display a wide variation in strength at laboratory strain rates and moderate melt fractions than Westerly granite and, hence, to give the appearance of an RCMP at lower melt fractions than the expected values of 45–50%. Dilatancy-hardening might also be partially responsible for the commonly observed phenomenon of an initial high yield strength followed by strain weakening that is commonly observed for partially molten granitic rocks.

Renner *et al.* (2000) also suggested that the persistence of relatively high strength of Westerly granite to higher melt fractions than for Delegate aplite might be due to dilatancy-hardening in a material with a higher melt viscosity. However, the different behavior of these two granites cannot be explained in this way because, in virtually all experiments on natural granites, solid–melt equilibrium was not attained, and during the course of each test the melt fraction increased by as much as several percent, thereby counteracting the effect of dilatancy on the melt pressure. By measuring melt fraction as a function of time, Rutter and Neumann (1995) estimated that at least a year would be required to attain an equilibrium melt fraction.

The "true" RCMP (due to the physical unlocking of the matrix of solid grains) is held widely to be of great importance for the behavior of partially molten rocks. The strength of migmatitic rocks in a tectonically active regime, even at melt fractions less than the RCMP, is expected to be very low compared to the intracrystalline-plastic flow strength of solid rocks. Thus, a migmatite that is trapped between impermeable solid crustal rock units above and below has effectively zero strength, and will be extremely effective as a zone of detachment, separating tectonic processes above from those below. Snoke *et al.* (1999) describe such a large-scale, melt-lubricated shear zone (1 km thick and extending 30 km along strike) in migmatitic rocks overlying a 10-km thick mafic sill of Permian age emplaced into lower crustal rocks in the Ivrea–Verbano zone in the Alps of northern Italy.

11.4.3 Microstructural considerations

What do the structure and microstructure of naturally deformed, partially molten rocks tell us about deformation mechanisms and the permeability of those rocks? Grain-shape preferred orientations are commonly seen in granitic rocks, such as the orientation of plates of idiomorphic feldspar crystals. The *shape* of idiomorphic crystals probably implies surface energy equilibration with respect to a melt phase, and the *preferred orientation* (of platy or prismatic crystals) suggests that some degree of granular flow had occurred. Such crystals also may demonstrate that diffusion-creep has occurred when they are seen to be interpenetrated or eroded in contact with neighboring crystals (Park & Means, 1996; Nicolas & Ildefonse, 1996; Rosenberg, 2001).

Granular flow is also implied in localized, melt-pervaded, shear zones in which there is evidence of dilatation in the form of a greater melt fraction than in the host rock. Granular flow is expected to be dilatant if diffusive mass transfer does not keep pace with the imposed rate of shear, or if deformation is accommodated by friction and fracturing of grains. The driving force for such dilatation is the shear stress on the aggregate, and because this is expected to be small (at low strain rates) compared to the total mean stress on the aggregate, the dilatation may occur only if the system is very nearly undrained. That is, the melt pressure is very nearly equal to the total mean stress; hence, the effective stresses are very small.

Recent detailed studies of melt distributions and interconnectivity on the grain scale and above have begun to demonstrate the ways that the formation and drainage of melt in natural anatectites at these scales is controlled by

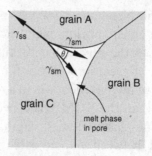

Fig. 11.13. Illustration of the control exerted by interfacial energy (γ_{sm} (solid–melt) or γ_{ss} (solid–solid)) on the shape of melt volumes (white) when at hydrostatic equilibrium. If $\theta > 60°$ melt is forced to reside at four-grain junctions or as isolated blebs along three-grain junctions. If $\theta < 60°$, continuous melt channels may exist along all three-grain edges, leading to permeability. If $\theta = 0°$, melt wets all grain interfaces. Note that for anisotropic materials, θ may vary with crystallite orientation.

lithological factors, local finite strain, or differential stress states (Brown *et al.*, 1999; Sawyer, 2001; Marchildon & Brown, 2002; Rosenberg & Handy, 2002; Holtzman *et al.*, 2003a, b). It may be unusual for such textures and micro-structures to be preserved at the grain scale in naturally partially melted rocks, which may be subject to later or penecontemporaneous microstructural readjustments.

Interconnectivity of liquid filled volume may be substantially enhanced in a localized shear zone, thereby providing an enhanced permeability pathway for melt through the granular matrix (Zimmerman *et al*, 1999; Bruhn *et al.*, 2000; Rosenberg & Handy, 2000; Rosenberg, 2001). This may be particularly impor-tant at low (<10 vol.%) melt fractions if: (1) equilibrium surface energy relations tend to develop between crystals and matrix, and (2) the equilibrium dihedral angle between melt and solids is relatively large, which tends to result in low permeability. Under hydrostatic (isotropic) stress, the microstructure of the partially molten aggregate is expected to tend towards an equilibrium in which the dihedral angle, θ, between two adjacent grains (Fig. 11.13) is determined by the relative interfacial energies between melt and solid (γ_{sm}) and between solid and solid (γ_{ss}):

$$\gamma_{sm}/\gamma_{ss} = 2\cos(\theta/2). \qquad (11.15)$$

Note that an apparent equilibrium dihedral angle of several tens of degrees does not necessarily preclude the stable existence of nanoscale melt films in a stressed interface (Clarke, 1987; Drury & Fitzgerald, 1996; Hess, 1994). For $\theta > 60°$, the melt is expected to be confined to four-grain corners, and three-grain edges will not be continuously wetted; hence, the melt phase will not be interconnected, and permeability will be zero. For smaller θ values, the melt phase will interconnect along the three-grain edges and permeability will be finite, increasing as the total melt fraction increases. This tends to be true for partially molten granitic rocks (e.g., Laporte, 1994). This rule may be relaxed somewhat by taking into account the anisotropy of surface energy that tends to characterize low-symmetry silicate structures. Waff and Faul (1992) point out that (010) faces in olivine are preferentially wetted by melt; hence, the permeability may be enhanced if intracrystalline plastic flow results in a crystallographic preferred orientation of the solid framework, or if crystals of platy habit become oriented during granular flow.

It should be noted that a crystallographic fabric may be produced by solid-state flow prior to melting, and there may not necessarily be an implication that deformation of the partially molten rock involved intracrystalline plastic flow of the solid framework. Although it is difficult to find unequivocal evidence that solid phases in a granitic rock underwent intracrystalline deformation in the presence of melt, Gapais and Barbarin (1986) and Rosenberg and Riller (2000) report careful microstructural studies that support such an inference.

Apart from studies on analog materials (Rosenberg & Handy, 2000; Rosenberg, 2001), experimental evidence from high-temperature–high-pressure studies for the focusing of melt into dilatant shear zones is presently only available for olivine aggregates pervaded with basaltic melt (Zimmerman *et al.*, 1999; Holtzman *et al.*, 2003a, b), olivine with albite melt and anorthite with basaltic melt (Holzman *et al.*, 2003a), or olivine with molten metals or metal sulphides (Bruhn *et al.*, 2000) deformed in forced shear zones. In all cases, interconnected sub-planar melt arrays formed in the sheared volume at about 10–30° to the shear direction (Fig. 11.14a, b), whereas isolated melt pockets tended to form in hydrostatically stressed samples. Holtzman *et al.* (2003b) found that the strain partitioning that accompanies melt segregation into shear bands influences the pattern of crystallographic-preferred orientation that developed as a result of the intracrystalline plastic flow of the olivine matrix. This has implications for the seismic anisotropy that might develop during flow of partially molten mantle rocks.

In axially symmetric compression experiments, Bussod and Christie (1991) and Daines and Kohlstedt (1997) noted a tendency for melt to become preferentially concentrated into grain boundaries oriented at 20–30° to the compression direction, and Jin *et al.* (1994) have observed that deformation of partially molten ultramafic rocks results in the break-up of isolated melt pockets, enhanced redistribution of melt, and spreading around grain

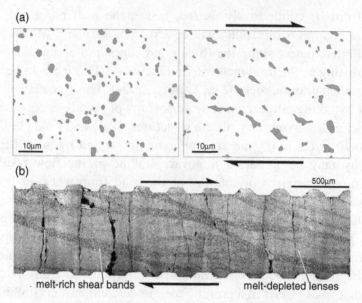

Fig. 11.14. (a) Tracings of optical micrographs comparing undeformed (left) and sheared (right) samples of hot-pressed olivine with a metal sulfide melt phase (7 vol.%, gray shaded) from experiments at 1250 °C under undrained conditions reported by Bruhn *et al.* (2000). The undeformed sample shows isolated melt pockets with large dihedral angles; hence, zero permeability. Deformation causes segregation of melt into fewer, larger, melt pockets, preferentially oriented about 20 °C to the shear direction, the existence of which implies deformation-enhanced connectivity. Reconstruction of the 3-D structure by serial sectioning showed these to be well-connected sheets of molten material extending into the third dimension. (b) SEM image of sheared fine-grained (3 μm) olivine + chromite matrix with 4% basaltic melt, showing segregation of melt (black) into dilatant shear bands oriented about 20° to the bulk shear direction, leaving lenses of melt-depleted matrix (shear strain = 3.5, flow stress 30–60 MPa). The vertical black cracks were induced during specimen recovery, and the stepping of the top and bottom of the sheared sample corresponds to steps in the forcing pistons made to minimize slip at the edges of the sample. (Photo courtesy of B. Holtzman.)

boundaries. These examples all involve dilatation of grain boundaries. Most experimental studies on the axisymmetric deformation of partially molten granitic rocks (e.g., Rutter & Neumann, 1995) at differential stresses sufficiently large to produce fracturing of grains, have noted the drainage of melt into such fractures or into similarly oriented dilated grain boundaries (Gleason *et al.*, 1999).

Both experimental studies and observations of dilatation and focusing of melt into localized shear zones (e.g., D'Lemos *et al.*, 1992; Vigneresse, 1995; Fig.11.15) demonstrate that granular flow under non-hydrostatic stress is

Fig. 11.15. Photograph of a natural migmatite showing preferential concentration of melt phase (now crystallized as fine-grained granite) into a ductile shear zone (origin of sample unknown). Diameter of lens cap is 5 cm.

important in nature, and that such localized deformation may be important in the drainage of melt and its accumulation as the first step in the formation of plutons. It is also clear that fracturing may be an important process for tapping melts. A three-dimensional network of fractures or of localized dilatant shear zones may enhance the effective permeability of a partially molten rock mass by several orders of magnitude (cf. Chapter 10).

11.4.4 Permeability of partially molten rocks

It is usual to assume that the factors governing the permeability of partially molten rocks are the same as those governing the permeability of solid rocks to water at low temperatures. In texturally equilibrated rock, flow may occur pervasively at the grain scale, along three-grain edges with the geometry of the conductive channels controlled by the wetting relations between solid and liquid phases, or individual grain boundaries may dilate under tectonic stresses. Partially molten rocks may also develop mesoscale fracture networks or other extensional failure features such as interboudin partitions that may become filled with melt. If they are interconnected, they may become very effective drainage pathways. For either case, simple permeability models may be erected, linking, on the one hand, intergranular porosity and grain size to permeability, or, on the other, crack porosity and average crack spacing and opening aperture (or other more sophisticated statistical descriptions of crack array geometry) to permeability.

A simple but widely used expression for grain scale permeability, k, is

$$k = d^2 \phi^r / C,$$ (11.16)

in which d is mean grain diameter, and C is a dimensionless number, the value of which depends on pore geometry, which may be derived theoretically or empirically. For the case of Fontainebleau sandstone, the data of Bourbie and Zinzner (1985) give $C \sim 400$ and $r = 3$. For flow through a network of planar veins, one may take as a starting-point the result for the mean flow velocity U for flow in a parallel-sided channel of width s under a pressure gradient dp/dx

$$U = \frac{-s^2}{12\eta} \frac{dp}{dx}.$$ (11.17)

For a parallel array of N channels per meter, this is analogous to Darcy's law with permeability given by

$$k = Ns^3 / 12.$$ (11.18)

The crack porosity of the network is Ns. A more complex arrangement of channels will affect the permeability by a numerical factor. For the same crack porosity, one wide crack is more effective than many narrow cracks.

It is instructive to compare grain scale permeability with that due to a crack network, for the same porosity. Assuming a grain size of 5 mm and a melt fraction, ϕ, of 10 vol.%, grain scale permeability is $6 \times 10^{-11} \, \mathrm{m}^{-2}$, whereas for a crack spacing of 500 mm, the crack permeability is $2 \times 10^{-5} \, \mathrm{m}^{-2}$. Even a low-density crack array forms a much more effective drainage network than flow through pore spaces. Once melt has been gathered into cracks and veins, the enhanced permeability of crack arrays allows even high viscosity melts to flow at significant rates.

11.4.5 *Extraction of granitic melts from their protoliths*

Brown *et al.* (1995) identify three principal contributions to the driving forces for melt extraction: (1) gravity driven compaction of the solid matrix, causing upward drainage of melt; (2) forces arising from volume changes associated with melting (expansion–contraction convection), and (3) melt segregation assisted by differential stress. Brown *et al.* (1995) are particularly concerned with the processes involved in local segregation of melt to form stromatic migmatites, in which melt lenses (represented by leucosomes) form parallel arrays within source material (mesosome), sometimes with mafic-rich selvages (melanosome) flanking the leucosome (see Chapter 10). The latter probably represent solid residue from partial melting reactions. In such rocks, the mesosome is often schistose; thus, the processes leading to the segregation of leucosomes and the interconnection of channels probably are influenced strongly by the mechanical anisotropy of the starting material, rather than dominantly by the applied stress (including melt pressure) state. At the other extreme, migmatites may be nearly isotropically textured, but transected by arrays of veinlets, shear zones, and other more or less irregular bodies in which melt has accumulated, which may represent fast drainage routes, ultimately feeding dikes that transport the magma to higher crustal levels. It will be argued that the development of closely spaced (of the order of 1 m) veinlets are an essential part of the ability of a partially molten granite protolith to drain its melt in a geologically reasonable time frame.

We will consider below the contributions of (1) and (3) above, but it should be noted that, in general terms, melting without added fluid phase causes a net volumetric expansion of the system, so that the excess melt volume must dilate the rock mass, creating spaces between grains, or forming mesoscopic crack arrays, including opening of pre-existing schistosity planes if present. Melting involving addition of fluid from outside, or from what is trapped in pre-existing pore spaces, results in net volume decrease, requiring some degree of compaction by permanent deformation of the framework of solid grains. Petrographically distinct layers or volumes will respond differently to these forces, leading to the potential for local segregations (process (2) above).

Brown *et al.* (1999) point out that compositions of granitic plutons suggests that the melting of 10–25 vol.% of the source region was required, implying that a pluton requires 4–10 times its own volume of protolith to have been tapped in the source region. By the time a source region has become exposed to inspection by geologists, it may have suffered any degree of drainage. Veinlets may be full, frozen in the act of supplying melt to a sink that has adventitiously been cut off, or most of the melt may have drained away, with concomitant collapse under gravity of most of the remaining melt-filled porosity and vein networks.

Production and accumulation of granitic melt are sequential processes. The slowest step will control the overall rate and affect the ultimate frozen appearance of the system. The rate of supply of heat to drive the endothermic melting reactions limits the maximum rate of production of the system. There is no requirement for a steady state to become established; thus, if the roof of the suprasolidus rock mass does not fail by fracture, and if lateral drainage pathways are also impermeable, melt fraction in intergranular spaces and in veinlets or other accumulations will increase until limited by available heat supply. In this totally undrained state, the rock will be so weak that tectonic stress gradients will be insignificant, but the rock mass will be able to respond in a very ductile fashion to small regional stresses, and may be a very effective detachment horizon in the lower crust. If the roof or sides of such a region should become breached, this region of protolith may then be able to supply large amounts of magma to higher levels very rapidly.

If drainage is permitted concurrently with melt production (at a rate determined by the heat flux) – and assuming: (1) porous flow from intergranular regions to collecting veinlets, followed by (2) channel flow along veinlets – the melt volume (and, hence, veinlet aperture dimensions) and pressure drops in each part of the system, will adjust spontaneously by inflation with melt or compaction of the solid framework until the volume averaged melt flux everywhere in the system is the same.

It is instructive to estimate the relative ease of gravity driven porous flow between grains compared to gravity driven flow within veinlets. Widespread use has been made of results derived theoretically by McKenzie (1984, 1985), which allow an estimate to be made of the time t_h required to reduce the melt filled porosity of a layer of thickness, h, by a factor e, and the amount of melt (equivalent to a layer of thickness h_m) extracted in time t_h. These are given by

$$t_h = \tau_0 h / \delta_c \quad \text{and} \quad h_m = h\phi(e-1)/e, \tag{11.19}$$

where δ_c is the compaction length, the height over which compaction of the solid matrix is occurring against an impermeable lower boundary (Fig. 11.16). δ_c is given by

$$\delta_c = (\mu k / \eta)^{1/2}, \tag{11.20}$$

in which μ is the effective viscosity of the solid matrix. At higher levels, the upward flux of expelled melt prevents compaction. The characteristic compaction time τ_0, is the time required to reduce the melt fraction by a factor e at the base of the compacting layer

$$\tau_0 = \delta_c / [w_0(1-\phi)], \tag{11.21}$$

in which w_0 is the relative velocity between melt and matrix, given by

$$w_0 = k(1-\phi)(p_s - p_f)g/\eta\phi. \tag{11.22}$$

These expressions allow us to compute, for example, the time required to extract, as granite melt by porous flow, 10 vol.% melt from a protolith under the influence of gravity. Figure 11.17 compares these times as a function of temperature and water content for: (1) porous flow with a grain size of 5 mm and a permeability calculated from Eq. (11.16), and (2) drainage of a network

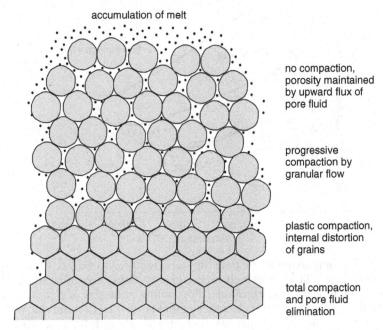

Fig. 11.16. Illustration of gravity driven extraction of melt and resultant compaction of the matrix of grains. Compaction is assumed to take place against an undeformable substrate, but everywhere the bulk stresses are hydrostatic (the lateral constraint locally equals the vertical pressure). The lowermost grains are permanently deformed either by intracrystalline plasticity or by diffusive mass transfer, to eliminate pore space. Above, there is a transition to pore space reduction by intergranular sliding. Above the compacted zone, the upward flux of melt prevents compaction, and melt eventually accumulates at the top of the section.

of veinlets that are 5 cm wide (on average) and 1 m apart. The heat source is assumed to be a mafic sill at 1200 °C lying about 500 m beneath. Melt fractions were calculated following Clemens and Vielzeuf (1987), but the water contents shown have only been assumed to affect melt viscosity, and their effect on equilibrium melt fraction beyond the point of initial melting has been neglected. It is also assumed, for simplicity, that the water content of the melt remains constant with progressive melting, whereas in reality the melt water content will decrease with progressive melting.

The effective permeability of the rock mass in case (2) is $\sim 10^6$ greater than in (1), and this accounts for the much faster rate of drainage. Except for very wet melts (>3 wt.% water), porous flow cannot lead to granitic melt extraction in geologically reasonable periods of time ($<10^7$ years, Wickham, 1987), whereas once the melt is collected into a network of channelways such as veins,

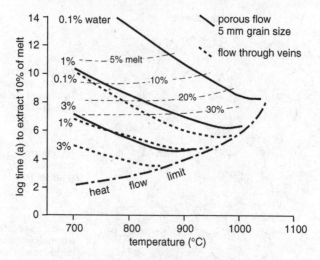

Fig. 11.17. Comparison of the time (in years) required to extract 10% of the volume of a granitic rock as melt (under gravity alone) according to: (1) the porous flow model for intergranular flow with a grain size of 5 mm (upper solid curves, labeled with melt water contents of 0.1, 1.0, and 3.0 wt.%, respectively), and (b) the porous flow model applied to a network of veinlets of volume fraction 5% (lower dashed curves for same range of melt water contents) (based on Rutter & Neumann, 1995). An upper limit to extraction rate is set by the rate that the latent heat of fusion may be supplied to the material (heat flow limit curve). Contours of melt volume fraction are also shown applying to the porous flow curves. Melt viscosities are based on the Hess and Dingwell (1996) description and melt fractions are calculated following Clemens and Vielzeuf (1987). Porous flow alone is too slow to extract melt in geologically realistic times, except for very wet melts.

interboudin partitions, or dilatant shear zones, this problem disappears. But the melt is generated at interfaces between grains on the grain scale, so porous flow is required at least initially to extract the melts into high conductivity channelways. Gravity induced vertical melt pressure gradients are too small ($\sim$0.003 MPa m^{-1}) to drive this intergranular flow of viscous melts.

It is in this initial process that tectonic deformation may play a part, provided differential stress gradients on the order of 1 MPa m^{-1} may arise. The melt in a veinlet cannot sustain differential stress (its viscosity is orders of magnitude less than that of the partially molten protolith for melt fractions less than 45 vol.%). We may imagine that the migmatitic rock mass is like a pile of grapes surrounded by interconnected "veinlets" of juice. Differential load may be transmitted to the grapes via their contiguous network of stationary and sliding contact surfaces, yet the entire mass is drained by the juice channels. Thus, the deviatoric stress gradient forcing juice from the grapes into the channels is of the order of the far-field applied stress divided by the grape diameter. In migmatite terms, this is the far-field differential stress (of the order of 1 MPa or less) divided by the channel separation (of the order of 1 m). Hence, the stronger the rock mass (i.e., at smaller melt fractions), the greater is the momentum that the tectonic stress may impart to the melt. The enhancing effect of a higher strength at lower melt fractions is partially offset by the lower permeability at lower melt fractions. Lower melt fraction also may imply lower temperature and, hence, higher melt viscosity that also helps offset the effect of the higher strength. For realistic variations of melt fraction and viscosity with temperature, increasing temperature always increases the rate of deformation-enhanced melt extraction.

Rutter and Neumann (1995) described a model for the expulsion of melt into veins in a simplified way. Melt-producing rock volumes of permeability k and mean diameter $2a$ (the separation between veinlets) are subject to a far-field differential stress σ_a. Assuming a steady state, so that the geometry remains constant over a period of time, the rate of production of melt $\dot{\phi}$ is the same everywhere and equals the rate of melt loss to the veins. Poisson's equation must then be satisfied for the pore pressure p:

$$\nabla^2 p = \phi\eta/k. \tag{11.23}$$

The solution to this equation (Rutter & Neumann, 1995) gives the radial distribution of melt pressure in the fluid filled pore spaces between the veinlets. By incorporating into the solution the requirements of mechanical equilibrium, and making the arbitrary simplifying assumption that local pore pressure is raised above the pressure in the veins by one-third the local differential stress (i.e., $A = {}^1\!/_3$ in Eq. (11.12)), the rate of stress-assisted supply of melt to the veins is given by

$$\dot{\phi} = 8k\sigma_a/3a^2\eta. \tag{11.24}$$

In this expression, $\dot{\phi}$ includes any time-dependent generation of melt according to the heat flux and the stress-induced compaction of pore spaces. The rate of loss of the melt is then limited by viscosity and permeability and driven by the stress-induced elevation of melt pressure. Thus, the flux is determined by the requirement to maintain a steady state to match the permeability and viscosity terms. In practice, the rheology of the solid matrix limits the rate at which it may collapse and expel melt. Modifying Eq. (11.24) to include, for example, a rheology determined by diffusion-accommodated grain boundary sliding, reduces somewhat the predicted melt expulsion rate, effectively by reducing the value of the empirical parameter A and making it temperature-dependent.

The two processes: (1) supply of melt to veins or other drainage channels, and (2) gravity driven drainage of those channels, are sequential processes, so the slower one controls the overall rate. In practice, the amount of inflation of the channel network and, hence, its permeability, is expected to adjust until the two rates become equal. Substituting $Ns^3/12$ (Eq. (11.18), for a simple representation of permeability due to flow through veinlets) into Eq. (11.22), the rate of melt separation under the influence of gravity may be written as

$$\dot{\phi} \approx Ns^3(\rho_s - \rho_f)g(1 - \phi)/12h\eta\phi, \qquad (11.25)$$

in which h is the height of the rock column through which the melt flows to its collection region. Equating this to the melt flux to the veinlets given by Eq. (11.24), the average width of the veinlets required to satisfy the stress assisted flux of melt from the regions between veins is

$$s = \sqrt[3]{32kh\phi\sigma_a/N(\rho_s - \rho_f)g(1 - \phi)a^2}. \qquad (11.26)$$

Figure 11.18 shows graphically how the time required to extract 10 vol.% of melt from a protolith varies with temperature for different melt water contents

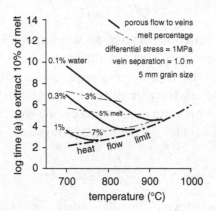

Fig. 11.18. Computed time (in years) to extract 10 vol.% of melt by shear-enhanced compaction from a hypothetical quartzofeldspathic protolith of grain diameter 5 mm, containing 0.1, 0.3, and 1.0 wt.% water. Melt is assumed to accumulate in veinlets, separated by 1 m, that inflate until they are sufficiently permeable to drain melt upwards by virtue of the density contrast between melt and solids. At short times, extraction rate is limited by heat flow. Applied differential stress is assumed to be 1 MPa. The long times required for dry melts combined with low temperatures can be reduced proportionally by increasing the differential stress and, hence, distortional strain rate. Contours are shown for melt fraction (percent). Melt viscosity is assumed to follow the Hess and Dingwell (1996) description. The combination of melt extraction by shear-enhanced compaction and gravity driven flow through veins allows even viscous melts to be extracted from their protoliths to form plutons in geologically realistic times.

at a constant differential stress of 1 MPa. The relationships between water content, temperature, melt viscosity, and melt fraction are as given by Eqs. (11.8), (11.9) and (11.10). The distortional part of the flow and the influence of local effective stress on permeability are neglected. Figure 11.19 shows graphically how veinlet thickness (Eq. (11.26)) varies for two cases: (1) where the rate of compaction retains the same value (10^{-13} s^{-1}) at all temperatures (in which case the stress decreases with temperature), and (2) where the applied stress retains the same value of 1 MPa at all temperatures. The range of vein thicknesses and separations predicted are reasonable when compared with the structure of migmatite complexes seen in the field (Chapter 10). The predicted time scales are also reasonable when compared to the expected duration of synorogenic magmatic cycles (<10 Ma). A minimum time limit is imposed on the melt extraction process because it cannot run faster than the

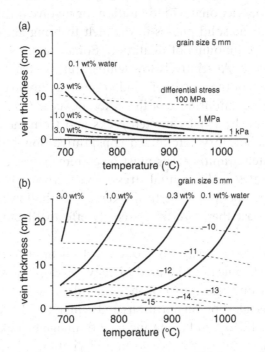

Fig. 11.19. Graphs showing the expected average thickness of veins in a vein network of average spacing 1 m in a protolith of mean grain size 5 mm for the water contents indicated, such that the rate of density driven flow through the veinlets equals the rate of supply of melt to the veins by shear-enhanced compaction. (a) is for a constant compaction rate of 10^{-13} s^{-1}, which corresponds to 10% volume loss from the protolith in 30 000 years. Contours of constant differential stress (dashed lines) are shown. (b) is for a constant applied differential stress of 1 MPa. Contours of constant log compactive strain rate (dashed lines) are shown.

rate at which heat is supplied for the melting reactions. It is also assumed that the dP/dT for the melting reaction will not lead to freezing on depressurization. Without the suggested two-stage melt separation process involving local non-hydrostatic stress gradients to push melt into veinlets, followed by gravitational separation in the veinlets, gravity driven porous flow as a separation mechanism would be slower by approximately the ratio of intergranular permeability to vein network permeability (i.e., about 10^5). This would require longer times than geologically available to separate melts of viscosities greater than about 10^6 Pa s.

11.5 Conclusions

The flow of partially molten rocks that gives rise to a granitic melt may involve granular flow, wherein solid crystals slide over one another. The accommodation process may be frictional with dilatation, or involve diffusive redistribution of material of the solid phases in the melt to bring about unlocking of asperities with lesser amounts of dilatation. Some fracturing of solid grains may also be produced. At relatively low temperatures, where the rock may be strong enough, flow of the matrix of solid grains may involve intracrystalline plasticity. The limited importance of the latter, deduced from the appearance of naturally deformed partially molten rocks and from modeling based on extrapolation of laboratory-determined constitutive flow laws, also tells us that partially molten granitic rocks must be extremely weak, able to flow at geological strain rates at differential stresses on the order of 1 MPa or less, according to the temperature. Vertical intervals in the lower crust that are partially molten must therefore be extremely effective tectonic detachment horizons.

The structure of natural partially molten rocks demonstrates that they are capable of being tectonically fractured or may develop cracks as a result of volume expansion on melting. Also, they may develop more "ductile" melt filled voids, such as interboudin partitions and vein networks, and shear zones. The latter may be effective at facilitating melt drainage by linking extensional fractures if granular flow involves substantial dilatancy. Experiments show that an increase in melt percentage causes dramatic weakening. Once the melt fraction in a shear zone becomes higher than the surroundings it would be expected to continue as a localized zone of weakness.

The density difference between melt and its protolith leads to a gravitational potential for the separation of melt by porous flow between grains. However, for viscous fluids such as granitic melts, homogeneous intergranular porous flow is too slow to cause segregation of granitic liquids over geologically reasonable periods of time. Provided *local* differential stress gradients may develop between veinlets and adjacent volumes of partially molten protolith, pore fluid pressure gradients may be induced that are hundreds to thousands of times greater than the gravitational potential gradient, and able to drive melt transport from intergranular spaces to the much more permeable vein network. Gravity driven drainage of the vein network then may separate the melt in a geologically reasonable time scale.

11.6 Acknowledgements

This work was supported by UK NERC grants GR9/2640 and GR3/13062. Experimental work was aided by experimental officer Robert Holloway. Ben Holtzman is thanked for providing Fig. 11.14b. Helpful and constructive reviews by Gayle Gleason and an anonymous reviewer led to considerable improvements to this chapter.

References

Arzi, A. A. (1978). Critical phenomena in the rheology of partially melted rocks. *Tectonophysics*, **44**, 173–84.

Auer, F., Berkhemer, H. and Oehlschlegel, G. (1981). Steady-state creep of fine grain granite at partial melting. *Journal of Geophysics*, **49**, 89–92.

Bourbie, T. and Zinzner, B. (1985). Hydraulic and acoustic properties as a function of porosity in Fontainebleau sandstone. *Journal of Geophysical Research*, **90**, 11 524–32.

Brace, W. F. and Martin, R. J. (1968). A test of the law of effective stress for crystalline rocks of low porosity. *International Journal of Rock Mechanics and Mining Sciences*, **5**, 415–26.

Brown, M., Averkin, Y. A., McLellan, E. L. and Sawyer, E. W. (1995). Melt segregation in migmatites. *Journal of Geophysical Research*, **100**, 15 665–79.

Brown, M. A., Brown, M., Carlson, W. D. and Denison, C. (1999). Topology of melt-flow networks in the deep crust: inferences from three-dimensional images of leucosome geometry in migmatites. *American Mineralogist*, **84**, 1793–818.

Bruhn, D., Groebner, N. and Kohlstedt, D. L. (2000). An interconnected network of core-forming melts produced by shear deformation. *Nature*, **403**, 883–6.

Bussod, G. Y. and Christie, J. M. (1991). Textural development and melt topology in spinel lherzolite experimentally deformed at hypersolidus conditions. *Journal of Petrology*, **32**, 17–39.

Chekmir, A. S. and Epel'baum, M. B. (1991). Diffusion in magmatic melts: new study. In *Physical Chemistry of Magmas*, ed. L. L. Perchuk and I. Kushiro. *Advances in Physical Geochemistry*, 9. New York: Springer-Verlag, pp. 99–119.

Clarke, D. R. (1987). On the equilibrium thickness of intergranular glass phases in ceramic materials. *Journal of the American Ceramic Society*, **70**, 15–22.

Clemens, J. D. and Vielzeuf, D. (1987). Constraints on melting and magma production in the crust. *Earth and Planetary Science Letters*, **86**, 287–306.

Curran, J. H. and Carroll, M. M. (1979). Shear stress enhancement of void compaction. *Journal of Geophysical Research*, **84**, 1105–12.

Daines, M. and Kohlstedt, D. L. (1997). Influence of deformation on melt topology in peridotites. *Journal of Geophysical Research*, **107**, 10 257–71.

Dell'Angelo, L. N. and Tullis, J. (1988). Experimental deformation of partially melted granitic aggregates. *Journal of Metamorphic Geology*, **6**, 495–516.

Dell'Angelo, L. N., Tullis, J. and Yund, R. A. (1987). Transition from dislocation creep to melt-enhanced diffusion creep in fine-grained granitic aggregates. *Tectonophysics*, **139**, 325–32.

Dimanov. A., Dresen, G. and Wirth, R. (1998). High-temperature creep of partially molten plagioclase aggregates. *Journal of Geophysical Research*, **103**, 9651–64.

D'Lemos, R. S., Brown, M. and Strachan, R. A. (1992). The relationship between granite and shear zones: magma generation, ascent and emplacement in a transpressional orogen. *Journal of the Geological Society London*, **149**, 487–90.

Drury, M. R. and Fitzgerald, J. D. (1996). Grain boundary melt films in experimentally deformed olivine-orthopyroxene rock; implications for melt distributions in upper mantle rocks. *Geophysical Research Letters*, **23**, 701–4.

Gapais, D. and Barbarin, B. (1986). Quartz fabric transition in a cooling syntectonic granite (Hermitage, France). *Tectonophysics*, **25**, 357–70.

Gleason, G. C., Bruce, V. and Green, H. W. (1999). Experimental investigation of melt topology in partially molten quartzo-feldspathic aggregates under hydrostatic and non-hydrostatic stress. *Journal of Metamorphic Geology*, **17**, 705–22.

Green, H. W. and Borch, R. S. (1990). High pressure and temperature deformation experiments in a liquid confining medium. In *The Brittle–Ductile Transition in Rocks*, ed. A. G. Duba, W. B. Durham, J. W. Handin and H. F. Wang. Geophysical Monograph 56, American Geophysical Union, pp. 195–200.

Hess, K.-U. and Dingwell, D. B. (1996). Viscosities of hydrous leucogranite melts: a non Arrhenian model. *American Mineralogist*, **81**, 1297–300.

Hess, P. C. (1994). Thermodynamics of thin films. *Journal of Geophysical Research*, **99**, 7219–29.

Hiraga, T., Anderson, I. M., Zimmerman, M. E., Mei, S. and Kohlstedt, D. L. (2002). Structure and chemistry of grain boundaries in deformed olivine + basalt and partially molten lherzolite aggregates: evidence of melt-free grain boundaries. *Contributions to Mineralogy and Petrology*, **144**, 163–75.

Hirth, G. and Kohlstedt, D. L. (1995a). Experimental constraints on the dynamics of the partially molten upper mantle: deformation in the diffusion creep regime. *Journal of Geophysical Research*, **100**, 1981–2001.

Hirth, G. and Kohlstedt, D. L. (1995b). Experimental constraints on the dynamics of the partially molten upper mantle: deformation in the dislocation creep regime. *Journal of Geophysical Research*, **100**, 15 441–9.

Holtzman, B. K., Groebner, N. J., Zimmerman, M. E., Ginsberg, S. B. and Kohlstedt, D. L. (2003a). Stress-driven melt segregation in partially molten rocks. *Geochemistry, Geophysics, Geosystems*, **4**, 8607, doi:10.1029/2001GC000258.

Holtzman, B. K., Kohlstedt, D. L., Zimmerman, M. E., Heidelbach, F., Hiraga, T. and Hustoft, J. (2003b). Melt segregation and strain partitioning: implications for seismic anisotropy and mantle flow. *Science*, **301**, 1227–30.

Jeffrey, D. J. and Acrivos, A. (1976). The rheological properties of suspensions of rigid particles. *Journal of the American Institute of Chemical Engineers*, **22**, 417–32.

Jin, Z.-M., Green, H. W. and Zhou, Y. (1994). Melt topology during dynamic partial melting of mantle peridotite. *Nature*, **372**, 164–7.

Laporte, D. (1994). Wetting behavior of partial melts during crustal anatexis: the distribution of hydrous silicic melts in polycrystalline aggregates of quartz. *Contributions to Mineralogy and Petrology*, **116**, 486–99.

Lejeune, A.-M. and Richet, P. (1995). Rheology of crystal-bearing silicate melts: an experimental study at high viscosities. *Journal of Geophysical Research*, **100**, 4215–29.

Luan, F.-C. and Paterson, M. S. (1992). Preparation and deformation of synthetic aggregates of quartz. *Journal of Geophysical Research*, 97, 301–20.

Marchildon, N. and Brown, M. (2002). Grain-scale melt distributions in two contact aureole rocks: implication for controls on melt localization and deformation. *Journal of Metamorphic Geology*, **20**, 381–96.

McKenzie, D. (1984). The generation and compaction of partially molten rock. *Journal of Petrology*, **25**, 713–65.

McKenzie, D. (1985). The extraction of magma from the crust and mantle. *Earth and Planetary Science Letters*, **74**, 81–91.

Mecklenburgh, J. and Rutter, E. H. (2003). On the rheology of partially molten synthetic granite. *Journal of Structural Geology*, **25**, 1575–85.

Mortensen, A. (1997). Kinetics of densification by solution-reprecipitation. *Acta Materialia*, **2**, 749–58.

Muir Wood, D. (1990). *Soil Behaviour and Critical State Soil Mechanics*. New York: Cambridge University Press.

Nicolas, A. and Ildefonse, B. (1996). Flow mechanism and viscosity in basaltic magma chambers. *Geophysical Research Letters*, **23**, 2013–16.

Paquet, J. and François, P. (1980). Experimental deformation of partially melted granitic rocks at 600° to 900°C and 250 MPa confining pressure. *Tectonophysics*, **68**, 131–46.

Paquet, J., François, P. and Nedelec, A. (1981). Effect of partial melting on rock deformation: experimental and natural evidences for rocks of granitic compositions. *Tectonophysics*, **78**, 545–65.

Park, Y. and Means, W. D. (1996). Crystal rotation and growth during grain flow in a deforming crystal mush. In *Evolution of Geological Structures in Micro- to Macro-Scales*, ed. S. Sengupta. London: Chapman and Hall, pp. 245–58.

Paterson, M. S., (1990). Rock deformation experimentation. In *The Brittle–ductile Transition in Rocks*, ed. A. G. Duba, W. B. Durham, J. W. Handin and H. F. Wang. American Geophysical Union Geophysical Monograph 56, pp. 187–94.

Paterson, M. S. (1995). A theory for granular flow accommodated by material transfer via an intergranular fluid. *Tectonophysics*, **245**, 135–52.

Paterson, M. S. (2001). A granular flow theory for the deformation of partially molten rock. *Tectonophysics*, **335**, 51–61.

Paterson, M. S. and Luan, F.-C. (1990). Quartzite rheology under geological conditions. In *Deformation Mechanisms, Rheology and Tectonics*, ed. R. J. Knipe and E. H. Rutter. Geological Society of London Special Publication 54, pp. 299–308.

Persikov, E. S., Bukhtiyarov, P. G. and Polskoy, S. F. (1990). The effects of volatiles on the properties of magmatic melts. *European Journal of Mineralogy*, **2**, 621–42.

Renner, J., Evans, B. and Hirth, G. (2000). On the rheologically critical melt percentage. *Earth and Planetary Science Letters*, **181**, 585–94.

Renner, J., Viskupic, K., Hirth, G. and Evans, B. (2003). Melt extraction from partially molten peridotites. *Geochemistry, Geophysics, Geosystems*, **4**, 8606, doi:10.1029/2002GC000369.

Roscoe, R. (1952). The viscosity of suspensions of rigid spheres. *British Journal of Applied Physics*, **3**, 267–9.

Rosenberg, C. L. (2001). Deformation of partially molten granite: a review and comparison of experimental and natural case studies. *International Journal of Earth Sciences (Geologisches Rundschau)*, **90**, 60–76.

Rosenberg, C. L. and Handy, M. R. (2000). Syntectonic melt pathways during simple shearing of a partially molten rock analogue (norcamphor-benzamide). *Journal of Geophysical Research*, **105**, 3135–49.

Rosenberg, C. L. and Handy, M. R. (2002). Mechanisms and orientation of melt segregation paths during pure shearing of a partially molten rock analogue (norcamphor-benzamide). *Journal of Structural Geology*, **23**, 1917–32.

Rosenberg, C. L. and Riller, U. (2000). Partial melt topology in statically and dynamically recrystallized granite. *Geology*, **28**, 7–10.

Rutter, E. H. (1972). Effects of strain rate changes on the strength and ductility of Solnhofen limestone at low temperature and confining pressures. *International Journal of Rock Mechanics and Mining Sciences*, **9**, 183–9.

Rutter, E. H. (1997). The influence of deformation on the extraction of crustal melts: a consideration of the role of melt-assisted granular flow. In *Enhanced Fluid Transport in the Earth's Crust*, ed. M. B. Holness. *Deformation*. London: Chapman and Hall, pp. 82–110.

Rutter, E. H. and Neumann, D. H. K. (1995). Experimental deformation of partially molten Westerly granite, with implications for the extraction of granitic magmas. *Journal of Geophysical Research*, **100**, 15 697–715.

Sawyer, E. W. (2001). Melt segregation in the continental crust: distribution and movement of melt in anatectic rocks. *Journal of Metamorphic Geology*, **19**, 291–310.

Skempton, A. W. (1954). The pore pressure coefficients A and B. *Géotechnique*, **4**, 143–7.

Smith, G. N. (1990). *Elements of Soil Mechanics*. Oxford: Blackwell.

Snoke, A. W., Kalakay, T. J., Quick, J. E. and Sinigoi, S. (1999). Development of a deep crustal shear zone in response to tectonic intrusion of mafic magma into the lower crust, Ivrea–Verbano zone, Italy. *Earth and Planetary Science Letters*, **166**, 31–45.

Tullis, T. E. and Tullis, J. (1986). Experimental rock deformation techniques. In *Mineral and Rock Deformation: Laboratory Studies*, ed. B. E. Hobbs and H. C. Heard. Geophysical Monograph 36, American Geophysical Union, pp. 297–324.

van der Molen, I. and Paterson, M. S. (1979). Experimental deformation of partially melted granite. *Contributions to Mineralogy and Petrology*, **70**, 299–318.

Vigneresse, J. L. (1995). Control of granite emplacement by regional deformation. *Tectonophysics*, **249**, 173–86.

Viskupic, K. M., Renner, J., Hirth, G. and Evans, B. (2001). Melt segregation from partially molten peridotites. *EOS, Transactions of the American Geophysical Union*, **82**, F1107–F8.

Waff, H. S. and Faul, U. H. (1992). Effects of crystalline anisotropy on fluid distribution in ultramafic partial melts. *Journal of Geophysical Research*, **97**, 9003–14.

Wen, S. and Nekvasil, H. (1994). Ideal associated solutions: application to the system albite-quartz-H_2O. *American Mineralogist*, **79**, 316–31.

Wickham, S. M. (1987). The segregation and emplacement of granitic magmas. *Journal of the Geological Society London*, **144**, 281–97.

Zimmerman, M. E., Zhang, S., Kohlstedt, D. L. and Karato, S. (1999). Melt distribution in mantle rocks deformed in shear. *Geophysical Research Letters*, **26**, 1505–8.

List of symbols used

σ_i	Principal stress component (Pa), $i = 1, 2,$ or 3.
ε_j	Conventional strain component, $j = 1, 2,$ or 3.
p	Pore fluid (melt) pressure (Pa).
A, B	Skempton pore pressure coefficients.
x, z	Distance along coordinate axis (m).
η	Melt viscosity (Pa s).
k	Matrix permeability (m^2).
ϕ	Porosity ($=$ melt fraction).
θ	Dihedral angle (degrees).
δ_c	Compaction distance (m).
w_0	Melt velocity (m s^{-1}).
τ_0	Compaction time (s).
h	Layer thickness and height of rock column (m).
μ	Effective viscosity of solid matrix (Pa s).
ρ_s	Solid density (kg m^{-3}).
ρ_f	Melt density (kg m^{-3}).
γ_{sm}	Melt–solid surface energy (J m^{-2}).
γ_{ss}	Solid–solid surface energy (J m^{-2}).
κ	Pore pressure multiplier in effective stress law.
s	Vein width (m).
N	Number of veins per meter (m^{-1}).
$2a$	Distance between veins (m).
R	Gas constant (J K^{-1} mol^{-1}).
T	Temperature (K).
T_C	Temperature (°C).
Ψ	Water content of melt (wt.%).
∇^2	Laplacian operator.
H	Activation enthalpy (J mol^{-1}).
d	Grain diameter (m).
g	Gravitational acceleration (ms^{-2}).
D	Diffusion coefficient (m^2 s^{-1}).
V	Solid phase molar volume (m^3).
U	Melt flow velocity (m s^{-1}).
c'	Equilibrium molar concentration.
n	Stress exponent in flow law.
r	Melt fraction exponent.
q	Exponent in Einstein–Roscoe equation.
a, b, c	Coefficients in Hess–Dingwell viscosity formula.
C_i, C'_i	Empirical constants in flow laws, $i = 1, 2,$ or 3.

12

Melt migration in the continental crust and generation of lower crustal permeability: inferences from modeling and experimental studies

TRACY RUSHMER AND STEPHEN A. MILLER

12.1 Introduction

The fundamental process by which the crust differentiates is through partial melting and the subsequent migration and emplacement of that melt into upper levels of the crust (e.g., Fyfe *et al.*, 1978; Chapter 9). On the grain scale, this basic process is one of melt segregation from a solid matrix and subsequent migration induced by an active driving force. However, the possible range in rates of melt migration, the type of melt transport paths, the volume of melt, and the subsequent impact on melt geochemistry and crustal rheology are subjects still under debate (Molnar, 1988; Clemens & Mawer, 1992; Paterson & Vernon, 1995; Royden, 1996; Weinberg, 1996; Vanderhaeghe & Teyssier, 1997, 2001; Chapters 13 & 14). In addition, these processes probably are dependent on tectonic setting and vary according to whether there is penetrative regional deformation. Deformation strongly influences the nature of permeable networks at the source (Laporte & Watson, 1995; Brown & Rushmer, 1997), but exactly how migration occurs requires not only a knowledge of the permeable network, but its development within the system as it responds to changing physical and/or chemical conditions.

This chapter addresses the importance of determining the mechanisms of fluid distribution and extraction at the grain scale for understanding crustal differentiation processes in different tectonic environments. A basis for understanding fluid migration in the crust is to determine the controls on the development of permeable networks. We provide a framework that addresses fluid migration in systems where high pore-fluid pressure is developed by mineral reaction, either by dehydration or melting. We present a conceptual model and simple model results to demonstrate a view in which melt migration is controlled by the nucleation, growth, and coalescence of small-scale

Evolution and Differentiation of the Continental Crust, ed. Michael Brown and Tracy Rushmer. Published by Cambridge University Press. © Cambridge University Press 2005.

microcracks that evolve to an interconnected permeable pathway for fast migration. In this conceptual model, fluid pressures resulting from melt or dehydration are sufficient to induce a small-scale hydrofracture, thus switching the local permeability from (essentially) zero to a very large value associated with flow through a rough fracture (Miller & Nur, 2000). We consider at this stage only reactions that generate fluid overpressures, thus only reactions with positive Clapeyron (dP/dT) slopes. This is a reasonable assumption because most common reactions in the crust that generate significant amounts of melt are fluid-absent (without the participation of a free volatile phase) and thus their solidii have variable, but overall positive, slopes in pressure–temperature space at typical lower crustal pressures (Thompson, 1990).

The nature of the growing crack network will depend on the volume change associated with the dehydration/melting reaction, the nature of the dehydration/melting reaction (e.g., the presence of solid–solution) and the ability to which the matrix may respond elastically to the deformation imposed by reaction (Connolly, 1997). The presence or absence of external deformation will also influence strongly the nature of permeable networks at the source (Davidson *et al.*, 1994; Brown, 1994; Holness, 1997). We address the development of reaction induced permeability in the mid and lower crust from modeling and experimental studies.

12.2 Development of a permeable network

12.2.1 Definition of porosity and permeability

Porosity and permeability are fundamental rock properties that have been the subject of numerous geological studies. *Porosity* in rocks refers to the ratio of pore volume to total volume of a given rock, whether these pores are interconnected or not. *Permeability* describes the ability of a rock to transmit a fluid and therefore is a measure of the relative ease of fluid flow through a rock (Guilbert & Park, 1986). The driving force for movement of fluid is usually a pressure gradient, which may be combined with buoyancy forces (Bredheoft & Norton, 1990; Jamtveit & Yardley, 1997). Rocks that have porosity may not be permeable, but permeability is required for fluid (melt) migration. The permeability of rocks may be developed through a variety of contributing effects. Guilbert and Park (1986) divide these effects into two categories: (1) primary permeability, and (2) secondary permeability. Primary permeability is intrinsic to the rock and created when the rock forms, and secondary is induced permeability, such as by deformation. Primary and secondary permeability may be supercapillary (>1 mm, unrestricted, turbulent or laminar), capillary (0.01–1.0 mm, restricted flow, diffusion, osmosis), and subcapillary (<0.01 mm, highly restricted flow, diffusion ≫ flow).

For melt segregation beginning at the grain-scale, we are concerned with permeability (both primary and secondary) at the capillary and subcapillary scale. We consider the permeability related to melting (and dehydration) reactions during periods of change in the lower crust to be a primary permeability and intrinsic to the rock. External deformation in addition to mineral reaction is considered to generate secondary permeability. Table 12.1 gives primary permeabilities in several typical crustal dehydration reactions and the quartz + muscovite melting reaction, and describes the nature of the permeability system. Development of primary permeability in the lower and middle crust is probably produced mainly by mineral reaction during changing conditions that encourage the production of a fluid (melt) pressure. Albeit transient, high permeability may develop during prograde metamorphism.

12.2.2 Understanding permeability development in lower crustal environments

The pressure of fluids produced in metamorphic reactions must necessarily be at least as high as the rock pressure. For example, if the fluid produced at individual reaction sites is less than the porosity created, the pore space collapses and brings the fluid to rock pressure. If more fluid is produced than porosity, then the fluid pressure exceeds the rock pressure and hydrofracture, and a region of high permeability may result. Even though dehydration (and dehydration-melting) reactions reduce the solid volume, total ΔV may be positive or negative with increasing pressure. In general, it is considered that positive ΔV reactions may cause fluid overpressure and hydrofracture, and negative ΔV reactions may cause underpressure, cannot induce fracture, and will restrict fluid (melt) migration. However, Connolly (1997) suggests that in fact the most important variables are connected porosity generated by the reduction in solid volume, and then the changes in rock and fluid pressure gradients combined with compaction. The magnitude of the ΔV is important in determining at what point embrittlement will occur.

Several important studies have investigated the nature of permeability development and fluid pressure during dehydration as a result of prograde metamorphism (e.g., Hanson, 1995; Ko *et al.*, 1997; Connolly, 1997; Hacker, 1997; Wong *et al.*, 1997; Simpson, 1999; Miller & Nur, 2000). Most early studies that incorporated modeling of dehydration-induced fluid pressures viewed fluid overpressure as part of a larger problem of compaction and fluid flow. Dehydration-induced fluid overpressure was incorporated as a

Table 12.1. *Primary permeabilities in some typical crustal dehydration reactions and the quartz + muscovite melting reaction and the nature of the permeability system. Permeabilities for various crustal rock types are given; beach sand is given for reference. Primary permeability is that which is present when the rock is formed. Secondary permeability is processes that create permeability after rock formation (e.g., deformation events)*

Permeabilities in representative rock types	Nature of permeability network (Guilbert & Park, 1986)	k (m^2)	Reference(s)
Gabbro	*Primary*: cooling cracks;		
	Secondary: expansion/contraction, thermal effects, deformation induced mineral cleavage	1.4×10^{-20}	Davis (1969)
Granite	Same as above	2.9×10^{-22} to 1.0×10^{-25}	Norton and Knapp (1977); Davis (1969)
Gneiss	*Primary*: Intragrain discontinuities such as twin planes;		
	Secondary: expansion/contraction, thermal effects, deformation-induced mineral cleavage, foliation planes	4.9×10^{-15} to 1.9×10^{-18}	Norton and Knapp (1977); Davis (1969)
Schist	Same as above	1.4×10^{-12} to 9.8×10^{-17}	Norton and Knapp (1977); Davis (1969)
Beach Sand	*Primary*: compaction features, channels	9.8×10^{-15} to 9.8×10^{-14}	Davis (1969)
Quartz Diorite (fractured)	*Secondary*: joints, shears, fractures	2.9×10^{-14} to 1.0×10^{-12}	Norton and Knapp (1977)
Reaction-produced permeability: **Dehydration:**			
Gypsum $\rightarrow$ bassanite + 3/2 H$_2$O	*Primary*: interconnected pore network through reduction of solid volume during reaction	1.5×10^{-20}	Estimated from experimentally based model of pore dilation and fluid transport during dehydration (see text) Wong *et al.* (1997)

Table 12.1. (cont.)

Permeabilities in representative rock types	Nature of permeability network (Guilbert & Park, 1986)	k (m^2)	Reference(s)
Muscovite + quartz → sillimanite + K feldspar + H$_2$O	*Primary*: Reaction-produced permeability, but nature of network unknown, but assumed to be similar to above for permeability estimates	1.6×10^{-18}	Estimates from Wong *et al.* (1997)
Antigortie → 18 forsterite + 4 talc + 27 H$_2$O	As above	2.8×10^{-18}	Estimates from Wong *et al.* (1997)
Biotite dehydration during regional metamorphism	As above	10×10^{-19} to 2.0×10^{-2}	Estimates from Connolly (1996); Wood and Orville (1992); Connolly and Thompson (1989)
Melting: Muscovite + 3.94 quartz → 8.23 melt + 0.03 biotite + 0.16 mullite	*Primary*: interconnected melt-filled microcracks from positive volume change and associated high dilational strain during reaction	1.0×10^{-13} to 1.0×10^{-15}	Calculated from development of crack network in experiments Connolly *et al.* (1997)

porosity production term in the fluid flow equations (e.g., Walder & Nur, 1984; Wong *et al.*, 1997). Modeling studies (e.g., Wong *et al.*, 1997) show that substantial fluid overpressure is possible but that compaction will reduce crustal permeability and porosity faster than metamorphic time scales, which was in agreement with the long-standing view that during regional metamorphism fluid pressure remains equal to the lithostatic pressures.

Connolly (1997) investigated numerically the evolution of dehydrating regions in the lower crust during this process and developed models for coupling strain and reaction rates to fluid pressure in reactions for both positive and negative total volume change (ΔV), and for both discontinuous and continuous metamorphic reactions. This study points out that, whereas it is thought that the ΔV of metamorphic reactions is of first-order importance in determining whether reactions will generate over-pressuring and embrittlement, the modeling shows that the most important variables are connected porosity, changes in rock and fluid pressure gradients, and compaction. The results from these studies, which are relevant to both metamorphic dehydration reactions and partial melting reactions in the mid to lower crust, suggest that in the lower crust there may exist zones or domains of high permeability. The domains are envisioned as being developed by interplay between permeability generation by reaction, and reduction by compaction. We describe below the primary permeability development model that uses mineral reactions to induce changes in local permeability and how this may produce domains of high permeability.

12.2.3 Cyclic primary permeability development during dehydration and partial melting

Conceptual and numerical models for cyclic permeability in the earthquake cycle (Miller *et al.*, 1996; Miller, 2002) were applied to cyclic permeability evolution in dehydrating systems that are characterized by positive volume change (Miller & Nur, 2000; Miller *et al.*, 2003). In this model, the dehydration or melting reaction produces enough fluid overpressure to hydrofracture the rock, thus creating small-scale changes in the local permeability. The overall permeability remains low because the evolving network is not sufficiently mature to provide a larger-scale pathway for flow. However, the large-scale permeability of the system may change drastically when isolated networks link at the percolation threshold to create an interconnected crack network.

For melt reaction with a positive Clapeyron slope, a fluid overpressure develops at the time of the phase change. Early in the melting process, melt occurs at isolated reaction nuclei with no connection to the overall permeability of the system. If the fluid overpressure is sufficient for hydro-fracture, then a crack is formed that increases the local permeability but not the global permeability of the melting horizon. The orientation of the induced fracture depends on the general tectonic setting, with fracture occurring in the direction of the maximum principal stress. As the system evolves, local fracture networks (developed at individual reaction sites) begin to link and increase the scale of the connected fracture network. Large-scale increases in the permeability of the melting horizon occur at the percolation threshold. That is, prior to the percolation threshold, a fully connected network is not formed and each 'pocket' of melt behaves independently of the overall system. At the percolation threshold, regions of the system are completely linked, and the macroscopic permeability is greatly enhanced (cf. Sawyer, 2001).

These concepts are illustrated using a simple cellular automaton model of the processes involved. A complete description of the model is found in Miller and Nur (2000) and Miller *et al.* (2003), and is outlined here. In a system with very small (negligible) permeability, the rate of pore pressure increase is related to the ratio of a source term to the porosity and compressibility of the fluid and rock. The source is a combination of the rate at which porosity is changing and the rate at which a direct fluid source is input into the system. For a melt reaction, the direct fluid source is greater than the porosity created, so overpressure develops and hydrofracture may occur. To model this, we treat permeability as a toggle switch, where permeability is zero before a fracture criterion is achieved, to a locally high value at fracture. Very high permeability is limited to the immediate neighborhood, and whether the scale of interaction is large or small depends on the previous history of connectivity.

Prior to the percolation threshold, the consequence of local permeability increases is limited to the immediate neighborhood (Fig. 12.1a). As the system evolves, hydrofracture at the local scale links with other developing systems and a structure of interconnected pockets emerges (Fig. 12.1b, c). Near the percolation threshold (Fig. 12.1d and enlarged in Fig. 12.2), a definite (complex) structure is apparent, and indicates regions that are hydraulically connected. When the system is hydraulically connected, it behaves as a unit where buoyancy forces and large-scale fluid flow may result. In this model, the driving force of the system is based on simple concepts that are also supported by natural observations on permeability evolution.

Figures 12.1 and 12.2 show that a complex network of possible fluid pathways is produced by overpressure during reaction. We may also investigate the development of porosity and permeability, to some extent, experimentally. The following section describes two experimental studies of the nucleation and

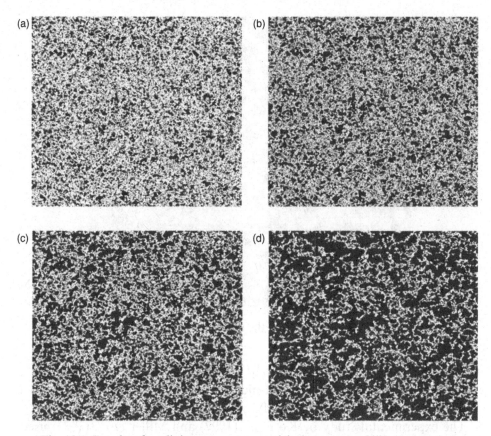

Fig. 12.1. Results of a cellular automaton model where permeability is treated as a toggle switch. In this 1000 × 1000 matrix, black regions correspond to interconnected networks. The evolution of the system approaching the percolation threshold (a–c) shows how the connectivity structure develops. At the percolation threshold (d), a clear interconnected system is evident.

growth of permeable pathways for fluid flow produced during reaction. Both studies used microstructural evidence for calculating and/or estimating ranges of permeability developed while the rock underwent reaction. The first study looks at dehydration of gypsum; the other is a fluid-absent partial melting study on muscovite-bearing quartzite.

12.2.4 Permeability development in experimental studies

The development of high permeability domains by mineral reaction is shown to be possible, as described above. Primary permeability development has been observed in dehydrating systems that maintain pore pressure excess during

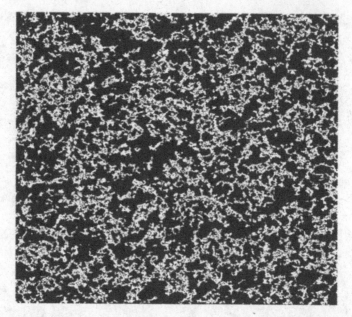

Fig. 12.2. Blow-up of Fig. 12.1d of the system at the percolation threshold.

mineral dehydration reactions (Ko *et al.*, 1997; Wong *et al.*, 1997; Miller *et al.*, 2003) and by the production of melt-induced cracks in partially molten rocks that are muscovite-bearing (Connolly *et al.*, 1997; Rushmer, 2001).

The experimental study by Ko *et al.* (1997) and Miller *et al.* (2003) used gypsum aggregates as a starting material and investigated the generation and maintenance of excess pore pressure during dehydration of the gypsum to bassanite plus water. The study used a triaxial deformation apparatus that was connected directly to a pore fluid system. Experiments were conducted under both drained conditions, where fluid expulsion was monitored, and under undrained conditions, where pore pressure increases were monitored. Microstructural analyses were conducted to examine the evolution of the pore network during dehydration of the initially low porosity, low permeability gypsum aggregate. The results of the analyses showed that after initial pore pressure increased due to fluid entrapment in unconnected pores, an interconnected pore network system developed and fluid readily escaped. Fluid expulsion rate then reduced when the reaction neared completion. The fluid expulsion rate itself was affected by temperature, pore pressure, effective pressure (the difference between applied confining pressure and pore pressure), differential stress, and pore compaction rates.

Microstructures examined in samples recovered at different stages in the experiments showed evolution from small, isolated, nuclei of bassanite on gypsum grain boundaries, and voids (pore space) due to decrease in solid volume, the aspect ratios of which are similar to the bassanite grains themselves. As the dehydration reaction continued, bassanite fibers, often several hundreds of microns in length, grew and eventually linked up in an interconnected network. The pores themselves mimicked in shape the bassanite fibers and linked up to form a path for fluid expulsion. By the end of the reaction, all the gypsum had been converted to form bassanite and void space, and there was no preferred orientation in the final product of bassanite fibers and elongated pores. The geometry of the pore space is complex.

These results provide insight into pore evolution during dehydration and show that sample-wide hydraulic conductivity is achieved when linking of void space occurs, similar to that observed in the cellular automaton model when hydrofractures on the local scale link with the other developing systems (e.g., growing neighbor bassanite fiber and void spaces) and a complex structure of interconnected pockets, or voids, is realized. However, the experimental data show that even though the porosity develops through a stage of initial pore pressure increase, interconnectivity may be achieved by linking of low aspect ratio voids, although whether this is achieved by small hydrofractures or by progressive coalescence is unknown.

The gypsum–bassanite study of Ko *et al.* (1997) also shows that there is a partitioning of liberated fluid during reaction between pores, which is directly expelled from the source. Effective pressure is suggested as an important control. The effective pressure will be determined by the strength of the solid portion of the rock. High effective pressure, present in rocks with weaker solid frameworks, will compact void space more successfully and squeeze out fluid, whereas low effective pressure, present in rocks that have strong solid frameworks, may preserve and may dilate pore space providing a site for fluid retainment within the rock. Clearly, the strength of the solid framework changes with reaction and may be weakened by the development of additional fine-grained reactant products (Brodie & Rutter, 1985). However, for first-order estimations of developing permeability during dehydration in the mid to lower crust, this conceptual model may be treated theoretically (Wong *et al.*, 1997) to provide estimates of permeability in other dehydrating systems.

Connolly *et al.* (1997) investigated the permeability change in a muscovite-bearing quartzite undergoing melting. The melting reaction creates a melt overpressure due to the volume change of the reaction ($0.021\,\mathrm{m}^3\,\mathrm{m}^{-3}$ of original starting material) and induces microcracking in the rock. The hydrostatic partial melting experimental study used cores of a muscovite-bearing quartzite, and in some experiments quartz sand was loaded in the capsules

along with the cores to act as a drain for excess melt and to test interconnectivity of the melt-lined fracture network. Experimental observations find
that the melting reaction begins at quartz–muscovite grain boundaries, and
melt-filled cracks are produced during the melting of muscovite plus quartz
producing melt, mullite, and biotite, even in experiments where muscovite was
incompletely reacted, due to the positive volume change associated with this
reaction.

The experiments documented that during early stages of the reaction, melt-
filled cracks were typically 1 μm wide and extended <150 μm from the
melting sites. As melting progressed, to approximately halfway to completion, the cracks have propagated further and reached lengths of several
100 μm with widths of 1–10 μm. With increasing time, there was no significant increase in microcrack density and a permeable network was established
and the network allowed melt to drain reducing melt overpressure. In all
experiments with the quartz sand trap, the sand was saturated with melt, and
there was no evidence of melt movement between the interface of the sample
and the capsule wall. In addition, the total crack length per unit area
decreased towards the sand trap, suggesting that melt had migrated along
the cracks into the sand trap and the crack network collapsed when melt
pressure was removed.

Figure 12.3 shows a photomicrograph of melt-filled microcracks developed
during melting reaction of muscovite + quartz. Observed crack density in the
muscovite + quartz experiments implied relatively high permeabilities by reac-
tion-produced microcracks, but did not allow direct estimation of the permeability. Therefore, further analysis was needed to calculate the permeability.
The microcrack distribution in the experiments showed a lack of preferred
orientation and a similarity in crack lengths in sections cut both orthogonal to,
and parallel to, the rock foliation. The statistical model of Dienes (1982) allows
calculation of the permeability based on a network of randomly oriented
penny-shaped cracks; calculations using this model suggested that microcrack
development during partial melting created significant transient permeability,
from $10^{-13.5}$ to $10^{-14.5}$ m^2, rather than what is typical for regional metamorphism, from 10^{-18} to 10^{-21} m^2 (Dipple & Ferry, 1992; Manning &
Ingebritsen, 1999). Although the textural observations of microcrack distribution are not as complex as indicated by the cellular automation model (Figs.
12.1 and 12.2), the crack density and calculated associated permeability suggest that interconnectivity has occurred allowing for the formation of a high-
permeability domain. Table 12.1 shows that primary permeability produced
during partial melting of this assemblage may equal that of fractured crystalline rocks.

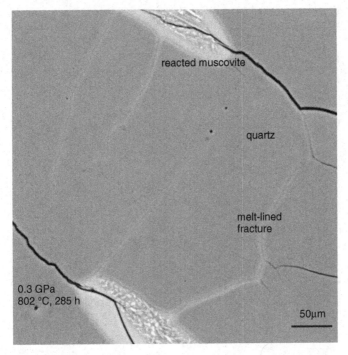

Fig. 12.3. Backscatter image of solidified muscovite + quartz melting experiment to show fracture network. Experiment in a high-pressure gas apparatus was run for 285 h 54 min at $P = 0.3$ GPa and $T = 802$ °C under fluid-absent conditions. The reacted muscovite is pseudomorphed by melt + mullite + biotite. Melt-lined fractures begin to connect melt pools. Dark cracks formed during the quench portion of the experiment and are not melt filled.

12.3 Dilational strain analysis – estimating reaction controlled permeability during partial melting

The volume change of the reaction may be converted to dilational strain associated with reaction by knowing the stoichiometry of the melting reaction and the modal abundance of the hydrous phase (Connolly *et al.*, 1997). Dilational strain analysis provides a method by which to compare different fluid-absent melting reactions and the possibility for increased permeability during melting (Rushmer, 2001). Results of partial melting experiments on a biotite gneiss and a muscovite + biotite schist, in addition to the muscovite + quartz study described above, were used to calculate volume change for a given partial melting reaction associated with general crustal anatexis reactions (Rushmer, 2001). Texturally, the results from experiments on biotite gneiss cores show that melt cracks are not produced during partial melting

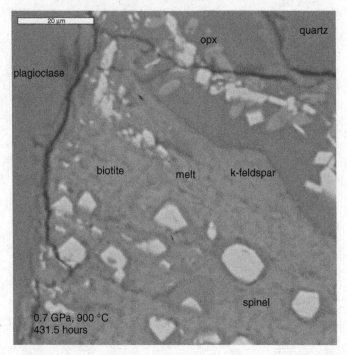

Fig. 12.4. Backscatter image of solidified partial melting experiment of a biotite gneiss under static conditions at $P = 0.7$ GPa, $T = 900$ °C for 431 h (Rushmer, 2001). The reaction shown here is: biotite + plagioclase + quartz → spinel + opx + Fe–Ti oxides + K-feldspar + melt. The stoichiometry of the reaction is given in Table 12.2, reaction (2). No melt fractures are developed as in Fig. 12.3; instead, melt lines grain boundaries and resides within the biotite with spinel.

as observed in the muscovite-bearing assemblages (Fig. 12.4). The calculations are carried out by first determining the stoichiometry of the melting reaction, then determining the overall volume change of the reaction and finally calculating the dilational strain associated with the reaction as described in Connolly *et al.* (1997). Table 12.2 shows the stoichiometry of some common melting reactions, the volume change associated with the reaction, calculated on a 1-O basis to be able to use the hydrous melt model from Ochs and Lange (1997), and the dilational strain in percent for a rock with a mode of 30% hydrous phase. Dilational strain is calculated by using the mode of the hydrous phase, which provides the amount of hydrous phase per cubic centimeter of rock and the volume change for the dehydration melting reaction.

The slope in *P–T* space from experimental studies may be used as a first approximation of the volume change associated with the melting reaction, as we assume entropy will be positive for all melting reactions from the

Table 12.2. *Mode of hydrous phase, stoichiometry, volume change and associated dilational strain for five crustal dehydration melting reactions. Volume change (ΔV) is calculated after converting to units to a 1-oxygen basis (see Rushmer 2001 for further details on units)*

Reaction	ΔV (1 oxygen)	Mode (vol.%)	Dilational strain
(1) 1 muscovite + 0.32 plagioclase + 0.36 quartz = 1.14 melt + 0.09 biotite + 0.22 K-feldspar 0.22 sillimanite (600 MPa and 820 °C; Patiño-Douce & Harris, 1998)	+2.64	30	+6.76%
(2) 1 biotite + 0.54 quartz + 0.053 plagioclase = 0.73 opx + 0.002 spl + 0.59 K-feldspar + 0.28 melt (at 700 MPa and 900 °C; Rushmer, 2001)	+0.30	30	+0.70%
(3) 1 biotite + 0.55 quartz + 0.13 plagioclase = 0.615 opx + 0.017 cord + 0.39 kspar + 0.63 melt (at 200 MPa and 825 °C; Montel & Vielzeuf, 1997)	+0.25	30	+0.58 %
(4) 1 biotite + 0.36 quartz + 0.198 plag = 0.565 opx 0.119 gt + 0.52 k-feldspar + 0.32 melt (at 800 MPa and 913 °C; Montel & Vielzeuf, 1997)	−0.67	30	−1.57%
(5) 0.75 biotite + 0.21 quartz + 0.48 plagioclase = 0.76 melt + 0.04 opx + 0.01 oxide + 0.13 gt + 0.06 cpx (at 1500 MPa and 950 °C; Patiño-Douce & Beard, 1995)	−5.23	30	−12.24%

Clapeyron equation, $dP/dT = \Delta S / \Delta V$. These slopes may be compared with the calculated volume change (e.g., the point at which the slopes go from positive to negative volume when the dense phase garnet becomes stable). From previous experimental work on biotite dehydration melting (e.g., Vielzeuf & Montel, 1994; Patiño-Douce & Beard, 1995; Stevens *et al.*, 1997; Montel & Vielzeuf, 1997), most slopes determined in *P–T* space suggest steep, but positive, volumes associated with reaction and then a backbend (shift from positive to negative volume change) with increasing pressures. Under moderate pressures, there are variations in Clapeyron slopes and Stevens *et al.* (1997) show a steep negative slope for the biotite dehydration melting in metagreywacke, and Patiño-Douce and Harris (1998) find that the dP/dT solidus slopes for their experimental results on biotite-bearing assemblages have much steeper slopes than the muscovite melting curves. Melting experiments on biotite gneiss cores show no evidence of melt-induced fractures, suggesting the volume change was less than that for muscovite melting (Fig. 12.3,

Table 12.2). The lower volume change (thus low dilational strain) during biotite melting may also represent an important "cut-off" point where the matrix is able to compensate elastically for the change in volume, whereas for muscovite melting this threshold is overstepped (Connolly *et al.*, 1997).

The results from the dilational strain analyses suggest that permeability will vary according to dehydration reaction, but as modeling has shown (Connolly *et al.*, 1997), there are additional factors to consider. The other important variables are connected porosity, changes in rock and fluid pressure gradients, and compaction. During fluid-absent melting induced by changing thermal gradients, there may exist domains of high permeability in the lower crust, but they are being developed by feedbacks between permeability generation by reaction and reduction by compaction with modifications by variations in rock and fluid pressure gradients. This may explain why field studies of grain and outcrop-scale distribution of melt produced during low dilational strain biotite fluid-absent melting do not necessarily show a trapping of melt at low melt fraction along grain boundaries during anatexis, as observed in the hydrostatic experimental studies (Sawyer, 2001). The presence of non-hydrostatic stress fields will create changes in rock and fluid (melt) pressure gradients, potentially driving melt into planes of foliation and branched grain boundary orientations. The negligible volume change associated with these reactions is not a first-order control on forming permeable pathways for melt escape during regional metamorphism and associated deformation. In the section below, we discuss the nature of the different fluid-absent reactions, low versus high dilational strain during deformation and the textural evolution of melt-present deformation features as partial melting initiates.

12.4 Development of secondary permeable melt networks in the crust and the role of deformation

Holyoke and Rushmer (2002) performed solid-media deformation experiments to assess the differences between the dehydration melting reactions in muscovite- and biotite-bearing natural rocks under conditions of applied stress, and to evaluate the implications for melt segregation during orogenesis. The differences in dilational strain for the two different crustal melting reactions described above suggest that melt pore pressure is developed rapidly in muscovite-bearing rocks undergoing dehydration melting, and that deformation will induce melt-enhanced embrittlement leading to cataclasis close to the solidus. Fluid-absent melting in muscovite-bearing rocks produces a change in ΔV over a narrow temperature range, essentially just above the solidus. This is in contrast to the other common crustal hydrous phases, such as biotite, which undergo dehydration melting continuously over a wider temperature interval (Patiño-Douce & Harris, 1998; Holyoke & Rushmer, 2002). Deformation concomitant with partial melting of low dilational strain assemblages may result in the onset of cataclasis being suppressed or reduced due to the slower melt generation rate and the negligible dilational strain associated with dehydration melting.

The specific melting reactions investigated experimentally under deformation were chosen to represent a high dilational strain reaction (muscovite + quartz + plagioclase → melt + biotite + aluminosilicate + K-feldspar, Reaction (1) in Table 12.2) and a low dilational strain reaction (biotite + plagioclase + quartz → orthopyroxene + spinel + K-feldspar + melt, Reaction (2) in Table 12.2).

Figure 12.5 (from Holyoke & Rushmer, 2002) summarizes the textural evolution in the different assemblages at the beginning of partial melting. The experimental results showed that as melt fraction reached ~5 vol.% in the biotite gneiss (Reaction (2)) there was melt flow along grain boundaries, a development of a disaggregated framework of quartz and plagioclase grains, and the presence of largely unfractured grains surrounded by melt. Some zones of grain size reduction were present and melt-assisted granular flow-accommodated deformation. Locally, there was development of high melt pore pressures, causing cataclasis. These processes together led to a mixture of melt-assisted granular flow and cataclastic flow as shown in Fig. 12.5. When compared to cataclastic zones observed in samples from the muscovite-bearing samples with similar melt fractions (Reaction (1)), the zones of cataclasis were sharply defined, grains were intensely fractured, and grain size was reduced (Figs. 12.5 and 12.6a, b).

Holyoke and Rushmer (2002) attributed the differences in microstructure development at low melt fractions in the muscovite-bearing assemblage and biotite gneiss to the contrasting behavior of the two different dehydration–melting reactions. The difference in dilational strain associated with melting may explain difference in textural evolution and behavior between the two rock types deforming in the presence of melt. At lower melt fractions, the low dilational strain present during the biotite fluid-absent melting will not produce enough pore pressure to induce fracturing, but may be enough to allow migration along grain boundaries. Holyoke and Rushmer (2002) observed ~1 vol.% melt in the biotite gneiss had migrated along foliation, even at the high strain rates used in experimental rock deformation studies.

The production of a secondary permeability – a highly interconnected network enhanced by rapid development of melt pore pressure during partial melting of a muscovite-bearing metapelite – may be a mechanism by which low-fraction melts may segregate, not necessarily requiring foliation. In lower

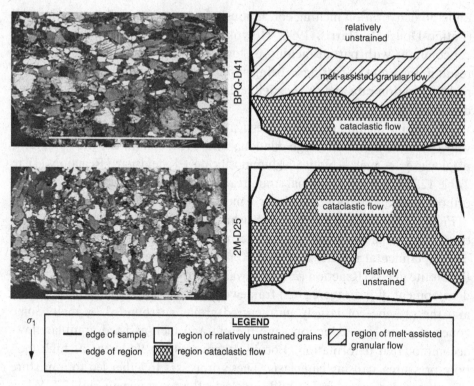

Fig. 12.5. Photomicrographs (crossed polarized light) to contrast examples of mixed melt-assisted granular flow and cataclasis in BPQ-D41 and purely cataclasis in 2M-D25 in the two different hydrous phase-bearing assemblages deformed at the same bulk strain rate and with the same melt fraction (5 vol.%). Scale bars in photomicrographs are 5 mm. Both of these experiments were conducted close to the solidus of the hydrous phase dehydration melting reaction of each respective assemblage (Holyoke & Rushmer, 2002). The upper photomicrograph and behavior region map show mixed behavior (melt-assisted granular flow and cataclasis) that developed during experiment BPQ-D41 (upper photomicrograph and drawing, $T = 930$ °C, $P_C = 1000$ MPa, ($\varepsilon = 13\%$, yield strength $= 105$ MPa). The region of primarily melt-assisted granular flow in BPQ has grains that are partially-reacted, few fractures, and melt has flowed along grain boundaries, disaggregating those grains. The region of cataclasis has intense grain fracturing (see legend for pattern descriptions). The cataclastic zone which developed in 2M-D25 (lower photomicrograph, and drawing $T = 740$ °C, $P_C = 700$ MPa, ($\varepsilon = 20\%$, yield strength $= 43$ MPa,) no longer contains unfractured grains in the extensive cataclastic zone.

crustal biotite-bearing assemblages, melt pathways are more controlled by pre-existing anisotropy in the rock as shown by Sawyer (2001). Unless pore pressure may be built up locally, melt-induced fractures may not be as easily formed.

Fig. 12.6. (a) Photomicrograph of BPQ-D45 ($T = 920$ °C, $P_C = 1000$ MPa, $\varepsilon = 13\%$) that exhibits a melt-assisted granular flow microstructure. This microstructure is where large, relatively unfractured grains with rounded edges, due to reaction, are surrounded by melt on all sides and the material has flowed as one mass towards the edges of the sample during deformation. (b) Photomicrograph of 2M-D23 ($T = 920$ °C, $P_C = 700$ MPa, $\varepsilon = 17\%$) that exhibits cataclastic flow. Cataclastic flow is used to describe regions that contain intensely fractured grains (region "a" in the photomicrograph) and grain-size reduction. Grains are surrounded by melt and have flowed in response to deformation (Holyoke & Rushmer, 2002). Cataclastic zone in region "a" is composed of intermixed quartz, feldspar, and glass (melt). "b" is a partially reacted muscovite grain adjacent to the cataclastic zone and is kinked and banded. Strain in this sample is accommodated by pervasive development of cataclasis and kinking of muscovite and biotite (Holyoke & Rushmer, 2002).

12.5 Applications to lower crustal environments

Experimental data gathered from static experiments suggest that in environments where tectonic activity is minimal, grain-scale fluid distribution may be viewed as a combination of planar features and cracks (Laporte & Watson, 1995; Laporte *et al.*, 1997). In this case, the permeability of the crust developed by mineral reactions may be viewed as primary and "reaction-controlled". Permeability produced by external mechanical processes, as would be expected in deformation environments, may be considered a secondary permeability or "mechanically controlled" (Oliver, 1996).

In active tectonic environments where deformation is present, strain is partitioned into weaker rock types or sites where partial melt is present, creating shear zones. Permeability has been estimated within ductile shear zones by both Geraud *et al.* (1995) and by Dipple and Ferry (1992), who used fluid–rock ratios and flow duration calculations, to be approximately from 10^{-15} to 10^{-16} m^2. These authors suggest that permeability may be strongly anisotropic. During partial melting, strain will be further partitioned into weak zones, and grain edges may become completely wetted by melt during deformation ("dynamic wetting", e.g., Jin *et al.*, 1994; Bai *et al.*, 1997). Under these conditions, microstructures suggest grain boundary migration and grain growth occur, but deformation may also be accommodated by melt-assisted granular flow (Rutter, 1997).

Deformation associated with external tectonic processes may induce melt segregation in partially molten source rocks in which melt under static conditions would not be able to become interconnected unless high melt fractions are reached (Petford *et al.*, 2000). Deformation-assisted melt segregation is likely to be common in active tectonic environments, and trace element geochemical signatures of low-melt fraction leucocratic granites have provided useful constraints on extraction rates during partial melting of the mid-crust (Harris *et al.*, 1995, 2000). Melt in this environment may move along shear zones, using them as pathways, because the shear zones have a higher permeability than the surrounding rock. Even more efficient melt segregation may occur at faster strain rates, where partial melting induces melt-enhanced cataclasis or embrittlement at pressures and temperatures at which rocks normally deform ductilely (Hollister & Crawford, 1986; Davidson *et al.*, 1994; Holyoke & Rushmer, 2002).

Orogenic evolution may very likely be tied to, and controlled by, crustal composition. For example, muscovite- and biotite-bearing assemblages comprise a significant portion of the continental crust and play an important role in the geochemical and rheological evolution of many orogenic belts. Pelitic assemblages are responsible for increasing the radiogenic heat budget, creating thermal anomalies, and their low solidus temperatures allow for extensive partial melting of the crust. As a result, orogenic belts containing significant amounts of mica-rich pelite are more susceptible to widespread, melt-induced weakening and eventual rheologic decoupling (e.g., Rey *et al.*, 2001).

Alternatively, orogenic belts dominated by mafic compositions may remain strong and vertically coupled. In Fiordland, New Zealand, interactions among magmatism, metamorphism, crustal melting, and deformation may be investigated at all crustal levels (e.g., Daczko *et al.*, 2001; Klepeis *et al.*, 2003). Here, observations suggest that the intermediate and mafic crustal composition, which dominates this orogen, influenced the mechanical evolution from lower to upper crustal levels (10–50 km paleodepths). This is most probably due to the low volumes of partial melt (due to high solidus of the mafic bulk composition), rapid crystallization of mantle-derived magmas, and a cold, dry, mafic lower crust that controlled the degree of vertical coupling and the partitioning of deformation. This produced the narrow, focused structural style of the orogen (Klepeis *et al.*, 2004).

In orogens bearing thick sequences of pelitic schists, fluid-absent melting reactions of muscovite and biotite will be the main reactions in which melt is produced, whereas in more mafic crustal belts, melting reactions will be dominated by hornblende, biotite to a lesser degree, and clinozoisite. How these reactions, which under fluid-absent conditions all have positive slopes in P–T space under crustal conditions, control the extent at which interconnectivity is achieved, is vital to understanding the different mechanisms of melt segregation at the grain scale. This is because successful melt segregation relies on the interplay between two main variables: (1) the melting reaction itself and subsequent grain-scale melt distribution, and (2) the tectonic environment. Both variables determine the melt fraction at which interconnectivity is achieved and the rate at which melt may segregate.

12.6 Conclusions

Dehydration melting in the crust has surprisingly different behaviors in regard to associated volume change. Models of dehydration reactions are useful in understanding the implications of different melting reactions in crustal differentiation. Dehydration melting reactions have different rates of pore pressure build-up and, depending on deformation environment, will in turn have distinct mechanisms for melt segregation. It may be useful to view the development of permeability of the melting horizon as an evolutionary process in which isolated pockets of melt behave independently prior to the percolation threshold. At the percolation threshold, the system behaves as a unit, and the large-scale permeability of the system can increase dramatically. Such behavior would have a strong impact melt on melt geochemistry and crustal rheology.

The implications of these results in general suggest that melt derived from muscovite-free assemblages may be trapped in the mid to lower crust at higher volumes than previously thought (Brown & Rushmer, 1997). This may provide a mechanism by which melt may be retained in the crust, potentially producing a weak mid to lower crust. The existence of a very weak mid to lower crust in orogenic belts may strongly affect the structural style of the crust, perhaps even allowing for decoupling of the upper crust from mantle motions (Huerta *et al.*, 1996; Axen *et al.*, 1998; Ellis *et al.*, 1998; Rey *et al.*, 2001; Klepeis & Clarke, 2004).

12.7 Acknowledgements

We acknowledge the support of NSF (EAR-0106241) and numerous helpful discussions with M. Brown, E. W. Sawyer, J. A. D. Connolly, G. Berganto, N. Petford, and K. Klepeis.

References

Axen, G. J., Selverstone, J., Byrne, T. and Fletcher, J. M. (1998). If the strong crust leads, will the weak crust follow? *GSA Today*, **8**, 1–8.

Bai, Q., Jin, Z.-M, and Green, H. W. II (1997). Experimental investigation of the rheology of partially molten peridotite at upper mantle pressures and temperatures. In *Deformation-enhanced Fluid Transport in the Earth's Crust and Mantle*, ed. M. B. Holness. The Mineralogical Society Series 8. London: Chapman and Hall, pp. 40–78.

Bredheoft, J. D. and Norton, D. L. (1990). *The Role of Fluids in Crustal Processes*. Wasington, DC: National Academy Press.

Brodie, K. H. and Rutter, E. H. (1985). On relationships between rock deformation and metamorphism, with special reference to the behavior of basic rocks. In *Metamorphic Reactions: Kinetics, Textures and Deformation, Advances in Physical Geochemistry*, Vol. 4, ed. A. B. Thompson and D. C. Rubie. New York: Springer-Verlag, pp. 239–64.

Brown, M. (1994). The generation, segregation, ascent and emplacement of granite magma: the migmatite-to-crustally-derived granite connection in thickened orogens. *Earth-Science Reviews*, **36**, 83–130.

Brown, M. and Rushmer, T. (1997). The role of deformation in the movement of granite melt: views from the laboratory and the field. In *Deformation-enhanced Fluid Transport in the Earth's Crust and Mantle*, ed. M. B. Holness. The Mineralogical Society Series 8. London: Chapman and Hall, pp. 111–44.

Clemens, J. D. and Mawer, C. K. (1992). Granitic magma transport by fracture propagation. *Tectonophysics*, **204**, 339–60.

Connolly, J. A. D. (1996). Mid-crustal focused fluid movement in the lower crust: thermal consequences and silica transport. In *Fluid Flow and Transport in Rocks: Mechanisms and Effects*, ed. B. Jamtviet and B. W. D. Yardley. London: Chapman and Hall, pp. 235–50.

Connolly, J. A. D. (1997). Devolatilization-generated fluid pressure and deformation-propagated fluid flow during prograde regional metamorphism. *Journal of Geophysical Research*, **102**, 18 149–73.

Connolly J. A. D. and Thompson, A. B. (1989). Fluid and enthalpy production during regional metamorphism. *Contributions to Mineralogy and Petrology*, **102**, 346–66.

Connolly, J. A. D., Holness, M. B., Rubie, D. and Rushmer, T. (1997). Reaction-induced microcracking: an experimental investigation of a mechanism for enhancing anatectic melt extraction. *Geology*, **25**, 591–4.

Daczko, N. R., Clarke, G. L. and Klepeis, K. A. (2001). Transformation of two-pyroxene hornblende granulite to garnet granulite involving simultaneous melting and fracturing of the lower crust, Fiordland, New Zealand. *Journal of Metamorphic Geology*, **19**, 549–62.

Davidson, C., Schmid, S. M. and Hollister, L. S. (1994). Role of melt during deformation in the deep crust. *Terra Nova*, **6**, 133–42.

Davis, S. N. (1969). Porosity and permeability of natural materials. In *Flow Through Porous Media*, ed. R. J. M. DeWiest. New York: Academic Press, pp. 54–89.

Dienes, J. K. (1982). Permeability, percolation and statistical crack mechanics. In *Issues in Rock Mechanics*, ed. R. E. Goodman and F. E. Heuze. New York: American Institute of Mineralogical, Metallurgical and Petroleum Engineering, pp. 86–94.

Dipple, G. M. and Ferry, J. M. (1992). Metasomatism and fluid flow in ductile fault zones. *Contributions to Mineralogy and Petrology*, **112**, 149–64.

Ellis, S., Beaumont, C., Jamieson, R. and Quinlan, G. (1998). Continental collision including a weak zone – the Vise model and its application to the Newfoundland Appalachians. *Canadian Journal of Earth Science*, **35**, 1323–46.

Fyfe, W. S., Price, N. J. and Thompson, A. B. (1978). *Fluids in the Earth's Crust*. New York: Elsevier.

Geraud, Y., Caron, J.-M. and Faure, P. (1995). Porosity network of a ductile shear zone. *Journal of Structural Geology*, **17**, 1757–69.

Guilbert, J. M. and Park, C. F. (1986). *The Geology of Ore Deposits*, New York: W. H. Freeman and Company.

Hacker, B. R. (1997). Diagenesis and fault valve seismicity of crustal faults. *Journal of Geophysical Research*, **102**, 24 459–67.

Hanson, R. B. (1995). The hydrodynamics of contact metamorphism. *Geological Society of American Bulletin*, **107**, 595–611.

Harris, N., Ayres, M. and Massey, J. (1995). Geochemistry of granitic melts produced during the incongruent melting of muscovite: implications for the extraction of Himalayan leucogranitic magmas. *Journal of Geophysical Research*, **100**, 15 767–77.

Harris, N., Vance, D. and Ayers, M. (2000). From sediment to granite: time scales of anatexis in the upper crust. *Chemical Geology*, **162**, 155–67.

Hollister, L. S. and Crawford, M. L. (1986). Melt-enhanced deformation: a major tectonic process. *Geology*, **14**, 558–61.

Holness, M. B. (1997). The permeability of non-deforming rock. In *Deformation-enhanced Fluid Transport in the Earth's Crust and Mantle*, ed. M. B. Holness. The Mineralogical Society Series 8. London: Chapman and Hall, pp. 9–34.

Holyoke, C. III and Rushmer, T. (2002). An experimental study of grain-scale melt segregation mechanisms in crustal rocks. *Journal of Metamorphic Geology*, **20**, 493–512.

Huerta, A. D., Royden, L. H. and Hodges, K. V. (1996). The interdependence of deformational and thermal processes in mountain belts. *Science*, **273**, 637–9.

Jamtveit, B. and Yardley, B. W. D. (1997). Fluid flow and transport in rocks: an overview. In *Fluid Flow and Transport in Rocks: Mechanisms and Effects*, ed. B. Jamtviet and B. W. D. Yardley. London: Chapman and Hall, pp. 1–14.

Jin, Z.-M., Green, H. W. II and Zhou, Y. (1994). Melt topology in partially molten mantle during ductile deformation. *Nature*, **372**, 164–7.

Klepeis, K. A. and Clarke, G. L. (2004). Evolution of an exposed lower crustal attachment zone in Fiordland, New Zealand. In *Vertical Coupling and Decoupling in the Lithosphere*, ed. J. Grocott, B. Tikoff, K. J. W. McCaffrey and G. Taylor. London: Geological Society Special Publication 227, pp. 197–230.

Klepeis, K. A, Clarke, G. L. and Rushmer, T. (2003). Magma transport and coupling between deformation and magmatism in the continental lithosphere. *GSA Today*, **13**(1), 4–11.

Klepeis, K. A., Clarke, G. L., Gehrels, G. and Vervoort, J. (2004). Processes controlling vertical coupling and decoupling between the upper and lower crust of orogens: results from Fiordland, New Zealand. *Journal of Structural Geology*, **26**(4), 765–91.

Ko, S.-C., Olgaard, D. L. and Wong, T.-F. (1997). Generation and maintenance of pore pressure excess in a dehydrating system, 1, Experimental and microstructural observations. *Journal of Geophysical Research*, **102**, 825–40.

Laporte, D. and Watson, E. B. (1995). Experimental and theoretical constraints on melt distribution in crustal sources – the effect of crystalline anisotropy on melt interconnectivity. *Chemical Geology*, **124**, 161–84.

Laporte, D., Rapaille, C. and Provost, A. (1997). Wetting angles, equilibrium melt geometry, and the permeability threshold of partially molten crustal protoliths. In *Granite: From Segregation of Melt to Emplacement Fabrics*, ed. J. L. Bouchez, D. H. W. Hutton and W. E. Stephens. Dordrecht: Kluwer Academic Publishers, pp. 31–54.

Manning, C. E. and Ingebritsen, S. E. (1999). Permeability of the continental crust: implications of geothermal data and metamorphic systems. *Reviews of Geophysics*, **37**, 127–50.

Miller, S. (2002). Properties of large ruptures and the dynamical influence of fluids on earthquakes and faulting. *Journal of Geophysical Research*, **107**, 2182, doi:10.1029/2000/JB000032.

Miller, S. and Nur, A. (2000). Permeability as a toggle-switch in fluid-controlled crustal processes. *Earth and Planetary Sciences*, **183**, 133–46.

Miller, S., Nur, A. and Olgaard, D. (1996). Earthquakes as a coupled shear stress–high pore pressure dynamical system. *Geophysical Research Letters*, **23**, 197–200.

Miller, S. A., van der Zee, W., Olgaard, D. L. and Connolly, J. A. D. (2003). A fluid-pressure-controlled feedback model of dehydration reactions: experiments, modeling, and application to subduction zones. *Tectonophysics*, **370**, 242–51.

Molnar, P. (1988). Continental tectonics in the aftermath of plate tectonics. *Nature*, **335**, 131–7.

Montel, J.-M. and Vielzeuf, D. (1997). Partial melting of metagreywackes, part II. Compositions of minerals and melts. *Contributions to Mineralogy and Petrology*, **128**, 176–96.

Norton, D. and Knapp, R. (1977). Transport phenomena in hydrothermal systems: nature of porosity. *American Journal of Science*, **277**, 913–81.

Ochs, F. A. and Lange, R. (1997). The partial molar volume, thermal expansively, and compressibility of H_2O in $NaAl-Si_3O_8$ liquid: new measurements and an internally consistent model. *Contributions to Mineralogy and Petrology*, **129**, 155–65.

Oliver, N. H. S. (1996). Review and classification of structural controls of fluid flow during regional metamorphism. *Journal of Metamorphic Geology*, **14**, 477–92.

Paterson, S. R. and Vernon, R. H. (1995). Bursting the bubble of ballooning plutons: a return to nested diapirs emplaced by multiple processes. *Geological Society of America Bulletin*, **107**, 1356–80.

Patiño-Douce, A. E. and Beard, J. S. (1995). Dehydration-melting of biotite gneiss and quartz amphibolite from 3 to 15 kbar. *Journal of Petrology*, **36**, 707–38.

Patiño-Douce, A. E. and Harris, N. (1998). Experimental constraints on Himalayan anatexis. *Journal of Petrology*, **39**, 689–710.

Petford, N., Cruden, A. R., McCaffrey, K. J. W. and Vigneresse, J.-L. (2000). Granite magma formation, transport and emplacement in the Earth's crust. *Nature*, **408**, 669–73.

Rey, P., Vanderhaeghe, O. and Teyssier, C. (2001). Gravitational collapse of continental crust: definitions, regimes, mechanisms and modes. *Tectonophysics*, **342**, 435–49.

Royden, L. (1996). Coupling and decoupling of crust and mantle in convergent orogens: implications for strain partitioning in the crust. *Journal of Geophysical Research*, **101**, 17 679–705.

Rushmer, T. (2001). Volume change during partial melting reactions: implications for melt extraction, melt geochemistry and crustal rheology. *Tectonophysics*, **34**, 2/3–4, 389–405.

Rutter, E. H. (1997). The influence of deformation on the extraction of crustal melts: a consideration of the role of melt-assisted granular flow. In *Deformation-enhanced Fluid Transport in the Earth's Crust and Mantle*, ed. M. B. Holness. Mineralogical Society Series 8. London: Chapman and Hall, pp. 82–110.

Sawyer, E. W. (2001). Melt segregation in the continental crust: distribution and movement of melt in anatectic rocks. *Journal of Metamorphic Geology*, **18**, 291–309.

Simpson, G. D. H. (1999). Evolution of strength and hydraulic connectivity during dehydration: results from a microcrack model. *Journal of Geophysical Research*, **104**, 10 467–81.

Stevens, G., Clemens, J. D. and Droop, G. T. R. (1997). Melt production during granulite-facies anatexis: experimental data from "primitive" metasedimentary protoliths. *Contributions to Mineralogy and Petrology*, **128**, 352–70.

Thompson, A. B. (1990). Heat, fluids and melting in the granulite facies. In *Granulites and Crustal Evolution*, ed. D. Vielzeuf and Ph. Vidal. *NATO* Advanced Study Institute Series C, 311. Dordrecht: Kluwer, pp. 37–57.

Vanderhaeghe, O. and Teyssier, C. (1997). Formation of the Shuswap metamorphic complex during late-orogenic collapse of the Canadian Cordillera: role of ductile thinning and partial melting of the mid- to lower crust. *Geodynamica Acta*, **10**, 41–58.

Vanderhaeghe, O. and Teyssier, C. (2001). Crustal-scale rheological transitions during late orogenic collapse. *Tectonophysics*, **335**, 211–28.

Vielzeuf, D. and Montel, J.-M. (1994). Partial melting of greywackes. 1. Fluid-absent experiments and phase relationships. *Contributions to Mineralogy and Petrology*, **117**, 375–93.

Walder J. and Nur, A. (1984). Porosity reduction and crustal pore pressure development, *Journal of Geophysical Research*, **89**, 11 539–48.

Weinberg, R. F. (1996). The ascent mechanism of felsic magmas: news and views. *Transactions of the Royal Society of Edinburgh: Earth Sciences*, **87**, 95–103.

Wong, T.-F., Ko, S.-C. and Olgaard, D. L. (1997). Generation and maintenance of pore pressure excess in a dehydrating system, 2, Theoretical analysis. *Journal of Geophysical Research*, **102**, 841–52.

Wood. B. J. and Orville, P. M. (1992). Volatile production and transport during regional metamorphism. *Contributions to Mineralogy and Petrology*, **79**, 252–7.

13

Emplacement and growth of plutons: implications for rates of melting and mass transfer in continental crust

ALEXANDER R. CRUDEN

13.1 Introduction

Granitic intrusions are important tectonic elements of the middle and upper crust of all orogens, and their emplacement represents a major heat and material transport process that has operated throughout the geological record. Despite the importance of granitic intrusions for understanding of crustal evolution, the processes of granite generation, ascent, and emplacement are enduring problems, although great advances have been made over the past 25 years (Pitcher, 1979; Brown, 2001; Petford *et al.*, 2000). One problem that has thwarted progress in this area of research is that each stage in the granite formation process has typically been regarded in isolation, with the demarcation between problems often falling along traditional disciplinary boundaries (e.g., melting and crystallization are petrological problems; ascent and emplacement are geodynamic and structural problems, etc.). The challenge for future research will be to develop integrated models that use an interdisciplinary approach to explore the linkages between all aspects of granitic magmatism (e.g., Petford *et al.*, 1997, 2000).

In this chapter I attempt to establish a link between the emplacement of granites in the middle and upper crust and the processes involved in the withdrawal of melt from partially molten source regions in the lower crust. Using geophysical and field data I first develop a geometric characterization of plutons and use this to evaluate the "room problem" for their emplacement. It is shown that plutons are geometrically self-similar, governed by an empirical power law scaling relationship. Field observations and constraints on the 3-D shape of plutons are used to suggest that most space for plutons is made by subsidence of their floors. Kinematic models of floor depression involving piston and cantilever mechanisms indicate that pluton emplacement and growth is a geologically rapid process, with typical sized intrusions forming

Evolution and Differentiation of the Continental Crust, ed. Michael Brown and Tracy Rushmer. Published by Cambridge University Press. © Cambridge University Press 2005.

over thousands to hundreds of thousands of years, at geologically reasonable strain rates.

Although displacements on active structures (faults and folds) for space creation (e.g., Hutton, 1988; Vigneresse *et al.*, 1999; Grocott & Taylor, 2002) are important local factors in helping to generate space for plutons, the presence of granitic intrusions in all tectonic regimes, together with recent spatial–statistical studies (Schmidt & Paterson, 2000), suggests that an independent or largely internal control operates to create space during granite emplacement (Cruden & McCaffrey, 2001). Further insights into the mechanisms of pluton growth are gained by comparing granitic intrusions to natural and experimental subsidence structures. Such a comparison leads to several predictions about the nature of the wall rock and internal structure of plutons, which may be used to both test and refine the floor depression models. Links between the pluton growth process and coeval regional deformation are also discussed.

The limiting controls on pluton growth, size, and spacing appear most likely to be related to the degree of melting and the mechanics of melt withdrawal in the lower crust. This is the focus of the second part of this chapter, in which I use the power law scaling of pluton size to place constraints on the distribution and degree of partial melting in granite magma source regions. Application of the results to two different granitic terrains illustrates the importance of tectonic setting in determining the spatial distribution of plutons. In the case of a continental magmatic arc, where a high degree of partial melting is focused in a narrow belt, the resulting plutons are closely spaced and probably tabular in form. Conversely, in orogens where areally extensive source regions with low degrees of partial melting occur, plutons are widely spaced and most likely are wedge-shaped. The time averaged flux of melt leaving the source regions in both cases is of the same order of magnitude, although consistently higher for magmatic arc settings, where thermal inputs into the base of the crust are also likely to be higher. Table 13.1 provides a summary of the parameters and their units employed in this chapter.

13.2 Shapes, external and internal structures of granitic plutons

Historically, granitic plutons have been viewed as areally extensive intrusions with steep sides that continue to great depth in the crust (e.g., Buddington, 1959; Paterson *et al.*, 1996; Miller & Paterson, 1999). Such a perception is largely a function of a sampling bias; erosion of a several kilometer thick, tabular body in an area of low to modest relief will tend to favor exposure of steep pluton walls. Preservation of roofs and dissection through floors is only

Table 13.1. *A summary of parameters*

Nomenclature		Units
a	slope of linear regression curve and/or slope of pluton growth curve on L versus T plots	
b	intercept of linear regression curve on L versus T plots	
g	gravitational acceleration	m s^{-2}
h	transient pluton thickness	m
l	thickness of compacting source region	m
r	radius of a circular pluton in map view	m
t	pluton filling time	s, yr
H_s	source region thickness	km
L	mean width of a pluton in map view	km
N	pluton spacing factor	
P_d	magma driving pressure	Pa
P_h	hydrostatic pressure	Pa
P_m	magma overpressure	Pa
P_v	viscous pressure drop	Pa
Q_A	volumetric magma ascent rate	m^3 s^{-1}
Q_E	volumetric pluton filling rate	m^3 s^{-1}
Q_W	volumetric melt withdrawal rate	m^3 s^{-1}
T	mean pluton thickness	km
V	pluton volume	km^3
V_m	melt volume generated in source region	km^3
V_s	total volume of source region	km^3
X_m	partial melt fraction in source region	
$\Delta\rho$	density contrast between crust and magma	kg m^3
λ	dike length	m
ω	dike width	m
μ	magma viscosity	Pa s
σ_H	horizontal tectonic stress	Pa
γ	bulk shear strain	
$\dot{\gamma}_c$	bulk shear strain rate in crust under pluton	s^{-1}
$\dot{\gamma}_f$	shear strain rate on pluton bounding fault	s^{-1}
$\dot{\varepsilon}$	pure shear strain rate in compacting source region	s^{-1}

likely in areas in which relief is greater than, or equal to, pluton thicknesses. This sampling problem is further compounded because uplift and erosion levels tend to stabilize close to the roofs of plutons (Leake & Cobbing, 1993; McCaffrey & Petford, 1997). However, a sufficient body of field- and geophysics-based data is now available in the literature to make several generalizations on the 3-D form of the majority of granitic plutons.

Because erosion is usually insufficient to expose both the floors and roofs of the majority of plutons, geophysical methods (e.g., gravity, magnetic or seismic) must be employed in order to estimate pluton thicknesses (e.g., Bott &

Smithson, 1967; Sweeney, 1976; Lynn *et al.*, 1981; Brun *et al.*, 1990; Evans *et al.*, 1994; Vigneresse, 1995; Amelgio & Vigneresse, 1999; Cruden *et al.*, 1999a; Roy & Clowes, 2000). Less frequently, tilted sections (e.g., John & Mukasa, 1990; Miller *et al.*, 1990), deep erosional dissection (e.g., Hamilton & Myers, 1967; Myers, 1975; Le Fort, 1981; Scaillet *et al.*, 1995; Rosenberg *et al.*, 1995; Skarmeta & Castelli, 1997; Grocott *et al.*, 1999) or analysis of structural patterns (e.g., Brun & Pons, 1981; Cruden *et al.*, 1999b) provide direct or projected estimates of pluton thicknesses. In these cases, both the roof and floor are commonly observed, whereas geophysical data typically only provide information on thickness from a sub-roof erosion level to the pluton floor. Despite this limitation, geophysical estimates of pluton thickness are similar in magnitude to field observations, as discussed in Section 13.2.9. Key observations on the nature of typical pluton structure are reviewed briefly in the following sections, with specific reference to four case studies (Figs. 13.1–13.4).

13.2.1 Dinkey Creek pluton, central Sierra Nevada batholith, California

The $\sim 800\,\mathrm{km}^2$, *c.* 102 Ma Dinkey Creek pluton (DCP; Fig. 13.1) is a well-exposed tonalite–granodiorite intrusion (Bateman, 1992), typical of the many hundreds of epizonal to mesozonal plutons that make up the Cordilleran batholiths of North and South America (e.g., Buddington, 1959; Hamilton & Myers, 1967; Pitcher, 1979). The pre-erosion thickness of the lobe structure in the southwest of the pluton is estimated to be between 900 and 3700 m, based on fluid-mechanical analysis of the magnetic fabric pattern, which also indicates that the pluton was fed by a northnorthwest-trending conduit (Cruden *et al.*, 1999b). Evidence for a relatively flat roof above the pluton is provided by remnants of metasedimentary rocks preserved at high elevations. Structure contours on the contacts of these roof pendants indicate that the roof now dips about 10° southwest. This is similar to the Cenozoic tilt of the adjacent Mount Givens pluton (MGP) determined by paleomagnetic studies (Gilder & McNulty, 1999), suggesting that the DCP roof was originally horizontal and located, on average, about 400 m above the present erosion surface. Geobarometry indicates that the pluton was emplaced at a paleodepth of about 15 km (Ague & Brimhall, 1988).

Repeated linear contact orientations, displacements of roof structure contours, and geometry of roof pendant contacts, suggest that the majority of the vertical contacts associated with the DCP, and irregularities in the roof, were controlled by pre-existing northwest-, northeast-, north- and east-trending faults and fractures, which acted as guides for vertical displacement of the

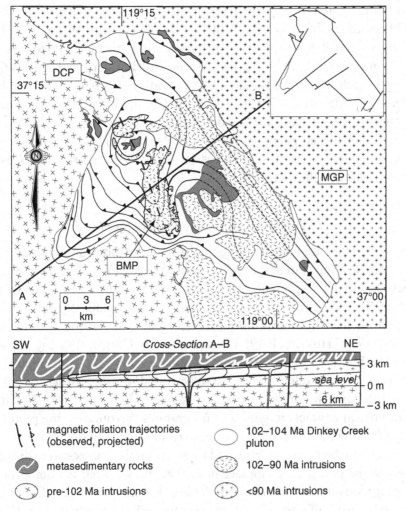

Fig. 13.1. Dinkey Creek pluton, central Sierra Nevada, California (after Bateman, 1992; Cruden *et al.*, 1999a). BMP = Bald Mountain pluton; DCP = Dinkey Creek pluton; MGP = Mount Givens pluton. Thick solid line in cross-section A–B decreasing in elevation from Northeast to Southwest is the present erosional surface; geology above this line was constructed using structure contours on roof pendants exposed at high elevations in the DCP; thickness of the DCP was determined by analysis of magnetic foliation data (Cruden *et al.*, 1999a). Northwest-trending dashed line with the DCP is the inferred location of the pluton feeder, based on the pattern of compositional zonation and foliation trajectories. Repetition of linear contact segments (mainly Northwest and Northeast) suggests pre-existing fracture control on location of pluton sides (inset).

pluton floor. Similarly oriented structures controlled the shape of the $40\,km^2$ Bald Mountain pluton and the $1500\,km^2$ MGP (Tobisch & Cruden, 1995; McNulty *et al.*, 2000). Evidence for pre-existing fracture control on sub-vertical pluton contacts has also been documented in Peru and northwestern Canada (Bussell, 1976; Dehls *et al.*, 1998). Field- and gravity-based studies in the Sierra Nevada batholith, California, the Coastal batholith, Peru, and the Patagonian Andes, Chile, have also determined pluton thicknesses in the range 1–3 km, similar to that proposed for the DCP (e.g., Hamilton & Myers, 1967; Myers, 1975; Coleman *et al.*, 1995; Sisson *et al.*, 1996; Skarmeta & Castelli, 1997; Haederle & Atherton, 2002).

13.2.2 Ljugaren granite, central Sweden

The *c.* 1700 Ma semi-circular Ljugaren granite intrudes amphibolite facies gneisses of the Svecokarelian orogen, Baltic Shield (Fig. 13.2). Its structural characteristics are representative of a post-tectonic mesozonal pluton (Buddington, 1959), emplaced into medium- to high-grade crystalline rocks. Such plutons have commonly been interpreted to be igneous diapirs (e.g., Sylvester, 1964; Holder, 1979; Bateman, 1984; Castro, 1986; Courrioux, 1987). The Ljugaren granite is bound to the west by the East Siljan fault, which is a Svecokarelian structure reactivated during formation of the Devonian Siljan Ring impact structure (Cruden & Aaro, 1992). Foliations and lithological contacts in gneisses and granites surrounding the pluton are bent into conformity with its steeply inward dipping contacts. Foliation trajectories are markedly asymmetric and suggest that the gneisses were displaced laterally to the east during emplacement. The granite itself is homogeneous, displays igneous microstructure and contains a weakly defined foliation concentric to its east half and a down-dip lineation where observed (Fig. 13.2).

The Ljugaren granite is associated with a residual gravity anomaly (>-6 mgal), centered on the west central part of the pluton within the East Siljan fault, which decreases outwards towards the margins (Cruden & Aaro, 1992). Model gravity profiles indicate that the bulk of the pluton is tabular and $<3\,km$ thick with a deeper ($\sim6\,km$) root on its west side (Fig. 13.2). Although the pluton roof is not exposed, regional considerations suggest that the granite spread beneath a cover sequence of syngenetic volcanic rocks. Examination of the structural pattern in combination with the gravity data suggests that the ductile wall rocks must have been displaced downwards as well as outwards during emplacement of the granite and that movement of magma was from west to east, emanating from a conduit coincident with the East Siljan fault (Cruden, 1998). Similar examples of asymmetric filling of a

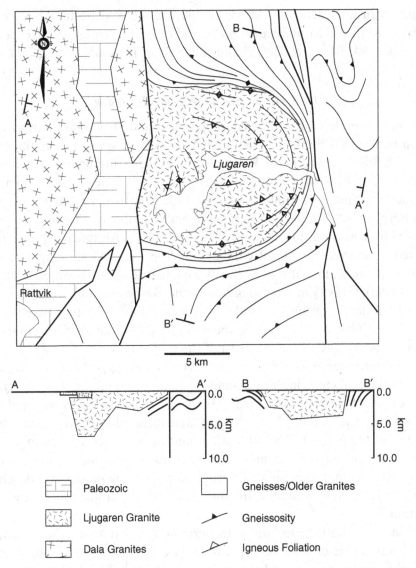

Fig. 13.2. Ljugaren granite, central Sweden (after Cruden & Aaro, 1992). Cross-sections A–A' and B–B' are based on 2.5-D forward models of residual gravity anomalies, combined with field-mapping data. Paleozoic sedimentary and Proterozoic igneous rocks west of the Ljugaren granite are exposed in the Devonian Siljan ring impact structure, which has modified the western margin of the pluton.

sill-like intrusion accompanied by ductile wall rock deflection have been documented in California, and western and central Canada (e.g., Cruden & Launeau, 1994; Brown & McClelland, 2000; Saint Blanquat *et al.*, 2001).

13.2.3 Gåsborn granite, central Sweden

A second example of a combined geophysical and structural analysis of a pluton is provided by the *c.* 1850–1650 Ma Gåsborn granite, Svecokarelian orogen, Baltic Shield, (Fig. 13.3), which has also been interpreted as a diapiric intrusion (Björk, 1986). Emplaced into greenschist facies metasedimentary and metavolcanic rocks and associated with a narrow (0–100 m) contact metamorphic aureole, the pluton is representative of a syn- to post-tectonic, epizonal to mesozonal intrusion (Buddington, 1959). Although the granite is exposed in an equant, almost circular, outcrop pattern, aeromagnetic and gravity data indicate that it is markedly asymmetric in cross-section (Cruden *et al.*, 1999a). Gravity models suggest that the Gåsborn granite consists of an up to 3 km deep, northwest–southeast trending root zone that overlies the trace of a major structural break in the wall rocks, a ~2-km thick mass west of the root, and a <1-km thick flap east of the root (Fig. 13.3).

The western margin of the pluton cuts across, and apparently deflects and overturns, the western limb of a tight, upright, regional syncline. A stratigraphic contact between metavolcanic units on the eastern side is deflected and truncated by the granite, but also dips under the pluton, steepening as it approaches the contact. The modest amount of wall-rock deflection on the western margin may be accounted for by regional post-emplacement transpressive strains that accumulated during cooling of the pluton and development of a northwest–southeast trending subvertical tectonic foliation within the granite.

The more dramatic deflection of stratigraphy adjacent to the eastern margin and its pluton-side-down deflection cannot be accounted for by regional post-emplacement strains, and has been attributed to the combined effects of lateral spreading of the pluton from the root zone and associated downfolding of wall rocks (Cruden *et al.*, 1999a). The Gåsborn granite is syntectonic in the sense that magma transport was channelled along a major regional structure, and emplacement and syn-cooling deformation occurred during a regime of regional transpression. However, space creation and the resulting 3-D form of the pluton appears to have been controlled by internal processes and variations in wall rock mechanical properties. This may be understood in terms of rates – ductile Svecokarelian strains accumulated over a time frame of millions of years, which is far longer than the time frame for pluton emplacement and

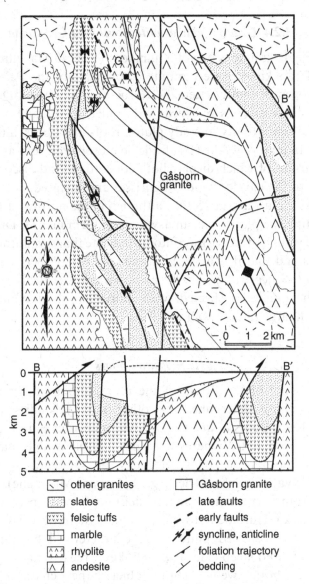

Fig. 13.3. Gåsborn granite, central Sweden (after Björk, 1986; Cruden *et al.*, 1999b). Cross-section B–B′ was constructed by integration of 2.5-D forward models of residual gravity anomalies and field data. G = Gåsborn; L = Långban.

crystallization (viz., Paterson & Tobisch, 1992). The structure of the Gåsborn granite and its relationship to regional ductile shear zones is similar to plutons in other transpressive orogens, such the New England Appalachians (Brown & Solar, 1998).

13.2.4 Graah Fjelde granite, southwest Greenland

The Graah Fjelde granite is one of a number of *c.* 1740 Ma, elliptical plutons composed of rapakivi-textured granite emplaced into the Paleoproterozoic Ketilidian orogen of south Greenland (Grocott *et al.*, 1999). Deep erosional dissection of these plutons by fjords with relief up to 2500 m provides an unprecedented opportunity to study pluton roofs and floors, although access is a problem. In map view, the Graah Fjelde pluton crops out as a 24 × 11 km body, the margins of which display both concordant and discordant contact relationships to the enveloping orthogneisses and metasediments (Fig. 13.4). Most contacts dip inward at 15–20°, although in the northwest the contact is steep, indicating the pluton has an asymmetric form in 3-D. Direct observations on fjord walls and structure contours indicate that the pluton is underlain by wall rocks and that it is exposed in the core of an open, basin-like fold, interpreted by Grocott *et al.* (1999) to be an emplacement-related feature. Internally, the granite is sheeted, with individual sheet boundaries being decorated with large rafts of wall rock.

13.2.5 Pluton floors

The characteristics of the floors of the four plutons described above are compared to those determined by other gravity studies in Fig. 13.5 (Vigneresse, 1995; Dehls *et al.*, 1998). Two first-order pluton floor geometries are observed (cf. Vigneresse *et al.*, 1999). Wedge-shaped plutons have one or more root zones and may be symmetric (e.g., Pontivy, Cabeza de Araya) or asymmetric (e.g., Ljugaren, Mortagne). Their floors dip inward from very shallow angles, defining broad open funnel shapes (e.g., Nordmarka–Hurdalen) to steep angles, defining carrot-like shapes (e.g., Ulu). Tablet shaped plutons are characterized by almost parallel roofs and floors and steep sides (e.g., Dinkey Creek, Graah Fjelde). Some plutons have both wedge and tablet-shape characteristics (e.g., Fichtelgebirge, Gåsborn). Almost all gravity studies find one or more funnel shaped root zones that are interpreted to be feeder structures (e.g., Ameglio & Vigneresse, 1999).

Field examples of the nature and geometry of pluton floors are relatively uncommon, for the reasons discussed above. However, limited observations in Greenland, North and South America and the Himalaya (e.g., Hamilton & Myers, 1974; Le Fort, 1981; Scaillet *et al.*, 1995; Hogan & Gilbert, 1997; Skarmeta & Castelli, 1997; Grocott *et al.*, 1999) are in general agreement with gravity models and seismic images of pluton floors. For example, several

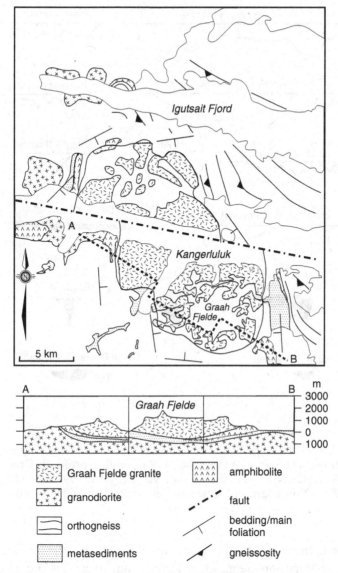

Fig. 13.4. Graah Fjelde granite, southeast Greenland (after Grocott *et al.*, 1999). White areas on land are ice-covered. Cross section A–B is based on field-mapping data and projection of observations on steep fjord walls into the line of section.

Proterozoic granites in South Greenland (e.g., Graah Fjelde and others, Grocott *et al.*, 1999) display well-exposed gently inward dipping bases that are commonly transgressive to wall-rock stratigraphy and in some cases show evidence for steepening towards a possible feeder zone (viz., Bridgewater *et al.*,

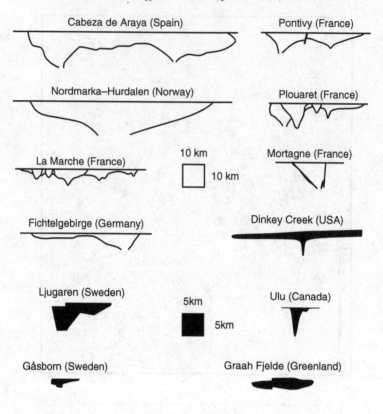

Fig. 13.5. Profiles of granitic intrusions determined from gravity and field observations. Data from Vigneresse (1995; unfilled profiles); Cruden and Aaro (1992); Dehls *et al.* (1998); Cruden *et al.* (1999a, 1999b) and Grocott *et al.* (1999). Note change of scale between unfilled and filled profiles.

1974). Syn-emplacement ductile fabrics are developed in the footwall rocks of some of these intrusions (e.g., Quernertoq) that overprint pre-emplacement regional fabrics. Ductile shear bands in the footwall rocks of the Quernertoq intrusion, together with the geometry of the contact, were previously interpreted to indicate emplacement into an active extensional ramp-flat structure (Hutton *et al.*, 1990). However, Grocott and others (see discussion in Hutton & Brown, 2000) consider these to be local structural features related to depression of the pluton floor. Similarly, Hamilton and Myers (1974) interpret map patterns associated with the base of the Boulder batholith, USA, to indicate that the intrusion is "bathtub-shaped" and that cuspate margin features represent lobes of granite that "flowed northward over the sinking floor."

Rosenberg *et al.* (1995) have mapped in 3-D the floor of the Bergell pluton in the central Alps. Although the floor has a gentle regional dip, it has a more complex geometry due to syn-emplacement folding.

In the Himalaya, the smaller Bhagirathi (Searle *et al.*, 1993) and Gangotri (Scaillet *et al.*, 1995) granites crop out as sub-horizontal lenticular bodies that may be controlled by flat-lying shear zones with syn-convergent, normal-sense displacements. The larger Manaslu granite appears to post-date displacement on the South Tibetan Detachment System and has a lower contact that is "rather flat and parallel to the main metamorphic cleavage, with a transition zone, a few hundred meters thick, where the gneisses and marbles are cross-cut by a network of sills and dykes" (Guillot *et al.*, 1993).

13.2.6 *Pluton roofs*

Paterson *et al.* (1996) have reviewed the characteristics of pluton roofs exposed in the Cordillera of North and South America. These roofs consistently show gentle dips to slightly domal morphologies and discordant contact relationships with pre-existing wall-rock structures (Fig. 13.6a, b). Furthermore, emplacement-related ductile strain in the wall rocks is typically absent to poorly developed and there is little evidence that the roofs have been lifted above their pre-emplacement position. Minor stoped blocks occur beneath the roof, and stoping is a likely candidate for generating the jagged profiles of the roofs (Fig. 13.6a, b), although the role of stoping as a major space-making mechanism is debatable, as discussed in Section 13.3. Other authors report more compelling evidence for upward displacements of pluton roofs (e.g., Morgan *et al.*, 1998a; Benn *et al.*, 1999; Grocott *et al.*, 1999). Hence, roof uplift may be an important contributor to the space-making process for some plutons, particularly in compressive and transpressive tectonic regimes, in which regional shortening may both aid in the roof-lifting process (e.g., Benn *et al.*, 1998) and act to squeeze magma upwards (e.g., Rosenberg *et al.*, 1995).

13.2.7 *Pluton sides*

Relatively undisturbed roofs, sharp transitions to steeply dipping walls, and the presence of either sharp wall-rock contacts or narrow strain aureoles with evidence for pluton-side down shear have been used by Paterson *et al.* (1996), Paterson and Miller (1998) and Miller and Paterson (1999) to argue that most space for emplacement of granites is due to downward transfer of material. Although these authors favor mechanisms such as stoping or return-flow

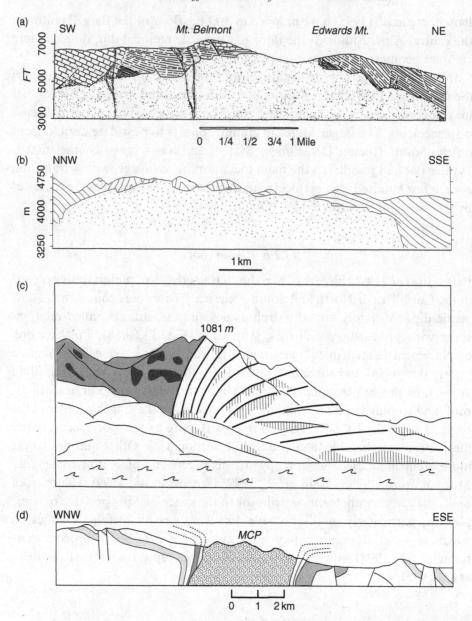

Fig. 13.6. Selected examples of pluton roofs (a, b) and sides (c, d). The roofs of the Marysville batholith (a) (Barrell, 1907) and Chita pluton (b) (Paterson *et al.*, 1996) are flat to bell-jar shaped and cut discordantly across pre-existing structures. Note also the sharp downturns to the pluton sides. In (c) and (d), two examples of pluton sides display well-defined downfolding, indicating downward displacement of wall rocks during pluton growth. In (c) (Quernertoq, Bridgewater *et al.*, 1974) gneissic layering is rotated towards the pluton margin, whereas in (d) (Marble Canyon plutons, California; Glazner & Miller 1997; Morgan *et al.*, 2000) the distortion of sedimentary formations defines down-folded wall rocks (rim monoclines).

during diapiric ascent, downward displacement and rotation of wall-rock structural markers and fabrics towards the margins of intrusions in Greenland, Sweden and North America (Fig. 13.6c, d) suggests that floor subsidence may be an important space-making process (Bridgewater *et al.*, 1974; Cruden, 1998; Benn *et al.*, 1999; Grocott *et al.*, 1999; Brown & McClelland, 2000; Potter & Paterson, 2000; Culshaw & Bhatnagar, 2001). Pluton-side down shear sense indicators and rollover of strata adjacent to some plutons have recently been ascribed to late-stage sinking of cooling magma bodies (Glazner & Miller, 1997; Sylvester, 1998; Morgan *et al.*, 2000). However, the observations around these intrusions could also be attributed to syn-emplacement floor subsidence, possibly accompanied by a component of lateral expansion of the pluton margins (see Section 13.3.9; Cruden, 1998), as exemplified by deflection of foliations around the Ljugaren granite (Fig. 13.2) and rotation of bedding–cleavage intersection lineations around the Eureka Valley–Joshua Flat–Bear Creek (EJB) pluton, California (Morgan *et al.*, 1998b). Large-scale tilting of roof pendants and wall rocks in the Sierra Nevada and Boulder batholiths has also been attributed to down drop of pluton floors during batholith growth and emplacement (Hamilton & Myers, 1974; Hamilton, 1988; Tobisch *et al.*, 2001).

13.2.8 *Internal structure of plutons*

Where not overprinted by tectonic strains (e.g., Fowler & Paterson, 1997), mineral and magnetic fabric studies of granites commonly reveal concentric magmatic foliation and lineation patterns that define one or more possible feeder zones (e.g., Bouchez, 1997; Cruden *et al.*, 1999b). These typically correspond to the root zones defined by gravity data where it is available (e.g., Benn *et al.*, 1999; Ameglio & Vigneresse, 1999). In some plutons, compositional zonation defines an annular pattern that is geometrically consistent with the presence of a feeder zone and indicates emplacement either as a series of nested pulses, or a progressive composition change during magma emplacement (e.g., Vigneresse & Bouchez, 1997; Cruden *et al.*, 1999b; Hecht & Vigneresse, 1999). Concentric structural patterns in plutons have also been delineated by finite strain analyses of deformed mafic enclaves and phenocryst distributions (e.g., Ardara granite, Holder, 1979; Molyneux & Hutton, 2000; Chinamora batholith, Ramsay, 1989). Finite strains are typically of the general flattening type and increase in magnitude from the core to the margin of the pluton. Such data have been used to support pluton emplacement by a ballooning mechanism, although the amount of space created by this process is controversial (i.e., between 30 and 80% see discussion in Molyneux & Hutton,

2000). As noted by Cruden (1998), the strain patterns in plutons ascribed to ballooning could also be produced by radial outward flow of magma from a central conduit into a horizontal widening and vertically thickening tabular body (see also Hunt *et al.*, 1953).

There is increasing evidence that some plutons, including those that are macroscopically homogeneous, are made up of many meter- to kilometer-scale sheets (e.g., McCaffrey, 1992; Everitt *et al.*, 1998; Cobbing, 1999; Grocott & Taylor, 2002). Detailed textural observations of intrusions in Maine, southwest Australia and southern New Zealand suggest that initially sub-horizontal sheets steepen with time during growth of a pluton (Wiebe & Collins, 1998). This is supported by U–Pb studies in the Coast Plutonic Complex, where plutons are interpreted to have grown from the floor upward by stacking of sheets and gradual subsidence and distortion of their floors (Brown & Walker, 1993; Brown & McClelland, 2000).

13.2.9 Empirical power law

McCaffrey and Petford (1997) proposed that both plutons and laccoliths display a scale-invariant relationship between their thickness, T, and width, L, that may be described by a simple power law of the form

$$T = bL^a. \tag{13.1}$$

Major axis regression on log T versus log L plots determined an intercept, b-value, of 0.12 (± 0.02) and slope, a, of 0.88 (± 0.10) for 135 laccoliths (Fig. 13.7). Although the fit for laccoliths seemed quite robust, the results of the analysis for plutons were based on only 21 observations (McCaffrey & Petford, 1997).

In order to better characterize the power law scaling of plutons, Cruden and McCaffrey (2001) utilized 66 studies of individual plutons, the majority (48) of these studies involved gravity surveying methods, and the remainder (18) employed field-based methods. They define the horizontal dimension of a pluton, L, as the equivalent diameter of a circle given by measurements of either the major and minor axes of elliptical bodies or their areas, and the thickness, T, is as the mean value where data are sufficient. The resulting data set (Fig. 13.7) spans over two orders of magnitude, from small plutons such as the Gåsborn granite ($L = 4.5$ km, $T = 1.6$ km; Cruden *et al.*, 1999a) to large-single plutons like the Lucerne granite ($L = 56$ km, $T = 4.5$ km; Sweeney, 1976) and composite batholiths such as the Sierra Nevada ($L = 600$ km, $T = 15$ km; Oliver, 1977). There is no obvious difference between pluton dimensions determined by field and geophysical methods, except for very large intrusions

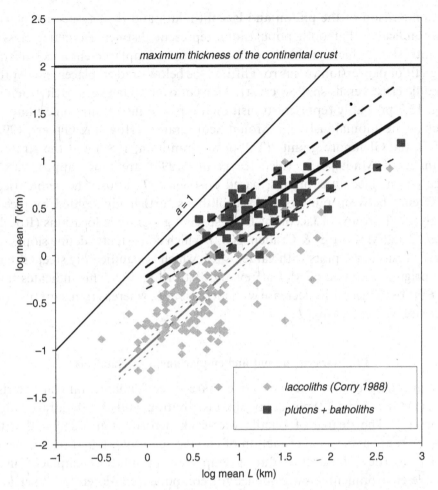

Fig. 13.7. Log thickness, *T*, versus log length, *L*, data for 66 plutons and 135 laccoliths as determined using geophysical methods and by field observations. Best-fit slopes (thick solid black and grey lines) and their 95th percentile confidence limits (dashed lines) were determined by reduced major axis regression (McCaffrey & Petford 1997; Cruden & McCaffrey 2001). The slope for a self-similar relationship between *T* and *L* ($a=1$) and the maximum thickness of continental crust (100 km) are shown for comparison. See text for discussion.

with $L > 50$ km, which are too large to provide field-based means for thickness estimation. Major axis regression of the pluton data yields a power law relationship (Eq. (13.1)) with a reasonable fit to the data characterized by an intercept value, $b = 0.6$ (± 0.15) and a slope, $a = 0.6$ (± 0.1). Error limits describe variation in the data at the 95th percentile level and define a field in log *T* versus log *L* space that incorporates most of the observations (Fig. 13.7).

Comparison of the pluton and laccolith data and their respective power laws indicates that both populations represent distinct geometric classes of intrusion, which probably reflects different emplacement mechanisms (e.g., floor depression versus roof lifting; see below) and emplacement depths (middle crust versus shallow crust). Areas of overlap between each group in Fig. 13.7 probably represent transitional types of intrusion. Some plutons, such as the anomalously thin Mount Scott granite (Hogan & Gilbert, 1995, 1997), the Glenmore granite (Talbot & Grantham, 1987) and the stratoid granites of Madagascar (Nédélec *et al.*, 1994) are more appropriately regarded as laccoliths or sills, based on their L/T ratios. The geometrical difference between plutons and laccoliths is further highlighted by recent analyses of groups of laccoliths from specific geographic locations (Rocchi *et al.*, 2002; McCaffrey & Cruden, 2002), which consistently define slopes on log T versus log L plots with values ~ 1.5, which is significantly steeper than the original estimate of McCaffrey and Petford (1997). This indicates that L/T ratios of laccoliths decrease with increasing L, whereas ratios for plutons increase with increasing L.

13.3 Granite ascent and emplacement mechanisms

The granite plutons reviewed above share three fundamental characteristics: (1) their shape; (2) power law size distribution, and (3) tabular to wedge geometry. The degree of ductile wall-rock deformation associated with intrusion appears to be a function of emplacement depth and the mechanical properties of the host during emplacement. Plutons emplaced in a ductile environment show evidence for components of lateral and vertical displacement of wall rocks, whereas those emplaced in brittle environments have involved only vertical translation of wall rock material, due to absence of wall-rock strains. Deeper-level, isolated granites with well-defined Bouguer gravity anomalies often have model shapes with flat to gently inward dipping floors and one or more roots, but no exposed roof. Higher-level intrusions preserve discordant flat roofs, steep sides and rare, but significant, evidence for gently inclined floors. Reasonable generalizations for many, but not all, granites are that: (1) they are emplaced as tabular- to wedge-shaped bodies with thicknesses ranging from 1 to 10 km; (2) they are fed by one or more vertical conduits, but the bulk of magma flow at the emplacement level is horizontal, and (3) the role of lateral displacement in creating space diminishes with decreasing ductility of the wall rocks. Any ascent and emplacement mechanism must account for these characteristics.

13.3.1 Diapirism

Tabular- to wedge-shaped bodies of granite may form by impingement and spreading of a diapir that has been arrested by a gently inclined mechanical barrier (e.g., Brun & Pons, 1981; Cruden, 1990) or by injection and subsequent thickening of a sill fed by a dike (Pollard & Johnson, 1973; Brisbin, 1986; Corry, 1988; McCaffrey & Petford, 1997). Evidence for lateral ductile displacement of wall rocks has been discussed above and is permissible evidence for spreading diapirs. However, it was also noted that the narrowness of the deformation aureole and the strain magnitude within it are unexpectedly low for a diapiric mechanism (Paterson *et al.*, 1991; Paterson & Fowler, 1993). Narrow strain aureoles around some plutons may still be accounted for by diapiric ascent if thermal softening (Marsh, 1982; Mahon *et al.*, 1988) and/or strain rate softening (Weinberg & Podlachikov, 1994) behavior occurs in the wall rocks. Although these models result in narrower strain aureoles than those predicted for diapiric ascent under isothermal, linear rheological conditions (Schmeling *et al.*, 1988), they will generate even higher strain intensities within the aureole, which have not been observed. More recently, ascent of "viscoelastic diapirs" has been proposed (Paterson & Miller, 1998), based on the analysis of dike propagation through a medium with variable viscous and elastic properties by Rubin (1993). Paterson and Miller (1998, their Fig. 2) envisage a spectrum from diapiric blobs with narrow ductile strain aureoles, through diapiric ridges with even narrower aureoles to discordant dikes. The problem of insufficient strain record still remains for diapirs and diapiric ridges, as does that of slow ascent rates that preclude long distances of magma ascent before freezing (Mahon *et al.*, 1988). Dike transport is discussed further in Section 13.3.3 below.

13.3.2 Stoping

A complete absence of wall-rock strain around higher-level discordant plutons precludes spreading of diapirs as a viable mechanism for their emplacement. Stoping is often invoked to explain the ascent and emplacement of such discordant intrusions (e.g., Barrell, 1907; Paterson *et al.*, 1996; Clarke *et al.*, 1998). However, energetic and thermal considerations suggest that stoping is unlikely to allow transport of granite magma more than a few thousand meters before freezing (Marsh, 1982, 1984). Furthermore, it is difficult to envision how magma ascending from its source by stoping would result in a tabular geometry at the level of final crystallization. Stoping is considered to be an important process in the modification of the roofs and walls of high-level intrusions, and although it may account for the last few hundreds of meters of magma ascent, it is probably unimportant for crustal-scale magma transport.

13.3.3 Channeled magma ascent

A long-standing objection to the dike transport of granite was that high-viscosity magma would freeze in the dike before a volume sufficient for filling a pluton could travel through it (Marsh, 1982). However, it is now realized that felsic melts have viscosities in the range 10^3–10^7 Pa s (Clemens & Petford, 1999; Dingwell, 1999). Consequently, dike transport of felsic magma may be sufficiently rapid to prevent freezing, and furthermore, dikes may fill pluton-sized upper crustal chambers over geologically rapid times (Petford *et al.*, 1993; Petford, 1996; Cruden, 1998). Alternative, but still rapid, mechanisms for channeled flow of low-viscosity magma are transport within mobile hydrofractures (Weertman, 1971; Bons & van Milligen, 2001) and pervasive flow (Brown, 1994; Collins & Sawyer, 1996; Weinberg & Searle, 1998; Weinberg, 1999). In the former, a penny-shaped, magma-filled crack ascends buoyantly by simultaneously propagating its tip and closing its tail. Dikes differ from hydrofractures in that they must stay open from top to bottom. Recent experimental work on hydrofractures suggests that they may allow step-wise accumulation of magma pulses into a pluton, leaving very little trace of their ascent in the underlying crust (Bons *et al.*, 2001). During pervasive flow, magma ascends upwards through an extensive network of channels that are formed by active deformation and kept open by a combination of dilational strain and magma pressure (Collins & Sawyer, 1996; Brown & Solar, 1998). This results in an intrusive migmatite terrain or injection complex that may be capped by tabular granites (e.g., Weinberg & Searle, 1998; Brown & Solar, 1999). Recent 1-D thermal modeling (Leitch & Weinberg, 2002) has shown that magma may flow pervasively as long as the host rock is above the magma solidus. Although incoming magma has the potential to push isotherms further up in the crust, transport by pervasive flow seems to be limited to only several kilometers in the lower to middle crust. Because dike transport lies in-between hydrofracturing and pervasive flow – in the sense that hydrofractures involve rapid, crust-traversing transport and pervasive flow involves dispersed, short range transport and connectivity between source and pluton, and because diking is relatively better understood process – I will assume this mechanism for granite magma transport below and in the following sections.

13.3.4 *Laccoliths versus lopoliths*

The driving force for dike transport is normally assumed to be buoyancy; both internal magmatic and tectonic over-pressuring (Robin & Cruden, 1994; Hogan *et al.*, 1998; Brown & Solar, 1998) are also attractive mechanisms for forcing magma up vertical conduits. Vertically propagating dikes must be arrested and thereafter be able to propagate horizontally in order to form a sill. Arresting mechanisms or "crustal magma traps" have been reviewed by Brisbin (1986), Corry (1988), Clemens and Mawer (1992), and Hogan *et al.* (1998) and include intersection with a freely slipping horizontal fracture, stopping of the propagating dike by a ductile horizon or a unit with high fracture toughness, and arrival at a level of neutral buoyancy.

Once an initial sill has formed, it may inflate, provided a sufficient magma pressure is available (Johnson & Pollard, 1973; Pollard & Johnson, 1973; Corry, 1988). Vertical inflation to plutonic dimensions may occur by roof lifting (i.e., laccolith emplacement), floor depression (i.e., lopolith emplacement), or a hybrid mechanism (Fig. 13.8; Cruden, 1998). The dynamics of laccolith emplacement by roof lifting are well established (Pollard & Johnson, 1973; Jackson & Pollard, 1988; Corry, 1988). Most models assume a two-stage process involving the formation of an initial sill following by vertical growth. Both stages are driven by the overpressure of magma at the emplacement site. The magma driving pressure, P_d is given by

$$P_d = P_h + P_m - P_v \pm \sigma_H \tag{13.2}$$

where P_h, P_m and P_v are the hydrostatic pressure, overpressure in the magma due to volatile release, and the viscous pressure drop, and σ_H is the magnitude of the tectonic stress normal to the dike wall (Table 13.1; Hogan *et al.*, 1998). Field and theoretical considerations show that vertical growth of laccoliths occurs by elastic, elastic-plastic or ductile bending of the roof rocks (e.g., Johnson & Pollard, 1973; Pollard & Johnson, 1973; Dixon & Simpson, 1987; Roman-Berdiel *et al.*, 1995), lifting of a piston by displacement on faults (e.g., Corry, 1988), or a combination of these mechanisms (Fig. 13.8; Jackson & Pollard, 1988). Note that "space creation" here is ultimately accommodated by surface uplift and subsequent erosion. The amount of vertical growth is a function of the horizontal cross-sectional area of the laccolith, the strength and effective thickness (which may be less than the true thickness if the rocks are mechanically anisotropic) of the roof rocks, and the available driving pressure (Pollard & Johnson, 1973; Dixon & Simpson, 1987). It appears that laccolith growth is self-limiting and rarely exceeds 2 km (Corry, 1988). The power law scaling of laccoliths is an expression of the limits placed on vertical growth, in

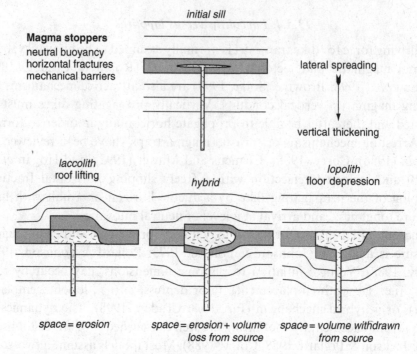

Fig. 13.8. Modes of tabular granite emplacement. On the left-hand side of each diagram, space for the granite (shaded) is created within the intruded unit (stippled) by offsets on faults, whereas space is created by ductile deformation on the right-hand side of each diagram.

which the force available for roof lifting is a function of the magma driving pressure and the area of the initial sill (McCaffrey & Petford, 1997). Further growth requires either rapid removal of the roof at the surface by erosion or gravity collapse, or simultaneous depression of its floor (Fig. 13.8).

With the exception of middle crustal plutons in which roof lifting was aided by tectonic compression discussed in Section 13.2.6, most studies consider laccoliths to be shallow-level intrusive phenomena, with all documented examples occurring at paleodepths <3 km (Corry, 1988). Experimental work suggests that the aspect ratios of laccoliths increase with depth. This is because horizontal sill propagation or growth is favored over roof lifting with increasing overburden thickness (Roman-Berdiel *et al.*, 1995). Corry (1988) proposes that at greater depths lopoliths form and that there is a continuous transition between intrusive styles from the epizone to the mesozone (Fig. 13.8). Although the depth of this transition is not well constrained it is noteworthy that the majority of the granitic plutons discussed in Section 13.2 were emplaced at paleodepths ≫3 km. The structural and geophysical attributes of these intrusions suggest that many plutons are geometrically similar to lopoliths (Fig. 13.8). However, the form and wall-rock structure of lopoliths are inferred to have formed by downward sagging of wall rocks after intrusion of the magma (Corry, 1988). The model proposed below is that space for incoming granitic magma is made by depression of the floor during intrusion.

13.3.5 *Pluton emplacement by floor depression*

Floor depression, or subsidence, has been considered as a possible space-making mechanism for granites for almost 100 years (Clough *et al.*, 1909; Cloos, 1923; Hamilton & Myers, 1967; Lipman, 1984). Most early models for whole-scale subsidence of pluton floors during emplacement invoked the presence of a magma chamber or reservoir in the lower crust (Branch, 1967; Whitney & Stormer, 1986) or in the crust underlying the growing intrusion (Myers, 1975; Bussell *et al.*, 1976; Pitcher, 1979). In such "cauldron subsidence" models, the intervening crust is assumed to drop down into an underlying magma chamber, usually guided by pre-existing fractures. Downward displacement of pluton floors by ductile flow mechanisms has also been discussed (Hamilton & Myers, 1967; Brown & Walker, 1993), possibly aided by isostatic depression of the Moho (Brown & McClelland, 2000). More recent models for floor depression take into account the notions that granitic magmas are probably extracted from partially molten lower crust (e.g., Brown, 1994; Thompson, 1999), that magma transport is probably channeled and rapid, and that the rates of melt extraction, ascent and emplacement must be balanced at the crustal scale (Cruden, 1998). The kinematics of this model are reviewed below.

Figure 13.9 illustrates schematically how pluton growth in the upper to middle crust by floor depression may be accommodated by volume loss due to melt withdrawal in the lower crust. The crustal column between the growing emplacement site and the source can be viewed as foundering into a deflating layer of partial melt, and the volume of melt lost from the lower crustal reservoir must be balanced by the volume emplaced in the pluton. Hence, the volumetric withdrawal rate in the source, Q_W, and the volumetric flux in the feeder dike (or dikes), Q_A, and the volumetric filling rate of the pluton, Q_E, must all be equal (Fig. 13.9a).

The two geometric scenarios considered here are discussed further in Section 13.4. In Fig. 13.9, a melt is extracted from a source volume restricted to the area directly beneath the pluton, which (arbitrarily) grows asymmetrically via a feeder on one side. Deflection of originally horizontal markers

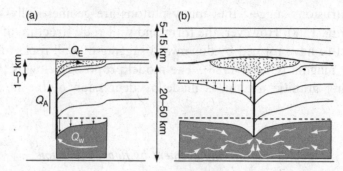

Fig. 13.9. Hypothetical scenarios for pluton emplacement by floor depression driven by withdrawal of melt from an underlying source region. In (a) the pluton (stippled) is asymmetric and fed by a conduit located on one side. Melt is withdrawn differentially from a partially molten source (shaded) the area of which is confined to the area below the pluton. Solid lines are initially horizontal markers. Dashed line above source indicates its original thickness. Thickness ranges of roof rocks, the pluton and the underlying crust are approximate and not shown to scale. Arrows indicate the flow of melt within the system, and the definitions of extraction (Q_W), ascent (Q_A) and emplacement (Q_E) flow rates. In (b) a symmetric, centrally fed, pluton is derived from a source region with horizontal dimension much greater than the final width of the pluton. The thin dashed line on the left below the pluton indicates the original position of a horizontal marker and arrows show its displacement during pluton growth. Lengths of white arrows in source indicate relative flow velocities schematically.

below the intrusion indicates that floor depression must be accommodated by deformation of the crust between the source and the growing pluton, and within the source itself. Mechanisms, structures, and rates involved in this deformation are discussed in Sections 13.3.6–13.3.10. In this model, and the one shown in Fig. 13.9b, the pluton is inferred to deepen towards the feeder, in accord with the floor geometry determined in many geophysical studies (Section 13.2.5; Fig. 13.5). Such geometry might arise due to differential melt withdrawal and resulting subsidence in the source. If melt extraction is driven primarily by density-driven flow into a conduit, then differential melt withdrawal will be a natural consequence of decreasing melt flow velocities away from the evacuation point. When melt is withdrawn from a source directly below the pluton (Fig. 13.9a), the degree of partial melting must be very high, or the thickness of the source must be substantially greater than the final thickness of the intrusion (see Section 13.4 for quantification of this statement). For the geometry sketched in Fig. 13.9a, about 25% of the original source volume must be extracted in order to grow the overlying pluton.

An alternative hypothetical source-pluton geometrical relationship is illustrated in Fig. 13.9b. Here, the intrusion grows symmetrically from a central feeder, and the source region has a much greater horizontal area than the final pluton. Because melt is being scavenged from a much larger source region, the degree of partial melting or the thickness of the source region potentially may be much less than the case shown in Fig. 13.9a. In this example, about 10% of the volume of the source has been removed to grow the overlying pluton (see also Section 13.4). One consequence of extracting melt from a source in which the horizontal area is much greater than the final intrusion is that deflation of the source must also lead to subsidence of the crust both below and adjacent to the pluton. This is illustrated by displacement of originally horizontal markers both above and below the intrusion (Fig. 13.9b). At the surface, this subsidence might be expressed by a regional reduction of relief that would otherwise be ascribed to extensional tectonics or isostatic effects.

Some authors have suggested that major pulses of granitic magmatism in some orogens were coeval with regional extension (e.g., Tobisch *et al.*, 1995; Collins, 2002). In the context of the model presented here (Fig. 13.9b), crustal extension would help the pluton formation processes in three ways: (1) it would help to accommodate regional subsidence caused by widespread melt withdrawal from the lower crust; (2) it would provide an additional driving force for melt extraction due to increased vertical stress acting on the source region, and (3) it would promote greater magma driving pressures due to reduction of σ_H across vertical conduits (Eq. (13.2); Hogan *et al.*, 1998).

13.3.6 Strain rates associated with floor depression

The deformation mechanisms that allow the downward transfer of material during pluton growth are not well constrained. However, given that this deformation occurs in the lower to middle crust (Fig. 13.9) of regions with high heat flow, it is reasonable to assume that the principal mechanism will be by ductile flow, aided by displacements on shear zones in regions of active regional deformation (e.g., Grocott & Taylor, 2002), (see Sections 13.3.9 and 13.3.10). Intuitively, rapid rates predicted for magma transport in dikes and filling of large tabular plutons (e.g., Petford, 1996) appear to be at variance with the concept of ductile flow of wall rocks, and therefore low strain rates of the middle and lower crust. Indeed, this problem is highlighted by the analysis of Johnson *et al.* (2001), who showed that wall rock strain rates due to outward expansion of spherical and spheroidal plutons ($L = 10$ km) being filled by a dike are of the order of from 10^{-6} to 10^{-12} s^{-1}, up to 6 orders of magnitude greater than the fastest accepted tectonic strain rates ($\sim 10^{-13}$ s^{-1}, Pfiffner & Ramsay, 1982).

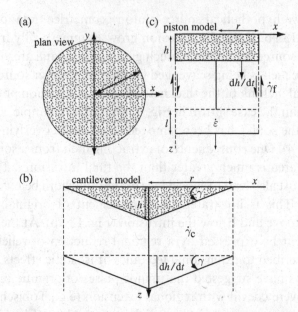

Fig. 13.10. Model geometries and parameters for calculation of strain rates in floor depression models. (a) Plan view of top of pluton (stippled) for both models. (b) Cantilever model. (c) Piston model. See text and Table 13.1 for definition of symbols.

This problem is evaluated below by considering an alternative mechanism for pluton growth by floor depression. Two simple end-member models are considered, the geometry of which are consistent with the wedge and tabular forms of plutons discussed in Section 13.2 (Fig. 13.10). Both models are circular in plan view (Fig. 13.10a) with radius r but differ in their cross-sectional geometry. It is also assumed in both cases that pluton inflation starts after the formation of an initial sill, although this may not be strictly valid in plutons with ductile wall rocks. The plutons are fed by a central conduit, which could correspond to a linear (dike) or point (pipe) source. The models may be easily formulated for asymmetric cases, in which magma is fed from one side (Cruden, 1998) and for plutons that are elliptical in plan view.

In the "cantilever" model, pluton growth occurs by tilting of the floor about a pivot point located at the perimeter of the pluton, farthest away from the feeder dike or pipe (Fig. 13.10b). This is similar to the trapdoor style of caldera growth (Lipman, 1997). The underlying crust deforms by bulk-progressive simple shear as it sinks into a region of partial melting in the lower crust. It is assumed that resistance to slip on the melt-filled conduit is negligible. In the "piston" model, growth occurs by depression of a horizontal floor, which is accommodated by a vertical, cylindrical, shear zone (or ring dike) at the perimeter of the pluton (Fig. 13.10c). The piston subsides into a partially molten source region at depth. In both cases, the pluton is filled at a rate Q_E, which is assumed to be constant. For the cantilever model

$$Q_E = \frac{dV}{dt} = \frac{\pi r^2}{3} \frac{dh}{dt},$$ (13.3)

where V is the pluton volume, and h is the depth of the floor adjacent to the feeder zone (Fig. 13.10b). The bulk shear strain in the underlying crust is

$$\gamma = \frac{h}{r},$$ (13.4)

and the shear strain rate is given by

$$\dot{\gamma}_c = \frac{1}{r} \frac{dh}{dt} = \frac{3Q_E}{\pi r^3}.$$ (13.5)

This is also the shear strain rate in the source region if it is assumed that the magma is withdrawn from an area of similar volume and shape in the lower crust (Fig. 13.9b). However, since melt percentages in the source are likely to be low, melt extraction and related deformation will be distributed over a much larger volume (Fig. 13.9b, Section 13.4). Hence, this shear strain rate is a maximum value for the source region. In the piston model

$$Q_E = \pi r^2 \frac{dh}{dt},$$ (13.6)

and the strain rate in the compacting source region is

$$\dot{\varepsilon} = \frac{1}{l} \frac{dh}{dt} = \frac{Q_E}{\pi r^2},$$ (13.7)

where l is the compaction length in the source (Fig. 13.10c). In the current geometry $l = h$, which implies that melt is extracted from an equivalent volume to that of the final pluton in the source. As in the cantilever model, this strain rate is also an upper bound on the deformation rate in the source region.

The variable of greatest uncertainty in the above equations is the rate of pluton filling, Q_E. A reasonable estimation may be proposed by equating Q_E with the volumetric flow rate in a feeder dike (or dikes) such that

$$Q_A = Q_E = \frac{\Delta\rho}{12\mu} g w^3 \lambda,$$ (13.8)

where $\Delta\rho$ is the density difference between crust and the magma, μ is magma viscosity, g is gravitational acceleration, w is the dike width, and λ is its length

(Table 13.1; Petford, 1996). Taking the following ranges in these parameters, $\Delta\rho =$ from 10 to 400 kg m^3, $\mu =$ from 10^4 to 10^8 Pa s, $\omega =$ from 1 to 10 m, and $\lambda =$ from 1 to 10 km, gives possible flow rates ranging from $<10^{-2}$ to $>10^3$ m^{3}s^{-1}. These values are in line with observed eruption and magma chamber filling rates in volcanic systems (e.g., Dzurisin et al., 1994; Rutherford & Gardner, 2000).

Using the above Q_E values as lower and upper bounds, strain rates required to fill plutons that grow to a final thickness, T, as predicted by the empirical power law (Section 13.2.9), have been determined in terms of the time required, t, and the pluton width (diameter), L, for both model geometries (Fig. 13.11). Strain rates fall within the range of from 10^{-9} to 10^{-16} s^{-1} to fill from 10 to 1000-km wide plutons in 10 to 10^8 yr. For a constant filling time, in the piston model the strain rate varies only as a function of Q_E (Fig. 13.11b), whereas in the cantilever model the strain rate is also dependent on pluton width (Fig. 13.11a), which is due to the rotational nature of deformation associated with the cantilever model.

Several limits may be placed on the application of these results to nature. Using the intrusions summarized in Section 13.2, pluton widths fall in the range L from $\approx$5 to 200 km. A lower bound on the filling time may be determined from estimates of solidification times of plutons, which are likely to vary from $\sim$1 yr to $\sim$1 Myr (e.g., Paterson & Tobisch, 1992). Upper bounds on the emplacement time may be made using average rates of tectonic defor- mation (10^{-13} to 10^{-15} s^{-1}; Pfiffner & Ramsay, 1982). Nyman et al. (1995) estimated strain rates in the Papoose Flat pluton wall rocks were between 10^{-12} and 10^{-13} s^{-1}, based on finite strain, thermobarometry and model cooling data, with emplacement occurring in $<80\,000$ yr. Faster strain rates on the order of 10^{-10} s^{-1} have been estimated for the growth of folds and displacement within shear zones (e.g., Paterson & Tobisch, 1992). A range of strain rates from 10^{-10} to 10^{-15} s^{-1} is also permissible on rheological grounds, given the likely temperature and stress variation in and above a granite source region.

In the piston model, estimates of the required strain rates on the bounding faults are considerably faster. For example, a 1-m wide fault requires a shear strain rate, $\dot{\gamma}_f$ of 10^{-6} s^{-1} for $\dot{\varepsilon} = 10^{-10}$ s^{-1}, and the same fault will deform with $\dot{\gamma}_f = 10^{-10}$ s^{-1} when $\dot{\varepsilon} = 10^{-13}$ s^{-1} (Cruden, 1998, Fig. 8). Such fast rates are reasonable if the fault is lubricated by magma (e.g., Spray, 1997). However, if it is not, this places a limit on the efficiency of piston sinking for space creation, and suggests that the cantilever mechanism may be energetically more favorable.

The analysis above demonstrates that considering floor depression as an alternative mechanism for pluton growth alleviates the problem of

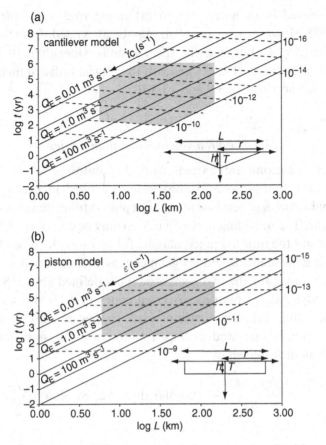

Fig. 13.11. Strain rates, filling rates, and times associated with growth of plutons by floor depression. Both nomograms show the time, *t* yr, required to fill plutons of a variety of widths, *L*, assuming that they grow to a final thickness governed by the empirical power law (Eqn. (13.1)) for a range of possible volumetric emplacement flow rates (Q_E m^{3}s^{-1}; solid lines). The corresponding strain rates ($\dot{\gamma}_c$ and $\dot{\varepsilon}$ s^{-1}) are indicated with dotted lines. (a) Cantilever model. (b). Piston model. Shaded areas indicate likely filling times and widths for tabular granite plutons, as discussed in the text.

unreasonably fast wall-rock strain rates predicted by models of lateral expansion of dike-fed plutons (Johnson *et al.*, 2001). Geologically rapid dike-fed pluton growth requires bulk strain rates in the underlying crust (cantilever model) or compacting source region (piston model) between 10^{-10} and 10^{-16} s^{-1}, which overlaps values estimated for orogens, pluton aureoles, folds, and shear zones (Pfiffner & Ramsay, 1982; Paterson & Tobisch, 1992; Nyman *et al.*, 1995). Faster strain rates (from 10^{-6} to 10^{-13} s^{-1}) are required if vertical growth is focused on a narrow (<10 m) single ring fault (piston model; Cruden, 1998). However, this problem may be alleviated if the fault is lubricated by magma (see above), if the shear zone is wider than 10 m, or if the downward displacement of rocks beneath the pluton is distributed on several shear zones (see Sections 13.3.9 and 13.3.10).

13.3.7 Pluton volumes and filling times

As noted earlier, a geometric characterization of pluton geometry, such as that provided by the power law relationship between T and L proposed above, may be used to make useful generalizations about pluton filling times, source region geometries, etc. The net filling time of a pluton may be determined, knowing its final volume and the time averaged magma filling rate (Q_E m^3 s^{-1}). Note that the volume is dependent on pluton geometry as well as absolute values of L and T. For example, for the two geometric types defined above (Section 13.2, Fig. 13.5) wedge-shaped plutons (inverted cones in 3-D) will have volumes two-thirds less than tabular intrusions (disk shapes in 3-D). Hence, for the same magma supply rate, wedge-shaped plutons will fill up faster than their tabular equivalents as given by

$$t = \frac{1000^3 \pi b L^{(a+2)}}{4 Q_E} \text{ (tabular disk-shaped plutons)}, \tag{13.9}$$

and

$$t = \frac{1000^3 \pi b L^{(a+2)}}{12 Q_E} \text{ (wedge-shaped plutons)}, \tag{13.10}$$

where t is filling time in seconds, L is in kilometers, and Q_E is in cubic meters per second (Table 13.1).

Using the Q_E values estimated in Section 13.3.6, the range of wedge and tablet-shaped pluton filling times is illustrated in Fig. 13.11. Note that these volumetric flow rates assume a continuous supply of magma into the pluton, whereas field evidence suggests that magma often arrives in batches that vary in thickness from meter- to decimeter-scale sheets to much larger pulses. If we assume that the time interval between each pulse is relatively short, then it is reasonable to equate the Q_E values employed as reasonable bounds on time-averaged supply rates. Pluton emplacement is potentially a geologically rapid process, with median filling times between 1000 and 10 000 yr for smaller plutons ($L = \sim 10$ km), between 10 000 and 100 000 yr for larger plutons ($L = \sim 50$ km) and $\gg 1$ Myr for batholiths (Fig. 13.11). If the time gap between pulses is long, then obviously an individual pluton could take much longer to form. For example, a 10-km wide, 2.4-km thick tabular intrusion made up of 200 12 m thick pulses each delivered at $Q_E = 1 \text{ m}^3 \text{ s}^{-1}$, could take a minimum of 6000 yr to fill if the time duration between each pulse was of the order of days to a few years. In this case, it takes 30 yr to emplace each pulse. If the time interval between injection of each pulse is increased to 1000 times longer than the emplacement time for each sheet, then the total pluton emplacement duration would be 6 Ma. Pluton construction durations between 6 and 8 Myr have been determined for construction of similarly sized plutons in the Coast Plutonic Complex using detailed geochronology (Brown & McClelland, 2000). Hence, although the ascent and emplacement (and withdrawal) of each pulse may be a rapid process, the net emplacement times for plutons shown in Fig. 13.11 should be regarded as upper bounds. Clearly, more detailed geochronological work is required to evaluate these rates and to put better constraints on time lags between pulses.

13.3.8 Structural accommodation of pluton floor subsidence and analog studies

The dimensional similarity displayed by plutons suggests that they must form by one or more scale-independent mechanisms. McCaffrey and Petford (1997) recognized three potential growth modes, characterized by the slope(s) of their growth curve(s) on log T versus log L plots: (1) self-similar growth ($a = 1$) in which the pluton horizontal–vertical dimensional ratios remain exactly the same with growth; (2) self-affine growth, where the ratio of horizontal–vertical dimension changes with time ($a > 1$ means vertical inflation > horizontal elongation; $a < 1$, horizontal elongation > vertical inflation), and (3) two-stage growth (Fig. 13.12). The observed power law relationship with $a \sim 0.6$ rules out self-similar growth, but cannot differentiate between self-affine growth from a very small, roughly equant "birth size" along a slope < 1 and two-stage growth involving formation of an initial sill with $L \gg T$ followed by vertical inflation along an $a > 1$ growth line (Fig. 13.12).

To address this problem and gain further insight into structural accommodation mechanisms involved in floor depression, Cruden and McCaffrey (2001) reviewed both natural and experimental surface subsidence features ranging in scales from centimeters (laboratory experiments), meters to hectometers (ice pits, ice cauldrons and sink holes), to kilometers (calderas) (see also Branney, 1995). All surface depressions formed due to physical removal of material from below are geometrically similar, lying on a curve with $a \sim 1.0$ in log T versus log

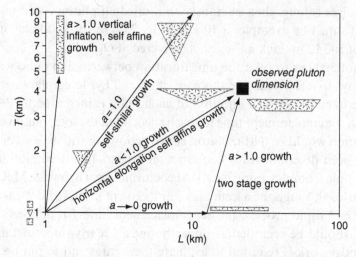

Fig. 13.12. Possible modes of pluton growth (after McCaffrey & Petford, 1997). A pluton with a given observed size (black square) may grow by either one-stage self-affine growth along a growth line with slope $a > 1.0$ or by a two stage mechanism involving initial lengthening along a growth line with $a \sim 0$ (sill injection), followed by vertical inflation along a growth line with $a > 1.0$.

L space. Plutons cluster in the same field as calderas, lending support to the proposal that they are also subsidence phenomena.

Analysis of laboratory subsidence experiments indicated that they grow in a similar fashion to the two-stage growth option illustrated in Fig. 13.12. The width of the structure at the end of the first stage is a function of the width of the source region and its depth. The second stage of growth follows a trajectory with a slope between $a = 6$ and ∞. If pluton formation by floor depression is regarded as an analogous process, this suggests that the power law scaling relationship (Section 13.2.9) is effectively a growth limit, in which the final intrusion thickness is dictated by the width of the initial sill achieved during the first stage and the amount of floor subsidence that is attainable. The initial sill width is a function of factors such as the emplacement depth, host rock rheology and driving pressure (Pollard & Johnson, 1973; Gudmundsson, 1990), which in turn is controlled by the distance between the source and emplacement site, source depth, magma physical properties, and tectonic environment (Eq. (13.2)). Limiting factors for the amount of vertical growth that is possible include the driving pressure and the amount of floor subsidence that may be achieved, which in turn will be controlled by crustal strength and the thickness and degree of partial melting in the source region (Section 13.4).

In addition to providing constraints on possible pluton growth modes, analog experiments also give us considerable insight into the detailed structural accommodation mechanisms that are likely to occur within the crust during floor depression. For example, Ramberg (1970) carried out a series of centrifuge experiments designed to study intrusion of magma under a wide range of conditions. The best-known of these are experiments on diapir formation in which a high-viscosity material ascended buoyantly through a stiffer overburden. However, Ramberg (1970) pointed out that with analog magma/wall rock viscosity contrasts close to 1, these experiments did not scale-up rheologically to nature. In a second set of experiments using much lower viscosity aqueous $KMnO_4$ solution as the analog magma and various putties as analog ductile and plastic wall rocks, Ramberg (1970) achieved a more realistic viscosity contrast of 10^{-10}. In his experiments M1 and M19, the buoyant $KMnO_4$ solution ascended through a horizontally layered overburden via a ladder-like series of dikes and sills controlled by the host anisotropy, and was eventually extruded at the free surface. In experiments I_1 to I_5 the model crust consisted of mechanically isotropic black modeling clay overlain by a stronger lid of white modeling clay and the $KMnO_4$ solution started in a cylindrical chamber at depth. In the centrifuge, the $KMnO_4$ solution chamber collapsed as the analog magma drained out via fractures to pond higher up beneath the strong barrier. The resulting "plutons" formed due to floor subsidence and have from wedge (Fig. 13.13a) to complex tabular (Fig. 13.13b) shapes in cross-section. Although Ramberg (1970) does not provide sufficient information for detailed geometrical and mechanical analyses of these experiments, we are provided with the only analog model studies to date of pluton growth by source region depletion and floor subsidence.

Several analog model studies have been carried out to study development of surface depressions formed in granular (Coulomb) materials associated with material withdrawal at depth. Sanford (1959) generated depressions by moving a rigid piston downward beneath a layer of sand. The resulting subsidence was accommodated by translation on normal faults and by downfolding. Marti *et al.* (1994) made scaled models of caldera formation by deflating spherical and elliptical balloons embedded in TiO_2 powder. In these experiments, flat floored, steep walled subsidence basins formed by down drop on both convex-up reverse faults and steeply inward-dipping normal faults. Mainly the size and shape of the underlying balloon and its depth controlled the size of the resulting depression. Ge and Jackson (1998) carried out a series of experiments designed to investigate surface subsidence structures caused by salt dissolution at depth. In these experiments, the shape, thickness, and burial depth of the analog salt material (silicone gum) was varied systematically, and

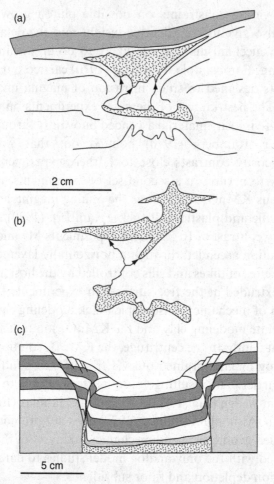

Fig. 13.13. (a) Line drawing of part of experiment I_5 of Ramberg (1970). Stippled areas contained $KMnO_4$ solution during experiment, with the lower area being a collapsed, initially cylindrical source and the upper area indicating the wedge-shaped pluton that formed by floor subsidence above. Arrows are inferred melt migration channels. Thin lines are representative structural markers in the host, indicating downfolding at the margins of the intrusion and substantial deformation below and within the deflated source. Gray layer above pluton is stronger material that prohibited further ascent of the $KMnO_4$ solution. (b) Line drawing of part of experiment I_1 of Ramberg (1970). An initially cylindrical chamber collapsed to form a complex tabular body above. No markers are visible in the host. (c) Line drawing of salt withdrawal experiment 250 of Ge and Jackson (1998). An initial analog salt (silicone gum; stippled) region with rectangular cross-section was drained from below, causing subsidence of an overlying sand pack. Subsidence was accommodated on both normal and reverse faults, indicated by displacement of passive markers (white and gray bands).

subsidence accommodation structures that formed in the overlying sand pack, as the salt analog was drained from the bottom, were documented by sectioning (Fig. 13.13c). Depressions formed above a wide tabular salt layer, drained from a central zone developed funnel geometries with gentle inward dips that grew by continuous deformation of the sand pack. Depressions that formed above laterally confined salt layers had flat-floored geometries that grew from an initial width similar to that of the underlying salt, then widened progressively as they deepened. In these experiments, floor subsidence was accommodated by displacements on both convex-up reverse faults and inward-dipping normal faults (Fig. 13.13c).

The experimental studies cited above indicate that growth of wedge and tabular shape plutons by floor depression is probably accommodated by a combination of bulk ductile flow, displacements on both normal and reverse shear zones, and downfolding of wall rocks adjacent to intrusion margins. In cases in which there is a large mechanical contrast between the intrusion and its wall rock, and/or when wall rocks have high strength, coupling between the growing pluton and its surroundings will be minimal, resulting in discordant margins. Under the opposite conditions, coupling between the pluton and its wall rocks will be significant, and downfolding and structural concordance is expected. Such a contrast in behaviors is observed along the contact of the South Mountain batholith (Culshaw & Bhatnagar, 2001). Where the granite is in contact with a heavily jointed slate unit, it is discordant. Contacts with a more massive sandstone unit are characterized by the development of a margin parallel foliation and downward rotation of pre-intrusive fold hinges towards the batholith.

The floor depression model presented here explicitly links the presence of downfolds or rim monoclines observed around many granitic intrusions (e.g., Bridgewater *et al.*, 1974; Cruden & Aaro, 1992; Glazner & Miller, 1997; Grocott *et al.*, 1999; Cruden *et al.*, 1999a; Morgan *et al.*, 2000; Potter & Paterson, 2000; Culshaw & Bhatnagar, 2001) to ductile coupling between a down-dropping floor and adjacent wall rocks during pluton growth. This model is therefore an alternative to that of Glazner and Miller (1997) who proposed that rim monoclines might form by late stage sinking of plutons after crystallization, but while their wall rocks were still thermally softened. This model is feasible if the crystallized pluton is denser than the surrounding rocks, as was the case in the area they studied. However, this does not hold in most of the other studies cited above, where the plutons must be less dense than their wall rocks, based on compositional criteria, negative Bouguer gravity anomalies, and petrophysical measurements. Furthermore, the timing of downfolding differs between the models. Glazner and Miller (1997) interpret the lack of solid-state strain in their plutonic rocks to indicate that rim monoclines form after the pluton had largely crystallized. In the floor depression model, the same observations imply that downfolding occurs before crystallization, when the magma has little or no strain memory. Hence, the kinematic record of floor down-drop will usually only be found in the wall rocks, unless passive markers, such as rotated compositional layers, are preserved in the pluton (e.g., Weibe & Collins, 1998; Grocott & Taylor, 2002).

13.3.9 *Links between pluton growth and regional deformation*

Although the above sections imply that pluton emplacement is an inherent consequence of melt withdrawal and transfer into the middle and upper crust (i.e., melt extraction, ascent, and emplacement may constitute a closed, internally driven system), numerous structural studies have demonstrated strong connections between regional deformation and granite intrusion processes (e.g., Pitcher, 1979; Castro, 1986; Hutton, 1988; Vigneresse, 1995; Brown & Solar, 1998; Saint Blanquat et al., 1998). Space-making mechanisms for granitic plutons have been linked to dilation within strike-slip faults (e.g., Guineberteau et al., 1987; Tikoff & Teyssier, 1992), reverse faults (e.g., Ingram & Hutton, 1994), normal faults (e.g., Hutton et al., 1990; Grocott et al., 1994), transpressional fault systems (e.g., McCaffrey, 1992; Benn et al., 1999), and folds (e.g., Schwerdtner, 1990; Roig et al., 1998). I propose that the conclusions of such studies are not at variance with the hypothesis that the granite emplacement process is driven primarily by melt withdrawal at depth and the need for melt to ascend to higher levels in the crust. Rather, the observed relationships between pluton emplacement and the development of regional structures is a natural consequence of the crustal and local scale interaction between vertical material transfer (i.e., magma up, wall rocks down) and tectonic deformation.

An illustrative case for such interactions is made by field-based observations in the Coastal Cordillera of North Chile where one or more margins of elongate Mesozoic plutons are frequently bound by high temperature mylonites associated with strands of the Atacama and Tigrillo fault systems (Grocott & Taylor, 2002). Detailed studies of the kinematics of these faults and the mylonite zones adjacent to and within the plutons indicate that space for granitic magmas was made by a combination of (mainly) floor depression and roof uplift, accommodated by syn-emplacement extensional to transtensional displacements on the regional fault systems. Based on the observed field relationships, Grocott and Taylor (2002) proposed

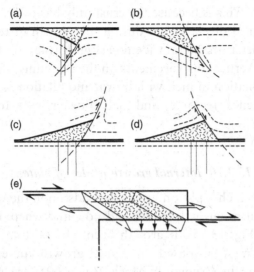

Fig. 13.14. Hypothetical links between pluton growth and displacements on active faults. Cross-sections (a-d) illustrate how displacements on active dip-slip faults (a and d, normal; b and c, reverse) may aid in the growth of a pluton (stippled) by floor depression (a and b) and roof uplift (c and d). (After Grocott & Taylor, 2002). In (a-d) thin dashed lines are originally horizontal marker horizons; thick black lines mark the initial intrusion level; and thin vertical lines represent possible magma feeders. Three-dimensional sketch (e) illustrates the structure of a pluton (stippled) emplaced in a transtensional step-over or releasing bend within a dextral strike-slip fault zone (thick black lines). Extensional strain (open arrows) within the step-over results in localization of magma and facilitates vertical pluton growth (thin arrows) due to floor depression.

several models that show how displacements on both active normal and reverse faults may accommodate pluton floor depression and roof uplift (Fig. 13.14a–d).

A similar approach may be applied to the case of plutons located in releasing bends of strike fault systems (e.g., Guineberteau *et al.*, 1987; Hutton, 1988). Although published studies have emphasized the role of lateral widening for creating space for magma, the primary space-making mechanism for intrusions in such settings may also be floor depression (Fig. 13.14e). The argument here is analogous to deposition in pull-apart basins, in which the dominant sediment accommodation mechanism involves subsidence of the basin floor (Sylvester, 1988). If magma emplacement in such settings is controlled primarily by vertical rather than horizontal displacements, then many of the objections to pluton emplacement in transtensional structures, based on rates of lateral widening and magma cooling rates (e.g., Yoshinobu *et al.*, 1998), become less severe. This is because the crust only has to accommodate a few kilometers of vertical displacement during the growth of the pluton, rather than tens of kilometers of lateral widening (Fig. 13.14e). In the case of pluton emplacement, the vertical displacements on the bounding faults are probably driven by a combination of melt withdrawal and dilation associated with the releasing bend. Hence, tectonics and melt transfer work together to create space for the pluton.

13.3.10 *Internal growth modes of plutons*

The processes by which a pluton grows depends upon the wall-rock accommodation mechanism, growth curve, and how and where magma is added to the intrusion (Fig. 13.15). If growth occurs by bottom-up growth, governed by an $a \approx 6$ curve (based on typical growth curves for subsidence structures reviewed in Cruden & McCaffrey, 2001; see also Fig. 13.12), then the resulting pluton will have a dish-shaped geometry (Fig. 13.15a). The internal structure will be massive if magma is added continually, or horizontally sheeted if it arrives in pulses (Fig. 13.15b). Note that in the latter case, evidence for sheeting may be destroyed by thermal- and/or gravity-induced flow if the rheological contrast between sheets is minimal, as would be expected for short time intervals between injection of pulses (Bergantz, 2000).

If growth occurs by top-down addition of magma following an $a \approx 6$ curve, then the resulting pluton will either have a bell-jar shape when a horizontal floor drops with time (Fig. 13.15c), or a wedge shape if the floor rotates downwards with time (Fig. 13.15d). In the latter scenario, the hinge point migrates outwards with time and the older intrusive phases will be exposed towards the center of the pluton after erosion. When growth occurs with $a \approx \infty$, then the resulting pluton will have either a tabular or wedge shape (Fig. 13.15e, 13.15d), depending on whether floor depression occurs by a piston-like or cantilever-type mechanism. If magma is always added to the top of the growing pluton, then piston-like growth results in a vertically stacked intrusion with young to old stratigraphy (Fig. 13.15e). Progressive $a \approx \infty$ growth of a wedge-shaped pluton results in an inward younging of material in map view and, if emplacement occurs by sheet addition, the sheet dip will increase with relative age (Fig. 13.15f; Weibe & Collins, 1998). Again, in each of these scenarios, the potential exists for homogenization and destruction of sheets if the time interval between delivery of pulses is short. In particular, gravity-induced mixing of sheets would be encouraged during progressive steepening of internal contacts (Bergantz, 2000).

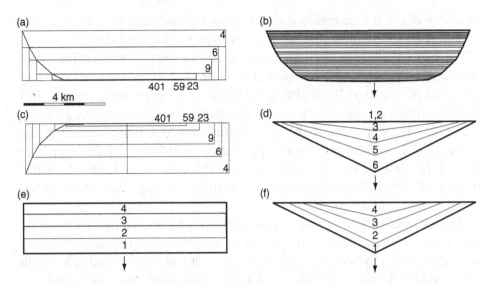

Fig. 13.15. Possible subsidence and growth modes of a 10-km wide pluton with resulting internal structural patterns. (a) Bottom-up growth by successive addition of sheets, its size of which follows a growth curve with a slope $a = 6.0$ (e.g., Fig. 13.12). The heavy black curve on left-hand side illustrates location of the pluton contact, defining a dish shape. Numbers on right-hand side indicate aspect ratio, L/T, at different stages of pluton growth (see Cruden & McCaffrey, 2001 for further details). (b) Internal structure (schematic) of a pluton that grows according to (a) by injection of many thin horizontal sheets. (c) Top-down growth by successive addition of sheets, the size of which (indicated by L/T ratios) follows an $a = 6.0$ growth curve. The resulting pluton will have a bell jar shape (heavy black curve on left side) and grows by stacking of horizontal sheets. (d) If the pluton in (c) forms by a cantilever mechanism, then it will grow progressively wider and deeper in the stages indicated, shown here for the case of basal accretion. Numbers indicate relative age of each pulse. In map view, such a pluton will have the oldest phase in the center and the youngest at the margin. (e) Growth of a tabular intrusion by a simple piston down-drop mechanism following an $a = \infty$ growth curve. Here, growth is from the base up (indicated by number order), resulting in a layered pluton with the youngest phase at the top. (f) Growth of a wedge-shaped pluton by a cantilever rotation of its floor following an $a = \infty$ growth line. Growth is from the base up, resulting in progressive steepening of the dip of older phases and a young (inner) to old (outer) phase distribution in map view.

13.4 Relationships between melt extraction and pluton growth

The observed power-law scaling of pluton dimensions (Section 13.2.9), together with the proposed growth mechanisms by floor subsidence (Section 13.3) suggest that the size and shape of plutons is ultimately controlled by the geometry of the source region and the amount of melt that may be extracted from it (Fig. 13.9). These relationships are explored further below.

For first-order geometrical analysis, we assume the following: (1) the final thickness and volume (V_p) of a pluton of width, L, may be predicted using an empirical power law; (2) the pluton volume is equal to the melt volume (V_m) extracted from the source region; (3) the melt volume extracted is given by $V_m = X_m \times V_s$, where X_m is the total partial melt fraction produced in the source, and V_s is the total volume of the source region involved in supplying melt to the pluton; (4) that the source region thickness $H_s > T$, and (5) melt is drained efficiently from the source region to be focused into one or several conduits that feed the pluton.

These simplifications ignore factors such as the likelihood that melt extraction is not 100% efficient, the degree of partial melting is unlikely to be uniform through the source region, and that melt will be produced at a certain rate as it is withdrawn. The assumption (2) – that melt volume delivered to a pluton is equal to the amount extracted from the source region – is particularly problematic because a significant amount of material may be lost during ascent due to fractional crystallization and other processes (Brown, 2001); hence, estimates of source volume dimensions given below should be regarded as lower bounds. However, given the current uncertainties in our understanding of the physicochemical processes occurring in partially molten source regions, the above assumptions are justifiable for the purposes of the first-order analysis below.

13.4.1 Influence of pluton shape

The simplest scenario occurs when melt extraction occurs from a region with the same areal extent as the final pluton (Fig. 13.9a). In this case, the required minimum source region thickness, H_s, is a function of the degree of partial melting and the pluton geometry and volume (Fig. 13.16). The range of source region thicknesses required to produce both tabular and wedge-shaped plutons when 25% partial melting occurs is shown in Fig. 13.17a. For a 25-km wide tabular pluton, minimum source region thicknesses range from about 9 to 28 km, with a median value of 17 km for the range of pluton thicknesses predicted by the empirical power law (T = from 2 to 7 km). Large plutons forming in this way either require very high degrees of partial melting or a source region that is over half the thickness of average continental crust. Because of their smaller volume, wedge-shaped plutons require much thinner source regions. A 25-km wide wedge-shaped pluton must be extracted from a 3–10-km thick (median 5.5 km), 25% partially molten source (Fig. 13.17a).

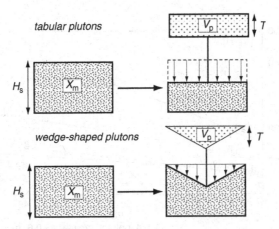

Fig. 13.16. Simple geometrical relationships between a deflating magma source region of initial thickness H_s and a growing pluton for which melt is extracted from an area confined to the region beneath the pluton. Tabular plutons (above) require two-thirds more melt extraction than wedge-shaped plutons (below). The total amount of melt available to feed a pluton with volume, V_p, is determined by the thickness, H_s, and melt fraction, X_m, in the source. Arrows indicate the vertical deflation of the source required to accommodate pluton growth.

For all wedge-shaped plutons >10 km wide, a 25% partially molten source must be >2 km thick.

13.4.2 Influence of pluton spacing

Melt extraction from source regions that "image" the final pluton requires either high degrees or very thick zones of partial melting. For lower degrees of partial melting or thinner source regions, melt extraction must occur from an area greater than that of the final pluton (Figs. 13.9b and 13.18). This is probably one of the reasons why plutons (and ultimately, volcanoes) are observed to occur with a characteristic spacing (e.g., Rickard & Ward, 1981; Pelletier, 1999). Taking a circular pluton with radius r, we may define the region below it from which melt is being scavenged as being wider than the pluton by a factor N. The radius of the melt extraction region is then $r(N + 1)$, and the margin to margin spacing between plutons of equal size would be $2Nr$ (Fig. 13.18). Melt is assumed to drain radially into a conduit feeding the pluton, in which case the size of each melt extraction domain, and ultimately the pluton spacing, is governed by the efficiency of the pumping mechanism draining melt. The situation is analogous to the siting of wells pumping an aquifer; their spacing is designed to minimize draw-down of the aquifer for a

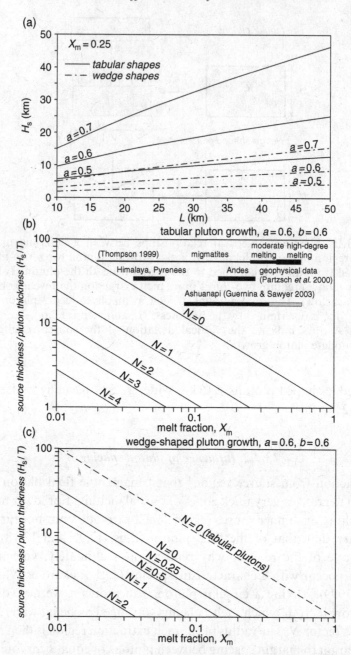

Fig. 13.17. (a) Minimum source region thicknesses, H_s, required to form tabular and wedge-shaped plutons at 25% partial melting (X_m, by volume) and melt extraction. (b) and (c) H_s/T *verses* X_m plots for tabular pluton growth (b) and wedge-shaped pluton growth (c). N is the pluton spacing factor (Fig. 13.18), and describes the size of the source region with respect

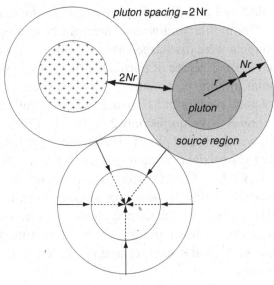

Fig. 13.18. Plan view illustrating relationships between the horizontal dimensions of plutons (radius r) and their respective source regions (radius $r(N+1)$). Pluton spacing ($2Nr$) is ultimately controlled by the efficiency of horizontal radial flow into a conduit system that drains the source and fills the resulting pluton.

given pumping rate (analogous to Q) and groundwater recharge rate (analogous to the rate of melt production) (e.g., Domenico & Schwartz, 1998). The spaced nature of plutons and volcanoes in many terrains suggests that melt extraction, ascent, and emplacement systems are strongly self-organized (Section 13.5).

The effects of varying N for both tabular and wedge-shaped plutons with $a = 0.6$ and $b = 0.6$ are summarized in Fig. 13.17b, c. We use the ratio H_s/T rather than H_s for comparative purposes, with $H_s/T \rightarrow 1$ being considered the optimum situation. $N = 0$ corresponds to the case in which the melt extraction domain is of equal area to the overlying pluton. Taking tabular

Figure 13.17. (cont.)

to the size of the pluton. Tabular pluton growth is favored when the degree of source region melting is high or when plutons are spaced far apart, whereas wedge-shaped pluton growth may occur at lower degrees of partial melting and closer pluton spacing. Ranges of degrees of partial melting as determined by geophysical, petrological, and experimental studies are shown for reference in (b). Note that the position of these ranges with respect to the H_s/T axis is arbitrary.

pluton growth with $N=0$ and $X_m = 1$ corresponds to the case in which the source region is 100% molten and would therefore be analogous to caldera formation if the magma were discharged at the surface. Increasing N dramatically reduces the required degree of partial melting for a desired H_s/T ratio. For example, tabular plutons being extracted from a source twice as thick as the resulting intrusions require 50% partial melting, while allowing $N=1$ drops this to 15% and $N=3$ requires only 2% (Fig. 13.17b).

The situation for wedge-shaped plutons is even more dramatic (Fig. 13.17c). For $H_s/T=2$, only 15% partial melting is required when $N=0$, and this drops to 3% when $N=1$. Wedge shaped pluton growth therefore will be favored when degrees of partial melting in the source are low. For the same degree of source region partial melting, wedge shaped plutons will tend to be larger and more closely spaced than tabular plutons. Tabular pluton growth requires either higher degrees of partial melting and/or the emplacement of small, widely spaced, plutons.

13.4.3 Other constraints on degrees of partial melting in granite source regions

Field, geophysical, and experimental approaches have been applied to estimate degrees of partial melting in potential granite source regions (Fig. 13.17b). The lowest estimates come from seismic and electrical measurements of active orogens, with estimated degrees of melting in the lower crust of the Himalaya and Pyrenees at about 4%, and in the central Andes of 15% (Partzsch *et al.*, 2000; Schilling & Partzsch, 2001). A review of experimental and thermodynamic data on partial melting of pelites and tonalites led Thompson (1999) to differentiate between three regimes. The migmatite regime represents 0–25% melting of pelitic source rocks under conditions of regulated moderate heat supply, and is less likely to produce granite plutons (but see Brown, 2001). Both moderate- (25–40%) and high-degree (40–60%) melting will yield granites (e.g., Guernina & Sawyer, 2003). Moderate-degree melting requires high geothermal gradients, high radioactive heat-producing layers or proximity to hot magmatic sills, while high-degree melting requires large volumes of pre-heated fertile crust and widespread mantle heat sources (Thompson, 1999).

The above estimates of degrees of partial melting may be compared to our evaluation of pluton growth and spacing (Fig. 13.17). Plutons exposed in eroded collisional orogens (e.g., Paterson & Schmidt, 1999; Schmidt & Paterson, 2000) and Archean granite-greenstone belts tend to have distinctly spaced distributions with distances between plutons between about $0.5r$ and $3r$. This pattern is consistent with lower degrees of partial melting observed in the Himalaya and Pyrenees (Partzsch *et al.*, 2000) and the moderate amounts of melting predicted by thermal models of tectonically thickened crust (Thompson, 1999). Plutons exposed in eroded continental magmatic arcs such as the Sierra Nevada batholith (e.g., Bateman, 1992) and Coast Plutonic Complex (e.g., Brown & McLelland, 2000) tend to form in discrete belts that migrate normal to the trend of the arc with time (due to either slab roll-back or slab-dip changes) (e.g., Tobisch *et al.*, 1995). Within each belt, individual plutons tend to be emplaced with spacing between $0r$ (i.e., touching) to $<1r$. Such a distribution is consistent with higher degrees of partial melting observed in the Andes (Schilling & Partzsch, 2001) and high amounts of melting predicted in thermal models involving large amounts of advected mantle heat (underplating; Thompson, 1999).

13.4.4 Estimates of degrees of partial melting and melt fluxes in the central Sierra Nevada and Superior Province

As a first-order attempt to evaluate the above concepts, we take the central Sierra Nevada batholith and the Archean Superior Province as environments where upper mantle and/or lower crustal melting occurred in focused versus widespread settings, respectively (Fig. 13.19). In each case, by determining the areas of all plutons, the total volume of intrusive material may be estimated using the empirical power law (Eqn (13.1), Section 13.2.9). This in turn yields an estimate of the magma production rate and the magma flux, given the time span of magmatic activity and assumptions about spatial distribution of melting in each region (Table 13.2). Areas of plutons in both cases were determined from digital images using a conventional image analysis program (Scion Image/NIH Image). Outlines of granites in the Superior Province study were obtained from Arc View shape files provided by Digital Geoscience at the Geological Survey of Canada CD-ROM Data Sampler (2001, Natural Resources Canada), at a spatial resolution of $4\,km\ pixel^{-1}$. Individual plutons in composite batholiths are not differentiated in the data, and are generally not available on regional geological maps of the Superior Province. Pluton outlines in the central Sierra Nevada batholith study were obtained by digitization of a 1:250 000 map (Bateman, 1992, Plate 1). In this case, individual plutons within this composite batholith are differentiated. The resulting spatial resolution in the digital data was $65\ m\ pixel^{-1}$.

Mesozoic Central Sierra Nevada batholith

As summarized by Bateman (1992), the central part of the Sierra Nevada batholith is probably one of the best-mapped, subdivided, and analyzed

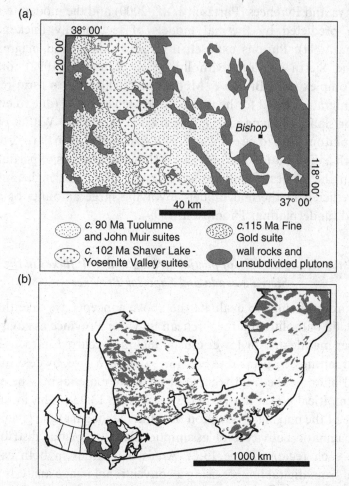

Fig. 13.19. (a) Distribution of granitic suites, wall rocks, and roof pendants in the Jurassic–Cretaceous central Sierra Nevada batholith, California (after Bateman, 1992). (b) Distribution of late granites in the Archean Superior Province, Canada (data from Digital Geoscience at the Geological Survey of Canada CD-ROM Data Sampler (2001, Natural Resources Canada).

portions of any composite arc batholith (Fig. 13.19a). The Sierra Nevada batholith is considered to be a product of eastward subduction of the oceanic Farallon plate beneath the continental North American plate in the Jurassic and Cretaceous (e.g., Bateman, 1992; Tobisch *et al.*, 1995). Geochemical and petrological studies of intrusive rocks of the batholith indicate that they were derived by partial melting of juvenile igneous and continental metasedimentary rocks at depths of about 40 km, probably in response to underplating of hot, subduction-related mafic melts (Bateman, 1992; Coleman *et al.*, 1995;

Table 13.2. *Central Sierra Nevada granites*

Suite	Fine Gold	Shaver Lake/ Yosemite	John Muir/ Tuolumne	Totals	Units
Duration	15	8	11	34	Myr
Area	3.1×10^3	1.8×10^3	4.3×10^3	9.2×10^3	km^2
Number of plutons	32	37	40	109	
Volume*	1.7×10^4	5.4×10^3	1.7×10^4	3.9×10^4	km^3
Equivalent thickness**	5.5	3.0	3.9	4.2	km
Production rate*	0.03	0.02	0.05	0.04	$m^3 s^{-1}$
Flux*	365	369	354	124	$m\,Myr^{-1}$

* Volumes, production rates, and fluxes based on tabular pluton shapes.

** Equivalent thickness of a single sheet of magma spread out over the total area of each suite.

Patiño-Douce & Beard, 1996; Sisson *et al.*, 1996). Emplacement depths of the magmas varied between 4 and 15 km (Ague & Brimhall, 1988), suggesting a source to pluton transport distance of between 25 and 36 km. The locus of magma emplacement in the central Sierra Nevada migrated from west to east during the middle Cretaceous, possibly in response to slab flattening and changes in the mantle wedge flow pattern (Tobisch *et al.*, 1995).

Three slightly overlapping northwest-trending (arc parallel) magmatic belts are defined in the central Sierra Nevada (Bateman, 1992): the 50–20-km wide, *c*. 120–105 Ma Fine Gold suite in the west; the 30–10-km wide, *c*. 105–97 Ma Shaver Lake – Yosemite Valley and Washburn Lake – Buena Vista Crest – Merced Peak suites in the middle; and the 40–50 km wide, *c*. 96–85 Ma John Muir and Tuolumne suites in the east (Fig. 13.19a). Each suite consists of discrete, cross-cutting (i.e., touching) plutons ranging in composition from gabbro to alkali feldspar granite, but are predominantly granodiorites and tonalites. Pre-batholithic to syn-batholithic metasedimentary and metavolcanic rocks underlie a relatively small area of the batholith, occurring as roof pendants at high elevations and occasional steeply inclined screens between plutons (Fig. 13.19a).

The highly focused character of intrusive activity in the Sierra Nevada batholith suggests that partial melting in the source region must have also occurred in a linear northwest-trending, 20–30-km wide zone in the lower crust during the formation of each suite. This is consistent with the focused nature of subduction zone magmatism and consequent advection of heat into the base of the crust. We may place limits on the generation of magmas in the lower crustal source region by estimating the volume of magma extracted and emplaced into the overlying batholith over a known period of time. This has been done for the central Sierra Nevada by determining the area of each pluton in each suite and then applying the empirical scaling law derived in Section 13.2.9 to estimate the volume of each pluton and of each suite. Based on limited geological observations (Coleman *et al.*, 1995; Sisson *et al*, 1996; Cruden *et al.*, 1999b) that suggest that Sierra Nevadan plutons have flat roofs and floors and steep sides, and comparison to field and geophysical studies of similar batholiths in South America (Myers, 1975; Pitcher, 1979; Skarmeta & Castelli, 1997; Haederle & Atherton, 2002), we assume that all of the plutons analyzed have tabular shapes. Hence, the resulting volumes, rates, and fluxes are upper bounds. We also assume that each pluton is fed by a conduit and grows by floor depression (see Section 13.3.5), probably guided by pre-existing fractures (e.g., Tobisch & Cruden, 1995; McNulty *et al.*, 2000; Petford *et al.*, 2000). Other mechanisms, such as roof uplift, stoping, tectonic dilation, and diapirism may well have also played a role (e.g., Paterson *et al.*, 1996), but they are not considered to be rate-limiting.

Results of the analysis of each suite are summarized in Table 13.2. Each suite is made up of 30–40 individual plutons, ranging in size from the very large Bass Lake tonalite (Fine Gold suite, area $= 2250 \text{ km}^2$, volume $= 14\ 720 \text{ km}^3$) and Mount Givens granodiorite (John Muir suite, area $= 1435 \text{ km}^2$, volume $= 8196 \text{ km}^3$) to small bodies with areas of $\sim 3 \text{ km}^2$ and volumes $\sim 2.6 \text{ km}^3$. The average pluton has an area of $\sim 100 \text{ km}^2$ and a volume of $\sim 250 \text{ km}^3$, corresponding to an average width, $L = 11.3 \text{ km}$ and thickness, $T = 2.6 \text{ km}$. However, most of the volume in the batholith resides in the few very large plutons in each suite (e.g., 88% of the volume of the Fine Gold suite occurs in the Bass Lake tonalite). The total volume of each suite ranges from $17\ 000 \text{ km}^3$ (Fine Gold) to 5400 km^3 (Shaver Lake–Yosemite Valley). These volumes would correspond to layers of intrusive material underlying the area of each suite with uniform thicknesses ranging from 5.5 km (Fine Gold) to 3.0 km (Shaver Lake–Yosemite Valley). This "equivalent thickness" (Table 13.2) is a scale-independent measure of the relative amount of material residing in each suite. It also corresponds to the minimum thickness of the underlying source region if it was 100% molten.

Taking the volume of each suite and dividing it by the time interval to form it gives an average production rate for each suite. Production rates in the central Sierra Nevada varied between 0.02 and 0.05 $\text{m}^3 \text{ s}^{-1}$, which is on the low end of magma extraction, ascent, and emplacement rates estimated in Section 13.3.6. A scale-independent measure of melt transfer from the source to the batholith is the flux, given by production rate/area. Fluxes for melt transfer in the central Sierra Nevada appear to have been similar for each suite at around 360 m Myr^{-1}.

Late granites of the Archean Superior Province

The Superior Province is the world's largest contiguous Archean craton (e.g., Windley, 1984; Card & Poulsen, 1998). It consists of alternating east–west-trending granite-greenstone, metasedimentary, metaplutonic, and high-grade metamorphic belts that formed and amalgamated between *c.* 3.1 and 2.6 Ga. Tectonic origins of Archean crust, including the Superior Province, are still debated, with interpretations ranging from uniformitarian accretionary plate tectonic models (e.g., Windley, 1984; Williams *et al.*, 1992; DeWit, 1998) to non-uniformitarian endogenic models (e.g., Hamilton, 1998; Zegers & van Keken, 2001). Like other Archean cratons, over 75% of the area of the Superior Province is underlain by granitic rocks (Fig. 13.19b; Williams *et al.*, 1992; Card *et al.*, 1998). Regardless of tectonic origin, it is therefore the most impressive "granite factory" in the geological record.

Granitic rocks in the Superior Province are subdivided into two major suites: the earlier tonalite–trondhjemite–granodiorite suite (TTG) and the later granite–granodiorite suite (Beakhouse & McNutt, 1991; Williams *et al.*, 1992; Card & Poulsen, 1998). The TTG suite is considered generally to be syn-tectonic with respect to the Kenoran orogeny between about 2710 and 2690 Ma (Card *et al.*, 1998). Ideas on the origin of the TTG magmas range from melting of a subducting oceanic slab to melting of lower crustal garnet amphi-bolite, possibly related to mantle plume activity or crustal delamination (see discussions in Smithies, 2000, and Zegers & van Keken, 2001).

The granite–granodiorite suite is considered to be post-tectonic and formed between about 2695 to 2655 Ma, with a 10–20 Myr time lag between the north-ern and southern Superior Province (Card *et al.*, 1998). A subduction origin is difficult to apply to these rocks because they clearly cross-cut early accretionary structures and boundaries and post date the major tectonothermal Kenoran orogeny. Furthermore, they were emplaced across the entire area of the Superior Province in a narrow time window of 40 ± 10 Myr. Together with late hydro-thermal activity in major faults, the granite–granodiorite suite appears to be a manifestation of a period of widespread lower crustal metamorphism (Moser *et al.*, 1996). This "Pan Superior" event represents a major thermal perturbation of the lower crust, resulting in re-melting of older TTGs to form the granite–granodiorite suite (e.g., Beakhouse & McNutt, 1991). Possible causes of this event are not well constrained, but could include thermal relaxation of the Kenoran orogeny or lithospheric delamination (e.g., Moser *et al.*, 1996).

Table 13.3. *Superior Province, late granites*

Total area of granites $= 5.4 \times 10^5$ km^2*	Tabular	Wedge	Units
Volume	5.8×10^6	1.9×10^6	km^3
Equivalent thickness (regional)**	3.1	1.0	km
Equivalent thickness (local)**	10.7	3.6	km
Production rate	4.6 ($\pm$1.5)	1.5 ($\pm$0.5)	m^3 s^{-1}
Flux (regional)***	76 ($\pm$26)	25 ($\pm$9)	m Myr^{-1}
Flux (local)**	267 ($\pm$89)	89 ($\pm$30)	m Myr^{-1}

* Total area of Superior Province analyzed $= 1.9 \times 10^6$ km^2.
** Equivalent thickness of a uniform layer of granite spread out over the total area of
   the Superior Province (regional) or over the total area of granite intrusions (local).
*** Flux based on magma production over 40 ± 10 Myr.

In contrast to the focused magmatism of the Sierra Nevada, the late granites
of the Superior Province therefore provide an example of widespread heating
and partial melting of the lower crust. The result is a highly dispersed pattern
of granite–granodiorite intrusions displaying a wide range of sizes, and some
possible clustering into belts, reflecting variations in the fertility of source
rocks, local tectonic setting, etc. (Fig. 13.19b). Table 13.3 summarizes results
of the analysis of the Superior Province data, for which volumes for both
tabular and wedge geometries are calculated. Although geophysical data are
not available for all 320 granite areas in Fig. 13.19b, published results of
gravity and seismic reflection studies consistently show that pluton floors in
the Superior Province define symmetrical to asymmetrical wedge shapes (e.g.,
Jackson *et al.*, 1994; Everitt *et al.*, 1998). As noted above, the data used for the
analysis here does not differentiate between individual plutons in composite
batholiths (i.e., many of the large regions in Fig. 13.19b) or the case of
touching, isolated, granites. This will result in an overestimation of the volume
of granite; hence, the results presented in Table 13.3 should be regarded as
first-order upper bounds.

The volume of late granite–granodiorite in the Superior Province varies
between two (wedge) and six (tabular) million cubic kilometers. Taking all of
this material and spreading it out evenly across the entire area of the Superior
Province would result in a layer of granite between 1 and 3 km thick (regional
equivalent thickness). This represents the minimum thickness of the magma
source region if melt was scavenged uniformly from a 100% molten lower
crust. If the magma was derived only from areas below outcropping intrusions,
then the local equivalent thickness and corresponding minimum source region
thickness would be between about 11 and 4 km, respectively (Table 13.3).

Magma production rates vary between 4.6 (tabular) and 1.5 (wedge) $m^3 s^{-1}$, which is the mid range of magma extraction, ascent, and emplacement rates estimated in Section 13.3.6.

Deriving magma from the entire lower crust of the Superior Province results in low fluxes (Table 13.3), ranging from about 100 m Myr^{-1} (tabular granites forming over 30 Myr) to 16 m Myr^{-1} (wedge granites forming over 50 Myr). If magma is derived only locally, then fluxes are higher approaching those for the Sierra Nevada in the case of tabular granites forming over 30 Ma (flux = 356 m Myr^{-1}, Table 13.3). Given the likely overestimation of the volume of granite and the evidence for wedge shaped pluton floors, magma flux in the Superior Province was probably towards the low end of the estimates in Table 13.3 (i.e., between about one-quarter and one-third of the flux in the Sierra Nevada). Preliminary analyses of the Archean Chilimanzi granites of Zimbabwe and the Devonian granites of New Hampshire and Maine gives similar equivalent thicknesses and fluxes to those determined here for the Superior Province.

Comparison of magma source regions below the Sierra Nevada batholith and the Superior Province

Greater magma fluxes estimated for the Sierra Nevadan intrusive suites, compared to those for late granites in the Superior Province are consistent with the general thermal regime expected for both environments. As discussed above, spatially and temporally focused advection of heat, driven by underplating of mantle-derived mafic magmas into the lower crust of a magmatic arc is likely to result in a higher upward flux of melt compared to widespread anatexis due to slow warming of the lower crust. The net production, as measured by the equivalent thickness is also substantially different, being higher in the central Sierra Nevada. Again, this is probably a consequence of heat focusing, resulting in higher degrees of partial melting and melt extraction, and the longer-term availability of melt, and is consistent with observations discussed in Section 13.4.3. The results (Tables 13.2 and 13.3) may also be used to provide constraints on likely source region thicknesses, H_s, and total partial melt fractions, X_m, in both areas (Fig. 13.20).

As noted above, the equivalent thickness of granite in the central Sierra Nevada is also the minimum thickness of the magma source region, if it was 100% molten. Figure 13.20 plots source region thickness values for an equivalent thickness of 3 km over a more reasonable range of bulk partial melting (5–37%). Even high degrees of partial melting ($\sim$35%) require the magma source to be between 10 and 15 km thick and extremely high degrees of partial melting (approaching 100%) are required for a source approaching the

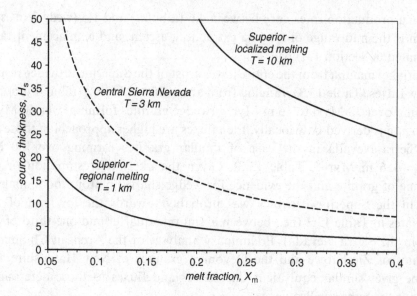

Fig. 13.20. Comparison of source thickness estimates, H_s for the central Sierra Nevada and Superior Province as a function of total melt fraction, X_m is the source and the mean thickness, T, of the resulting plutons.

equivalent thickness of 3 km. This supports the notion suggested in Section 13.4.3 that the formation of closely spaced, tabular plutons, as in the central Sierra Nevada, will require high degrees of partial melting and a thick magma source region. One way partly to offset these requirements would be provided if the lower crust were replenished continuously with new material throughout the melt-producing phase of the source. Strongly focused, sustained heating, coupled with continuous underplating of mantle-derived material in continental magmatic arc settings appears to be able to provide these conditions.

Two end-member scenarios for the late granites of the Superior Province are plotted in Fig. 13.20. The curve for $T = 10$ km, corresponds to derivation of all magma from below exposed intrusions, assuming that they have tabular shapes. Under these conditions, moderate degrees of partial melting ($\sim 25\%$) require the source region to be much thicker than the present thickness of the Superior Province crust. Even extremely high degrees of partial melting require a source approaching half the thickness of the crust. Deriving melt from the entire lower crust of the Superior Province to produce wedge-shaped granites allows for a much thinner source region at low to moderate degrees of melting: 25% melting requires a 5 km thick source, and 7.5% melting requires a 10-km thick source (Fig. 13.20). Given the fact that melting over such a large area was probably heterogeneous and diachronous as well as the likelihood of both wedge and tabular shaped plutons, the situation for the Superior Province is probably somewhere between the $T=1$ km and $T=4$ km curves. Nevertheless, these results support the proposal in Section 13.4.3 that low degrees of widespread partial melting will favor the production of spaced, wedge-shaped plutons.

The size–frequency relationships of plutons in both the Superior Province and the Sierra Nevada batholith show a power law distribution. Although the subject of work in progress, this relationship, together with other spatial statistical measures, suggests that magmatism in both regions is self-organized (see also Section 13.4.2). Similar relationships have been demonstrated by Pelletier (1999) for Mesozoic and Cenozoic volcanic rocks in the North American Cordillera. Recent experimental work suggests that both temporal and spatial self-organization may be a consequence of melting and melt extraction processes (Bons & van Milligen, 2001).

13.5 Conclusions

Two end-member geometrical types of pluton (tabular and wedge shapes) have been identified, the dimensions of which are characterized by an empirical power law. Although aided by other mechanisms not modeled here, many plutons appear to grow predominantly by floor subsidence, and the limiting factors on their ultimate size are the amount of melt that may be withdrawn from a lower crustal source, the thickness of the source, and the spacing between zones or conduits that channel magma vertically to supply the intrusion. Pluton growth is potentially very fast (thousands to hundreds of thousands of years) if magma is delivered continuously or in pulses separated by short time intervals (days to years). However, if the time scale for pulse addition is greater than pulse emplacement time, then plutons may take several million years to form.

A number of different structural arrangements of phases within plutons are predicted depending on the growth mode (tabular versus wedge) and the nature of magma arrival in the intrusion (continuous versus pulsed; top down versus base up). If homogenization of pulses has not occurred, it is therefore possible to evaluate the growth history of a specific pluton by detailed field and geochronological analysis. Likewise, pluton spacing potentially may be used to estimate the size, shape, and degree of partial melting within a magma source region. However, more rigorous constraints on the relationships between the amount of melt generated versus the amount supplied to the pluton, as well as source region melt migration and focusing mechanisms, are required to solve this problem satisfactorily. Despite these limitations, geophysical, field, and petrological estimates of the degrees of partial melting in the lower crust of modern collisional orogens and continental magmatic arcs compare favorably to those currently predicted from the spacing of plutons in ancient unroofed batholiths.

Tectonic setting appears to have a major influence on the spatial distribution and 3-D form of plutons. Local structural controls (faults and folds) on pluton emplacement identified in numerous field-based studies are an expected consequence of the interaction between active regional deformation and (predominantly) vertical pluton growth, melt transfer, and source withdrawal. Preliminary indications suggest that granitic magma systems are strongly self-organized from the bottom up. The challenge for future research will be to develop internally consistent models that integrate heat transfer, melting, melt extraction, ascent and emplacement mechanisms that may both predict and explain the nature and causes of this self-organization.

13.6 Acknowledgements

I have been supported by Operating and Lithoprobe grants from the Natural Sciences and Engineering Research Council of Canada. Fruitful discussions on granites over the last decade or so with J. Grocott, K. McCaffrey, N. Petford, P.-Y. Robin, C. Talbot, and O. Tobisch have helped me to clarify many of the ideas presented in this chapter. Thorough, constructive, and critical reviews by J. Hogan, S. Morgan and R. Law led to substantial improvements in the manuscript.

References

Ague, J. J. and Brimhall, G. H. (1988). Regional variations in bulk chemistry, mineralogy, and the compositions of mafic and accessory minerals in the batholiths of California. *Geological Society of America Bulletin*, **100**, 891–911.

Ameglio, L. and Vigneresse, J. -L. (1999). Geophysical imaging of the shape of granitic intrusions at depth: a review. In *Understanding Granites: Integrating New and Classical Techniques*, ed. A. Castro, C. Fernandez and J. -L. Vigneresse. Geological Society, London, Special Publication 168, pp. 39–54.

Barrell, J. (1907). *Geology of the Marysville Mining District, Montana*. US Geological Survey Professional Paper 57.

Bateman, P. C. (1992). *Plutonism in the Central Part of the Sierra Nevada Batholith, California*. US Geological Survey Professional Paper 1483.

Bateman, R. (1984) On the role of diapirism in the segregation, ascent and final emplacement of granitoid magmas. *Tectonophysics*, **110**, 211–31.

Beakhouse, G. P. and McNutt, R. H. (1991). Contrasting types of Late Archean plutonic rocks in northwestern Ontario: implications for crustal evolution in the Superior Province. *Precambrian Research*, **49**, 141–66.

Benn, K., Odonne, F. and de Saint Blanquat, M. (1998). Pluton emplacement during transpression in brittle crust: new views from analogue experiments. *Geology*, **26**, 1079–82.

Benn, K., Roest, W. R., Rochette, P., Evans, N. G. and Pignotta, G. S. (1999). Geophysical and structural signatures of syntectonic batholith construction: the South Mountain Batholith, Meguma Terrane, Nova Scotia. *Geophysical Journal International*, **136**, 144–58.

Bergantz, G. W. (2000). On the dynamics of magma mixing by reintrusion: implications for pluton assembly processes. *Journal of Structural Geology*, **22**, 1297–1309.

Björk, L. (1986). Beskrivning till bergrundskartan Filipstad NV (with English Summary). *Sveriges Geologiska Undersökning*, **Af147**.

Bons, P. D., Dougherty-Page, J. and Elburg, M. A. (2001). Stepwise accumulation and ascent of magmas. *Journal of Metamorphic Geology*, **19**, 627–33.

Bons, P. D. and van Milligen, B. P. (2001). New experiment to model self-organized critical transport and accumulation of melt and hydrocarbons from their source rocks. *Geology*, **29**, 919–22.

Bott, M. H. P. and Smithson, S. B. (1967). Gravity investigations of subsurface shape and mass distributions of granite batholiths. *Geological Society of America Bulletin*, **78**, 859–78.

Bouchez, J. L. (1997). Granite is never isotropic: an introduction to AMS studies of granitic rocks. In *Granite: From Segregation of Melt to Emplacement Fabrics*, ed. J. L. Bouchez, D. H. W. Hutton and E. Stephens. Dordrecht: Kluwer Academic Publishers, pp. 95–112.

Branch, C. D. (1967). The source of eruption for pyroclastic flows: caldrons or calderas? *Bulletin of Volcanology*, **30**, 41–53.

Branney, M. J. (1995). Downsag and extension at calderas: new perspectives on collapse geometries from ice-melt, mining, and volcanic subsidence. *Bulletin of Volcanology*, **57**, 303–18.

Bridgewater, D., Sutton, J. and Watterson, J. (1974). Crustal downfolding associated with igneous activity. *Tectonophysics*, **21**, 57–77.

Brisbin, W. C. (1986). Mechanics of pegmatite intrusion. *American Mineralogist*, **71**, 644–51.

Brown, E. H. and McClelland, W. C. (2000). Pluton emplacement by sheeting and vertical ballooning in part of the southeast Coast Plutonic Complex, British Columbia. *Geological Society of America Bulletin*, **112**, 708–19.

Brown, E. H. and Walker, N. W. (1993). A magma loading model for Barrovian metamorphism in the southeast Coast Plutonic Complex, British Columbia and Washington. *Geological Society of America Bulletin*, **105**, 479–500.

Brown, M. (1994). The generation, segregation, ascent and emplacement of granite magma: the migmatite-to-crustally-derived granite connection in thickened orogens. *Earth Science Reviews*, **36**, 83–130.

Brown, M. (2001). Crustal melting and granite magmatism: key issues. *Physics and Chemistry of the Earth*, **26**, 201–12.

Brown, M. and Solar, G. S. (1998). Granite ascent and emplacement during contractional deformation in convergent orogens. *Journal of Structural Geology*, **20**, 1365–93.

Brown, M. and Solar, G. S. (1999). The mechanism of ascent and emplacement of granite magma during transpression: a syntectonic granite paradigm. *Tectonophysics*, **312**, 1–33.

Brun, J. P. and Pons, J. (1981). Strain patterns of pluton emplacement in a crust undergoing non-coaxial deformation, Sierra Morena, S. Spain. *Journal of Structural Geology*, **3**, 219– 29.

Brun, J. P., Gapais, D. and Cogne, J. P. (1990). The Flamanville granite (northwest France): an unequivocal example of a syntectonically expanding pluton. *Geological Journal*, **25**, 271–86.

Buddington, A. F. (1959). Granite emplacement with special reference to North America. *Geological Society of America Bulletin*, **70**, 671–747.

Bussell, M. A. (1976). Fracture control of high-level plutonic contacts in the Coastal Batholith of Peru. *Proceedings of the Geological Association*, **87**, 237–46.

Bussell, M. A., Pitcher, W. S. and Wilson, P. A. (1976). Ring complexes of the Peruvian Coastal Batholith: a long-standing subvolcanic regime. *Canadian Journal of Earth Science*, **13**, 1020–30.

Card, K. D. and Poulsen, K. H. (1998). Geology and mineral deposits of the Superior province of the Canadian Shield. In *Geology of the Precambrian Superior and Grenville Provinces and Precambrian Fossils of North America*, ed. S. B. Lucas and M. R. St. Onge. Geological Society of America, C1, pp. 13–204.

Card, K. D., Frith, R. A., Poulsen, K. H. and Ciesielski, A. (1998). *Lithotectonic Map of the Superior Province, Canada and Adjacent Parts of the United States of America*. Geological Survey of Canada, Map 1948A, scale 1:2 000 000.

Castro, A. (1986). Structural pattern and ascent model in the central Extramadura batholith, Hercynian belt, Spain. *Journal of Structural Geology*, **8**, 633–45.

Clarke, D. B., Henry, A. S. and White, M. A. (1998). Exploding xenoliths and the absence of "elephant's graveyards" in granite batholiths. *Journal of Structural Geology*, **20**, 1325–43.

Clemens, J. D. and Mawer, C. K. (1992). Granitic magma transport by fracture propagation. *Tectonophysics*, **204**, 339–60.

Clemens, J. D. and Petford, N. (1999). Granitic melt viscosity and silicic magma dynamics in contracting tectonic settings. *Journal of the Geological Society London*, **156**, 1057–60.

Cloos, H. (1923). Das batholithenproblem. *Forschritte Geologie und Paleontologie*, **1**, 1–80.

Clough, C. T., Maufe, H. B. and Bailey, E. B. (1909). The cauldron-subsidence of Glen Coe, and the associated igneous phenomena. *Journal of the Geological Society, London*, **65**, 611–78.

Cobbing, E. J. (1999). The Coastal Batholith and other aspects of Andean magmatism in Peru. In *Understanding Granites: Integrating New and Classical Techniques*, ed. A. Castro, C. Fernandez and J. -L. Vigneresse. Geological Society, London, Special Publication 168, pp. 111–22.

Coleman, D. S., Glazner, A. F., Miller, J. S., Bradford, K. J., Frost, T. P., Joye, J. L. and Bachl, C. A. (1995). Exposure of a late Cretaceous layered mafic-felsic magma system in the central Sierra Nevada batholith, California. *Contributions to Mineralogy and Petrology*, **120**, 129–36.

Collins, W. J. (2002). Hot orogens, tectonic switching and creation of continental crust. *Geology*, **30**, 535–8.

Collins, W. J. and Sawyer, E. W. (1996). Pervasive granitoid magma transfer through the lower-middle crust during non-coaxial compressional deformation. *Journal of Metamorphic Geology*, **14**, 565–79.

Corry, C. E. (1988). *Laccoliths: Mechanics of Emplacement and Growth*. Geological Society of America, Special Publication 220.

Courrioux, G. (1987). Oblique diapirism: the Criffel granodiorite/granite zoned pluton (southwest Scotland). *Journal of Structural Geology*, **9**, 87–103.

Cruden, A. R. (1990). Flow and fabric development during the diapiric rise of magma. *Journal of Geology*, **98**, 681–98.

Cruden, A. R. (1998). On the emplacement of tabular granites. *Journal of the Geological Society London*, **155**, 853–62.

Cruden, A. R. and Aaro, S. (1992). The Ljugaren granite massif, Dalarna, central Sweden. *Geologiska Föreningens i Stockholm Förhandlingar*, **114**, 209–25.

Cruden, A. R. and Launeau, P. (1994). Structure, magnetic fabric and emplacement of the Archean Lebel Stock, S. W. Abitibi Greenstone belt. *Journal of Structural Geology*, **16**, 677–91.

Cruden, A. R. and McCaffrey, K. J. W. (2001). Growth of plutons by floor subsidence: implications for rates of emplacement, intrusion spacing and melt-extraction mechanisms. *Physics and Chemistry of the Earth, Part A, Solid Earth and Geodesy*, **26**, 303–15.

Cruden, A. R., Sjöstrom, H. and Aaro, S. (1999a). Structure and geophysics of the Gåsborn granite, Central Sweden: an example of fracture-fed asymmetric pluton emplacement. In *Understanding Granite: Integrating New and Classical Techniques*, ed. A. Castro, C. Fernandez and J.-L. Vigneresse. Geological Society, London, Special Publication 168, pp. 141–60.

Cruden, A. R., Tobisch, O. T. and Launeau, P. (1999b). Magnetic fabric evidence for conduit-fed emplacement of a tabular intrusion: Dinkey Creek pluton, central Sierra Nevada batholith, California. *Journal of Geophysical Research*, **104**, 10 511–30.

Culshaw, N. and Bhatnagar, P. (2001). The interplay of regional structure and emplacement mechanisms at the contact of the South Mountain batholith, Nova Scotia: floor-down or wall-up? *Canadian Journal of Earth Sciences*, **38**, 1285–99.

Dehls, J. D., Cruden, A. R. and Vigneresse, J.-L. (1998). Fracture control of late-Archean pluton emplacement in the Northern Slave Province, Canada. *Journal of Structural Geology*, **20**, 1145–54.

DeWit, M. J. (1998). On Archean granites, greenstones, cratons and tectonics: does the evidence demand a verdict. *Precambrian Research*, **91**, 181–226.

Dingwell, D. B. (1999). Granitic melt viscosities. In *Understanding Granites: Integrating New and Classical Techniques*, ed. A. Castro, C. Fernandez and J.-L. Vigneresse. Geological Society, London, Special Publication 168, pp. 27–38.

Dixon, J. M. and Simpson, D. G. (1987). Centrifuge modeling of laccolith intrusion. *Journal of Structural Geology*, **9**, 87–103.

Domenico, P. A. and Schwartz, F. W. (1998). *Physical and Chemical Hydrogeology*. New York: Wiley Interscience.

Dzurisin, D., Yamashita, K. M. and Kleinman, J. W. (1994). Mechanisms of crustal uplift and subsidence at the Yellowstone caldera, Wyoming. *Bulletin of Volcanology*, **56**, 261–70.

Evans, D. J., Rowley, W. J., Chadwick, R. A., Kimbell, G. S. and Millward, D. (1994) Seismic reflection data and the internal structure of the Lake District batholith, Cumbria, northern England. *Proceedings of the Yorkshire Geological Society*, **50**, 11–24.

Everitt, R., Brown, A., Ejeckam, R., Sikorsky, R. and Woodcock, D. (1998). Litho-structural layering within the Archean Lac du Bonnet Batholith, at AECL's Underground Research Laboratory, Southeastern Manitoba. *Journal of Structural Geology*, **20**, 1291–1304.

Fowler, T. K. Jr. and Paterson, S. R. (1997). Timing and nature of magmatic fabrics from structural relations around stoped blocks. *Journal of Structural Geology*, **19**, 209–24.

Ge, H. and Jackson, M. P. A. (1998). Physical modeling studies of structures formed by salt withdrawal: implications for deformation caused by salt dissolution. *American Association of Petroleum Geologists Bulletin*, **82**, 228–50.

Gilder, S. and McNulty, B. A. (1999). Tectonic exhumation and tilting of the Mount Givens pluton, central Sierra Nevada, California. *Geology*, **27**, 919–22.

Glazner, A. F. and Miller, D. M. (1997). Late-stage sinking of plutons. *Geology*, **25**, 1099–102.

Grocott, J. and Taylor, G. K. (2002). Magmatic arc fault systems, deformation partitioning and emplacement of granitic complexes in the Coastal Cordillera, north Chilean Andes (25° 30'S to 27° 00'S). *Journal of the Geological Society London*, **159**, 425–42.

Grocott, J., Brown, M., Dallmeyer, R. D., Taylor, G. K. and Treloar, P. J. (1994). Mechanisms of continental growth in extensional arcs: an example from the Andean plate-boundary zone. *Geology*, **22**, 391–4.

Grocott, J., Garde, A., Chadwick, B., Cruden, A. R. and Swager, C. (1999). Emplacement of Rapakivi granite and syenite by floor depression and roof uplift in the Paleoproterozoic Ketilidian orogen, South Greenland. *Journal of the Geological Society London*, **156**, 15–24.

Gudmundsson, A. (1990). Emplacement of dikes, sills and crustal magma chambers at divergent plate boundaries. *Tectonophysics*, **176**, 257–75.

Guernina, S. and Sawyer, E. W. (2003). Large-scale melt-depletion in granulite terranes: an example from the Archean Ashuanipi Subprovince of Quebec. *Journal of Metamorphic Geology*, **21**, 181–201.

Guillot, S., Pecher, A., Rochette, P. and Le Fort, P. (1993). The emplacement of the Manaslu granite of central Nepal: field and magnetic susceptibility constraints. In *Himalayan Tectonics*, ed. P. J. Treloar and M. P. Searle. Geological Society Special Publication 74, pp. 413–28.

Guineberteau, B., Bouchez, J. L. and Vigneresse, J.-L. (1987). The Mortagne granite pluton (France) emplaced by pull-apart along a shear zone: structural and gravimetric arguments and regional implications. *Geological Society of America Bulletin*, **99**, 763–70.

Haederle, M. and Atherton, M. P. (2002). Shape and intrusion of the Coastal Batholith, Peru. *Tectonophysics*, **345**, 17–28.

Hamilton, W. (1988). Tectonic setting and variations with depth of some Cretaceous and Cenozoic structural and magmatic systems of the Western United States. In *Metamorphism and Crustal Evolution of the Western United States: Rubey Volume 7*, ed. W. G. Ernst. New Jersey: Prentice Hall, pp. 1–40.

Hamilton, W. B. (1998). Archean magmatism and deformation were not products of plate tectonics. *Precambrian Research*, **91**, 143–79.

Hamilton, W. and Myers, W. B. (1967). *The Nature of Batholiths*. US Geological Survey Professional Paper 554(c), pp. 1–30.

Hamilton, W. and Myers, W. B. (1974). Nature of the Boulder batholith of Montana. *Geological Society of America Bulletin*, **85**, 365–78.

Hecht, L. and Vigneresse, J.-L. (1999). A multidisciplinary approach combining geochemical, gravity and structural data: implications for pluton emplacement and zonation. In *Understanding Granites: Integrating New and Classical*

Techniques, ed. A. Castro, C. Fernandez and J.-L. Vigneresse. Geological Society, London, Special Publication 168, pp. 95–110.

Hogan, J. P. and Gilbert, M. C. (1995). The A-type Mount Scott granite sheet: importance of crustal magma traps. *Journal of Geophysical Research*, **100**, 15 779–92.

Hogan, J. P. and Gilbert, M. C. (1997). Intrusive style of A-type sheet granites in a rift environment: the Southern Oklahoma Aulacogen. In *Middle Proterozoic to Cambrian Rifting, Central North America*, ed. R. W. Ojakangas, A. B. Dickas and J. C. Green. Geological Society of America, Special Paper 312, pp. 299–311.

Hogan, J. P., Price, J. D. and Gilbert, M. C. (1998). Magma traps and driving pressure: consequences for pluton shape and emplacement in an extensional regime. *Journal of Structural Geology*, **20**, 1155–68.

Holder, M. T. (1979). An emplacement mechanism for post-tectonic granite and its implication for geochemical features. In *Origin of Granite Batholiths: Geochemical Evidence*, ed. M. Atherton and J. Tarney. Orpington: Shiva, pp. 116–28.

Hunt, C. B., Averitt, P. and Miller, R. L. (1953). *Geology and Geography of the Henry Mountains Region*, Utah. US Geological Survey Professional Paper, **228**.

Hutton, D. H. W. (1988). Granite emplacement mechanisms and tectonic controls: inferences from deformation studies. *Transactions of the Royal Society of Edinburgh: Earth Sciences*, **79**, 245–55.

Hutton, D. H. W. and Brown, P. E. (2000). Discussion of: emplacement of Rapakivi granite and syenite by floor depression and roof uplift in the Paleoproterozoic Ketilidian orogen, South Greenland. *Journal of the Geological Society, London*, **157**, 701–4.

Hutton, D. H. W., Dempster, T. J., Brown, P. E. and Becker, S. M. (1990). A new mechanism of granite emplacement: intrusion into active extensional shear zones. *Nature*, **343**, 452–5.

Ingram, G. I. and Hutton, D. H. W. (1994). The great tonalite sill: emplacement into a contractional shear zone and implications for late Cretaceous to early Eocene tectonics in southeastern Alaska and British Columbia. *Geological Society of America Bulletin*, **106**, 715–28.

Jackson, M. D. and Pollard, D. D. (1988). The laccolith-stock controversy: new results from the southern Henry Mountains, Utah. *Geological Society of America Bulletin*, **100**, 117–39.

Jackson, S. L., Cruden, A. R., White, D. and Milkereit, B. (1994). A seismic reflection based regional cross section of the southern Abitibi greenstone belt. *Canadian Journal of Earth Sciences*, **32**, 135–48.

John, B. E. and Mukasa, S. B. (1990). Footwall rocks of the mid-Tertiary Chemehuevi detachment fault: a window into the middle crust in the Southern Cordillera. *Journal of Geophysical Research*, **95**, 463–85.

Johnson, A. M. and Pollard, D. D. (1973). Mechanics of growth of some laccolithic intrusions in the Henry Mountains, Utah, I. Field observations, Gilbert's model, physical properties and flow of the magma. *Tectonophysics*, **18**, 261–309.

Johnson, S. E., Albertz, M. and Paterson, S. R. (2001). Growth rates of dike-fed plutons: are they compatible with observations in the middle and upper crust? *Geology*, **29**, 727–30.

Leake, B. E. and Cobbing, J. (1993). Transient and long-term correspondence of erosion level and the tops of granite plutons. *Scottish Journal of Geology*, **29**, 177–82.

Le Fort, P. (1981). Manaslu leucogranite: a collision signature of the Himalaya a model for its genesis and emplacement. *Journal of Geophysical Research*, **86**, 10 545–68.

Leitch, A. M. and Weinberg, R. F. (2002). Modeling granite migration by mesoscale pervasive flow. *Earth and Planetary Science Letters*, **200**, 131–46.

Lipman, P. W. (1984). The roots of ash flow calderas in western North America: windows into the tops of granitic batholiths. *Journal of Geophysical Research*, **89**, 8801–41.

Lipman, P. W. (1997). Subsidence of ash-flow calderas: relation to caldera size and magma-chamber geometry. *Bulletin of Volcanology*, **59**, 198–218.

Lynn, H. B., Hale, L. D. and Thompson, G. A. (1981). Seismic reflections from the basal contacts of batholiths. *Journal of Geophysical Research*, **86**, 10 633–8.

Mahon, K. I., Harrison, T. M. and Drew, D. A. (1988). Ascent of a granitoid diapir in a temperature varying medium. *Journal of Geophysical Research*, **93**, 1174–88.

Marsh, B. D. (1982). On the mechanics of igneous diapirism, stoping and zone melting. *American Journal of Science*, **282**, 808–55.

Marsh, B. D. (1984). Mechanics and energetics of magma formation and ascension. In *Explosive Volcanism: Inception, Evolution and Hazards*, ed. F. R. Boyd, Jr. Washington: National Academy Press, pp. 67–83.

Marti, J., Ablay, G. J., Redshaw, L. T. and Sparks, R. S. J. (1994). Experimental studies of collapse calderas. *Journal of the Geological Society London*, **151**, 919–29.

McCaffrey, K. J. W. (1992). Igneous emplacement in a transpressive shear zone: Ox Mountains igneous complex. *Journal of the Geological Society London*, **149**, 221–35.

McCaffrey, K. J. W. & Cruden, A. R. (2002). Dimensional data and growth models for intrusions. In *First International Workshop: Physical Geology of Subvolcanic Systems – Laccoliths, Sills & Dykes (LASI)*, ed. C. Breitkreuz, A. Mock and N. Petford. Wissenschaftliche Mitteilungen der Bergakademie Freiberg, Vol. 20, pp. 37–9.

McCaffrey, K. J. W. and Petford, N. (1997). Are granitic intrusions scale-invariant? *Journal of the Geological Society London*, **154**, 1–4.

McNulty, B., Tobisch, O. T., Cruden, A. R. and Gilder, S. (2000). Multi-stage emplacement of the Mt. Givens pluton, central Sierra Nevada, California. *Geological Society of America Bulletin*, **112**, 119–35.

Miller, C. F., Wooden, J. L., Bennett, V. C., Wright, J. E., Solomon, G. C. and Hurst, R. W. (1990). *Petrogenesis of the Composite Peraluminous–Metaluminous Old Woman–Piute Range Batholith, Southeastern California Isotopic Constraints*. Geological Society of America Memoir 174, pp. 99–109.

Miller, R. B. and Paterson, S. R. (1999). In defense of magmatic diapirs. *Journal of Structural Geology*, **21**, 1161–73.

Molyneux, S. J. and Hutton, D. H. W. (2000). Evidence for significant granite space creation by the ballooning mechanism: the example of the Ardara pluton, Ireland. *Geological Society of America Bulletin*, **112**, 1543–58.

Morgan, S. S., Law, R. D. and Nyman, N. W. (1998a). Laccolith-like emplacement model for the Papoose Flat pluton based on porphyroblast-matrix analysis. *Geological Society of America Bulletin*, **110**, 96–110.

Morgan, S. S., Law, R. D. & de Saint Blanquat, M. (1998b). Aureole deformation associated with inflation of the concordant Eureka Valley–Joshua Flat–Bear Creek composite pluton, central White-Inyo Range, eastern California. In *1998*

Field Guidebook, ed. R. Behl. Geological Society of America, 94th Annual Meeting of Cordilleran Section, pp. 5.1–5.30.

Morgan, S. S., Law, R. D. and Saint Blanquat, M. (2000). Papoose Flat, Eureka Valley-Joshua Flat-Bear Creek, and Sage Hen Flat plutons: examples of rising, sinking, and cookie cutter plutons in the central White-Inyo Range, eastern California. In *Great Basin and Sierra Nevada: Boulder, Colorado*, ed. D. R. Lageson, S. G. Peters and M. M. Lahren. Geological Society of America Field Guide 2, pp. 189–204.

Moser, D. E., Heaman, L. M., Krogh, T. E. and Hanes, J. A. (1996). Intracrustal extension of an Archean orogen revealed using single-grain U–Pb zircon geochronology. *Tectonics*, **15**, 1093–109.

Myers, J. S. (1975). Cauldron subsidence and fluidization: mechanisms of intrusion of the Coastal Batholith of Peru into its own volcanic ejecta. *Geological Society of America Bulletin*, **86**, 1209–20.

Nédélec, A., Paquette, J.-L., Bouchez, J.-L., Olivier, P. and Ralison, B. (1994). Stratoid granites of Madagascar: structure and position in the Panafrican orogeny. *Geodynamica Acta*, **7**, 48–56.

Nyman, M. W., Law, R. D. and Morgan, S. S. (1995). Conditions of contact metamorphism, Papoose Flat Pluton, eastern California, USA: implications for cooling and strain histories. *Journal of Metamorphic Geology*, **13**, 627–43.

Oliver, H. W. (1977). Gravity and magnetic investigations of the Sierra Nevada batholith, California. *Geological Society of America Bulletin*, **88**, 445–61.

Partzsch G. M., Schilling F. R. and Arndt, J. (2000). The influence of partial melting on the electrical behavior of crustal rocks: laboratory examinations, model calculations and geological interpretations. *Tectonophysics*, **317**, 189–203.

Paterson, S. R. and Fowler, T. K. Jr. (1993). Re-examining pluton emplacement processes. *Journal of Structural Geology*, **15**, 191–206.

Paterson, S. R. and Miller, R. B. (1998). Mid-crustal magmatic sheets in the Cascades Mountains, Washington: implications for magma ascent. *Journal of Structural Geology*, **20**, 1345–63.

Paterson, S. R. and Schmidt, K. L. (1999). Is there a close spatial relationship between faults and plutons? *Journal of Structural Geology*, **21**, 1131–42.

Paterson, S. R. and Tobisch, O. T. (1992). Rates of processes in magmatic arcs: implications for the timing and nature of pluton emplacement and wall rock deformation. *Journal of Structural Geology*, **14**, 291–300.

Paterson, S. R., Vernon, R. H. and Fowler, T. K. Jr. (1991). Aureole tectonics. In *Contact Metamorphism*, ed. D. M. Kerrick. *Reviews in Mineralogy*, Vol. 26. Washington: Mineralogical Society of America, pp. 673–722.

Paterson, S. R., Fowler, T. K. Jr. and Miller, R. B. (1996). Pluton emplacement in arcs: a crustal-scale exchange process. *Transactions of the Royal Society of Edinburgh: Earth Sciences*, **87**, 115–23.

Patiño-Douce, A. E. and Beard, J. S. (1996). Effects of P, $f(O_2)$ and Mg/Fe ratio on dehydration melting of model greywackes. *Journal of Petrology*, **37**, 999–1024.

Pelletier, J. D. (1999). Statistical self-similarity magmatism and volcanism. *Journal of Geophysical Research*, **104**, 15 425–38.

Petford, N. (1996). Dykes or diapirs? *Transactions of the Royal Society of Edinburgh: Earth Sciences*, **87**, 105–14.

Petford, N., Kerr, R. C. and Lister, J. R. (1993). Dike transport of granitoid magmas. *Geology*, **21**, 845–48.

Petford, N., Clemens. J. D. and Vigneresse, J. -L. (1997). The application of information theory to the formation of granitic rocks. In *Granite: From Segregation of*

Melt to Emplacement Fabrics, ed. J. L. Bouchez, D. H. W. Hutton and E. Stephens. Dordrecht: Kluwer Academic Publishers, pp. 3–10.

Petford, N., Cruden, A. R., McCaffrey, K. J. W. and Vigneresse, J.-L. (2000). Dynamics of granitic magma formation, transport and emplacement in the Earth's crust. *Nature*, **408**, 669–73.

Pfiffner, O. A. and Ramsay, J. G. (1982). Constraints on geologic strain rates: arguments from finite strains of naturally deformed rocks. *Journal of Geophysical Research*, **87**, 311–21.

Pitcher, W. S. (1979). The nature, ascent and emplacement of granitic magmas. *Journal of the Geological Society London*, **136**, 627–62.

Pollard, D. D. and Johnson, A. M. (1973). Mechanics of growth of some laccolithic intrusions in the Henry Mountains, Utah, II. Bending and failure of overburden layers and sill formation. *Tectonophysics*, **18**, 311–54.

Potter, M. E. and Paterson, S. R. (2000). Rolling-hinge model for downward flow of host rock during emplacement of the Pachalka pluton, Mojave Desert, southeastern California. *Geological Society of America Abstracts with Programs*, **32**, abstract 52968.

Ramberg, H. (1970). Model studies in relation to plutonic bodies. In: *Mechanism of Igneous Intrusion*, ed. G. Newall and N. Rast. *Special Issue of the Geological Journal*, **2**, 261–86.

Ramsay, J. G. (1989). Emplacement kinematics of a granite diapir: the Chinamora batholith, Zimbabwe. *Journal of Structural Geology*, **11**, 191–209.

Rickard, M. J. and Ward, P. (1981). Paleozoic crustal thickness in the southern part of the Lachlan orogen deduced from volcano- and pluton-spacing geometry. *Journal of the Geological Society of Australia*, **28**, 19–32.

Robin, P.-Y. F. and Cruden, A. R. (1994). Strain and vorticity patterns in ideally ductile transpression zones. *Journal of Structural Geology*, **16**, 447–66.

Rocchi, S., Westerman, D. S., Dini, A., Innocenti, F. and Tonarini, S. (2002). Two-stage growth of laccoliths at Elba Island, Italy. *Geology*, **30**, 983–6.

Roig, J.-Y., Faure, M. and Truffert, C. (1998). Folding and granite emplacement inferred from structural, strain, TEM and gravimetric analyses: the case of the Tulle antiform, SW French Massif Central. *Journal of Structural Geology*, **20**, 1169–89.

Roman-Berdiel, T., Gapais, D. and Brun, J. P. (1995). Analogue models of laccolith formation. *Journal of Structural Geology*, **17**, 1337–46.

Rosenberg, C. L., Berger, A. and Schmid, S. M. (1995). Observations from the floor of a granitoid pluton: inferences on the driving force of final emplacement. *Geology*, **23**, 443–6.

Roy, B. and Clowes, R. M. (2000). Seismic and potential field imaging of the Guichon Creek batholith, British Columbia, Canada, to delineate structures hosting porphyry copper deposits. *Geophysics*, **65**, 1418–34.

Rubin, A. M. (1993). Dykes *vs.* diapirs in viscoelastic rock. *Earth and Planetary Science Letters*, **119**, 641–59.

Rutherford, M. J. and Gardner, J. F. (2000). Rates of magma ascent. In *Encyclopedia of Volcanoes*, ed. H. Sigurdsson. San Diego: Academic Press, pp. 207–17.

Saint Blanquat, M., Tikoff, B., Teyssier, C. and Vigneresse, J.-L. (1998). Transpressional kinematics and magmatic arcs. In *Continental Transpressional and Transtensional Tectonics*, ed. R. E. Holdsworth, R. A. Strachan and J. F. Dewey. Geological Society, London, Special Publication 135, pp. 327–40.

Saint Blanquat, M., Law, R. D., Bouchez, J.-L. and Morgan, S. S. (2001). Internal structure and emplacement of the Papoose Flat pluton: an integrated structural, petrographic, and magnetic susceptibility study. *Geological Society of America Bulletin*, **113**, 976–95.

Sanford, A. R. (1959). Analytical and experimental study of simple geological structures. *Bulletin of the Geological Society of America*, **70**, 19–52.

Sawyer, E. W., Dombrowski, C. and Collins, W. J. (1999). Movement of melt during synchronous regional deformation and granulite-facies anatexis: an example from the Wuluma Hills, Central Australia. In *Understanding Granites: Integrating New and Classical Techniques*, ed. A. Castro, C. Fernandez and J.-L. Vigneresse. Geological Society, London, Special Publication 168, pp. 221–38.

Scaillet, B., Pêcher, A., Rochette, P. and Champenois, M. (1995). The Gangotri granite (Garhwal Himalaya): laccolithic emplacement in an extending collisional belt. *Journal of Geophysical Research*, **100**, 585–607.

Schilling F. R. and Partzsch, G. M. (2001). Quantifying partial melt portion in the crust beneath the central Andes and the Tibetan Plateau. *Physics and Chemistry of the Earth, Part A, Solid Earth and Geodesy*, **26**, 239–46.

Schmeling H., Cruden, A. R. and Marquart, G. (1988). Finite deformation in and around a fluid sphere moving through a viscous medium: implications for diapiric ascent. *Tectonophysics*, **149**, 17–34.

Schmidt, K. L. and Paterson, S. R. (2000). Analyses fail to find coupling between deformation and magmatism. *Eos, Transactions of the American Geophysical Union*, **81**, 197–203.

Schwerdtner, W. M. (1990). Structural tests of diapir hypotheses in Archean crust of Ontario. *Canadian Journal of Earth Sciences*, **27**, 387–402.

Searle, M. P., Metcalf, R. P., Rex, A. J. and Norry, M. J. (1993). Field relations, petrogenesis and emplacement of the Bhangirathi leucogranite, Garhwal Himalaya. In *Himalayan Tectonics*, ed. P. J. Treloar and M. P. Searle. Geological Society Special Publication 74, pp. 429–44.

Sisson, T. W., Grove, T. L. and Coleman, D. S. (1996). Hornblende gabbro sill complex at Onion Valley, California, and a mixing origin for the Sierra Nevada batholith. *Contributions to Mineralogy and Petrology*, **126**, 81–108.

Skarmeta, J. J. and Castelli, J. C. (1997). Intrusión sintectónica del granito de las Torres del Paine, Andes Patagónicos de Chile. *Revista Geológica de Chile*, **24**, 55–74.

Smithies, R. H. (2000). The Archean tonalite–trondhjemite–granodiorite (TTG) series is not an analogue of Cenozoic adakite. *Earth and Planetary Science Letters*, **182**, 115–25.

Solar, G. S. and Brown, M. (2001). Petrogenesis of migmatites in Maine, USA: possible source of peraluminous leucogranite in plutons. *Journal of Petrology*, **42**, 789–823.

Spray, J. (1997). Superfaults. *Geology*, **25**, 579–82.

Sweeney, J. F. (1976). Subsurface distribution of granitic rocks, south-central Maine. *Geological Society of America Bulletin*, **87**, 241–9.

Sylvester, A. G. (1964). The Precambrian rocks of the Telemark area in south central Norway, III, geology of the Vrådal granite, *Norsk Geologisk Tidsskrift*, **44**, 445–82.

Sylvester, A. G. (1988). Strike-slip faults. *Geological Society of America Bulletin*, **100**, 1666–1703.

Sylvester, A. G. (1998). Magma mixing, structure, and re-evaluation of the emplacement mechanisms of Vrådal pluton, central Telemark, southern Norway. *Norsk Geologisk Tidsskrift*, **78**, 259–76.

Talbot, C. J. and Grantham, G. H. (1987). The Proterozoic intrusion and deformation of deep crustal 'sills' along the south coast of Natal. *South African Journal of Geology*, **90**, 520–38.

Thompson, A. B., (1999). Some time–space relationships for crustal melting and granitic intrusion at various depths. In *Understanding Granites: Integrating New and Classical Techniques*, ed. A. Castro, C. Fernandez and J.-L. Vigneresse. Geological Society, London, Special Publication 168, pp. 7–26.

Tikoff, B. and Teyssier, C. (1992). Crustal-scale, en-echelon "*P*-shear" tensional bridges: a possible solution to the batholithic room problem. *Geology*, **20**, 927–30.

Tobisch, O. T. and Cruden, A. R. (1995). Fracture-controlled magma conduits in an obliquely convergent magmatic arc. *Geology*, **23**, 941–4.

Tobisch, O. T., Saleeby, J. B., Renne, P. R., McNulty, B. A. and Tong, W. (1995). Variations in deformation fields during development of a large volume magmatic arc, central Sierra Nevada, California. *Geological Society of America Bulletin*, **107**, 148–66.

Tobisch, O. T., Fiske, R. S., Sorenson, S. S., Saleeby, J. B. and Holt, E. (2001). Steep tilting of metavolcanic rocks by multiple mechanisms, central Sierra Nevada, California. *Geological Society of America Bulletin*, **112**, 1043–58.

Vigneresse, J.-L. (1995). Control of granite emplacement by regional deformation. *Tectonophysics*, **249**, 173–86.

Vigneresse, J.-L. and Bouchez, J. L. (1997). Successive granitic magma batches during pluton emplacement: the case of Cabeza de Araya (Spain). *Journal of Petrology*, **38**, 1767–76.

Vigneresse, J.-L., Tikoff, B. and Ameglio, L. (1999). Modification of the regional stress field by magma intrusion and formation of tabular granitic plutons. *Tectonophysics*, **302**, 203–24.

Weertman, J. (1971). Theory of water-filled crevasses in glaciers applied to vertical magma transport beneath ocean ridges. *Journal of Geophysical Research*, **76**, 1171–83.

Weinberg, R. F. (1999). Mesoscale pervasive felsic magma migration: alternatives to dyking. *Lithos*, **46**, 393–410.

Weinberg, R. F. and Podlachikov, Y. (1994). Diapiric ascent of magmas through power law crust and mantle. *Journal of Geophysical Research*, **99**, 9543–59.

Weinberg, R. F. and Searle, M. P. (1998). The Pangong injection complex, Indian Karakoram: a case of pervasive granite flow through hot viscous crust. *Journal of the Geological Society London*, **155**, 883–91.

Whitney, J. A. and Stormer, J. C. Jr. (1986). Model for the intrusion of batholiths associated with the eruption of large-volume ash-flow tuffs. *Science*, **231**, 483–5.

Wiebe, R. A. and Collins, W. J. (1998). Depositional features and stratigraphic sections in granitic plutons: implications for the emplacement and crystallization of granitic magma. *Journal of Structural Geology*, **20**, 1273–89.

Williams, H. R., Stott, G. M., Thurston, P. C., Sutcliffe, R. H., Bennet, G., Easton, R. M. and Armstrong, D. K. (1992). Tectonic evolution of Ontario: summary and synthesis. In *Geology of Ontario*. Ontario Geological Survey Special Volume 4, Part 2, pp. 1255–332.

Windley, B. F. (1984). *The Evolving Continents*, 2nd edn. New York: John Wiley & Sons.

Yoshinobu, A. S., Okaya, D. and Paterson, S. R. (1998). Modeling the thermal evolution of fault-controlled magma emplacement models: implications for the solidification of granitoid plutons. *Journal of Structural Geology*, **20**, 1205–18.

Zegers, T. E. and van Keken, P. E. (2001). Middle Archean continent formation by crustal delamination. *Geology*, **29**, 1083–6.

14

Elements of a modeling approach to the physical controls on crustal differentiation

GEORGE W. BERGANTZ AND SCOTT A. BARBOZA

14.1 Introduction

Crustal differentiation is a progressive, and sometimes repeated, process of phase change, transport, and emplacement (Brown, 1994). For example, the upward migration of melts depleted in Y and the heavy rare earth elements (HREEs), which are generated during high-grade metamorphism, tends to stratify the concentration of large ion lithophile elements within the crust, concentrating them in the upper crust (Chapter 4). However, as with many persistent issues in crustal differentiation, the appropriate length scales of the compositional gradients and rates of the transport processes are unknown.

The essential process in crustal differentiation is the relative motion between phases; without relative motion there will be no differentiation, only bulk transport. Advective processes govern crustal-scale differentiation, as diffusion is not capable of transporting sufficient mass, but there are a number of questions that remain unanswered about these processes. What are the controls on relative motion? What are the mechanics of phase gathering and dispersal? To what degree are putative thermodynamic volumes controlled by rheological contrasts and reaction rates? All geological flow is formally chaotic in the sense of Aref (1990) and Ottino (1989), and includes crossing scales of influence (Petford *et al.*, 1997; Brown & Solar, 1998a, b; Yuen *et al.*, 2000), but are there underlying simplifications that are useful for geological applications?

Relatively simple, but interrelated, deterministic laws for which the dynamics may be quite complex govern geological systems. This yields systems possessing a dual nature; there is a tendency to disorder, but that is offset by the underlying mechanical deterministic templates. It is our view that these deterministic templates reflect multiphase, multiscale processes. Our objective in this final chapter is to review modeling strategies for addressing the

Evolution and Differentiation of the Continental Crust, ed. Michael Brown and Tracy Rushmer. Published by Cambridge University Press. © Cambridge University Press 2005.

transport mechanics of these multiphase, multiscale processes and to offer guidelines and recommendations for their quantitative assessment.

When coupled with deterministic models, both ergodic theory and chaotic dynamics provide possible approaches to constructing explanations that may bridge the heterogeneous scales of the information content of natural examples. Our emphasis is largely on the fluid-dominated regime of crustal differentiation; we will not discuss physical processes that have been addressed adequately elsewhere (e.g., Holness, 1997) and in earlier chapters. Hence, we will focus on methodologies and physical considerations that have not been elaborated previously, and that the student of crustal processes may find of some use.

One new but widely applied paradigm for multiphase systems invokes the notion of "granularity." In the broadest sense, granularity refers to the structure of the information content of a data set. It may also refer to the indeterminacy, or limits of resolution, of both space and time in a data set. Granularity in our context refers to the hierarchy of size, or scale, or levels of detail, that is intrinsic to geological data sets. It is important to appreciate that spatial granularity scales, such as the size, composition, and distribution of leucosomes in a migmatite, may have very different temporal granularity. Thus, the definition of proximity may take on different scales in both space and time.

This is implicit in the recognition that the geological systems that lead to crustal differentiation are open systems, and that the information content of an outcrop may exhibit deformation of both spatial and temporal elements. For example, a leucosome may be the result of intermittent melt migration and the composition may not reflect any simple relationship to the surrounding melanosome (e.g., Sawyer, 1999), or the thickness of a dike while magma is flowing may be greater than the measured thickness in the field, or the thermal conditions during partial melting may be overprinted by subsequent high-temperature events. Other examples include the plutonic expression of magma mixing and mingling, where what one observes in the field is the complex superposition of events that represent the very last and least dynamic state of the system (e.g., Bergantz, 2000). Thus, geological samples may be "over-sampled" with respect to temporal, kinematic, or compositional information in a way that is difficult to assess.

Data granularity is an explicit expression of phase-relative motion and, hence, multiphase processes. Each of these processes has characteristic length and time scales, and it is the nonlinear interactions and feedback among them that give rise to the dissipative structures that are recorded as spatial variations in composition and kinematic character of a rock body. These elements compose the "memory" of a rock body, and the spatial distribution of "memory" is not likely to be linearly related to time, especially in bodies with isotherms that are associated with profound rheological transitions (e.g., Vigneresse *et al.*, 1996).

14.2 Some general principles of multiphase flow

Practically all geological systems are multiphase, multicomponent systems. A phase is a physically distinct portion of matter, and a mixture is any multiphase ensemble. For example, magma may be composed of numerous solid phases, a melt phase and perhaps a separate volatile phase, all of which may move with non-zero phase and relative velocities. If the melt phase is volumetrically dominant, it would be called the "carrier phase," and we might consider the mixture a dilute suspension. This simplifies the physical description of the multiphase mixture considerably, as the interphase coupling is not important in describing the bulk flow. The other extreme would be flow through a fixed porous medium, in which the velocity of the solid phase usually is taken to be small or zero and the velocity of the fluid phase is non-zero. In this case, the solid phase is volumetrically dominant. Developing a continuum description of multiphase heat and mass transfer for all possible conditions, from melt to solid dominated, has been especially challenging due to the change in the intrinsic averaging scales that define the continuum.

14.3 Suspensions and slurries

14.3.1 Definitions

It is apparent that a dispersed-phase mixture is not a continuum in the molecular limit, and one must develop carefully the definition of a material "point" so that the differential equations of transport may be applied to the mixture (Poletto & Joseph, 1995). The definition of mixture quantities requires the definition of a stationary average volume ΔV_o

$$\Delta V_o = N_m L^3, \tag{14.1}$$

where N_m is the minimum number of particles for a stationary average, and L is the particle spacing. If L is taken to be 10 particle diameters, the particle diameters to be 0.5 mm and N_m to be 10^3, the stationary average volume ΔV_o is $0.125 \, \mathrm{m}^3$, or a volume 0.5 m on a side. If this dimension is of the same order of magnitude of the system scale, defined to be the scale that describes some distinct region of homogeneity, then the mixture cannot be treated as a continuum. Considering that many migmatite bodies have measurable variability at or below this scale, attempts to integrate kinematic and compositional elements with continuum theory require considerable care. Equation (14.1) provides a measure of what might be considered geological "noise" relative to a model that represents averaged quantities, and highlights the difficulties of invoking the continuum assumption for many geological applications.

Equation (14.1) motivates the definition of the phase volume fraction, ε_p:

$$\varepsilon_p = \lim_{\Delta V \to \Delta V_0} \frac{\Delta V_p}{\Delta V} \tag{14.2}$$

for a phase volume of ΔV_p. Conservation of mass requires that the volume fraction of n phases sum to unity

$$\sum_{i=1}^{n} \varepsilon_i = 1. \tag{14.3}$$

The mechanics of the relative motion between melt, cumulate, and residual phases determine the length and time scales of crustal differentiation. Such mixtures are described as suspensions or slurries, depending on the degree to which the presence of particles influences the rheology or the exchange of momentum (Roberts & Loper, 1987; Marsh, 1988; Iverson, 1997). The mechanics of a dilute suspension are governed largely by the physical properties of the melt. With increasing solid fraction, the suspension may undergo a transition to a slurry, the mechanics of which are dominated by the transmission of stress and/or momentum between the separate phases. Kinematic and dynamic criteria provide alternative quantitative definitions for the suspension to slurry transition.

The kinematic criterion follows from the rule-of-thumb that particles that are closer than two to three diameters experience mutual hydrodynamic interaction and so cannot be considered to be isolated elements, and so the mixture is not a dilute suspension. One implication of this is that simple drag laws, such as Stokes' law are not applicable without some correction. If it is assumed that the particles are equally spaced, the particle diameter, D, the particle spacing, L, and the volume fraction are related by

$$\frac{L}{D} = \left(\frac{\pi}{6\varepsilon_p} \right)^{1/3}, \tag{14.4}$$

and so for a particle volume fraction of 0.1, the particle spacing ratio is about 1.7, and the particles cannot be treated as individual elements. Therefore, the mechanics of a multiphase mixture is influenced by particle collisions or hydrodynamic interactions even at particle volume fractions of less than 10%, and the multiphase mixture cannot be considered as a simple collection of single particles.

A highly simplified dynamic criterion for the suspension-to-slurry transition may be obtained by the consideration of whether the mixture is dilute or dense. In a dilute mixture, the momentum transfer is strictly between the particle and the drag and lift forces of the fluid, and particle–particle interactions are negligible. One measure of mechanical dilution is the particle response time. The particle response time is the time necessary for the fluid to influence the particle trajectory. For example, if the particle response time to the fluid forces is shorter than the time between particle collisions, the fluid is dilute because particles are dominantly responding to fluid forces. Conversely, if the particle–particle collisions happen so rapidly that the particle cannot respond to fluid forces before the next collision, the fluid is considered to be a slurry.

The particle response time τ_M to the fluid forces in the particle sub-inertial or Stokes regime range (note that the fluid may still be turbulent), is given by (Marble, 1970; Crowe *et al.*, 1996)

$$\tau_M = \frac{\Delta\rho D^2}{18\mu},\tag{14.5}$$

where $\Delta\rho$ is the particle–fluid density contrast, D the particle diameter and μ the dynamic viscosity of the carrier phase. The time between particle collisions is

$$\tau_C = \frac{1}{n\pi D^2 \nu_r},\tag{14.6}$$

where n is the number density of particles, and ν_r the relative velocity between particles. Thus, the requirement for a mixture to be dilute is

$$\frac{\tau_M}{\tau_C}\tag{14.7}$$

For a relative velocity of unity, which is probably much higher than what one might have in practice, and the relatively high viscosities, the criterion for a dilute condition might be satisfied even to high particle volume fractions in most geological applications. A more complete dimensional and scaling analysis of momentum transport at high particle volume fraction provides additional dynamic criteria for distinguishing multiphase flow regimes (Iverson, 1997).

14.3.2 Characteristic scales of crystal sorting and sedimentation

Many expressions of transport of multiphase mixtures may be addressed by consideration of the particle momentum response time. The equation of motion for a non-steady, single particle is given by the Basset–Boussinesq–Oseen equation (BBO), which is a complex integro-differential equation (Stock, 1996; Kim *et al.*, 1998). However, if one is interested in the time-averaged behavior, and if the fluctuations in the flow-field are not rapid, many of the complications of the BBO equation may be neglected (Vojir & Michaelides, 1994). With these caveats in mind, we will focus on the (pseudo) steady-state drag term, as the integrated influence of this term dominates the particle–fluid interaction.

First, we introduce the characteristic time scale for the flow

$$\tau_F = \frac{L_O}{U_O}, \tag{14.8}$$

where U_O and L_O are the local characteristic velocity and length scale of the flow (but also see Crowe, 2000). The ratio of the particle response time to the characteristic time scale of the fluid motion is called the Stokes number, *Sk*, and is given by

$$\frac{\tau_M}{\tau_F} = Sk, Sk = \frac{\Delta\rho D^2}{18\mu} \frac{U_O}{L_O} \tag{14.9}$$

(Ling *et al.*, 1998). If the Stokes number is much less than 1, the particle relaxation time is small relative to the time it experiences the fluid forces, velocity equilibrium will be immediately attained, and the particle dispersion will be the same as the fluid dispersion (Yang *et al.*, 2000). Conversely, if the Stokes number is large, the particles do not have sufficient time to respond to fluid forces, and will move through the flow field without being entrained by the flow. When the Stokes number is near unity, the fluid has sufficient viscosity to spin the particle to significant angular velocities without suppressing the radial trajectory. Thus, the particle relaxation time is matched to the eddy rotation time, and the particles circulate on the margins of the eddy. This provides a mechanism for the de-mixing of an initially well-mixed multiphase fluid, and in fact may lead to particle sorting and selective sedimentation even in turbulent flow (Zarrebini & Cardoso, 1998).

Gravity forces introduce another time scale in the description of the particle motion, which leads to the scaled particle settling time

$$\tau_S = \frac{18\mu L_O}{\Delta\rho D^2 g}, \tag{14.10}$$

which is the time it takes the particle to settle through an eddy of diameter L_O. This time scale provides a partial measure of the time a particle has available for chemical reaction within a dynamic eddy, the state of which is defined by an average chemical potential. One important difference between the particle response time scale, τ_M and the particle settling time scale, τ_S, is that the particle response time scale represents the time scale of unsteady motions. The particle settling time scale is that associated with the steady forcing of gravity (Raju & Meiburg, 1995; Burgisser & Bergantz, 2002). The ratio of the particle settling time and the scaled flow time is

$$\frac{\tau_S}{\tau_F} = \frac{Fr^2}{Sk}, Fr = \frac{U_O}{\sqrt{Log}}, \tag{14.11}$$

where Fr is the Froude number. Neglecting some terms for brevity of discussion, the BBO equation may now be written as

$$\frac{d\mathbf{v}_p}{dt} = \frac{f}{Sk}\left[\mathbf{u}(\mathbf{X}, t)|_{x=x_p} - \mathbf{v}_p(t)\right] + \frac{1}{Fr^2}\mathbf{e}_g, \tag{14.12}$$

where $\mathbf{u}$ is the fluid velocity evaluated at the particle, $\mathbf{v}_p$ the particle velocity, and $\mathbf{e}_g$ the unit direction of the acceleration of gravity. The term, f, provides for departures from Stokes drag at higher-particle Reynolds number, or for more involved corrections related to turbulence intensity, non-spherical particles and other factors.

The notion of particle response time may be extended to heat and mass transfer as well. As with the case of non-steady motion, a complete description of the non-steady heat and mass transfer between a particle and the carrier (fluid) phase yields a complex integro-differential equation (Michaelides & Feng, 1994; Feng & Michaelides, 2000). However, a useful characteristic time scale for the particle thermal response time to changes in the fluid temperature is

$$\tau_T = \frac{\rho_P c_P D^2}{12k_c} \tag{14.13}$$

where ρ_P is the particle density, c_P the particle specific heat capacity and k_c the thermal conductivity of the carrier phase.

14.3.3 *Phase segregation by shear*

Particle dispersal and gathering takes place in both rectilinear and curvilinear flow at low carrier-phase Reynolds number, where there is little time-dependence to the flow. There are many physical manifestations of this process, such as the resuspension of a particles that have settled, or the migration of particles out of regions of high shear as observed in conduit and viscometer flow (Karnis *et al.*, 1966; McTigue *et al.*, 1986; Leighton & Acrivos, 1987; Averbakh *et al.*, 1997; Lyon & Leal, 1997; Marchioro & Acrivos, 2001). Two theories have been used to illuminate these physical processes: (1) the Lagrangian approach, such as the Stokesian dynamics (Brady & Bossis, 1988; Singh & Nott, 2000; Marchioro & Acrivos, 2001), or (2) the lattice Boltzmann approach (Aidun *et al.*, 1998), and the effective continuum (non-local) models where the movement of individual particles is represented by changes in the local particle volume fraction.

Of the continuum models, there are generally two approaches to the physical description of the migration of particles: (1) the diffusive flux model of Leighton and Acrivos (1987), subsequently extended by Phillips *et al.* (1992), and (2) the granular "temperature," or suspension balance models, of Jenkins and McTigue (1990) and Nott and Brady (1994). The diffusive flux model is a phenomenological model that attempts to describe the non-local particle migration perpendicular to the shear plane as a diffusion process, and a constitutive equation is derived for the particle flux. In this model, the shear-induced particle diffusivity is linearly proportional to the local shear rate. The description of particle segregation from the initial inlet conditions to a final steady configuration is given by a particle volume fraction conservation balance, where the particle segregation flux, J, is explicitly decomposed into contributions from gravitational and hydrodynamic forces, and spatial variations in viscosity as a function of local particle volume fraction (Fang & Phan-Thien, 1995; Subia *et al.*, 1998; Shauly *et al.*, 2000). The explicit forms of the contributions to the general model for the particle segregation J are complex, and the interested reader is directed to the references given above. Despite the fact that the phenomenological diffusive flux approach is ad hoc, it does provide an accessible, empirically based model for particle migration in simple shear flow that is amenable to numerical implementation.

The granular temperature model of Jenkins and McTigue (1990) is based on the notion of a scalar migration potential associated with just the particle phase. The suspension balance model of Nott and Brady (1994), while also invoking the notion of particle temperature and, hence, pressure, includes both phases. However, both models have been difficult to use and implement. One notable exception is the study of Petford and Koenders (1998), which employs the model of Jenkins and McTigue (1990) to assess the possibility of particle migration in the flow of silicic dikes. However, it has been concluded subsequently that the definition of a granular temperature as a scalar variable is not very realistic, as the particle velocity fluctuations are highly anisotropic (Shapley *et al.*, 2002, 2004). This does not mean that these models have no utility for geological applications, only that they are currently most useful in the high-Peclet number limit where putative particle temperatures have a negligible contribution to the potential energy of the transport.

An extension of the so-called "suspension balance approach" offers a more fundamental theory of particle segregation (Morris & Boulay, 1999), and does not require ad hoc tuning of field variables as in the diffusion flux model, nor invoke (inspired but) model-based notions such as granular temperature. It is a rheologically based approach, and has the advantage over the diffusive flux model that anisotropy of normal stresses is explicitly modeled and its role as the cause of particle segregation is explained. This approach satisfactorily recovers many of the experimental results, and illuminates how the normal stresses fall-off quadratically as the particle volume fraction becomes small. In fact, it has been difficult experimentally to demonstrate particle migration for very dilute systems where the particle volume fraction is less than 10% (Hampton *et al.*, 1997). Thus, we would expect that particle segregation during flow will be more important at higher solid fraction, such as one might obtain in diatexites. The implementation of the model is rather complicated, and the needed experimental work to provide the transport coefficients for the normal and shear stress rheology and the particle hindrance function has yet to be done.

Of central interest to geological applications is the time scale over which unmixing, or particle migration, may occur. Most theories and experiments predict that the length scale, L, over which an initially well-mixed material will un-mix as a consequence of particle migration, will scale as (Nott & Brady, 1994; Phan-Thien & Fang, 1996; Hampton *et al.*, 1997)

$$L \approx \frac{W^3}{a^2}(1.1), \tag{14.14}$$

where W is the width of the channel, and a is the particle radius. For a diatexite or dike in simple pressure flow (i.e., 0.5 m wide, and a residual particle radius of 5 mm) the length scale to a steady particle profile is 5 km. Thus, despite being in a state of very low Reynolds number, the material being transported will have a complex history. This indicates that assuming steady conditions in the modeling of a multiphase conduit requires some justification. If the channels do not have a simple geometry, although the flow is laminar, it will be chaotic (Jones *et al.*, 1989; Wang *et al.*, 1990). This may complicate the assessment of steady-state conditions further. In addition, all the models described above are for conditions of mono-disperse mixtures, which are unlikely to be the norm in practice. For poly-disperse mixtures, more complex arrangements may result (Shauly *et al.*, 2000), and the first-order physical models are not yet available that include normal stresses.

14.4 Mixing and compositional heterogeneity

One of the most challenging aspects in the study of crustal differentiation is the association of upper-crustal products with their source areas. This is partly a consequence of the fact that these source areas are rarely exposed, but is also due to open-systems processes of differentiation and mixing that produce intermittent homogenization and fractionation, which may mask source area diagnostics. These processes are mixing in the sense that conditions change from one state of simplicity to another (Tavare, 1986). For example, source regions may be regionally simple in the statistical sense of bulk composition, mode, kinematic fabric, etc., and the ultimate product of crustal differentiation (a granite, for example) might manifest similar scales of statistical uniformity. But this simplicity is the result of complex open system processes and dissipative structures, so how may one see through the low-energy final products and characterize the complex spatial and temporal patterns? This motivates us to consider some of the physical controls on the time and space scales of mixing in multiphase reacting flow.

In the discussion below, we consider only mixing that is a consequence of advection. The role of velocity is two-fold, it increases the surface area by stretching, and aids diffusion by transporting folded and stretched elements. We do not explicitly consider the mixing that results from diffusion alone across a solid, bounding body. These are controlled by the diffusivity of compositional elements, which is much less than the thermal diffusivity, which is often rate-controlling. We will also not consider the mixing of granular systems, where the momentum exchange and dissipative structures are dominated by particle–particle friction. These processes have been considered by Khakhar *et al.* (1999), but are not obviously relevant to melt-present advective transport dominated by buoyancy changes in a gravity field.

Mixing may be regarded generally as consisting of three distinct stages: (1) initial interpenetration or injection; (2) stirring, and (3) mixing by molecular diffusion. The interpenetration stage initializes the macroscopic bodies of a composite system with a distinct volume, surface, and position. This stage may also provide for the potential energy for subsequent internal mixing, the initial distribution of surface area, and so transports extensive properties associated with distinct, pre-mixed volumes.

One may imagine two end-members to this process. The first is where the interpenetration process brings pre-mixed volumes together with sharp boundaries. This typically happens if there is a contrast in viscosity and thermophysical properties, and if the Reynolds number is low, which may often be the case in diatexites or dikes (Petford, 1996). This could form distinct islands of unmixed material, which would then undergo blending by repeated stretching. Another end-member is where the interpenetration process is intimately accompanied by mixing, and the stirring and interpenetration stages are virtually indistinguishable. This is called "stationary entrainment" and requires nearly equal viscosities and the rapid, local, conversion of potential energy to kinetic energy at high Reynolds number conditions, such as that documented by Linden *et al.* (1994). Examples of both end-members are illustrated in the flow of multiphase plumes by Bergantz and Ni (1999).

The dissipative structures that do the work of mixing, such as Taylor layers (Broadwell & Mungal, 1991), are essentially transient and difficult to quantify after the process is completed. In the context of mantle convection there have been significant efforts to elucidate the mixing of passive features (Hoffman & McKenzie, 1985). These modeling efforts demonstrate that the mixing times and time-dependent spatial scales are sensitive to both the kinematics of the flow and the rheology, and are difficult to generalize. For example, models of Rayleigh–Bernard convection under mantle conditions reveal that Newtonian rheology produces more rapid progress to mixing than does a non-Newtonian material (Ten *et al.*, 1997). Forward modeling provides valuable insights into specific conditions of mixing and has significant heuristic value, but does not usually have generic and widespread applicability due to the wide range of possible and chaotic conditions. For example, this approach generally does not relate a general metric of mixing efficiency to a particular velocity field.

This motivates some definitions of the state of mixing, as proposed by Danckwerts (1952) and implemented in a geological context by Oldenburg *et al.* (1989), which is recommended as a succinct introduction. Two measures of mixing are the scale and intensity of segregation, as long as there are no persistent large-scale heterogeneities. The scale of segregation is associated with a two-point correlation function and is related to the average traversal length through the phases. This function gives the probability that two volume elements of a concentration, C, will be correlated

$$R(|\mathbf{r}|) = \frac{\overline{C_i(x + r) \cdot C_i(x)}}{\overline{C'_i}^2}, \qquad (14.15)$$

where C'_i is the deviation from the system average concentration. Note that for large values of $|\mathbf{r}|$, $R(|\mathbf{r}|)$ tends to zero. If periodic structures are present, $R(|\mathbf{r}|)$ will not tend to zero, but will have a periodic variation in $|\mathbf{r}|$ and the scale of segregation is defined by the correlation function

$$\xi = \int_0^\infty R(|\mathbf{r}|) d|\mathbf{r}|. \qquad (14.16)$$

Another measure of goodness-of-mixing is the intensity of segregation, which indicates how sharp the compositional variations are. If diffusion is unimportant, the intensity of segregation is at a maximum, regardless of the spatial distribution of such compositional elements, and complete homogenization would produce a value of zero. For a two-component mixture of A and B it is defined as

$$I = \frac{\overline{C'_A}^2}{C'_A{}^2 C'_B{}^2},\tag{14.17}$$

where I is the deviation of the concentrations in the segregated phases from the mean. These measures may be applied to the interpretation of outcrop scale systems. They provide a basis for comparison with forward models, and interpretation of scales of data, as statistical measures from, for example, crystal-scale features, may be assessed on the scales of meters to kilometers if outcrop is sufficiently extensive. The measures of spatial statistical significance are also present in the application of the variogram, a widely used element of geostatistics.

The potential energy for mixing may arise as a consequence of forcing by boundary conditions, or the decay of an unstable initial condition, or both; an example is mixing on reintrusion, which is likely to be important in geological systems (Jellinek & Kerr, 1999; Bergantz, 2000; Eichelberger *et al.*, 2000). Numerical experiments of transient, buoyant, multiphase cavity flow quantify the progress of mixing in terms of the Lyapunov exponent, σ_A (Williams, 1997), which is defined as

$$\sigma_A = \lim_{t \to \infty} \frac{1}{t} \ln \frac{\mathrm{d}A(t)}{\mathrm{d}A(0)},\tag{14.18}$$

where $A(0)$ is the initial area and $A(t)$ is the area at a time after the start of stirring. A positive value of σ_A indicates an exponential increase of the surface area with time, and the reduction in the thickness of any region of distinct composition (the striation thickness). A kinematic description of the Lyapunov exponent is that it is the long-time average of the strain rate.

For a Reynolds number, Re, that is greater than $10^3 - 10^4$, turbulent stirring produces what is known as "Eulerian chaos." This critical value of the Reynolds number is called the "mixing transition" (Breidenthal, 1981), and is associated with the notion of turbulence being fully developed where many scales of the flow are present. This critical value for the mixing transition is remarkably robust, being approximately the same for many different geometries and flow styles. The efficiency of stirring at high Re is the result of straining and rapid exponential area increase in the smallest eddies of the turbulent flow. Mixing takes place by entrainment and nearly simultaneous dispersion. Hence, the rate-limiting aspects to the mixing are the time scales, usually the rotation period, of the largest eddies as they are the dynamic entities that dominate the entrainment (Brown & Roshko, 1974).

A salient point is that the striation thickness will decrease as

$$\frac{V}{A} \approx \eta e^{-\sigma_A t}; \quad \sigma_A \approx 0.3 e_K, \tag{14.19}$$

where η is the characteristic length of the Kolmogorov microscale, and e_K the strain rate at this length scale. The Lyapunov exponent for these kinds of flows is always positive and the flow is formally chaotic. This is the kind of mixing that is most common in engineering and environmental flows and a thorough treatment is given by Baldyga and Bourne (1999). For a Re of 10^2, the flow is in the convective regime, and the Lyapunov exponent is of 10^{-2}, indicating a flow that is transitional to chaotic. For a Re of 10^{-1}, the near absence of non-linear convective forces leads to a vanishing Lyapunov exponent, and the kinematics of mixing may change fundamentally.

Ultimately, the volume of perfectly mixed material is controlled by molecular diffusion. This diffusion process manifests the usual scaling and the volume, V, of mixed material after a time, t, is given by the proportionality

$$V \approx A\sqrt{Dt}, \tag{14.20}$$

where D is the diffusivity of the mixed component and A the surface area between mixing volumes. As D is a (non-constant) material property, the only way to increase the efficiency of mixing is to increase the surface area by stirring. This motivates the introduction of the striation thickness as one metric for the progress of mixing (Ottino, 1989), V/A. For volume-conserving incompressible materials, this is the scaled thickness of the unmixed volumes while stirring proceeds, and complete mixing is optimized in a system in which deformation reduces this length scale.

Viscosity contrasts are eventually rate-limiting for the efficiency of stirring, and these contrasts are determined by the thermal "lifetime" of the overall process. Thus, to be effective, the stirring must decrease the striation thickness of the pre-mixed volumes such that the thermal macro-scale of the process is equal to the time scale for compositional homogenization by diffusion

$$\frac{L^2}{(V/A)^2} \approx \frac{\kappa}{D}, \tag{14.21}$$

where L is the length scale associated with the overall thermal anomaly, and κ the thermal diffusivity. The ratio of the diffusivities will be of the order of from 10^4 to 10^8 depending on the definition of the components. For a diffusivity ratio of 10^6 and a thermal length scale of 1 km, the striation thickness must be reduced to 1 mm for molecular mixing to be achieved. This length scale may be smaller than the size of the crystals in multiphase flow, and illustrates the requirements and difficulty of complete mixing at the crystal scale.

Mixing of silicate melt and crystals may involve materials with differing viscosities and densities (Campbell & Turner, 1986; Sparks & Marshall, 1986; Jellinek & Kerr, 1999). The basic physics of mixing of variable property mixtures of magmas is complicated by the fact that temperature changes induce rheological changes through crystallization and exsolution of volatiles. None the less, it is instructive to consider some of the fundamentals of the variable property mixing of single-phase fluids (Burmester *et al.*, 1992; Rielly & Burmester, 1994; Jellinek & Kerr, 1999).

Consider, as a simple example, the intrusion of material into a magma accumulation site or chamber that is already in a state of unsteady motion. It has been shown that the efficiency of mixing depends on the ambient kinetic energy of the host mixture, the viscosity contrast, the density contrast, and the method of addition of the two materials. Mixing when high viscosity materials are added to a fluid environment in motion may only be effective if the inertial shear stresses associated with local unsteady motion may overcome the viscous stresses in the added material that resist stretching and entrainment. Thus, only the large eddies may distort the viscous additions leading to local break-up at large scales. It has been confirmed experimentally that the viscosity ratio is unimportant if

$$Re\frac{\mu_1}{\mu_2} \geq O(10^2), \tag{14.22}$$

where Re is the host environment Reynolds number, μ_1 the host viscosity, and μ_2 the viscosity of the addition. For values of this parameter group below this threshold, the mixing time may increase by orders of magnitude and the mixing will not be self-similar with respect to the Reynolds number.

The method of addition is also important, as that influences the surface area and hence the stresses. For example, material injected as a stream has a smaller characteristic length scale and will mix more readily than material added as clumps. The location of addition will influence the mixing as well unless the host is in a state of isotropic turbulence, which is unlikely in natural examples. Dense material added at the top in the form of a stream will be more likely to mix than material added near the floor, as entrainment into down-going (multiphase?) plumes is likely.

Density differences also influence mixing – if added material is heavier, it has a tendency to sediment. This is a function of the Richardson number, Ri, which is proportional to the density contrast, or reduced gravity, between the added and resident materials. Assuming that the length scale in the Richardson number is that of the added material, and Ri is large, added material will sediment and tend to stay on the bottom, extending any delays in mixing associated with viscosity contrasts. Thus, effective mixing requires small density and viscosity contrasts, and addition at small length scales near the top of the system.

The "mixing" of immiscible fluids manifests two principle stages as a function of the capillary number, Ca, which is the ratio of the shear stress, τ, transmitted by the carrier phase and the resisting force, σ, associated with the drop interfacial tension and radius, R,

$$Ca = \frac{\tau R}{\sigma}. \tag{14.23}$$

When Ca is greater than unity under conditions of equal viscosity, the mixing process is called "distributive mixing" and drops are extended affinely. As mixing proceeds, Ca tends to the critical value, disturbances grow, and break-up of the disperse phase occurs. This is called "dispersive mixing." If the viscosity of the dispersed phase is less than the carrier phase, the micro-rheological processes and transition to dispersive mixing occur faster. The critical value of the capillary number for the transition between distributed and dispersive mixing is a function of the viscosity ratio and the flow kinematics, with the most effective dispersion not necessarily associated with a viscosity ratio of unity (Grace, 1982).

14.5 Heat transfer, melting, and rheological models

Changes in chemical potential are required for the production of a melt phase. These changes are usually the result of increasing temperature or the addition of volatiles or some combination of both. Estimating the role of fluid infiltration is hampered by the fact that it is often difficult to estimate the volatile content of newly added magmatic material into the crust, or to quantify the physical processes associated with volatile generation and transfer (cf. Holness & Clemens, 1999). Conversely, the consequences of heat transfer, often the rate-controlling step in crustal differentiation (Petford *et al.*, 1997; Harris *et al.*, 2000), are easier to estimate and predict.

We begin by noting that the discussion of heat transfer should first be developed explicitly in terms of enthalpy transfer, as the heat equation is based on the conservation of an extensive quantity, and secondly in terms of rates. It is incorrect to do a simple thermal balance, a popular but ad hoc calculation, which relates temperature changes between two bodies by simply equating their specific enthalpies and mass to an equilibrium temperature. In conjugate heat transfer (Bergantz, 1992), temperature changes are a function of the rates of heat transfer. The initial total enthalpy contrast may have no relation to the maximum temperatures experienced during the prograde step of a thermal event.

A simple example from conduction will suffice. Consider two semi-infinite regions in thermal contact. If these two regions were at constant temperature initially, the maximum temperature at their contact will be roughly the mean temperature regardless of how long the heat transfer takes place. Thus, the maximum temperature is not related to the actual amount of enthalpy exchanged. However, the integrated amount of heat transferred will depend on the initial specific enthalpy and the mass. So for a partial melting event, the amount of melt produced will depend on the magnitude of the enthalpy difference, but the maximum degree of melting will be determined only by the magnitude and spatial scale of the initial temperature contrast.

It is the interdependence between the phase relations and the temperature that determines the compositional spectrum of melts produced. In the simplest case, there is no melt transport during melting, and the phase relations and temperature equation (with initial and boundary conditions) provide a complete description of the system (Bergantz, 1990). It is important to appreciate that thermodynamic statements (e.g., the functional relationship between the enthalpy and temperature, or temperature and phase volume fraction) appear explicitly only as a prescribed closure condition. If the system is open, then one must have a complete description of the compositional phase space, as open system transport will change the progress of melting by changing the bulk composition of the system. One strategy for parameterization of the phase relation linked to a transport model is exemplified in Barboza and Bergantz (1997).

During a prograde thermal event, the maximum temperature dictates one of the compositional extremes and the maximum amount of melt. Melt composition and volume fraction dominate the rheology of the partially molten region and, hence, the transport styles and rates. There is an explicit feedback between rates of melting, maximum temperatures, and rheological states. This has been explored in detail for crustal melting associated with intrusion of mafic material into the crust (Barboza & Bergantz, 1996, 1997, 1998). These studies indicate that the changes in rheological conditions associated with partial melting had a larger influence on the extent and progress of the melting than did changes in the phase diagram, or melt-fraction interval. Invoking assumptions that would optimize the conditions of enhanced heat transfer, these authors concluded that high degrees of melting, (e.g., greater than 40 vol.%) were not likely under most crustal conditions. This is in agreement with the estimated compositions of melts produced by natural protoliths, where greater amounts of partial melting produces melts that do not resemble natural examples (Beard & Lofgren, 1989).

Of course, if the mean temperatures of the host and intruding material produce melt fractions above this value (e.g., under conditions of a very steep geotherm) then even conduction would produce a mobile material above a critical melt fraction. However, it is notable that these conditions of rapid and extensive melting are rarely (ever?) reported in the rock record (Barton *et al.*, 1991). Hence, based on an overwhelming number of natural examples, as well as model analyses, magma once chambered, rarely produces thermal effects greater than that predicted assuming conductive heat transfer. The counter-examples, where magma has demonstrably been flowing in a dike or volcanic neck, often do melt the margins; see references in Knesel and Davidson (1999) and Davies and Tommasini (2000).

The rheological complexity of multiphase magmatic mixtures has been discussed by Vigneresse *et al.* (1996), Bagdassarov and Dorfman (1998), Vigneresse and Tikoff (1999) and Renner *et al.* (2000), and a comprehensive review of the rheology of suspensions may be found in Liu and Masliyah (1996). The notion that there is a single rheological critical melt fraction (RCMF) that limits melt transport is only applicable under nearly closed-system conditions, which may not occur often in natural samples (Sawyer, 1994). Rheological experiments do not reveal a single value for a RCMF (Rushmer, 1995; Rutter & Neumann, 1995; Bagdassarov & Dorfman, 1998) leaving it uncertain if there is a comprehensive constitutive model. In a compelling set of descriptive arguments combining outcrop-scale observations and experiments from material science, the notion of a RCMF has been extended by Vigneresse and Tickoff (1999) to include percolation thresholds to allow for phase-relative motion. While these distinctions lack a detailed theoretical description, they provide a very useful basis for developing more sophisticated models of multiphase mechanics at low to moderate local melt fraction.

However, one should be cautious about applying a comprehensive rheological model to multiphase systems subject to diverse kinematics (Iverson & Vallance, 2001). Buoyant flows involve both simple and pure shear; hence, constitutive equations must be developed for the entire range of flow kinematics. In addition, rheological models of multiphase mixtures are impli- citly dependent on the scale of averaging, or the scale of the stationary volume for the continuum. This sets the scales for any transport theory, whether formal or "back of the envelope;" this is discussed in more detail below.

A transport theory requires a statement of momentum balance, and if there is phase-relative motion, a momentum balance must be provided for each phase. Under these circumstances, the viscosity of the constituent phases of the mixture and multiphase drag interactions must be considered explicitly. The viscosity is then different in each transport equation and reverts to the literal definition, that it represents the drag of a phase on itself. Models for mixture viscosity that attempt to incorporate the physics of phase interaction by defining a "mixture" viscosity require a condition of no phase-relative motion. This means that there is only one local velocity for all phases. In our previous efforts in which a mixture rheological model was introduced (Barboza & Bergantz, 1997, 1998), we made it explicitly clear that there was no phase-relative motion, and this is a limitation of those works. Hence, the interpretation of the dynamics of outcrop-scale features manifesting prior melt-present conditions should be tempered with the notion that the invoca- tion of a constitutive relationship requires an appreciation for the implicit assumption of length scales and possibility of phase-relative motion.

One consequence of partial melting is the development of porosity (Bergantz, 1990; Lupulescu & Watson, 1999) and the progressive development of melt transport networks (Brown, 1994; Sawyer, 2001). The interplay between the progress of melting, the development of a melt network, deforma- tion, and melt migration, is complex (Brown & Solar, 1998a, 1998b; Marchildon & Brown, 2002, 2003; Guernina & Sawyer, 2003). Sawyer (2001) suggests that there are two end-member types of melt channel networks: (1) a melt-draining network that allows for within-layer melt redistribution, and (2) melt-transfer networks that allow melt to move across and out of its source layer. The melt-draining network moves melt from sites of incipient melting into a set of dynamic, hierarchical structures, the granularity of which is controlled by the foliation planes.

Rates of melt migration have been estimated based on the degree of chemi- cal equilibration between leucosomes and protolith (Sawyer, 1991, 1994; Harris *et al.*, 2000). It has suggested that melt generation and deformation- assisted segregation may take place on the order of 10^2 years, although such rates have yet to be independently conformed based on an internally consistent physical model and consideration of a field study.

There is uncertainty as to the importance of the magnitude of the volume change and the appropriate form for the resulting permeability tensor for melt migration, especially at water under-saturated conditions. Positive volume changes associated with muscovite dehydration melting may be around 2% (Patiño-Douce & Johnston, 1991; Connolly *et al.*, 1997; Rushmer & Brearley, 1998). This positive volume change may induce microfracturing, which may provide a means of generating permeability (Watt *et al.*, 2000), and Connolly *et al.* (1997) have estimated that the permeability may be 10^{-14} m^2 and isotropic. In contrast, Rushmer and Brearley (1998) caution that biotite dehydration may have small or negative volume changes, although Guernina and Sawyer (2003) document efficient melt removal, at up to 40% of melting, during regional biotite dehydration. Thus, it may be difficult to generalize the role of dehydration reactions in the production of porosity.

The cellular automaton approach provides one very promising, quasi-physical approach to exploring the means by which drainage networks develop (Miller & Nur, 2000). Although notable progress towards a master continuum model has been made (Connolly, 1997), a first-principles, comprehensive mechanical model for the progressive formation of kilometer-scale drainage networks that provides for melt to accumulate and migrate remains to be developed. It is often assumed that the Blake–Kozeny–Carman equation relating porosity to permeability is applicable. However, whether the permeability is anisotropic and what is the form of the tortuosity are both uncertain.

Recalling our discussion of granularity of data, there is always uncertainty as to whether regional suites of leucosomes were all present at the same time, and that their size reflects some measure of conditions at the time of putative melt transport. Two studies have considered natural examples, and one further study was motivated by metallurgical applications. In a novel study, Tanner (1999) concluded that the distribution of leucosomes was scale-invariant (fractal), allowing generalization to a simple transport model. He concluded that the melt could escape by buoyant flow alone, a conclusion that might warrant further consideration in light of a more complex transport model. Brown *et al.* (1999) conclude that melt channels were straight and uniform, suggesting that the flow may be regarded as a Darcy-like system on a regional basis. In an experimental study of the (dendritic) crystallization of metals, Nielsen *et al.* (1999) concluded that the Blake–Kozeny–Carman equation gave good agreement with experiments. The ability of melt to migrate during deformation also depends on the viscosity, which will vary as a function of melt composition as melting proceeds (Renner *et al.*, 2000).

In a study of pelite melting, Barboza and Bergantz (1996) considered the ratio of the permeability to the viscosity as a measure of likelihood of melt migration (Fig. 14.1.) and concluded that a melt fraction between 0.055 and 0.145 was more likely to be extracted than higher melt fractions. Although the

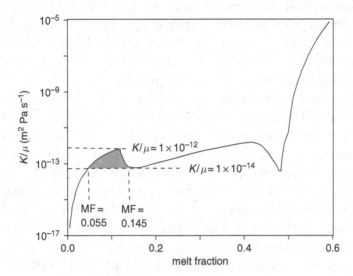

Fig. 14.1. Propensity for melt migration for pelite melting. The vertical axis is the ratio of the permeability divided by the melt viscosity. Higher values indicate a higher propensity for melt movement. Note that the values reached between melt fraction of 0.05 and 0.15 are not exceeded until after 40% of the pelite has been melted. Stippled region is the low melt fraction window of extractable melt.

permeability was not as high, the melt had a distinctly lower viscosity; hence, the ratio was greater. In summary, successful application of the methods used by Brown *et al.* (1999) and Tanner (1999), coupled with field observations (Sawyer, 1999, 2001; Marchildon & Brown, 2002, 2003) holds promise that more realistic models for permeability will emerge.

14.6 Transport models

Continuum macro-modeling may play an important role in illuminating aspects of the melt generation and transport, especially where the absence of real-time measurements makes direct verification difficult. We wish to distinguish our use of a continuum "macro-model" from, for example, a model for a mineral solution or a constitutive equation. Macro-models provide a means of isolating processes of interest, of exploring the behavior of systems characterized by extreme non-linearity, and at the very least are potent heuristic tools for developing intuition and insight. It is often said that one must start out with simple models to understand more complex systems. Of course, this cannot be literally true, as there is no guarantee that the super-position of the non-linear, perhaps simpler, processes will yield a system that resembles in any way a combination of the added elements for coupled, non-linear systems. However, in our experience, working with simpler model systems is invaluable in that intuition is developed, experimental procedure refined, and confidence emerges. Additionally, models may provide significant insight into the construction of negative tests for a proposed process.

Models generally come in one of three classes (Kleinstreuer, 2003): analytical mathematical models, in which notions like a "semi-infinite strip" may be posited; experimental models, in which analog materials are arranged to explore kinematic states and process scaling; and, computer models, in which transport equations are discretized. Each of these models yields different information content at different scales of resolution as a function of their inherent observability. For example, the observability of an analog experiment might be related to the optics of the camera system, while the observability of the numerical experiment is dictated by the round-off error and the assumptions in the governing equations and discretization schemes (Roache, 1998). Thus, one must choose a model approach that represents some mixture of appropriate scale of observation, available resources and convenience.

In principle, one should be able to formulate a multiphase flow system in terms of the local and instantaneous variables. However, that necessitates recasting the system as a deforming multiboundary problem, and accounting for many of the micro-scale interactions, the forms of which may be unknown and that are rarely important at the macro scale. As a result, local averaging is used to define macro-scale variables, and some combination of the mixture or Eulerian or Lagrangian approaches is commonly used.

The mixture approach is an implicit multiphase model in which the field variables of all the phases are represented as being the same in any stationary volume. It is assumed that there is one "local" temperature for the multiphase mixture at every point, and that the local velocity is single-valued. Hence, if the system is in suspension flow, both the melt phase and the particle phases have the same velocity, or, if in the Darcy regime, the solid phases have zero velocity and the local velocity is the Darcy velocity of the melt phase (Oldenburg & Spera, 1992; Barboza & Bergantz, 1998). But regardless, there is only one local velocity. The advantages of the mixture approach are that the system dimensionality is reduced as there is virtually only one phase, with the construction of rather ad hoc constitutive equations to capture any multiphase interactions implicitly. It is also easier to use from the standpoint of algorithmic design, as standard techniques for single-phase flow computational fluid dynamics may be used. The disadvantages are the inability to model relative motion between phases, although at low Stokes number this is not a severe limitation. However, if one is interested in possible relative motion in flow of a complex mixture like a diatexite, then the mixture approach may not be appropriate.

In the multiphase Eulerian approach, each of the phases is assumed to be a continuum, and the governing equations are assumed to hold at every point in the domain. This yields an elliptic system of equations in boundary value form. When averaged, information regarding the behavior of individual particles is lost, as the governing equations are cast in terms of the local volume fraction for each variable, which is identical to the existence probability of the phase occurring at that point. Assumptions are made regarding grain shape and size, and from the local volume fraction an interfacial area may be calculated. This is then used to calculate the drag and scalar transport. This approach is also called the "two-fluid" or Eulerian/Eulerian method, as it assumes that every phase is a virtual fluid even though one might be dealing with a particle–fluid mixture. This leads to some especially ad hoc constitutive relations, particularly in the scalar transport equations. For example, if one has a dispersed particle phase, what is the actual meaning of the within phase "thermal conductivity" if none of the elements of that phase are touching? There are a variety of constitutive functions to accommodate the interphase drag for the entire range of 0–100% liquid, with none seeming especially better than any other; hence, we recommend Agarwal and O'Neill (1988).

The Eulerian method is also subject to numerical diffusion, as discrete particles are represented by a field variable, the particle volume fraction. The numerical integration of field variables may lead to smearing of the interfaces as a consequence of numerical diffusion that occurs during numerical integration, and one must choose an advection scheme that aggressively reduces numerical diffusion (Andrews, 1995). However, the Eulerian method has been used widely in engineering design, and is the most commonly used multiphase theory in practice. It is reasonably robust when applied with awareness of its assumptions (Lahey & Bertodano, 1991). Further, elaboration of the nature of the Eulerian approach may be found in Durst *et al.* (1984) and Ni and Beckermann (1991), and geological examples in Bergantz and Ni (1999) and Valentine (1994).

Another approach for the physical description of a multiphase system would be to identify every occurrence of every phase, and follow their respective motions while accounting for all possible phase macro-interactions, phase change and properties. This is called the Lagrangian approach that yields a parabolic system and has the advantage of being straightforward, as the modeling of discrete particle interactions and heat transfer is mathematically simple. One distinct advantage is that particles may be tracked, and relative motion explicitly modeled. However, at the very least, the number of degrees of freedom increases in direct proportion to the number of particles. For example, there would be about 10^7 crystals in a cubic meter of magma with 10 vol. % crystals, each with a diameter of 1 cm. And if the flow is fully turbulent, the number of degrees of freedom to describe the flow of the melt-phase varies as $Re^{9/4}$ where *Re* is the Reynolds number. Hence, for a Reynolds number of 10^5, the melt phase alone requires the solution of 10^{11} linear equations. It is obvious that a strict Langrangian approach is not feasible for turbulent or chaotic flow.

One refinement is to combine both Eulerian and Lagrangian methods. At moderate loadings of particles, the melt phase may be modeled as a Eulerian continuum, and the particles tracked individually as in the Lagrangian method. If the particle loadings are high, and the flow is coupled, one must employ iterative techniques involving cycling between the integration of the particle path and the state of the carrier phase as each depends on the other. Although this may slow down the calculations considerably, it is perhaps the most promising approach for the numerical investigation of diatexites. Sokolichin *et al.* (1997) have performed a comparison of the methods and the interested reader is directed there for further elaboration.

However, this approach is still not entirely satisfactory in that geological systems may be chaotic (i.e., small local changes may induce very different outcomes. Hence, any fully deterministic approaches like those described above will represent only one realization, and a strongly averaged one that is in effect a low-pass filter with regard to the spatial and temporal granularity of observations. One approach is the composite continuum–stochastic algorithm (Hersum & Bergantz, 2000) that will allow one to assess a range of possible outcomes or responses in a Monte Carlo sense. The new element of the algorithm is multigridding, in which hierarchical scales of gridding are used in space, time and methodology. By this we mean that a continuum, determi-nistic formulation of heat and mass transfer is developed in the usual way at averaging scales that are much larger than the grain scale. This reflects the controls that originate on the basic and usual transport constitutive laws. The finer scales, or sub-grids, are modeled using a stochastic approach to relate the distribution and transport of melt using either a cellular automaton approach or even simpler stochastic sweeping over a range of possible responses and outcomes. This approach allows one to assess the sensitivity to reasonable geological variability at a number of distinct scales and at each time step.

14.7 Closure

Crustal differentiation by melt generation and migration is a process with extreme granularity. The "memory" of these processes in the rock record is both fractal and incomplete, with some events being over-represented. Rock memory is inevitably controlled by rheological transitions, which may vary at all scales and in highly local way throughout prograde-to-retrograde tectonic cycles. Rheological patchiness, representing in some combination local, but perhaps subtle, variations in protolith mode, state of hydration, and mineral fabric, controls the dynamics of melt transport and mixing, which are a consequence of the fundamentally multiphase character of crustal differentiation.

The absence of a comprehensive dynamic model for the interplay between the rate of change of chemical potential that induces melting, and the temporal and length scales of melt collection and transport, is the result of the local self organization that typifies multiphase dynamics. The transitions between end-member states such as heterogeneous porous flow, to transport of melt and residue in sheets and dikes, may be triggered by local stress focusing, reflecting non-general protolith conditions. The potentially rapid and local changes in the multiphase continuum may preclude the use of any approach based on field theories. Alternatively, a hierarchical approach, in which field theories are used to represent the energy and mass fluxes on average scales but are coupled to a stochastic or cellular automaton approach locally, may provide more insight in the range of possible kinematic responses and fluxes of energy and mass. The point is that, given the nature of observability between model approaches and natural examples, it does not seem likely that one will ever be able to test either a general modeling approach, or a particular geological case study with a detailed forward model.

References

Agarwal, P. K. and O'Neill, B. K. (1988). Transport phenomena in multi-particle system-1. Pressure drop and friction factors: unifying the hydraulic-radius and submerged-object approaches. *Chemical Engineering Science*, **43**, 2487–99.

Aidun, C. K., Lu, Y. and Ding, E. J. (1998). Direct analysis of articulate suspensions with inertia using the discrete Boltzmann equation. *Journal of Fluid Mechanics*, **373**, 287–311.

Andrews, M. J. (1995). Accurate computation of convective transport in transient two-phase flow. *International Journal of Numerical Methods Fluids*, **21**, 205–22.

Aref, H. (1990). Chaotic advection of fluid particles. *Philosophical Transactions of the Royal Society, London* A, **333**, 273–88.

Averbakh, A., Shauly, A., Nir, A. and Semiat, R. (1997). Slow viscous flows of highly concentrated suspensions part 1: Laserdoppler velocimetry in rectangular ducts. *International Journal of Multiphase Flow*, **23**, 409–24.

Bagdassarov, N. and Dorfman, A. (1998). Granite rheology: magma flow and migration. *Journal of the Geological Society London*, **155**, 863–72.

Baldyga, J. and Bourne, J. R. (1999). *Turbulent Mixing and Chemical Reaction*. New York: John Wiley & Sons.

Barboza, S. A. and Bergantz, G. W. (1996). Dynamic model of dehydration melting motivated by a natural analogue: applications to the Ivrea–Verbano zone, northern Italy. *Transactions of the Royal Society of Edinburgh*, **87**, 23–31.

Barboza, S. A. and Bergantz, G. W. (1997). Melt productivity and rheology: complementary influences on the progress of melting. *Numerical Heat Transfer*, **31A**, 375–92.

Barboza, S. A. and Bergantz, G. W. (1998). Rheological transitions and the progress of melting of crustal rocks. *Earth and Planetary Science Letters*, **158**, 19–29.

Barton, M. D., Staude, J.-M., Snow, E. A. and Johnson, D. A. (1991). Aureole systematics. In *Contact Metamorphism*, ed. D. M. Kerrick. Washington: Mineralogical Society of America, pp. 723–847.

Beard, J. S. and Lofgren, G. E. (1989). Effects of water on the composition of partial melts of greenstone and amphibolite. *Science*, **244**, 195–7.

Bergantz, G. W. (1990). Melt fraction diagrams: the link between chemical and transport models. In *Modern Methods of Igneous Petrology: Understanding Magmatic Processes*, ed. J. Nicholls and J. K. Russell. Washington: Mineralogical Society of America, pp. 240–57.

Bergantz, G. W. (1992). Conjugate solidification and melting in multicomponent open and closed systems. *International Journal of Heat Mass Transfer*, **35**, 533–43.

Bergantz, G. W. (2000). On the dynamics of magma mixing by reintrusion: implications for pluton assembly processes. *Journal of Structural Geology*, **22**, 1297–309.

Bergantz, G. W. and Ni, J. (1999). A numerical study of sedimentation by dripping instabilities in viscous fluids. *International Journal of Multiphase Flow*, **25**, 307–20.

Brady, J. F. and Bossis, G. (1988). Stokesian dynamics. *Annual Review of Fluid Mechanics*, **20**, 111–57.

Breidenthal, R. E. (1981). Structure in turbulent mixing layers and wakes using a chemical reaction. *Journal of Fluid Mechanics*, **109**, 1–24.

Broadwell, J. E. and Mungal, M. G. (1991). Large-scale structures and molecular mixing. *Physics of Fluids A*, **3**, 1193–206.

Brown, G. L. and Roshko, A. (1974). On density effects and large structure in turbulent mixing layers. *Journal of Fluid Mechanics*, **64**, 775–816.

Brown, M. (1994). The generation, segregation, ascent and emplacement of granite magma: the migmatite-to-crustally derived granite connection in thickened orogens. *Earth Science Reviews*, **36**, 83–130.

Brown, M. and Solar, G. S. (1998a). Shear zone systems and melts: feedback relations and self-organization in orogenic belts. *Journal of Structural Geology*, **20**, 211–27.

Brown, M. and Solar, G. S. (1998b). Granite ascent and emplacement during contractional deformation in convergent orogens. *Journal of Structural Geology*, **20**, 1365–93.

Brown, M. A., Brown, M., Carlson, W. D. and Denison, C. (1999). Topology of syntectonic melt-flow networks in the deep crust: inferences from three-dimensional images of leucosome geometry in migmatites. *American Mineralogist*, **84**, 1793–818.

Burgisser, A. and Bergantz, G. W. (2002). Reconciling pyroclastic flow and surge: the multiphase physics of pyroclastic density currents. *Earth and Planetary Science Letters*, **202**, 405–18.

Burmester, S. S. H., Rielly, C. D. and Edwards, M. F. (1992). The mixing of miscible liquids with large density differences in density and viscosity. In *Fluid*

Mechanics of Mixing, ed. R. King. Dordrecht: Kluwer Academic Publishers, pp. 83–90.

Campbell, I. H. and Turner, J. S. (1986). The influence of viscosity on fountains in magma chambers. *Journal of Petrology*, **27**, 1–30.

Connolly, J. A. (1997). Devolatilization-generated fluid pressures and deformation-propagated fluid flow during prograde regional metamorphism. *Journal of Geophysical Research*, **102**, 18 149–73.

Connolly, J. A. D., Holness, M. B., Rubie, D. C. and Rushmer, T. (1997). Reaction-induced microcracking: an experimental investigation of a mechanism for enhancing anatectic melt extraction. *Geology*, **25**, 591–4.

Crowe, C. T. (2000). On models for turbulence modulation in fluid particle flows. *International Journal of Multiphase Flow*, **26**, 719–27.

Crowe, C. T., Troutt, T. R. and Chung, J. N. (1996). Numerical models for two-phase turbulent flows. *Annual Reviews of Fluid Mechanics*, **28**, 11–43.

Danckwerts, P. V. (1952). The definition and measurement of some characteristics of mixtures. *Applied Science Research*, **A3**, 279–96.

Davies, G. and Tommasini, S. (2000). Isotopic disequilibrium during rapid crustal anatexis: implications for petrogenetic studies of magmatic processes. *Chemical Geology*, **162**, 169–91.

Durst, F., Milojevic, D. and Schonung, B. (1984). Eulerian and Lagrangian predictions of particulate two-phase flows: a numerical study. *Applied Mathematical Modeling*, **8**, 101–15.

Eichelberger, J. C., Chertkoff, D. G., Dreher, S. T. and Nye, C. J. (2000). Magmas in collision: rethinking chemical zonation in silicic magmas. *Geology*, **28**, 603–6.

Fang, Z. and Phan-Thien, N. (1995). Numerical simulation of particle migration in concentrated suspensions by a finite volume method. *Journal of Non-Newtonian Fluid Mechanics*, **58**, 67–81.

Feng, Z.-G. and Michaelides, E. E. (2000). A numerical study on the transient heat transfer from a sphere at high Reynolds and Peclet numbers. *International Journal of Heat Mass Transfer*, **43**, 219–29.

Grace, H. P. (1982). Dispersion phenomena in high viscosity immiscible fluid systems and application of static mixers as dispersion devices in such systems. *Chemical Engineering Communication*, **14**, 225–77.

Guernina, S. and Sawyer, E. W. (2003). Large-scale melt-depletion in granulite terranes: an example from the Archean Ashuanipi Subprovince of Quebec. *Journal of Metamorphic Geology*, **21**, 181–201.

Hampton, R. E., Mammoli, A. A., Graham, A. L. and Tetlow, N. (1997). Migration of particles undergoing pressure-driven flow in a circular conduit. *Journal of Rheology*, **41**, 621–40.

Harris, N., Vance, D. and Ayres, M. (2000). From sediment to granite: timescales of anatexis in the upper crust. *Chemical Geology*, **162**, 155–67.

Hersum, T. and Bergantz, G. W. (2000). The direct numerical solution of progressive crystal formation in solidifying magmas using a multi-scale algorithm. *Eos, Transactions of the American Geophysical Union*, **81**, F1293.

Hoffman, N. R. A. and McKenzie, D. P. (1985). The destruction of geochemical heterogeneities by deferential fluid motions during mantle convection. *Geophysical Journal of the Royal Astronomical Society*, **82**, 163–206.

Holness, M. B. (1997). *Deformation-Enhanced Fluid Transport in the Earth's Crust and Mantle*. The Mineralogical Society Series, 8, ed. A. P. Jones. London: Chapman and Hall.

Holness, M. B. and Clemens, J. D. (1999). Partial melting of the Appin Quartzite driven by fracture-controlled H_2O infiltration in the aureole of the Ballachulish Igneous Complex, Scottish Highlands. *Contributions to Mineralogy and Petrology*, **136**, 154–68.

Iverson, R. M. (1997). The physics of debris flows. *Reviews in Geophysics*, **35**, 245–96.

Iverson, R. M. and Vallance, J. W. (2001). New views of granular mass flow. *Geology*, **29**, 115–18.

Jellinek, A. M. and Kerr, R. C. (1999). Mixing and compositional stratification produced by natural convection 2. Applications to the differentiation of basaltic and silicic magma chambers and komatiite lava flows. *Journal of Geophysical Research*, **104**, 7203–18.

Jenkins, J. T. and McTigue, D. M. (1990). Transport processes in concentrated suspensions: the role of particle fluctuations. In *Two-phase Flows and Waves*, ed. D. D. Joseph and D. G. Schaffer. New York: Springer-Verlag, pp. 70–9.

Jones, S. W., Thomas, O. M. and Aref, H. (1989). Chaotic advection by laminar flow in a twisted pipe. *Journal of Fluid Mechanics*, **209**, 335–57.

Karnis, A., Goldsmith, H. L. and Mason, S. G. (1966). The kinetics of flowing dispersions I. Concentrated suspensions of rigid particles. *Journal of Colloidal Interface Science*, **22**, 531–53.

Khakhar, D. V., McCarthy, J. J., Gilchrist, J. F. and Ottino, J. M. (1999). Chaotic mixing of granular materials in two dimensional tumbling mixers. *Chaos*, **9**, 195–205.

Kim, I., Elghobashi, S. and Sirignano, W. A. (1998). On the equation of spherical particle motion: effect of Reynolds and acceleration numbers. *Journal of Fluid Mechanics*, **367**, 221–53.

Kleinstreuer, C. (2003). *Two-phase Flow: Theory and Applications*. New York: Taylor & Francis.

Knesel, K. M. and Davidson, J. P. (1999). Sr isotope systematics during melt generation by intrusion of basalt into continental crust. *Contributions to Mineralogy and Petrology*, **136**, 285–95.

Lahey, R. T. and Bertodano, M. L. (1991). The prediction of phase distribution using the two-fluid model. *ASME/JSME Thermal Engineering Proceedings*, 193–200.

Leighton, D. and Acrivos, A. (1987). The shear-induced migration of particles in concentrated suspensions. *Journal of Fluid Mechanics*, **181**, 415–39.

Linden, P. F., Redondo, J. M. and Youngs, D. L. (1994). Molecular mixing in Rayleigh–Taylor instability. *Journal of Fluid Mechanics*, **265**, 97–124.

Ling, W., Chung, J. N., Troutt, T. R. and Crowe, C. T. (1998). Direct numerical simulation of a three-dimensional temporal mixing layer with particle dispersion. *Journal of Fluid Mechanics*, **358**, 61–85.

Liu, S. and Masliyah, J. H. (1996). Rheology of suspensions. In *Suspensions: Fundamentals and Applications in the Petroleum Industry*, ed. L. L. Schramm. Washington, DC: American Chemical Society, pp. 107–76.

Lupulescu, A. and Watson, E. B. (1999). Low melt fraction connectivity of granitic and tonalitic melts in a mafic crustal rock at 800 °C and 1 Gpa. *Contributions to Mineralogy and Petrology*, **134**, 202–16.

Lyon, M. K. and Leal, L. G. (1997). An experimental study of the motion of concentrated suspensions in two-dimensional channel flow. I Monodisperse systems. *Journal of Fluid Mechanics*, **363**, 25–56.

Marble, F. E. (1970). Dynamics of dusty gases. *Annual Review of Fluid Mechanics*, **2**, 397–446.

Marchildon, N. and Brown, M. (2002). Grain-scale melt distribution in two contact aureole rocks: implications for controls on melt localization and deformation. *Journal of Metamorphic Geology*, **20**, 381–96.

Marchildon, N. and Brown, M. (2003). Spatial distribution of melt-bearing structures in anatectic rocks from southern Brittany, France: implications for melt transfer at grain to orogen-scale. *Tectonophysics*, **364**, 215–35.

Marchioro, M. and Acrivos, A. (2001). Shear-induced particle diffusivities from numerical simulations. *Journal of Fluid Mechanics*, **443**, 101–28.

Marsh, B. D. (1988). Crystal capture, sorting and retention in convecting magma. *Geological Society of America Bulletin*, **100**, 1720–37.

McTigue, D. F., Givler, R. C. and Nunziato, J. W. (1986). Rheological effects of nonuniform particle distributions in dilute suspensions. *Journal of Rheology*, **30**, 1053–76.

Michaelides, E. E. and Feng, Z.-G. (1994). Heat transfer from a rigid sphere in a nonuniform flow and temperature field. *International Journal of Heat Mass Transfer*, **37**, 2069–76.

Miller, S. and Nur, A. (2000). Permeability as a toggle switch in fluid-controlled crustal processes. *Earth and Planetary Science Letters*, **183**, 133–46.

Morris, J. and Boulay, F. (1999). Curvilinear flows of noncolloidal suspensions: the role of normal stresses. *Journal of Rheology*, **43**, 1213–37.

Ni, J. and Beckermann, C. (1991). A volume-averaged two-phase model for transport phenomena during solidification. *Metallurgical Transactions B*, **22B**, 349–61.

Nielsen, O., Amberg, L., Mo, A. and Thevik, H. (1999). Experimental determination of mushy zone permeability in aluminum–copper alloys with equiaxed microstructures. *Metallurgical Materials Transactions A*, **30A**, 2455–62.

Nott, P. R. and Brady, J. F. (1994). Pressure-driven flow of suspensions: simulation and theory. *Journal of Fluid Mechanics*, **275**, 157–99.

Oldenburg, C. M. and Spera, F. J. (1992). Hybrid model for solidification and convection. *Numerical Heat Transfer*, **B21**, 217–29.

Oldenburg, C. M., Spera, F. J., Yuen, D. A. and Sewell, G. (1989). Dynamic mixing in magma bodies: theory, simulations and implications. *Journal of Geophysical Research*, **94**, 9215–36.

Ottino, J. M. (1989). *The Kinematics of Mixing: Stretching, Chaos and Transport*. Cambridge: Cambridge University Press.

Patiño-Douce, A. E. and Johnston, A. D. (1991). Phase equilibria and melt productivity in the pelitic system: implications for the origin of peraluminous granitoids and aluminous granulites. *Contributions to Mineralogy and Petrology*, **107**, 202–18.

Petford, N. (1996). Dikes or diapirs. *Transactions of the Royal Society of Edinburgh*, **87**, 105–14.

Petford, N. and Koenders, M. A. (1998). Granular flow and viscous fluctuations in low Bagnold number granitic magmas. *Journal of the Geological Society London*, **155**, 873–81.

Petford, N., Clemens, J. D. and Vigneresse, J. L. (1997). Application of information theory to the formation of granitic rocks. In *Granite: From Segregation of Melt to Emplacement Fabrics*, ed. J. L. Bouchez, D. W. H. Hutton and W. E. Stephens. Dordrecht: Kluwer, pp. 3–10.

Phan-Thien, N. and Fang, Z. (1996). Entrance length and pulsatile flows of a model concentrated suspension. *Journal of Rheology*, **40**, 521–9.

Phillips, R. J., Armstrong, R. C., Brown, R. A., Graham, A. and Abbott, J. R. (1992). A constitutive model for concentrated suspensions that accounts for shear-induced particle migration. *Physics Fluids A*, **4**, 30–40.

Poletto, M. and Joseph, D. D. (1995). Effective density and viscosity of a suspension. *Journal of Rheology*, **39**, 323–43.

Raju, N. and Meiburg, E. (1995). The accumulation and dispersion of heavy particles in forced two-dimensional mixing layers. Part 2. The effect of gravity. *Physics Fluids*, **7**, 1241–64.

Renner, J., Evans, B. and Hirth, G. (2000). On the rheologically critical melt fraction. *Earth and Planetary Science Letters*, **181**, 585–94.

Rielly, C. D. and Burmester, S. S. H. (1994). Homogenization of liquids with different densities and viscosities. In *Industrial Mixing and Technology: Chemical and Biological Applications*, ed. G. B. Tatterson. American Institute of Chemical Engineers Symposium Series, 90, Event 299, pp. 171–85.

Roache, P. J. (1998). *Verification and Validation in Computational Science and Engineering*. Albuquerque: Hermosa Publishers.

Roberts, P. H. and Loper, D. E. (1987). Dynamical processes in slurries. In *Structure and Dynamics of Partially Solidified Systems*, ed. D. E. Loper. Dordrecht: Martinus Nijhoff, pp. 229–90.

Rushmer, T. (1995). An experimental deformation of partially molten amphibolite: application to low-melt fraction segregation. *Journal of Geophysical Research*, **100**, 15 681–95.

Rushmer, T. and Brearley, A. J. (1998). Don't speak volumes about melt segregation: the role of mineral reactions in melt extraction and melt geochemistry. *Eos, Transactions of the American Geophysical Union*, **79**, F1020.

Rutter, E. H. and Neumann, D. H. (1995). Experimental deformation of partially molten Westerly granite under fluid-absent conditions, with implications for the extraction of granitic magmas. *Journal of Geophysical Research*, **100**, 15 697–715.

Sawyer, E. W. (1991). Disequilibrium melting and the rate of melt residuum separation during migmatization of mafic rocks from the Grenville Front, Quebec. *Journal of Petrology*, **32**, 701–38.

Sawyer, E. W. (1994). Melt segregation in the continental crust. *Geology*, **22**, 1019–22.

Sawyer, E. W. (1999). Criteria for the recognition of partial melting. *Physics and Chemistry of the Earth*, **24**, 269–79.

Sawyer, E. W. (2001). Melt segregation in the continental crust: distribution and movement of melt in anatectic rocks. *Journal of Metamorphic Geology*, **19**, 291–309.

Shapley, N., Armstrong, R. C. and Brown, R. A. (2002). Laser Doppler velocimetry measurements of particle velocity fluctuations in a concentrated suspension. *Journal of Rheology*, **46**, 241–72.

Shapley, N., Brown, R. A. and Armstrong, R. C. (2004). Evaluation of particle migration models based on laser Doppler velocimetry measurements in concentrated suspensions. *Journal of Rheology*, **48**, 255–79.

Shauly, A., Wachs, A. and Nir, A. (2000). Shear-induced particle resuspension in settling polydisperse concentrated suspension. *International Journal of Multiphase Flow*, **26**, 1–15.

Singh, A. and Nott, P. R. (2000). Normal stresses and microstructure in bounded sheared suspensions via Stokesian Dynamics simulations. *Journal of Fluid Mechanics*, **412**, 279–301.

Sokolichin, A., Eigenberger, G., Lapin, A. and Lubbert, A. (1997). Dynamic numerical simulation of gas–liquid two-phase flows: Euler/Euler versus Euler/ Lagrange. *Chemical Engineering Science*, **52**, 611–26.

Sparks, R. S. J. and Marshall, L. (1986). Thermal and mechanical constraints on mixing between mafic and silicic magmas. *Journal of Volcanology and Geothermal Research*, **29**, 99–124.

Stock, D. E. (1996). Particle dispersion in flowing gases – 1994 Freeman scholar lecture. *Journal of Fluids Engineering*, **118**, 417.

Subia, S. R., Ingber, M. S., Mondy, L. A., Altobelli, S. A. and Graham, A. L. (1998). Modeling of concentrated suspensions using a continuum constitutive equation. *Journal of Fluid Mechanics*, **373**, 193–219.

Tanner, D. C. (1999). The scale-invariant nature of migmatite from the Oberpflaz, NE Bavaria and its significance for melt transport. *Tectonophysics*, **302**, 297–305.

Tavare, N. S. (1986). Mixing in continuous crystallizers. *American Institute of Chemical Engineers Journal*, **32**, 705–32.

Ten, A., Yuen, D. A., Podladchikov, Y. Y., Larsen, T. B., Pachepsky, E. and Malevsky, A. V. (1997). Fractal features in mixing of non-Newtonian and Newtonian mantle convection. *Earth and Planetary Science Letters*, **146**, 401–14.

Valentine, G. A. (1994). Multifield governing equations for magma dynamics. *Geophysical Astrophysical Fluid Dynamics*, **78**, 193–210.

Vigneresse, J. L. and Tikoff, B. (1999). Strain partitioning during partial melting and crystallizing felsic magmas. *Tectonophysics*, **312**, 117–32.

Vigneresse, J. L., Barbey, P. and Cuney, M. (1996). Rheological transitions during partial melting and crystallization with application to the felsic magma segregation and transfer. *Journal of Petrology*, **37**, 1579–600.

Vojir, D. J. and Michaelides, E. E. (1994). The effect of the history term on the motion of rigid spheres in a viscous fluid. *International Journal of Multiphase Flow*, **20**, 547–56.

Wang, L.-P., Burton, T. D. and Stock, D. E. (1990). Chaotic dynamics of heavy particle dispersion: fractal dimension versus dispersion coefficients. *Physics Fluids A*, **2**, 1305–8.

Watt, G. R., Oliver, N. H. S. and Griffin, B. J. (2000). Evidence for reaction-induced microcracking in granulite facies migmatites. *Geology*, **28**, 327–30.

Williams, G. P. (1997). *Chaos Theory Tamed*. London: Taylor & Francis.

Yang, Y., Crowe, C. T., Chung, J. N. and Troutt, T. R. (2000). Experiments on particle dispersion in a plane wake. *International Journal of Multiphase Flow*, **26**, 1583–607.

Yuen, D., Vincent, A. P., Bergeron, S. Y., Dubuffet, F., Ten, A. A., Steinbach, V. C. and Starin, E. (2000). Crossing of scales and non-linearities in geophysical processes. In *Problems in Geophysics in the New Millennium*, ed. E. Boschi, G. Ekstrom and A. Morelli. Bologua: Editrice Compositori, pp. 403–62.

Zarrebini, M. and Cardoso, S. S. S. (1998). Particle-size separation by a turbulent plume. *American Institute of Chemical Engineers Research Event*, 1–7.

Index

Andes 159–64
andesite model for formation of continental crust
        (*see also* geochemistry) 100, 102, 103, 139
Archean crust (*see also* geochemistry, sedimentary
        rocks) 196
    granitoid composition 197–8
        tonalite–trondhjemite–granodiorite (TTG)
                suite 9, 196, 198, 200–8, 210–11, 212,
                213–20, 316
        sedimentary rock compositions (*see also*
                sedimentary rocks) 196–7
        tonalite–trondhjemite–granodiorite (TTG)
                model for Archean crustal genesis
            problems 213–20
                experiments vs. nature 213–14
                older crustal component 217–19; slab fluids
                        and Rb-depletion 219–20
                sanukitoid 214–17
arcs (*see also* geochemistry) 2, 3, 4, 6, 8, 15, 54–5,
        139, 141
    mass fluxes:
        continental arcs 145–7
        island arcs 143–5, 150–2

Canadian shield
    East Athabasca mylonite triangle 236–52
continental growth models (*see also* geochemistry)
        28
    Andesite 100, 102, 103, 139
    arc 2, 3, 4, 6, 8, 15, 54–5, 139, 141
        crustal recycling 6
        mass fluxes: continental arcs 145–7; island
                arcs 143–5, 150–2
    cumulate 10
    felsic 102–3
    high-Mg andesite 155, 161
    oceanic plateau (including accretion) 8, 15, 140
    plume 8

differentiation of continental crust (*see also*
        geochemistry) 6, 7, 9–10, 28–9
    fractional crystallization
        cumulate formation 147, 149, 150, 156–8

hybridization models 158–9
    AFC (assimilation – fractional crystallization)
            158, 161
    MASH (melting, assimilation, storage and
            homogenization) 158, 161
mass transfer 331–2
    advective fluid flow 362
    diffusional fluid flow 362
mechanisms of melt extraction and ascent (*see
            also* metamorphism) 2, 3, 13–14, 333, 339,
            416–24, 430–1, 539–42
    channel flow 339, 340, 346, 355, 418
    deformation band networks 364–6
    diapirism 340
    dike 339, 341, 342, 345, 366, 368, 369, 370, 417
    ductile fracture 339, 355, 361, 366, 368–70
    porous flow 339, 340, 352, 355, 361–6,
            418–24
    shear enhanced compaction 364
melt ascent 333, 338–9, 354, 366–70
melting (*see also* geochemistry, metamorphism,
            rheology) 2, 3, 10, 11, 153, 158, 296–7
    melt flow networks 415
    volume change 334–5, 431
melt percolation threshold (*see also* rheology)
            366, 367
physical controls on differentiation
    crystal sorting 524–6
    heat transfer 534–9
    melting 534–9
    mixing 529–34
    multiphase flow 522; crystal sorting 524–6;
            phase segregation by shear 526–8;
            suspensions and slurries 522–8
    rheology 534–9
self-organization 333–4
transport of melt (*see also* plutons) 2, 14, 16,
            539–42

fluids (*see* metamorphism)

geochemistry of continental crust
    changes with time

continental growth models 7, 28
 rates of growth 141–4
continental crust composition 100–10, 135–9
 Archean vs. post-Archean
  (Archean-Proterozoic) 110–12,
  114–24, 154–5, 178–9
 average 174–80; lower continental crust 177–8
crust-mantle reservoirs and fluxes 189–95
 recent crustal growth and mantle fluxes 193–5
differentiation processes
 fractional crystallization: cumulate formation
  147, 149, 150, 156–8; intra-crustal
  melting 153, 158
early continental crust 195–9
 Archean 196
 granitoid composition 197–8
 sedimentary rock composition 196–7
formation of early continental crust 200–12
 adakite 210–11
 eclogite chemistry 207–8
 experimental evidence 202–6
 field evidence 202
 slab melting 212
 thermal modeling 208–10
 tonalite–trondhjemite–granodiorite (TTG)
  model for Archean crustal genesis:
  problems 213–20; experiments vs.
  nature 213–14; sanukitoid 214–17;
  older crustal component 217–19; slab
  fluids and Rb-depletion 219–20
 trace element modeling 206–7
growth of continental crust 7, 28, 181–9
 juvenile crust 270–4
 mass balance 2, 4
 mass fluxes: continental arcs 145–7; island
  arcs 143–5, 150–2
 no growth models 183–6; crustal recycling
  184–5; freeboard arguments 183;
  isotopic evidence 185; trace element
  ratios 184
 oceanic plateau accretion 8, 15, 140
 progressive growth models 186–9
 recycling 2, 3, 8
heat-producing elements (HPEs; *see* heat
  production) 92, 103
models for formation of continental crust (*see
  also* differentiation) 28
 andesite 100, 102, 103
 arc 6, 8
 crustal recycling 6
 cumulate 10
 felsic 102–3
 high-Mg andesite 155, 161, 214–17
 multi-component mixing model 220–2, 231–3
 oceanic plateau (including accretion) 8, 15, 140
 plume 8, 101–2
secular variations 110–25, 153, 154–5
 incompatible elements 118–20
 rare earth elements 116–18
 Th, U and Pb isotopes 120–4
upper continental crust 93, 124–5, 135

geophysics of continental lithosphere
 Conrad discontinuity 30
 continental lithosphere 21–2
 continental subdivisions 2, 3, 4, 5
  arc 54–5
  craton 31, 32–5, 41–3, 47, 89, 90
  extensional province 30, 34, 46, 47–8, 49–50
  orogen (orogenic belt) 29; orogenic plateau 29
  platform (*see* shield)
  rift 29, 48–9
  shield 29
 crustal reflectivity 41, 49–50
 crustal structure 2, 3
 crustal thickness and velocity structure 29–30,
  39–50
 detachment (*see also* orogen) 34, 35, 43, 46, 52
 finite element models 34–5
 heat flow (*see* thermal structure, heat production)
  23, 178–9
 lithosphere (continental) 21–2
 mantle dynamics (convection) 49, 51–2
 Moho (*see also* seismic velocity) 28, 30–1, 53–4
 orogen (orogenic belt) 29, 46–8
  plateau, orogenic 46
 plate tectonics 21
 potential field studies (gravity, magnetics,
  magnetotelluric) 22–3
 rheology 2, 3, 5, 10–11, 31, 32–5, 67, 68, 82–7, 88,
  268, 288, 337, 362–3, 421, 448–9
  Brace–Goetze rheology 82–4, 88
  brittle (semi-, failure, frictional) 32, 285
  brittle–ductile transition 32, 34, 269, 289
  Byerlee's law 32
  constitutive laws for minerals (*see also* flow
   laws) 32, 395
  decoupling 288–9
  ductile (plastic flow) 32, 49, 51–2
  strength envelope 32, 34, 82, 115–24
 seismic studies (lithoprobe, INDEPTH,
  continental dynamics) 25
 seismic velocity 23, 25–8
  Moho 1, 2, 3, 10, 28, 30–1, 43, 46, 53–4, 67, 80,
   83, 124–5, 135, 150, 153
 seismicity 269
 seismogenic zone 34
 seismology 23, 111
  active source (receiver function) 24, 25
  earthquake seismology (passive, natural) 24
  reflection (teleseismic arrays, wide-angle) 23,
   24, 46
  refraction 23, 24
 tectosphere 31, 43, 47, 52
 thermal structure 36–9
  Archean 38
  conductivity 36
  heat flow (crust, mantle, global data, surface,
   see also heat production) 36, 37, 52
  heat generation/production (*see also* heat
   production) 6, 36–7, 52
  Proterozoic 38
 tomography 23, 25

Hadean 7
  magma ocean 7
heat flow (*see also* geophysics, geochemistry, heat
       production) 6, 89–90, 92, 101–2, 103–7,
       112–15, 178–9
  continental 104–7, 111
  global 103, 107
heat production (*see also* geophysics) 6, 36–7, 52
  crustal 107
  distribution of HPEs 67–8, 68–74, 75, 83–6, 87–8
  exponential model 69–71
  heat producing elements (HPEs) 6
  heat sources 2, 3, 6
  homogeneous model 71
  redistribution of HPEs 68, 75, 88
  tectonic modification 74–5, 87
  thermal modeling 75–81

metamorphism and tectonics
  anatexis (*see also* partial melting) 2, 3, 10, 11, 153,
       158, 296–7
  fluids 11–13
  granulite facies 301–4
  lower continental crust 220–2, 231–3
    East Athabasca mylonite triangle 236–52;
       exhumed lower crust 235, 253–5;
       exhumation 235, 254–5; isobaric
       cooling 255; how representative
       257–9; nature of the lower continental
       crust 255–7
    geophysical characteristics 233
    high-pressure terranes 234–6; isobaric
       cooling 235, 253–5
    samples (xenoliths) 234
  melt
    extraction 2, 13–14, 16, 332, 334–9, 345–52, 354,
       360–70; melt flow network 345, 366
    segregation 13–14, 448, 449
    transport 2, 14, 16, 539–42
  Middle continental crust 268–70
    brittle–ductile transition 269
    geophysical characteristics 268–9
    rheology 268
    sutures 282–4
    tectonics 278, 287
  migmatite (*see also* anatexis) 334–9, 357–8,
       416–17, 420–1
    granite connection 357–8
    leucosome, structural control 336–9
    microstructures 335–6
  partial melting (*see also* differentiation) 11, 296–7
    experimental studies 305–19; lithological
       mixtures 317–19; metagreywackes
       310–12; metapelites 307–9; mafic
       rocks 312–14
    fertility of source rocks 320
    high-K calk-alkaline granitic magmas 319–20
    quartzofeldspathic rocks 312–14
    reactions 297–300, 304–5; fluid-absent
       (dehydration melting) 11, 304–5, 334–5
    tonalite–trondhjemite–granodiorite suite 316

volume change 334–5
pressure–temperature (–time-deformation)
    paths 341
  clockwise 341
  decompression 277
  isobaric cooling 276, 287
  looping 274–7, 287–8
  syntectonic plutons 277
Moho (*see also* geophysics) 1, 2, 3, 10, 28, 30–1, 43,
    46, 53–4, 67, 80, 83, 124–5, 135, 150, 153

orogens 10–11
  accretionary 10
  collisional 10–11

permeability 12, 415–16, 431–40, 444
  dilational strain 441–4
  experimental studies (*see also* porosity) 437–40
  fracture network 436
  lower crust 432–5, 448–9
  modeling, numerical 435
  partially molten rocks 435, 439–40, 441, 444–6
  percolation threshold 435, 436
  primary permeability 431, 435–7, 448
  secondary permeability 431, 444–6, 448
plate tectonics 21
  plate boundaries, convergent 10
  subduction 1, 2, 3
plume 2, 3, 5, 8, 140, 140–1
pluton 14–15
  ascent and emplacement (*see also* differentiation)
      472–92
    channel/dike 474
    diapirism 473
    floor depression 477–9; strain rates 479–84
    laccoliths vs. lopoliths 475–7
    stoping 473–4
    structural accommodation 485
    volumes and filling times 484–5
  floors 464–7
  growth (*see also* differentiation)
    internal growth modes 492
    melt extraction, relationship with 455–6,
      493–507
    partial melting 498; melt flux 499–507
    pluton shape, influence of 494–5
    pluton spacing, influence of 495–8
    regional deformation 490–2
  laccoliths 475–7
  lopoliths 475–7
  roofs 467
  shape 456–72
    internal structures 469–70
    power-law (empirical) 470–2
    tabular (tablet-shaped) 464, 497–8
    wedge (wedge-shaped) 464, 498
  sides 467–9
  source regions, melt flux 499–507
  spacing 495–8
  thickness 457–8
  volumes and filling times 484–5

plutonism 277–9, 285–7
  calc-alkaline 278
  syntectonic 278–9
porosity 384, 406, 431, 444
  experimental studies (*see also* permeability) 437–40
  pore pressure, modeling evolution of 435
Proterozoic
  Southwestern USA 269, 270–4

rheology (*see also* geophysics) 2, 3, 5, 10–11, 31,
    32–5, 67, 68, 82–7, 88, 268, 288, 337, 362–3,
    421, 448–9
  Brace–Goetze rheology 82–4, 88
  brittle–ductile (viscous) transition 32, 34, 269,
    280, 281, 289, 360
  deformation mechanisms 386–90
    accommodation processes 388–9
    fracturing 387
    granular flow 387–8, 396–402, 445, 448
    intracrystalline plastic deformation 387
  experimental studies
    diffusion-accommodated granular flow
        396–402, 421
    methods 390–2
    partially molten systems 392–4; strength
        394–6
  flow laws (*see also* constitutive laws for minerals)
    395
  mechanical behavior 385–6, 403–7
  microstructures 411–15

rheologically critical melt percentage (RCMP)
    407–11
strain localization 363
strength envelope 32, 34, 82, 279–82

secular
  analysis 1
  change 7, 9, 15, 153, 154–5, 178–9, 181–9,
    195–9, 200
  evolution 1, 2, 3
  variations 110–25
    incompatible elements 118–20
    rare earth elements (REE) 116–18
    Th, U and Pb isotopes 120–4
sedimentary rocks 93–8
  rare earth elements (REE) composition 94–5
  secular variations 110–25
    incompatible elements 118–20
    rare earth elements (REE) 116–18
    Th, U and Pb isotopes 120–4
  sedimentary recycling 124–5, 135
  upper crustal elemental abundances from 93–4,
    95–8, 108
self-organization (self-organized criticality) 1,
    333–4, 371–2
supercontinents 8

tonalite–trondhjemite–granodiorite (TTG) suite 9,
    110, 196, 198, 200–2, 202–8, 210–11, 212,
    213–20, 316